Pathways to Water Sector Decarbonization, Carbon Capture and Utilization

Pathways to Water Sector Decarbonization, Carbon Capture and Utilization

Edited by
Zhiyong Jason Ren and Krishna Pagilla

IWA PUBLISHING

Published by **IWA Publishing**
Unit 104–105, Export Building
1 Clove Crescent
London E14 2BA, UK
Telephone: +44 (0)20 7654 5500
Fax: +44 (0)20 7654 5555
Email: publications@iwap.co.uk
Web: www.iwapublishing.com

First published 2022

British Library Cataloguing in Publication Data
A CIP catalogue record for this book is available from the British Library

ISBN: 9781789061789 (Paperback)
ISBN: 9781789061796 (eBook)
ISBN: 9781789061802 (ePUB)

This eBook was made Open Access in March 2022

Contents

About the Editors xiii

List of Contributors xv

Foreword by Kala Vairavamoorthy xix

Foreword by Art K.Umble xxi

Preface xxiii

Chapter 1
Toward a net zero circular water economy ***1***
Zhiyong Jason Ren, Jerald L. Schnoor and Krishna R. Pagilla

1.1 The Water Sector and the Challenges and Opportunities on Decarbonization 1
1.2 Pathways Toward Water and Wastewater Decarbonization 4
1.2.1 Decarbonization requires a better understanding of emission baseline 4
1.2.2 Decarbonization requires a combination of approaches and collaborations among stakeholders 5
1.2.3 Processes and technologies that enable energy and resource recovery 6
1.2.4 Processes and technologies that enable additional benefits of carbon capture and utilization, and watershed management 9
1.2.5 Case studies on utility decarbonization practice 11
1.3 The Paradigm Change for a Net Zero Circular Water Economy 12
References 13

Chapter 2
What can we learn from decarbonization of the energy sector? ***15***
A. J. Simon and Seth W. Snyder

2.1 Introduction: Energy and Water: Similarities, Differences, and a Complex Relationship 15
2.1.1 The energy-water nexus 16
2.1.2 Differences in scale 17
2.1.3 The carbon-water nexus 18

2.2 Decarbonization of the Energy Sector 18
2.3 A Framework for Sustainability for Energy and Water 20
2.4 The Pace of Decarbonization 21
2.4.1 Residential and commercial equipment 24
2.4.2 Transportation equipment 24
2.4.3 Utility equipment 24
2.4.4 Integration 25
2.5 Case Studies 25
2.5.1 Energy efficient lighting 25
2.5.2 Electric vehicles 26
2.5.3 Cellulosic biomass 27
2.5.4 Wind and solar 28
Acknowledgements 29
References 30

Chapter 3
Greenhouse gases in the urban water cycle **33**
Andrew Shaw, Kubeshnee Chetty, Tak Fan Chan, Elena Lindsey, Anjana Kadava and Ben Stevenson

3.1 Introduction 33
3.1.1 Overview of the urban water cycle 33
3.1.2 Definition of scope 1, 2 and 3 emissions 35
3.1.3 Water footprint and carbon footprint 35
3.2 Greenhouse Gasses in the Water Cycle 35
3.2.1 Scope 1 – direct emissions – from own and controlled sources 37
3.2.2 Scope 2 – GHGs from energy use 38
3.2.3 Scope 3 – indirect emissions from other activities 39
3.2.4 Carbon sequestration and mitigation 40
3.3 Protocols 40
3.3.1 International protocols 40
3.3.2 Regional protocols 41
3.4 Methods of GHG Quantification 43
3.4.1 Emission factors 43
3.4.2 Direct measurement 44
3.4.3 Models 45
3.4.4 Quantification method selection 45
3.5 A Framework for Carbon Footprint Analysis 46
3.5.1 A roadmap to reducing carbon footprint in the water cycle 47
References 48

Chapter 4
Operational optimization and control strategies for decarbonization in WRRFs **51**
Krishna R. Pagilla

4.1 Introduction 51
4.2 Optimization Strategies at the Process/Operation Level 52
4.2.1 Wastewater pumping 54
4.2.2 Secondary treatment 55
4.2.3 Sludge treatment 57

4.3 Operational Strategies at the Whole Facility Level . . . 60
4.4 Pathways for Decarbonization and Future Perspectives . . . 62
References . . . 64

Chapter 5
Energy and resource recovery using the anaerobic digestion platform . . . **67**
Prathap Parameswaran, Jessica A. Deaver, Sudeep C. Popat, Vikas Khanna, Madison Kratzer and Mel Harclerode

5.1 Current State of the Art for Anaerobic Digestion in Municipal Wastewater Resource Recovery Facilities . . . 67
5.2 Need for Sludge Pretreatment to Enhance Viability of Anaerobic Digestion . . . 70
5.3 Diversifying Portfolio of Anaerobic Digestion at Municipal Wastewater Facilities – The Advent of Anaerobic co-digestion . . . 71
5.3.1 Theoretical basis/substrates used . . . 71
5.3.2 Challenges of ACoD . . . 71
5.3.3 Current research on ACoD . . . 72
5.4 Enhancing the Value of the Produced Biogas Through Co-generation and Further Purification to Natural Gas for Pipeline Delivery . . . 72
5.4.1 Biomethane for combined heat and power production . . . 72
5.4.2 Biomethane for electricity generation . . . 72
5.4.3 Biomethane for upgrading and pipeline delivery . . . 73
5.4.4 Biomethane for transportation . . . 73
5.4.5 Biogas to valuable chemicals . . . 73
5.5 Altering the AD Platform for Higher Organic Carbon Product Capture Coupled with Water Reuse and Nutrient Recovery . . . 73
5.6 Energy Management in Anaerobic Digestion for Overall Energy Neutrality or Energy Positive Treatment – The Case for Direct Anaerobic Treatment Through AnMBRs . . . 74
5.7 Techno-economic and Life Cycle Assessments for Shaping the Future of Anaerobic Digestion . . . 75
5.8 Future Strategies and Roadmaps to Decarbonization . . . 77
5.9 Anaerobic Digestion Techniques for Achievement of a Circular Economy . . . 78
Acknowledgements . . . 78
References . . . 78

Chapter 6
Carbon valorization using the microbial electrochemical technology platform . . . **83**
Jayesh M. Sonawane, Zhiyong Jason Ren and Deepak Pant

6.1 Introduction . . . 83
6.2 The Principles of Microbial Electrochemical Carbon Valorization . . . 84
6.2.1 Biocatalytic CO_2 capture and conversion to organic chemicals in MES and EF . . . 84
6.2.2 CO_2 capture and mineralization in MECC . . . 85
6.3 Valorization of Carbon Compounds by MES . . . 86
6.3.1 Methane or acetic acid production in MES . . . 87
6.3.2 Role of hydrogen in MES . . . 88
6.3.3 CO_2 valorization potential from the MES platform . . . 89
6.4 Valorization of Carbon Compounds by Electrofermentation . . . 90
6.4.1 Mechanisms of anodic and cathodic EFs . . . 91
6.4.2 Synergy between EF and anaerobic digestion . . . 91

6.5 CO_2 Mineralization in MECC ... 93
6.6 Outlook ... 94
References ... 95

Chapter 7
Decarbonization potentials in nitrogen management ... **99**
Kester McCullough, Stephanie Klaus and Charles Bott

7.1 Introduction ... 99
7.1.1 Carbon footprint costs of nitrogen removal ... 101
7.2 Carbon Removal/Diversion in Primary Treatment ... 103
7.3 Approaches for Carbon Efficient Nitrogen Removal ... 103
7.3.1 Traditional nitrification/denitrification ... 103
7.3.2 Nitrite shunt and PNA (NOB out-selection) ... 104
7.3.3 Partial denitrification/anammox ... 105
7.3.4 Aeration, alkalinity, and COD requirements for mainstream nitrogen removal technologies ... 105
7.3.5 Process control ... 107
7.4 Shortcut Nitrogen Removal Implementations ... 108
7.4.1 Sidestream treatment ... 108
7.4.2 Mainstream PNA/nitrite shunt ... 108
7.4.3 Mainstream PdNA ... 108
7.4.4 Partial denitrification/anammox (PdNA) case study ... 108
7.5 Conclusions and Future Outlook ... 109
References ... 111

Chapter 8
Decarbonization potentials in phosphorus management in the water sector ... **115**
Annalisa Onnis-Hayden, Dongqi Wang, Ali Akbari, Mi Nguyen and April Z. Gu

8.1 Overview of Global Phosphorus Consumption and Demand in Relation to Water Sustainability ... 115
8.1.1 P management and decarbonization potential pathways ... 117
8.1.2 Phosphorus management and policy: Current status and practice ... 118
8.2 Direct Decarbonization and Indirect Carbon Reduction Strategies from Point Source and Non-point Sources of Phosphorus ... 120
8.2.1 Phosphorus in agricultural waste streams ... 120
8.2.2 Phosphorus in industrial effluent: Best management practices for decarbonization ... 123
8.2.3 Phosphorus in domestic waste streams ... 123
8.2.4 Phosphorus in urban runoff and best management practices for decarbonization ... 126
8.3 Decarbonization in Phosphorus Removal and Recovery Processes ... 126
8.3.1 Carbon requirements in enhanced biological phosphorus removal processes ... 126
8.3.2 Carbon footprint reducing via operational strategies for EBPR ... 130
8.3.3 Carbon footprint reducing via new pathway/process for EBPR ... 131
8.3.4 Additional technologies for phosphorus removal and recovery from wastewater streams with carbon footprint reduction potential ... 132
8.4 Quantification of Decarbonization Potential from Phosphorus Removal and Recovery Processes ... 135
8.4.1 LCA studies for the quantification of decarbonization potential for non-point sources ... 135

8.4.2 LCA studies for P removal and recovery processes in WWTPs and quantification of decarbonization potential 136
8.5 Future Outlook and Research Needs 136
References 137

Chapter 9
Decarbonization potentials using photobiological systems 143
Lara Méndez, Cristian A. Sepúlveda-Muñoz, María del Rosario Rodero, Ignacio de Godos and Raúl Muñoz

9.1 Introduction 143
9.2 Photosynthetic Wastewater Treatment 144
9.2.1 Microalgae 144
9.2.2 Purple phototrophic bacteria 148
9.2.3 Photobioreactors for wastewater treatment and resource recovery 151
9.3 Enhanced Biogas Production from Microalgae and PPB for WWT Decarbonization 154
9.4 CO_2 Capture and Biogas Upgrading Using Photosynthetic Systems During WWT 157
Acknowledgement 161
References 161

Chapter 10
Sludge management and utilization for decarbonization 171
Meltem Urgun-Demirtas, Rachel Dalke and Krishna R. Pagilla

10.1 Overview of Current Sludge Management Practices 171
10.2 Sludge to Energy/Products 173
10.3 Land Application and Dedicated Landfilling 176
10.4 Reclamation of Brownfields 177
10.5 Composting 177
10.6 Resource Recovery 178
10.7 Sludge Stabilziation for Removal of Emerging Contaminants 179
10.8 Centralized vs. Distributed Sludge Management Practices 179
10.9 Environmental and Economic Life Cycle Assessment of Sludge Management Technologies 179
10.10 Implementation: Challenges and Opportunities, Regulatory and Social Issues 180
10.11 Future Strategies and Roadmaps 180
References 182

Chapter 11
Decarbonization potentials in intensified water and wastewater systems using membrane-related technologies 187
Boyan Xu, Shujuan Huang, Chuansheng Wang, Tze Chiang Albert Ng and How Yong Ng

11.1 Introduction 187
11.2 Membrane Strategies for Decarbonization in Wastewater Treatment and Resource Recovery 189
11.2.1 Aerobic granular sludge membrane bioreactors (AGMBRs) 189
11.2.2 Algae membrane bioreactors (A-MBRs) 190

11.2.3 Anaerobic membrane bioreactor (AnMBRs) ... 191
11.2.4 Membrane biofilm reactors (MBfRs) ... 192
11.2.5 Forward osmosis (FO) integrated processes for wastewater treatment/reclamation and resource recovery ... 193
11.3 Potential Membrane Strategies for Decarbonization in Desalination ... 196
11.3.1 Pressure retarded osmosis (PRO) in desalination for power generation ... 196
11.3.2 Forward osmosis-reverse osmosis (FO-RO) hybrid for desalination and wastewater concentration ... 198
11.3.3 Forward osmosis-membrane distillation hybrid (FO-MD) for desalination ... 199
11.4 Conclusions ... 200
References ... 200

Chapter 12
Natural treatment systems and integrated watershed management for decarbonization ... **207**
Hannah R. Molitor and Jerald L. Schnoor

12.1 Introduction ... 207
12.2 Natural Treatment Systems ... 209
12.2.1 Constructed wetlands ... 211
12.2.2 Treatment lagoons ... 212
12.2.3 Bioremediation and biofiltration ... 213
12.2.4 Microalgal cultivation ... 214
12.2.5 Land treatment systems ... 214
12.3 Case Study: Carbon Sequestration and Agricultural Runoff Treatment through Phytoremediation ... 217
12.4 Case Study: Power Plant Flue Gas and Fertilizer Wastewater Treatment by Nutritious Microalgae ... 218
12.4.1 Scaling-up microalgal cultivation ... 219
12.4.2 GHG and land footprints ... 220
12.4.3 Microalgal end products ... 221
12.5 Conclusions and Outlook ... 221
References ... 223

Chapter 13
Microbial electrochemical communication in carbon and electron flow for CO_2 methanation ... **227**
Aijie Wang, Bo Wang, Zechong Guo, Weiwei Cai and Wenzong Liu

13.1 Introduction ... 227
13.2 Carbon Conversion and Electron Flow to Electromethanogenesis ... 228
13.2.1 Functional communities and genes involved in carbon conversion ... 229
13.2.2 Organic conversion for CH_4 production under electrochemistry regulation ... 230
13.3 Upgraded CH_4 Production from Electrochemically Enhanced Anaerpbic Digestion ... 232
13.3.1 Hytrogentrophic methanogenesis pathway in anaerobic digestion ... 232
13.3.2 Microbial community evolution in cathode biofilm for methanation ... 233
13.3.3 Microbial network of electrochemically enhanced AD ... 235
13.4 CO_2 Methanation Driven by Solar-powered Bioelectrochemical System ... 237
13.4.1 Solar intermittent driven-power accelerates bioelectrochemical performances ... 237
13.4.2 Intermittent electro field mediates mutualistic interspecies electron transfer ... 240

13.5 Challenges and Perspectives. 241
Acknowledgements 241
References. 242

Chapter 14
***Thermal energy from wastewater* 249**
James McQuarrie

14.1 Introduction 249
14.2 Wastewater as a Thermal Energy Source 251
14.3 Integration of Thermal Energy Recovery from Wastewater with Modern District Energy Systems 254
14.4 Assessing the Technical Feasibility of Thermal Energy Recovery from Wastewater. 256
14.5 Adoption of Thermal Energy Recovery from Wastewater 257
14.6 Opportunities and Barriers to Thermal Energy from Wastewater. 259
14.6.1 Defining strategic planning 259
14.6.2 Demand and resource mapping 260
14.6.3 Technical feasibility. 261
14.6.4 Regulatory and financing frameworks 261
References. 263

Chapter 15
***Concept wastewater treatment plants in China* 265**
Jiuhui Qu, Hongqiang Ren, Hongchen Wang, Kaijun Wang, Gang Yu, Bing Ke, Han-Qing Yu, Xingcan Zheng and Ji Li

15.1 Introduction 265
15.2 The Current Challenges of China's Urban Wastewater Treatment Facilities 266
15.3 Wastewater Concept Plant Provides a Vision and Example for Future Development. 267
15.4 The Next Paradigm of Wastewater Treatment 272
References. 272

Chapter 16
***Data science tools to enable decarbonized water and wastewater treatment systems* 275**
Kathryn B. Newhart, Amanda S. Hering and Tzahi Y. Cath

16.1 Introduction 275
16.2 Principle Data Science Tools 277
16.2.1 Data preparation 277
16.2.2 Measuring accuracy. 279
16.2.3 Dimension reduction. 280
16.2.4 Principal component analysis. 281
16.2.5 Prediction and forecasting 284
16.2.6 Optimization 288
16.3 Data Science Applications to Select Treatment Systems. 290
16.3.1 Pump optimization 290
16.3.2 Nitrification 293
16.3.3 Anaerobic digestion. 294

16.4 Recommendations for Full-scale Implementation 296
16.5 Conclusions 297
Acknowledgements 297
References 297

Chapter 17
***Decarbonization policies and water sector opportunities* 303**
Jason A. Turgeon, Steven A. Conrad and Peter A. Vanrolleghem

17.1 Introduction 303
17.2 Core Concepts 303
17.2.1 Concept 1: In the scale of national or global policies the sector's energy use is relatively small, but there are other resources contained in wastewater worth considering 303
17.2.2 Concept 2: There is no overarching policy that mandates decarbonization in the water sector globally 306
17.2.3 Concept 3: Seek out co-benefits with other policy areas 309
17.2.4 Concept 4: Water reuse and energy recovery may be more beneficial in a distributed setting where the water and energy are recovered at/near the point of generation 312
17.2.5 Concept 5: Change is constant. We need water policy and technological platforms that can more easily adapt to change 315
17.3 Conclusion and Policy Suggestions 316
17.4 Additional Resources 317
References 318

Chapter 18
***Outlook for the carbon-negative circular water economy* 321**
Glen T. Daigger

18.1 Introduction 321
18.2 Resource Recovery 322
18.2.1 Historical perspective 322
18.2.2 Value hierarchies 323
18.2.3 Advancing markets and products 325
18.3 Accelerating Transitions 326
18.4 The Path Forward 329
References 331

Index 333

About the Editors

Dr. Zhiyong Jason Ren (@zjasonren) is a Professor in the Department of Civil and Environmental Engineering and the Andlinger Center for Energy and the Environment at Princeton University. He is the Associate Director for Research for the Andlinger Center and served as the Acting Director in 2020–2021. He is also a visiting scientist at the National Renewable Energy Laboratory.

Dr. Ren leads the Princeton WET Lab (Water & Energy Technologies) with research and teaching focusing on water & wastewater sector decarbonization and digitalization. His group uses electrochemistry, microbiology, and data science tools to understand the fundamental determining factors and develops models and technologies for resource recovery during environmental processes such as wastewater treatment, water desalination, environmental remediation, and carbon capture and utilization.

Dr. Ren has received numerous recognitions including the Paul L. Busch Award from the Water Research Foundation (2021), the Walter L. Huber Research Prize from the American Society of Civil Engineers (2020), the ISMET Innovation Award (2020), the Nanova/CAPEES Frontier in Research Award (2017), and the New Inventor of the Year Award (2015).

Dr. Ren has productive collaborations with water utilities and consulting companies. The partnerships have yielded the First Place Prize in Intelligent Water Systems Challenge (2021), successful technology demonstrations, and a startup company co-founded with students. Dr. Ren and his group have published nearly 200 peer reviewed articles in leading research journals, and their findings were featured by *National Public Radio, Forbes, and Scientific American*.

Dr. Ren serves on the National Steering Committee for Water Resource Recovery. He is a Topic Editor for *Environmental Science & Technology (ES&T) and ES&T Letters (ACS)* and an Associate Editor for *Environmental Science: Water Research Technology* (*RSC*). He received his Ph.D. in environmental engineering from Penn State University after spending several years in consulting industry.

Dr. Krishna Pagilla is the Ralph E. and Rose A. Hoeper Engineering Professor, Foundation Professor, and Chair of Civil and Environmental Engineering Department at the University of Nevada, Reno. He is also the Director of Nevada Water Innovation Institute, a university-utility collaboration to meet regional water technology and development needs, and to drive leading edge research and innovation in the water sector. Dr. Pagilla's expertise is in water quality, water resource recovery, indirect potable water reuse, and environmental biotechnology. More recently, his focus is on water reclamation for indirect potable reuse through ozone-biologically active carbon filtration-based treatment systems, wastewater-based epidemiology to determine community prevalence of COVID-19 and SARS-CoV-2 variants of concern, and decarbonization of water reclamation facilities.

Dr. Pagilla received numerous awards from national and international organizations, and is a Fellow of Water Environment Federation (WEF), International Water Association (IWA), and the American Society of Civil Engineers. Dr. Pagilla received the Engelbrecht International Service Award (2021), McKee Groundwater Protection, Restoration, or Sustainable Use Award (2020), Thomas R. Camp Applied Research Award (2013), and Fair Distinguished Engineering Educator Award (2013), and Harrison Prescott Eddy Medal for Outstanding Applied Research on Wastewater Principles and Processes (2011) from the Water Environment Federation (WEF). He received the Bill Boyle Outstanding Educator Award (2012) from the Central States Water Environment Association.

Dr. Pagilla held numerous leadership positions in the water profession. He served as President of Illinois Water Environment Association (2012–13), as President of Nevada Water Environment Association (2020–21), and as the Chair of the USA National Committee of IWA (2016–19). He currently serves as an International Juror for the Stockholm Junior Water Prize Competition conducted by the Stockholm International Water Institute, and as a member of the USEPA Science Advisory Board.

Dr. Pagilla holds a PhD degree in Civil/Environmental Engineering from the University of California (Berkeley, CA, USA) and MS and BE degrees in Civil/Environmental Engineering. Dr. Pagilla is a Registered Professional Engineer (PE) in Illinois and California, USA and Board-Certified Environmental Engineer (BCEE) of the American Academy of Environmental Engineers and Scientists.

List of Contributors

Ali Akbari
Cornell University, NY, USA

Charles Bott
Hampton Roads Sanitation District, VA, USA

Weiwei Cai
Beijing Jiaotong University, Beijing, China

Tzahi Y. Cath
Colorado School of Mines, CO, USA

Tak Fan Chan
Binnies, Surrey, United Kingdom

Kubeshnee Chetty
Binnies, Surrey, United Kingdom

Steven A. Conrad
Colorado State University, CO, USA

Glen T. Daigger
University of Michigan, MI, USA

Rachel Dalke
Argonne National Laboratory, IL, USA

Jessica A. Deaver
Clemson University, SC, USA

Ignacio de Godos
University of Valladolid, Valladolid, Spain

April Z Gu
Cornell University, NY, USA

Zechong Guo
Jiangsu University of Science and Technology, Zhenjiang, China

Mel Harclerode
CDM Smith Inc., NJ, USA

Amanda S. Hering
Baylor University, TX, USA

Shujuan Huang,
National University of Singapore, Singapore

Anjana Kadava
Black & Veatch Corporation, TX, USA

Bing Ke
Ministry of Science and Technology, Beijing, China

Vikas Khanna
University of Pittsburgh, PA, USA

Stephanie Klaus
Hampton Roads Sanitation District, VA, USA

Madison Kratzer
University of Pittsburgh, PA, USA

Ji Li
Jiangnan University, Wuxi, China

Elena Lindsey
Binnies, Surrey, United Kingdom

Wenzong Liu
Harbin Institute of Technology Shenzhen, Shenzhen, China

Kester McCullough
Cornell University & Hampton Roads Sanitation District, USA

James McQuarrie
Tetra Tech Engineering Services Company, CO, USA

Lara Méndez
University of Valladolid, Valladolid, Spain

Hannah R. Molitor
University of Iowa, IA, USA

Raúl Muñoz
University of Valladolid, Valladolid, Spain

Kathryn B. Newhart
United States Military Academy, NY, USA

How Yong Ng
National University of Singapore, Singapore

Tze Chiang Albert Ng
National University of Singapore, Singapore

Mi Nguyen
Cornell University, NY, USA

Annalisa Onnis-Hayden
Northeastern University, MA, USA

Krishna R. Pagilla
University of Nevada Reno, NV, USA

Deepak Pant
Flemish Institute for Technological Research (VITO), Boeretang, Belgium

Prathap Parameswaran,
Kansas State University, KS, USA

Sudeep C. Popat
Clemson University, SC, USA

Jiuhui Qu
Tsinghua University, Beijing, China

Hongqiang Ren
Nanjing University, Nanjing, China

Zhiyong Jason Ren
Princeton University, NJ, USA

María del Rosario Rodero
University of Valladolid, Valladolid, Spain

Jerald L. Schnoor
University of Iowa, IA, USA

Cristian A. Sepúlveda-Muñoz
University of Valladolid, Valladolid, Spain

Andrew Shaw
Black & Veatch Corporation, TX, USA

A. J. Simon
Lawrence Livermore National Laboratory, CA, USA

Seth W. Snyder
Idaho National Laboratory & Northwestern University, USA

Jayesh M. Sonawane
Laval University, Quebec, Canada

Ben Stevenson
Black & Veatch Ltd, Glasgow, United Kingdom

Jason A. Turgeon
United States Environmental Protection Agency, MA, USA

Art K. Umble
Stantec, Denver, CO, USA

Meltem Urgun-Demirtas
Argonne National Laboratory, IL, USA

Kala Vairavamoorthy
International Water Association, United Kingdom

Peter A. Vanrolleghem
Université Laval, Quebec, Canada

Aijie Wang
Harbin Institute of Technology Shenzhen, Shenzhen, China

Bo Wang
Aarhus University, Aarhus, Denmark

Chuansheng Wang
National University of Singapore, Singapore

Dongqi Wang
Xi'an University of Technology, Shaanxi, China

Hongchen Wang
Renmin University of China, Beijing, China

Kaijun Wang
Tsinghua University, Beijing, China

Dru Whitlock
Stantec, Salt Lake City, Utah, USA

Boyan Xu
National University of Singapore, Singapore

Han-Qing Yu
University of Science and Technology of China, Hefei, China

Gang Yu
Tsinghua University, Beijing, China

Xingcan Zheng
North China Municipal Engineering Design & Research Institute, Tianjin, China

Foreword

Kala Vairavamoorthy

Decarbonization is a topic and reality whose time has come. The water sector is committed to meeting the needs of society, from water supply and sanitation through to protection of the natural environment. Those needs now extend to contributing, where possible, to combating climate change, and to demonstrating leadership on how to best use the world's precious resources. This book supports progress on both of these fronts.

Great practical progress is already being made in cities and utilities around the world with actions aimed at achieving net zero carbon emissions. This progress combines a drive for greater efficiency in our current approaches to managing water with the development and deployment of new processes and technologies opening alternative approaches to resource use.

While technology breakthroughs and innovative designs can help us respond to some of these impacts, this needs to be coupled with comprehensive system change. Water is a system of systems, and decarbonization approaches need to be implemented and coordinated across these systems – at the basin level, the city level and the utility level. More than simply improving the performance and efficiency of the component parts, change is needed at a system level too. Therefore, such progress is not something those in the water sector can deliver fully by working alone. It requires cooperation and partnerships, in which the water sector can demonstrate leadership as a facilitator.

This book, written by experts – both water engineers and scientists – and specifically addressing decarbonization pathways for the water sector, is much needed. It combines the foundations, evidence and vision to support and stimulate practical progress.

This book can inspire the Global South. Here we see the biggest opportunities to reimagine how we do water, as many of these places are starting from scratch and are on the cusp of making huge investments in water infrastructure and services. It would be wonderful if this could be done in a low carbon way – with energy neutral, efficient and productive use of water, maximizing the capture of value from water and waste streams. Much as the Global South 'leapfrogged' fixed wires of communications infrastructure, so they can avoid the slow, costly, mechanistic, and heavy legacy of high carbon, centralized systems – by moving to low carbon, off-grid, distributed, flexible and circular systems.

The scope, content and intentions of the book align with the outlook of the International Water Association (IWA). The water sector, especially the utilities at its heart, faces multiple challenges in meeting the current and future needs of society. IWA supports the journey ahead, providing leadership and generating and sharing knowledge to enable action. IWA does this through initiatives such as

Climate Smart Utilities and Digital Water Program as well as others aimed at reframing thinking at the basin and the city level, as well as promoting the adoption of nature-based solutions.

The Climate Smart Utilities initiative encourages utilities to become leaders in climate mitigation, by providing them with approaches and tools to assess, monitor and reduce their greenhouse emissions while enhancing their ability to adapt to climate change. Meanwhile, our Digital Water Program is enabling utilities and its customers to transition towards a new low-carbon paradigm, where data-driven models can help integrate and optimize smart pumps, valves, sensors and actuators in order to maximize service levels while minimizing carbon footprint.

The latest developments in the field of decarbonization are brought together in this book thanks to the input of an outstanding array of authors. Their expertise spans the diversity of this important topic area, which touches all parts of the water sector, with each chapter drawing on a great depth of expertise.

These contributions have been brought together expertly by the Editors, Zhiyong Jason Ren and Krishna Pagilla. Their deep appreciation of this field and their awareness that now is such an important time for these perspectives to be presented underpin the value of this book.

The book also highlights the need for leadership, and itself represents a contribution to leadership in the sector, helping lay out a path ahead. Here, I see that the focus on decarbonization represents an opportunity of fundamental importance – both to the sector and to the world at large. To date, our economies have been built as high carbon economies and an expectation of a ready supply of water. There is a window of opportunity for water to be at the heart of the low carbon economy – one that shifts away from use of fossil fuels and at the same time recognizes our resource limitations and adopts a circular approach. It is an opportunity for those in the sector to show leadership, and this book provides valuable tools to do just that.

Kala Vairavamoorthy, PhD
Executive Director, International Water Association

Foreword

Art K. Umble

Few would argue that economic, scientific and socio-cultural advances achieved since the Industrial Revolution have greatly enhanced the quality of life for a large number of our world's population. The environmental costs of these achievements, however, are now beginning to manifest in ways which threaten those same life qualities we all have come to take for granted. The evidence is clear: the direct impacts on our physical world from climate change are on the rise, threatening our quality of life. Perhaps the greatest challenge climate change poses for humanity is not in discovering and implementing solutions that slow and reverse the rate of changing climate, but in loosening our collective grip on the comfort of our modern conveniences. This mindset hinders our progress toward effective change.

The dynamics of a changing climate are pressing down hard on the water industry. We must think differently about pathways to real solutions, the lifespan of solutions, and the resiliency of solutions. This thinking requires intensifying our efforts in identifying and accelerating innovative water treatment and distribution and wastewater collections and treatment technologies, coupled with tactical applied research, and forging strategic relationships with organizations, including regulators, to partner with us in implementing solutions. This book provided exactly these pathways.

Climate change is pushing us to put carbon management front and center of every solution approach in the water space (i.e., stepping down our carbon footprints to net-neutral and ultimately to net-zero operation). Though the water sector may need solutions oriented toward adaptation for the short term to bolster resiliency, emphasis needs to be on outcomes that result in long-term mitigation measures that ensure a sustainable future of our water environment–and the whole planet. All this means recognizing water's role in the management of all primary resources that support energy, agriculture, mineral mining, manufacturing, and construction economy sectors. It's also about reducing and capturing wastes for product reuse, remanufacturing and recycling, reducing operational carbon emissions and offsetting emissions through sequestration. These actions outline the circular economy, the core of a sustainable future.

In many ways, the global water sector is the epitome of an ultimate circular economy model. It is no secret that the changing climate significantly impacts the entire hydrologic cycle, from rainfall to drought, from declining aquifers to rising sea levels, from soil erosion to declining water quality, and more. The first and foremost mission of the water industry is to supply, treat and distribute safe, potable water to protect public health and sustain life itself, but in the context of climate change, water is a "product" that every living thing depends on for survival, so the water industry is uniquely positioned to effectively enact circular economic principles and lead by example on mitigating climate change through decarbonization.

Our time to implement circularity in the water industry to abate the impacts of climate change is running short. A business-as-usual mindset is unacceptable. The status quo must be shattered. The thought-provoking ideas and potential pathways contained in this book provide a framework for which circularity within the water industry can be a reality. The authors and the editors have provided the start and a route map well. The industry must act now.

Art K. Umble, PhD, PE, BCEE, F.WEF
Senior Vice President, Stantec
Director, Stantec Institute for Water Technology & Policy

doi: 10.2166/9781789061796

Preface

The water sector is in the middle of a paradigm shift from focusing on treatment and meeting discharge permit limits to integrated operation that also enables a circular water economy via water reuse, resource recovery, and system level planning and operation. While the sector has gone through different stages of such revolution, from improving energy efficiency to recovering renewable energy and resources, when it comes to the next step of achieving carbon neutral or negative emissions, it falls behind other infrastructure sectors such as energy and transportation. Decarbonization refers to the reduction of carbon footprint of an industry and in the long run creates a circular economy with integrated solutions for carbon management. The water sector carries tremendous potential to decarbonize, from technological advancements to operational optimization, to policy and behavioral changes.

This book aims to fill an important gap for different stakeholders to gain knowledge and skills in this area and equip the water community to further decarbonize the industry and build a low carbon society and economy. The book goes beyond technology overviews; rather it aims to provide a system level blueprint or pathways for decarbonization, carbon capture, and utilization in the water sector. We hope that this book will become an inspiration to develop practices and solutions that will drive innovation in the water sector for decarbonization. Here is a snapshot of what you will find in the book.

The first section of the book lays out a framework on the state-of-the-art in water sector carbon footprints. **Chapter 1** provides an overview of the challenges and opportunities on water sector decarbonization, and it summarizes the needs and approaches to achieve the net zero carbon goals. **Chapter 2** offers a comprehensive review on the pathways other infrastructure sectors (e.g., energy, transportation) explored and identifies the synergies and examples for the water sector to consider. **Chapter 3** discusses the different scopes of greenhouse gas (GHG) emissions associated with the urban water cycles and provides an overview of the carbon footprint accounting methods and protocols.

The second section of the book provides reviews and details on different processes and technologies that enable decarbonization, carbon capture, and utilization. The experts in each respective field offer deep insights on how the approach has been used to increase energy efficiency, reduce carbon footprint, recover resources, and capture and valorize GHG while maintaining treatment goals. **Chapter 4** starts with the easily implemented methods associated with operations at pumping, preliminary, primary, secondary, advanced, and sludge treatment level within a water resource recovery facility (WRRF). **Chapter 5** focuses on the energy and resource recovery from the commonly used anaerobic digestion (AD) platform, which also includes emerging processes such as anaerobic membrane bioreactors (AnMBR) and thermal hydrolysis. **Chapter 6** explores opportunities for renewable energy production and carbon valorization using the new microbial electrochemical technology (MET) platform. **Chapter 7** and **Chapter 8** investigate the critical considerations and tremendous potentials on decarbonization during nitrogen and phosphorus removal and recovery processes, respectively. **Chapter 9** describes the increasing popular photobiological systems using microalgae and cyanobacteria for CO_2 capture and conversion. **Chapter 10** focuses on sludge management and utilization using AD, compositing, incineration, as well as emerging processes like hydrothermal liquefaction (HTL). **Chapter 11** discusses several novel membrane technologies that enable process intensification with reduced energy consumption, and it provides an overview of system integration and optimization. **Chapter 12** analyzes the advantages and challenges of natural treatment systems and their potential for integrated watershed management and decarbonization in the context of One Water. **Chapter 13** discusses the fundamental carbon and electron flows occurring in anaerobic bioconversion systems for CO_2 capture and conversion to value-added organic chemicals. Lastly, **Chapter 14** assesses the potential in recovering the abundant low quality thermal energy from wastewater and its feasibility of integration with district heating.

The third section of the book covers the broader prospects in water sector decarbonization in the context of policy making, intelligent water systems, as well as case studies. **Chapter 15** describes several "wastewater concept plants" designed and built in China in recent years as examples for the next generation of WRRFs for integrated waste management and resource recovery. **Chapter 16** introduces the modern data science tools including statistical and machine learning methods that can be used for decarbonization. **Chapter 17** provides a critical analysis of local and national polices that will impact the efforts and highlights the need to seek multiple benefits whenever possible. Lastly, the concluding **Chapter 18** summarizes the evolution of the missions of water management, the need of concerted efforts to move the industry forward, and the tangible benefits such endeavors will make to the society, the economy, and the environment.

This book can be a reference book and textbook for undergraduate and graduate students, researchers, practitioners, consultants, and policy makers, and it will provide practical guidance for stakeholders to analyze and implement decarbonization measures in their professions. The goal is to provide pathways for decarbonization from various perspectives. We are confident that the readers of this book will be inspired to seek innovation in water management while achieving decarbonization.

We want to thank the authors and contributors for their time in the writing of this book and their dedications in advancing the understanding and contributing to the grand mission of water sector decarbonization. Their expertise and knowledge have been generously shared in this book. We also want to thank The International Water Association, Knowledge Unlatched, Princeton University, and University of Nevada Reno for their generous support to make this book accessible to many readers.

Zhiyong Jason Ren,
Princeton University

Krishna R Pagilla,
University of Nevada, Reno

doi: 10.2166/9781789061796

Chapter 1

Toward a net zero circular water economy

Zhiyong Jason Ren[1], Jerald L. Schnoor[2] and Krishna R. Pagilla[3]*

[1]*Department of Civil and Environmental Engineering & Andlinger Center for Energy and the Environment, Princeton University, Princeton, NJ, USA*
[2]*Department of Civil and Environmental Engineering, University of Iowa, Iowa City, IA, USA*
[3]*Department of Civil and Environmental Engineering, University of Nevada Reno, Reno, NV, USA*
**Correspondence: zjren@princeton.edu*

1.1 THE WATER SECTOR AND THE CHALLENGES AND OPPORTUNITIES ON DECARBONIZATION

Water underpins every aspect of life, and the water industry is the guardian of human society's and environmental water needs. From ancient Rome's aqueducts to modern water networks, the water sector has been playing a critical role in civilization and paving a pathway to a more sustainable and prospective world (Sedlak, 2014). Each nation's critical water infrastructure relies on the smooth operation of water and wastewater systems of different sizes. Water utilities treat and deliver billions of liters of water to homes and industries every day, and wastewater utilities collect and treat the generated wastewater to ensure the effluent can be safely discharged or reused. Acknowledging the interconnected nature of water management, the emerging 'OneWater' framework offers a holistic and integrated approach to consider all water resources from surface water, ground water, stormwater, potable water, wastewater, and recycled water as one to achieve reliable, sustainable, and resilient water systems (Figure 1.1).

However, the water sector is facing exacerbated challenges caused by climate change: extreme weather events, frequent floods or prolonged droughts, degradation of water quality, along with aging infrastructure and population redistribution. The industry needs a paradigm shift from focusing on water treatment and supply from scarce natural water resources, wastewater collection and treatment to meet discharge permits, and processing of residuals from these operations, to integrated water management to enable a low-carbon circular water economy using the OneWater concept. The goal should be on overall sustainability including energy, greenhouse gas (GHG) emissions, resource recovery, water resiliency, and socio-economic impacts in water management.

Water industry is energy- and material-intensive in procurement, production, and renewal of used water. The International Energy Agency (IEA) estimates ~4% of the world's electricity use goes toward moving and treating water and wastewater, and electricity consumption in the water sector is projected to increase by 80% over the next 25 years despite improvements in energy efficiency (Figure 1.2) (IEA, 2017). In search of more water, lower quality water sources, including non-traditional water sources,

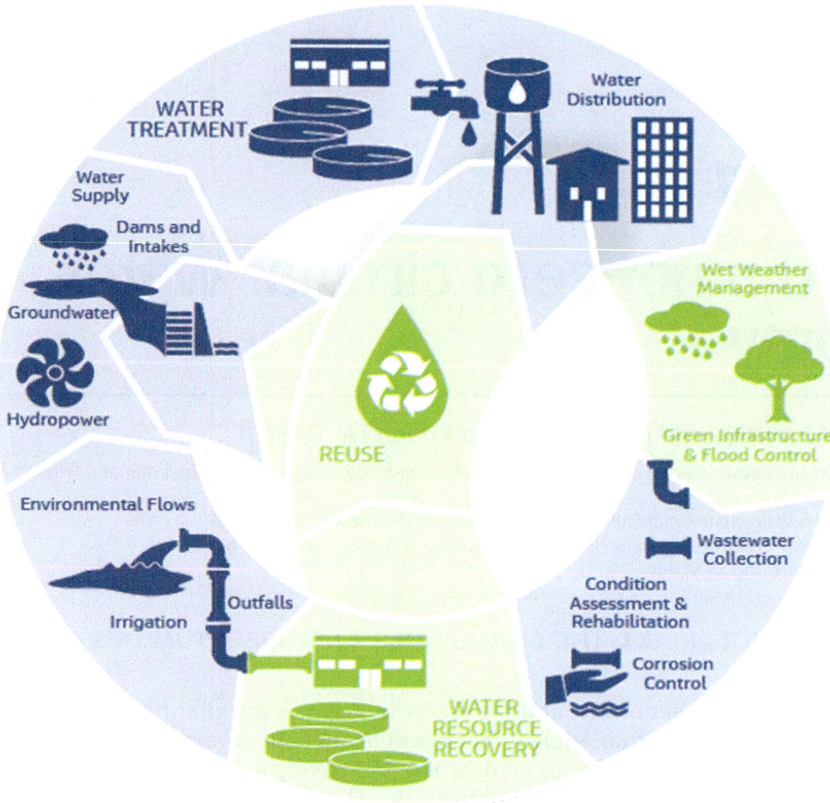

Figure 1.1 The OneWater framework for the urban water cycle (image adopted from jacobs.com, 2020).

are being considered for both water extraction and supply. This further adds to the energy intensity of water treatment and wastewater reclamation. Water, stormwater, wastewater collection, transport, and treatment use tremendous amounts of concrete, metal, and plastic materials, all of which are associated with non-renewable materials and energy sources or with intensive energy footprints.

Water and wastewater utilities typically spend 10–35% of their operational costs on energy, which is mostly generated from fossil fuel sources (IKI, 2020). This can account for as much as 40% of municipal energy use in some cities, and it mirrors the GHG emissions from the water sector. In water treatment facilities and supply systems, both scope 2 and scope 3 emissions are more significant than scope 1 (direct) emissions, while the direct non-biogenic GHG emissions within a wastewater treatment facility are scope 1 emissions. In addition, emissions include those related to imported electrical and thermal energy (scope 2) and other indirect emissions associated with the production and transportation of chemicals and fuels, waste disposal, as well as contracted services in both water and wastewater facilities (scope 3) (UNEP, 2017). However, different from other infrastructure sectors like energy and transportation where the primary GHG source is fossil-based CO_2, the direct release of CO_2 from wastewater via organic degradation is largely considered carbon neutral due to its biogenic nature, despite evidence of some carbon being of fossil origin (Griffith *et al.*, 2009). Instead, the non-CO_2 direct emissions (primarily CH_4, N_2O) from collection systems and treatment facilities are of significant concern because such GHGs are many times (28–298×) stronger in global warming potential (GWP) than CO_2 over 100 years (scope 1) (Figure 1.3) (Lu *et al.*, 2018). Currently, wastewater

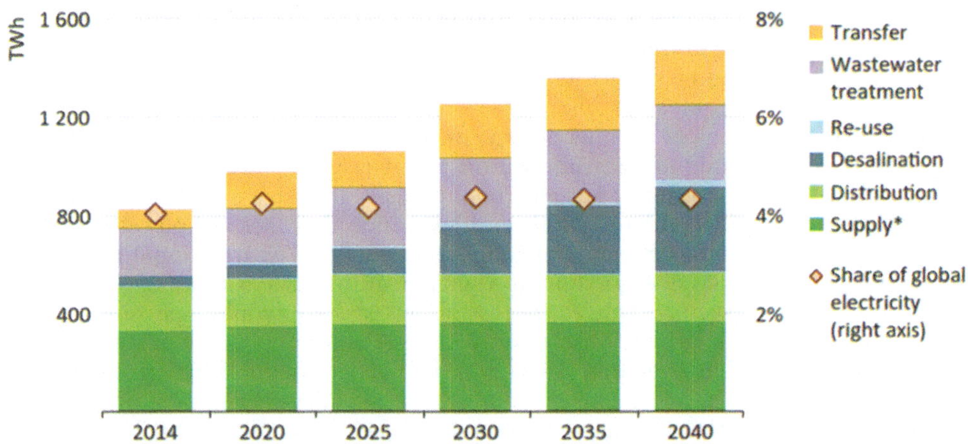

Figure 1.2 Global energy uses by different water-related activities (IEA, 2017).

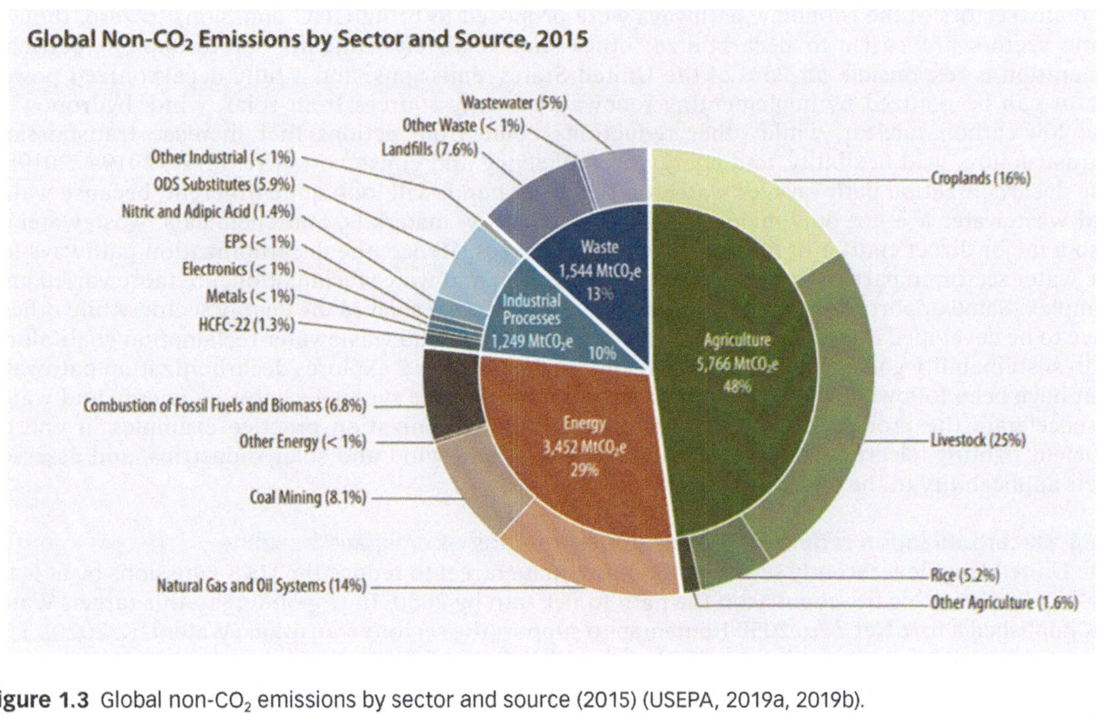

Figure 1.3 Global non-CO_2 emissions by sector and source (2015) (USEPA, 2019a, 2019b).

accounts for ~5% of the global total non-CO_2 GHG emissions, but the impacts are expected to be much higher with CH_4 emission control becoming a top priority in the next decade. Methane has >80 times the warming power of CO_2 over the first 20 years, and it only lasts 12–15 years in the atmosphere (Saunois *et al.*, 2020), so cutting methane emissions is the fastest opportunity to slow the rate of global warming and get to a net-zero emissions scenario by the mid-century.

In order to meet the 1.5°C global warming target in the Paris Agreement, many countries, cities, and industries have made commitments to move to net zero emissions by 2030–2060. That means GHG emissions would have to be dramatically reduced, and any remaining emissions will need to be balanced by absorbing an equivalent amount from the atmosphere (negative emissions or offsets). While many infrastructure sectors such as energy, transportation, and building systems have been extensively studied with regards to decarbonization pathways, the water sector is lagging behind. The water sector has not been considered as a carbon intensive industry, and because most water and wastewater utilities are heavily regulated public entities, hence, they lack the power to control prices that allow them to rationalize investments for long-term benefits. The traditional public perception is that water and wastewater service is a 'human right', meaning price is more of a function of cost, instead of value as in other sectors. Due to increasing water scarcity and environmental pollution in different parts of the world, such a 'water should be free' concept is being challenged, and many opportunities have emerged to overcome such hurdles by developing win-win solutions such as generating 'green' revenues via energy and resource recovery, developing new policies on carbon credits, and transforming empirical practice to data-driven decision making that improves efficiency and reduces cost.

1.2 PATHWAYS TOWARD WATER AND WASTEWATER DECARBONIZATION

In many sectors of the economy, pathways were proposed to bring GHG emissions to zero, though some sectors are easier to decarbonize, others are relatively difficult. For example, electricity generation is responsible for 25% of the United States' emissions, and a fully decarbonized power sector can be realized by implementing renewable energy sources from solar, wind, hydropower, and low-carbon nuclear, while other reductions come from actions that increase transmission infrastructure, grid flexibility, and energy use efficiency and conservation (USEPA, 2019a, 2019b). The decarbonization pathways for water, on the other hand, will look quite different, because water and wastewater are not only major consumers of energy, materials, and chemicals, wastewater is also a major direct emitter of non-CO_2 greenhouse gases. Hence, the decarbonization pathways for the water sector, in particular the wastewater collection and water reclamation, are more varied and complex. Some opportunities exist to learn from other sectors such as the energy sector, while others have to be developed through careful consideration of water and wastewater reclamation goals along with sustainability goals including net zero emissions. Chapter 2 explores decarbonization pathways that have been followed by the energy sector and identifies the synergies between energy and water to accelerate the process. It also discusses several decarbonization practice examples in energy efficient lighting, electric vehicles, cellulosic biomass, and wind and solar industries, and assesses their applicability to the water sector.

1.2.1 Decarbonization requires a better understanding of emission baseline

The United Kingdom recently set an ambitious climate target to reduce the UK's emissions by at least 68% by 2030, setting the country on the path to net zero by 2050. In responding to this target, Water UK published a first Net Zero 2030 Routemap to support the sector's transition (Water UK, 2020). The water companies constitute about a third of the UK's GHGs from industrial and waste management processes, but with the right support in place, it is possible this sector can become one of the most cost-effective sectors to become carbon neutral, even carbon negative, and this is a vision shared by many water professionals around the globe.

Figure 1.4 depicts the reference baseline emissions of the UK water sector in 2018–2019. This representative figure shows that the main sector emissions are attributed to CO_2 primarily from grid electricity, and CH_4 and N_2O emissions from wastewater and sludge treatment processes. These emissions can be offset by the purchase of green electricity, as well as the generation of renewable energy such as biomethane or combined heat and power.

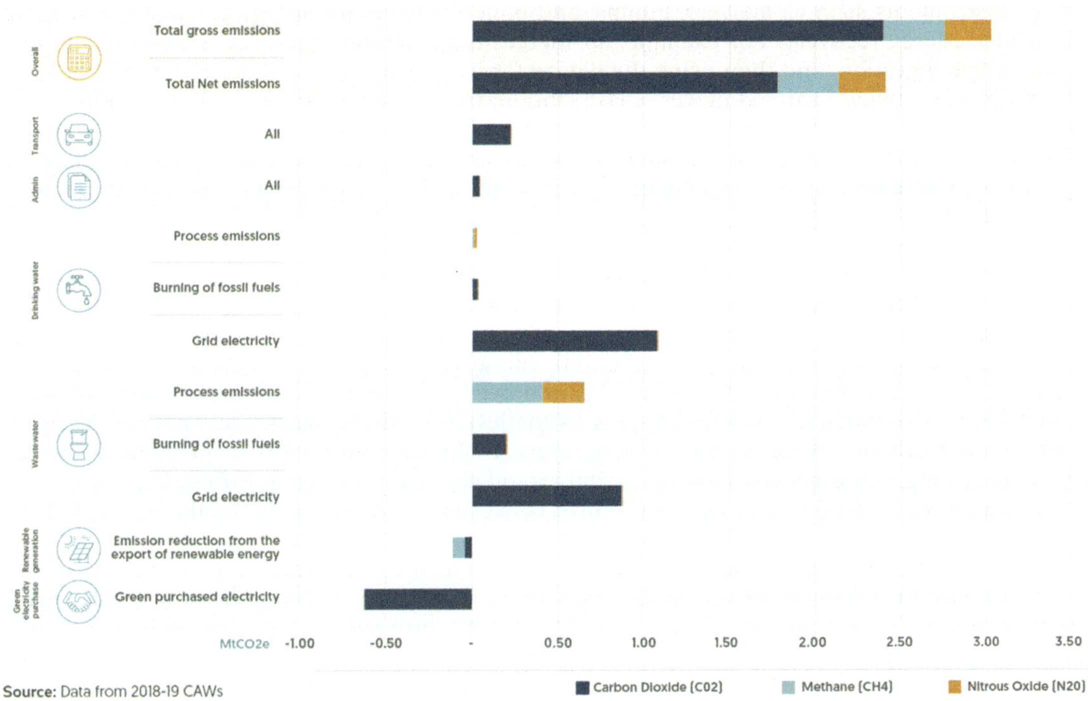

Figure 1.4 The reference baseline emissions of the UK water sector in 2018–2019 (Water UK, 2020).

While scope 2 emissions associated with grid electricity used in water/wastewater treatment and conveyance is relatively straightforward, larger uncertainties exist in estimating scope 1 direct emissions from treatment processes and conveyance systems. Generalized emission factor (EF) based estimates oversimplify the situation and do not reflect the reality of differences in actual emissions between various plants and processes, while localized flux chamber methods generate significant variations, sometimes as much as 3–4 orders of magnitude different (Delre *et al.*, 2017; Vasilaki *et al.*, 2019). Further research is required to establish a better scientific basis for the sector specific emission factors. Chapter 3 discusses the different scopes of GHG emissions associated with the urban water cycle. It also provides a comprehensive framework for carrying out a carbon footprint assessment along with an overview of available and relevant protocols and methods for assessing GHG emissions for the water sector.

1.2.2 Decarbonization requires a combination of approaches and collaborations among stakeholders

There is no single solution or method that can achieve net zero on its own, rather a combination of approaches and collaborations between different stakeholders, including those from academia, utilities, consulting firms, technology companies, government agencies, investors, and end users are critically needed. Decarbonization requires the advancement of science and technology, and it also calls for new policies, practices, and even behavioral changes toward sustainable living and practice.

Many utilities have launched programs that can reduce emissions immediately. These programs include transitioning operation and maintenance vehicle fleets away from fossil fuels by electrifying light duty vehicles and replacing heavy duty vehicles with alternative fuels such as renewable natural gas. It also includes purchasing green electricity from solar and wind sources, replacing energy intensive instruments such as blowers, pumps, and boilers with more energy efficient ones, as well as renewable energy recovery. For example, an increasingly common practice is to produce biogas from anaerobic digestion, and then using the biogas for heat, electricity or cogeneration of heat and electricity in a combined heat and power (CHP) station. In fact, the water sector is expanding to add waste materials such as food waste and other organic wastes as feedstocks for anaerobic digestion to produce biogas that can enable the facilities to be energy neutral or energy positive. Resource recovery from wastewater in terms of energy, nutrients (N and P) and water has resulted in renaming of wastewater treatment facilities into water resource recovery facilities (WRRFs). The embedded value of these recovered resources provides offsets for carbon emissions and hence can also contribute to net zero outcomes. Furthermore, new and innovative solutions such as carbon financing and real-time energy and chemical audits have also begun to make an impact.

In addition to treatment operations, the conveyance and storage of water and wastewater requires attention to reduce material consumption, energy demand, and construction of unnecessary infrastructure. Smart water meter installation to improve leak detection and reduce water use would also contribute to sustainable water management and lifestyle changes. Wastewater collection systems that minimize methane emissions due to anaerobic or septic conditions through both design and operational strategies are needed and are at embryonic stage and growing. Strategic evaluations of water and wastewater facilities siting to minimize conveyance energy needs should be coupled with centralized versus decentralized treatment facilities scenarios. The best decarbonization pathways are possible and provide maximum benefits when water is seen as OneWater in the urban watershed.

The water sector is heavily driven by a wide variety of policies. These policies not only include regulatory mandates on water quality and public health protection, but also include directives from different government agencies to ensure water affordability, social equity, ecological diversity, and infrastructure resiliency. Heavy investments in both financial and human capitals have gone into building and governing these systems as stove-piped services. There is no overarching policy that mandates decarbonization in the water sector globally, but seeking out co-benefits with other policy areas such as recovering local energy and resources, building multi-functional facilities, and protecting community assets would provide more feasible solutions. Continued progress will require strategic planning that begins at the local level and grows to global initiatives. This is especially the case with decarbonization. Otherwise, local action may focus almost solely on infrastructure resilience while the primary overlying issue of climate change requires decarbonization at the local level. In the last few decades, water and wastewater utilities have increasingly embraced the role and the value contribution to rate payers through performance-based operations. These efforts help reduce resource utilization and recover resources that hold local value, and such value-proposition provides utilities with better financial health and community support in making decisions that lead toward decarbonization. Chapter 17 discusses several concepts in policy making that impact efforts to decarbonize the water sector and highlights the need to seek multiple benefits whenever possible.

1.2.3 Processes and technologies that enable energy and resource recovery

Wastewater contains a significant amount of heat, chemical, and hydraulic energy that in total is estimated to be many times that required to treat the wastewater. Therefore, it is absolutely feasible to make WRRFs energy neutral or even positive. Figure 1.5 depicts a generalized view of energy flows at a WRRF. Technologies enable utilities to minimize the resources utilized to treat wastewater via process intensification, low energy treatment, and reduced chemical use. Leading utilities also maximize the extent of resources recovered through the treatment process via biosolids land application, nutrient recovery, biogas utilization, and water reclamation. A variety of technologies have been developed to make such operations a reality, and these technologies include but are not limited to anaerobic digestion, microbial electrochemistry, photobiological systems, advanced nitrogen and phosphorus management, process intensification using membranes and other technologies, as well as heat/pressure recovery and processes that significantly increase energy efficiency.

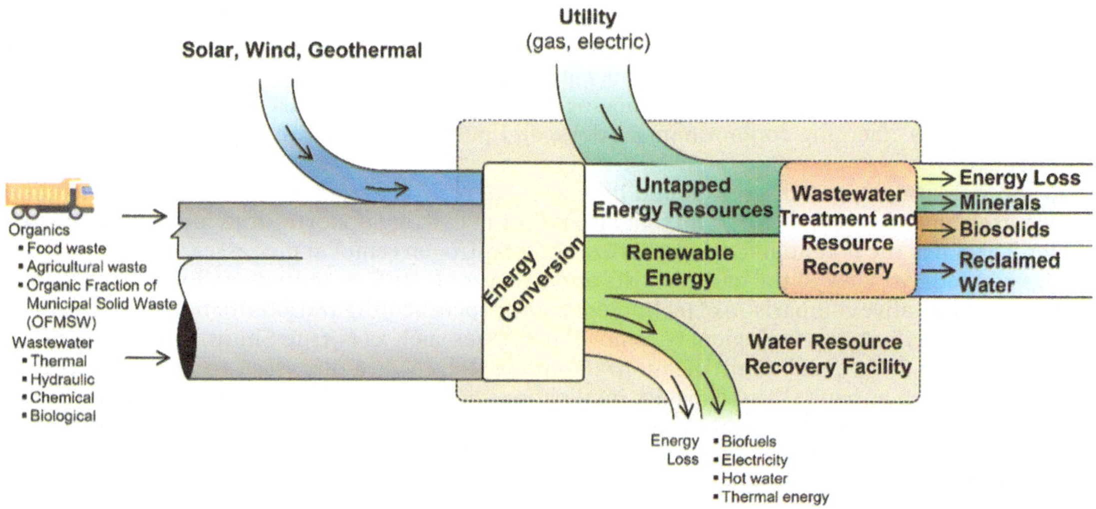

Figure 1.5 A generalized view of energy flows at a water resource recovery plant (WRRF) (revised from WEF (2019)).

The 'low-hanging fruit' for utility decarbonization is operations optimization. For an existing facility, the carbon footprint lies in operational activities and, thus, the decarbonization potential as well. Chapter 4 discusses the current opportunities for decarbonization at pumping, preliminary, primary, secondary, advanced, and sludge treatment level within a WRRF. For example, influent wastewater pumping has tremendous potential for reducing energy use, and data driven strategies using fuzzy logic, data mining, and bench-marking provide good tools to reduce specific energy and to improve energy savings (Torregrossa *et al.*, 2017). Similarly, aeration is the single largest source of energy use in most plants, and alternative diffusers and control systems taking advantage of the newer developments in membranes and online sensing have been utilized to make aeration energy efficient. At the whole plant level, different strategies can be employed in addition to individual unit level optimizations. For example, plant level benchmarking with comparable facilities to identify opportunities, or optimization of plant capacity utilization by arranging peak flow and load management, can be utilized. As chemicals use make up a significant portion of a WRRF's carbon footprint, reduction in chemical use through operational optimization is highly feasible for decarbonization and also results in operational costs reduction. Personnel training is another key area that can have a significant impact. With clear demonstration of cost savings and performance enhancement through best practices, operators will make direct contributions to decarbonization.

Anaerobic digestion (AD) is a model technology used in WRRFs to break down and stabilize wastewater sludge and to generate biogas and nutrient-rich effluent and biosolids. AD has been a central part of energy and resource recovery in the wastewater industry, and a suite of new processes and technologies have been developed to enhance sludge conversion, improve biogas production, upgrade biogas to higher-value products, and increase biosolid quality and applicability. Chapter 5 summarizes the current knowledge of the AD platform for energy and resource recovery, including the emerging trend of sludge pre-treatment using thermal, chemical, mechanical, and electrical methods to increase sludge degradability, sludge co-digestion with food waste and other organic waste for improved biogas yields and biosolids quality. It also discusses newly developed processes such as anaerobic membrane bioreactors (AnMBR), thermal hydrolysis, and volatile organic acids production using the AD platform. Chapter 10 furthers the discussion of AD in the broader perspective of sludge management, and it also discusses other practices including land application, composting, incineration, and landfilling. Emerging AD alternatives such as hydrothermal liquefaction (HTL) is also discussed to explore new pathways for sludge valorization and resource recovery. In addition, the chapter provides insights on how current management practices help decarbonization, the role of biosolids management strategies in achieving the decarbonization targets of utilities, and how challenges such as emerging contaminants, odors, and public scrutiny can be addressed in meeting such targets.

Advanced treatment for nutrients (N and P) removal or recovery has been a major driver increasing energy and chemical use in WRRFs, and that also points to existing opportunities for decarbonization. Chapter 7 explores the potential for decarbonization of nitrogen removal processes within WRRFs. It provides a broad overview of the carbon costs and decarbonization potentials, followed by a detailed review and quantitative comparison of technologies and process configurations for both sidestream and mainstream contexts. Novel biological N removal processes such as Nitrite Shunt, partial nitritation/anammox (PNA), and partial denitrification/anammox (PdNA) are discussed in detail, and critical considerations such as capital and capacity implications are evaluated. Similarly, Chapter 8 provides a comprehensive overview on phosphorus management and its potentials in decarbonization. It points out that sustainable P management requires a multi-tiered approach, and the cost and environmental consequences are likely to increase from higher-level management strategies for utilization efficiency improvements to lower levels of contaminant treatment and recovery. Direct decarbonization can be realized via reduced carbon input using enhanced biological P removal or even by implementing concurrent carbon sequestration methods. Indirect reduction relies on strategies that reduce carbon footprints throughout the life cycle of a given process, for example, by reducing chemical and energy demand or transportation. Both chapters also provide case studies of WRRFs implementing these strategies and note that whole plant level optimization is needed to coordinate C, N, P removal and recovery in order to accomplish decarbonization of the sector.

Membrane processes are playing critical roles in water engineering, from producing high quality water products to improving wastewater treatment efficiency and reducing spatial footprints. Membrane-based processes combine reaction with separation, and thus deliver high levels of process intensification. However, membrane separation operation generally consumes large amounts of energy and requires chemical cleaning, so it is important to evaluate the decarbonization potentials related to membrane operation. Chapter 11 discusses several novel membrane technologies that reduce energy consumption with high decarbonization potential. These include aerobic granular sludge membrane bioreactors (AGMBRs), algae membrane bioreactors (A-MBRs), anaerobic membrane bioreactors (AnMBRs), membrane biofilm reactors (MBfRs) and forward osmosis (FO) integrated processes. The chapter also includes membrane technologies for sustainable desalination, consisting of pressure retarded osmosis (PRO), forward osmosis-reverse osmosis (FO-RO) hybrid and forward osmosis-membrane distillation hybrid (FO-MD) methods. The benefits and challenges of these technologies are summarized, and research directions towards practical implementations for energy savings and low carbon footprints are provided.

While many technologies focus on converting the chemical energy in wastewater to usable forms like H_2, CH_4 or direct current, an even larger untapped area is the thermal energy embedded in wastewater. Discharged water with elevated temperature from commercial and industrial buildings, residential hot showers, dishwashing, clothes washers, and other appliances results in the embedding of substantial quantities of thermal energy in wastewater. Chapter 14 accordingly discusses this largely untapped energy source and its applications. It assesses the technical feasibility of thermal energy recovery from wastewater and envisions the possibility of integrating wastewater thermal energy recovery with district heating (DH) and district energy systems (DES) using heat pumps. The chapter also discusses the opportunities and barriers to thermal energy from wastewater, including but not limited to strategic planning, demand and resource mapping, technical feasibility, and regulatory and financial frameworks.

1.2.4 Processes and technologies that enable additional benefits of carbon capture and utilization, and watershed management

Tremendous progress on decarbonization has been made by increasing energy efficiency and recovering renewable energy, but these methods only reduce fossil fuel consumption and its associated carbon emissions. Considering the vast amount of wastewater generated each year (~1000 km^3 per year worldwide) and its positive correlation with population and industrial activities, wastewater treatment may even become carbon negative by capturing external CO_2 and CH_4 sources and converting them into value-added products. Because such practice can occur within existing wastewater infrastructure during treatment, no additional land or transportation would be required for such operations.

Natural treatment systems (NTS) utilize and enhance natural processes involving vegetation, soil and water, and the associated microbial ecosystems, and they are effective in advanced treatment, emerging contaminant removal, stormwater management, biomass production, recreational and educational services, and overall integrated watershed/sewershed management. NTS requires little mechanical or technological input to function, making them less chemical or energy intensive. Furthermore, these phyto- and microbial-based systems provide a wide range of carbon capture profiles depending on the level of treatment, seasonal variation, and system variation. Chapter 12 describes natural treatment technologies, their advantages and disadvantages, and their potential to decarbonize the water sector when incorporated in integrated watershed management. Under such practice, nutrients, energy, and water are recovered from wastewater, and GHG emissions are mitigated. By reclaiming and reusing the water for non-potable purposes such as irrigation/fertigation, aquifer recharge, graywater applications, or even for direct potable reuse, source water withdrawal can be reduced, aquifers are replenished, and a harmony between water demand and supply can be achieved. The chapter also provides two engineering case studies on the benefits of phytoremediation for carbon sequestration and agriculture runoff treatment, as well as using microalgae for combined power plant flue gas and fertilizer wastewater treatment.

Phototrophic microorganisms like microalgae or purple photosynthetic bacteria present a unique pathway for decarbonization, as they fix CO_2 during autotrophic growth while assimilating nutrients (N and P) in wastewater. Therefore, phototropic treatment systems are complementary to carbon-focused treatment processes like AD or microbial electrochemistry. Chapter 9 describes the decarbonization potentials using photobiological treatment systems. For example, microalgae have been widely studied for CO_2 capture and utilization, including extensive research on their use in large-scale (>5000 acre) cultivation systems to produce feedstocks for biofuels and bioproducts. When they are used in wastewater treatment, they are operated at an accelerated rate compared to terrestrial plants, and they can be integrated with AD to provide additional substrate and to condition biosolids, capture and utilize CO_2 and upgrade biogas into biomethane, and even use H_2S as an electron donor. Recent advances in photobioreactor design have boosted the biodegradation potential of photobiological-based systems, while lowering their energy demand, and they are critically discussed in this chapter.

Another promising technology platform for simultaneous wastewater treatment, resource recovery, and carbon capture and utilization is microbial electrochemical technology (MET). MET offers an extremely flexible platform for both oxidation and reduction reaction-oriented processes. The MET systems share one common principle in the anode chamber, in which biodegradable substrates are oxidized and generate electrical current. The current can be captured directly for electricity generation (microbial fuel cells, MFCs) or used to produce H_2 and other value-added chemicals (microbial electrolysis cells, MECs). In addition, such electrons from organic waste carbon can also be used in the cathode chamber to reduce CO_2 and generate organic or inorganic compounds, achieving double benefits of carbon capture and valorization. Chapter 6 presents the principles and popular MET carbon capture processes and discusses the variety of products and systems that have been developed. Microbial electrosynthesis converts CO_2 into organic compounds such as carboxylic acids and CH_4, while microbial electrolytic carbon capture mineralizes CO_2 into carbonate products. In addition, electro-fermentation (EF) uses electrochemistry to influence microbial metabolism and regulate fermentation pathways to valorize organic waste carbon to higher-value products, and many consider it to be an electrochemically enhanced AD system. A unique feature of MET is the complementary nature with anaerobic fermentation or digestion, in which synergistic interspecies electron transfer can occur between microbes to facilitate electro-methanogenesis or electro-acetogenesis. Chapter 13 focuses on the fundamental carbon and electron flows in such anaerobic bioconversion systems for CO_2 capture and conversion to value-added organic chemicals.

While transforming wastewater treatment to carbon-neutral/negative or even revenue-positive takes concerted efforts from stake holders, studies have demonstrated the potential benefits of implementing new processes and technologies. Figure 1.6 demonstrates a hypothetical process combination of microbial carbon capture cell (MECC) and microalgae reactor to replace the traditional anaerobic/anoxic/aerobic activated sludge process. The MECC specializes on organic carbon removal, while microalgae is effective in nutrient removal. Moreover, they both demonstrated excellent carbon capture and utilization (CCU) capability, with MECC converting CO_2 into carbonate mineral accompanied with high rate H_2 production, and microalgae captures CO_2 as biomass, which subsequently can be converted to biofuels or biochar (Lu *et al.*, 2018). Preliminary quantitative analyses using the US and China as examples showed that instead of being a net emitter of GHGs, a net of up to 112 (median; 5th–9th percentile range of 84–145; USA) and 75 (57–97; China) $MtCO_2e$ can be captured and converted to valued-added products. Among these negative emissions, approximately 41–56 and 47–58% are attributed to CCU during organic and nutrient removal, respectively; and −2–2% is credited to avoided consumption of fossil energy during CCU (avoided aeration, etc.; the negative value stems from uncertainty analysis). In terms of economic benefits, while the proposed system is likely to have even greater Capex and Opex costs than conventional processes, the recovered mineral and biofuel products may create 8.7 (6.9–10.9) and 5.6 (4.4–6.9) billion dollars in value per year for the US and China, respectively. Additionally, carbon capture credits in the two countries could also mobilize 4.5 (3.3–6.2) and 1.0 (0.7–1.5) billion dollars for US and China wastewater industries, correspondingly. These estimates demonstrate that the wastewater industry can become a significant contributor of negative carbon emissions, though significant technology development and testing are needed since neither process has been demonstrated in full scale.

Water utilities collect and store massive amounts of data to provide a reliable and efficient service. The data collected not only include water quantity and quality data in every treatment facility but also

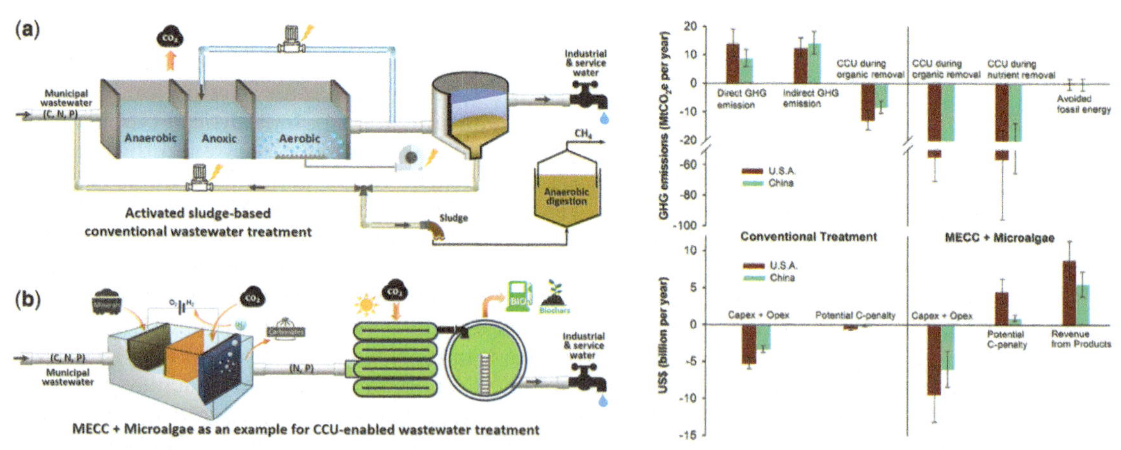

Figure 1.6 Preliminary estimates of carbon capture and utilization benefits from an example integrated MECC+microalgae process compared with conventional activated sludge process: (a) Conventional anaerobic/anoxic/aerobic activated sludge process for simultaneous carbon and nutrient removal; (b) Integrated MECC and microalgae cultivation for carbon and nutrient removal with resource recovery and CCU; (c) The preliminary estimates of CCU potential and economic impacts when the conventional process is compared with the MECC+algae process.

consists of real-time data from flow monitors, and rain and stream gauges across the watershed, as well as data at endpoints and in water/sewer pipelines. Data has become an essential asset for utilities and will play increasingly critical roles for utilities of the future. Accordingly, Chapter 16 introduces the modern data-driven modeling (DDM) including statistical and machine learning methods and uses specific examples to demonstrate how these tools can be used within a larger decarbonization strategy. The chapter explains data preparation, common DDM methods, and metrics for comparing different models. It also analyzes unit processes and how data-driven process optimization can become the proverbial 'low risk, high return' approach for carbon and cost reduction.

1.2.5 Case studies on utility decarbonization practice

Many leading utilities have established plans for energy- and climate neutrality within the next decade(s), and there are numerous case studies and best practices to follow. For example, the Strass im Zillertal plant in Austria has been a model on continued process optimization that accomplished the goal of producing more electricity than it consumes. The Strass plant provides two-stage biological treatment (A/B plants) to treat organic loads varying from 60 000 to 250 000 population equivalents (weekly average), and they implemented an SBR deammonification process to further reduce energy and carbon requirements for nitrogen removal. It also enhanced digester gas utilization via cogeneration and boosted electrical efficiency by more than 20%. They contributed the success to a highly educated workforce, high level of automation, the use of advanced analysis tools, and the ability to quantify gains (Wett *et al.*, 2007). Another example is VCS Denmark, which is the oldest water utility in Denmark and has been managing and treating wastewater in the Odense metro area since 1907. VCS Denmark has been energy neutral since 2019, which was accomplished by maximizing primary treatment efficiency, process intensification, advanced monitoring, control, and energy efficient equipment selection. They also partnered with district energy companies to install heat pump stations to draw thermal energy in the wastewater effluent for local district heating.

East Bay Municipal Utility District (EBMUD) in California was the first WRRF in North America to become energy positive in 2012. In 2020, it approved an ambitious plan to become carbon neutral for water operations by 2030. EBMUD runs a successful program on co-digestion, in which biodegradable wastes in sewage, food scraps and grease from local restaurants, plus waste streams from wineries and poultry farms, are mixed together for anaerobic digestion. The increased biogas production saves the district approximately $3 million each year by reducing electric power demand. The excess renewable energy is sold back to the electrical grid to cut fossil fuel use and GHG emissions, and provides savings for ratepayers. To meet the carbon neutrality goal, the district also completed several new kilowatt photovoltaic systems, switched all passenger vehicles to hybrid or electric, and are converting heavy-duty vehicles to renewable diesel.

China has the world's largest and still growing wastewater sector. The total number of wastewater treatment plants in Chinese cities increased by 10-fold, from 481 to 5640, during the period from 2000 to 2018 (Qu *et al.*, 2019). As Chapter 15 explains, in 2014, China formed the Concept WWTP Committee (CCWC), which gathered global insights and worked with domestic partners and launched the 'Concept Plant' project that aimed to build future treatment plants with integrated missions in 'sustainable water quality, resource recovery, energy neutrality, and environmental friendliness'. The first concept plant, Sui County No.3 WWTP, started to operate in 2019 with a designed flowrate of 20 000 m^3/day and serving a population of 900 000. The plant includes a liquid treatment area, an organic waste processing area, a constructed wetland, agriculture and sponge city demonstration areas, and an office building and education center. In 2021, another state-of-the-art Yixing Concept Plant went into operation (Figure 1.7). The plant consists of a water purification center with a capacity of 20 000 m^3/day, a production-oriented R&D center, and an organic co-processing center that convert sludge, kitchen waste, and agriculture waste to energy and fertilizer. The CCWC plans to build ~100 Concept Plants in the coming 5–8 years with designs incorporating considerations of geographical differences, capacities, treatment priorities, and integrated operation goals.

Figure 1.7 The design view and aerial view (insert) of the Yixing Concept Wastewater Resource Recovery Factory in Jiangsu, China (Qu *et al.*, 2022). The factory started operation in October 2021.

1.3 THE PARADIGM CHANGE FOR A NET ZERO CIRCULAR WATER ECONOMY

'Water management is a path, not a destination.' The same philosophy needs to apply to decarbonization of the water. It is how we pursue decarbonization, identify multiple pathways, try and refine them, and scale them to the entire water sector that will lead to positive outcomes. Former IWA President, Prof. Glen Daigger summarizes a water professional's core mission in the concluding Chapter 18. The evolution in the governance and infrastructure of a water utility reflects the progression of investments to address the most pressing needs for its community. The mission of water management has grown through time, from the initial focus on reliable water supply and prevention of the spread of diseases to include multifaceted objectives on water resource recovery, water/sewershed management, public health protection, and sustainable development to not only consider economics but also environmental and societal impacts. The momentum for change is accelerating as we transition from the current linear economy to a circular one, and the water sector can and should provide leadership in providing essential public services.

The topics in this book articulate many opportunities currently available to the water sector to decarbonize and transition into a circular water economy. Numerous innovators and early adopters are investigating options, conducting trials, and executing projects to increase energy efficiency, reduce carbon footprints, and recover resources from the OneWater cycle. There is no good or bad product, but there can be a right or wrong product for a specific utility. Resources need to be recovered and carbon footprints need to be reduced, but we also need to recognize that unless the products have sufficient market demand, a valid value proposition, and tangible benefits to the society from decarbonization, any changes made will not be sustainable in the long term.

There is no single solution that achieves net zero on its own, so it is imperative that all stakeholders work together and collectively transform how the water sector plans, invests, and operates with balanced near- and long-term goals in mind. The pathways laid out in the following chapters are developed by visioning the possible net zero futures for the sector as well as individual utilities, and they provide critical insights for the industry to move towards a circular economy that ensures sustainability for future generations.

REFERENCES

Delre A., Mønster J. and Scheutz C. (2017). Greenhouse gas emission quantification from wastewater treatment plants, using a tracer gas dispersion method. *Science of the Total Environment*, **605–606**, 258–268, https://doi.org/10.1016/j.scitotenv.2017.06.177

Griffith D. R., Barnes R. T. and Raymond P. A. (2009). Inputs of fossil carbon from wastewater treatment plants to US rivers and oceans. *Environmental Science & Technology*, **43**(15), 5647–5651, https://doi.org/10.1021/es9004043

IEA (2017). Water-Energy Nexus. IEA, Paris. Available at: https://www.iea.org/reports/water-energy-nexus (accessed 18 November 2021).

IKI (2020). Water Companies on The Way to CO_2 Neutrality. Available at: https://www.international-climate-initiative.com/en/news/article/watercompanies_on_the_way_to_co2_neutrality (accessed 18 October 2021).

Jacobs (2020). One Water. Available at: https://www.jacobs.com/solutions/water/one-water (accessed 18 November 2021).

Lu L., Guest J. S., Peters C. A., Zhu X., Rau G. H. and Ren Z. J. (2018). Wastewater treatment for carbon capture and utilization. *Nature Sustainability*, **1**(12), 750–758, https://doi.org/10.1038/s41893-018-0187-9

Qu J., Wang H., Wang K., Yu G., Ke B., Yu H. Q. and Gong H. (2019). Municipal wastewater treatment in China: development history and future perspectives. *Frontiers of Environmental Science & Engineering*, **13**(6), 1–7.

Qu J., Ren H., Wang H., Wang K., Yu G., Ke B. and Li J. (2022). China launched the first wastewater resource recovery factory in Yixing. *Frontiers of Environmental Science & Engineering*, **16**(1), 1–2, https://doi.org/10.1007/s11783-021-1429-z

Saunois M., Stavert A. R., Poulter B., Bousquet P., Canadell J. G., Jackson R. B. and Zhuang Q. (2020). The global methane budget 2000–2017. *Earth System Science Data*, **12**(3), 1561–1623, https://doi.org/10.5194/essd-12-1561-2020

Sedlak D. (2014). Water 4.0. Yale University Press, New Haven, CT, USA.

Torregrossa D., Hansen H., Hernandez-Sancho F., Cornelissen A., Schutz G. and Leopold U. (2017). A data-driven methodology to support pump performance analysis and energy efficiency optimization in waste water treatment plants. *Applied Energy*, **208**, 1430–1440, https://doi.org/10.1016/j.apenergy.2017.09.012

UNEP (2017). The Emissions Gap Report 2017: A UN Environment Synthesis Report. UN Environment, Kenya.

USEPA (2019a). Global Non-CO_2 Greenhouse Gas Emission Projections & Mitigation 2015–2050. USEPA, USA.

USEPA (2019b). Sources of Greenhouse Gas Emissions. Available at: https://www.epa.gov/ghgemissions/sources-greenhouse-gas-emissions (accessed 18 October 2021).

Vasilaki V., Massara T. M., Stanchev P., Fatone F. and Katsou E. (2019). A decade of nitrous oxide (N_2O) monitoring in full-scale wastewater treatment processes: a critical review. *Water Research*, **161**, 392–412, https://doi.org/10.1016/j.watres.2019.04.022

Water Environment Federation (2019). ReNEW water project: resource recovery to fuel and grow a circular economy. WEF, USA.

Water UK (2020). Net Zero 2030 Routemap. Available at: https://www.water.org.uk/routemap2030 (accessed 18 October 2021).

Wett B., Buchauer K. and Fimml C. (2007). Energy self-sufficiency as a feasible concept for wastewater treatment systems. 4th IWA Leading-edge Conference and Exhibition on Water and Wastewater Technologies, IWA, United Kingdom.

doi: 10.2166/9781789061796

Chapter 2

What can we learn from decarbonization of the energy sector?

A. J. Simon[1]* *and Seth W. Snyder*[2,3]

[1]Lawrence Livermore National Laboratory, Livermore, CA, USA
[2]Idaho National Laboratory, Idaho Falls, ID, USA
[3]Northwestern University, Evanston, IL, USA
**Correspondence: simon19@llnl.gov*

2.1 INTRODUCTION: ENERGY AND WATER: SIMILARITIES, DIFFERENCES, AND A COMPLEX RELATIONSHIP

Decarbonizing water and wastewater treatment is an enormous challenge, but it is substantially smaller, in total carbon emissions, than decarbonizing the energy sector. When planning, executing, and assessing strategies for decarbonizing the water sector, water experts should partner with the energy sector and heed that sector's lessons-learned in its ongoing process of decarbonization. In the energy sector, decarbonization pathways can be as simple as a supply-side technology that converts fuel to electricity more efficiently, reducing net carbon emissions for every kilowatt-hour generated. The pathways can be much more complex, however, as is the case with the demand-side reordering of behavior as seen with online shopping or working from home during a public health crisis. Both of those pathways reduce demand for private-vehicle fuel and shift some work, and associated carbon emissions, to other parts of the economy. This chapter explores decarbonization pathways that have been followed by the energy sector and assesses their applicability to the water sector.

Energy and water are two infrastructure sectors that are coupled in multiple ways. This chapter is not intended to quantitatively assess the opportunities for energy and water to contribute to each others' decarbonization. For a full understanding of issues and opportunities at the 'energy-water nexus', the reader is advised to review the literature (DOE, 2014; EPRI, 2013; Gleick, 1994; Greenberg *et al.*, 2017; Grubert & Sanders, 2018). The most important concepts that frame the decarbonization opportunities are as follows:

- Water and wastewater treatment are generally consumers of energy for aeration, pumping and heating. That energy use is associated with the carbon emissions that result from electricity production or onsite usage of natural gas. Energy sector decarbonization can contribute to decarbonization of the water sector.
- Wastewater is a carrier of carbon, most of which is oxidized to CO_2 during wastewater treatment or in the environment after it is discharged. Intercepting and permanently immobilizing this carbon is a major opportunity for the water sector to reduce its gross emissions.

- Anaerobic digestion of wastewater fractionates the embedded carbon into methane and CO_2. Methane has a radiative forcing factor 28 times greater than CO_2 over 100 years (and 86 times greater over 20 years), resulting in substantial short-term climate impacts if the methane is not captured and combusted (Roy *et al.*, 2015).
- Innovations at the unit process level are the most straightforward way to manage the carbon intensity of the water sector.
- Structural change in the water economy (more efficient water use, potable and non-potable reuse, etc.) may also affect the sector's carbon intensity.

2.1.1 The energy-water nexus

The energy-water nexus refers to the coupling between the energy and water sectors. The water sector requires a significant amount of energy to operate and presents opportunities for energy recovery and electricity generation. Similarly, the energy sector requires significant amounts of water to operate, and also presents opportunities for water treatment and delivery. At the energy-water nexus, changes to one sector may affect the economic and environmental sustainability of the other. Modern water systems use exogenous energy to acquire, convey, purify, distribute, collect, treat, and dispose of water. To the extent that this exogenous energy is electric (power for pumps and blowers), the carbon intensity of the water system is coupled to the carbon intensity of electricity supply. Decarbonization of the electricity supply is already underway as coal-fired electricity generation is being replaced by natural gas (which has roughly one-half the carbon intensity as coal), and as more electricity is generated from non-fuel resources such as solar and wind.

A more thorough accounting of the challenges and opportunities at the energy-water nexus is available in the literature (see above, including works cited in those references), and some of the major interactions are briefly summarized here:

- Water acquisition, treatment and distribution requires electricity for pumping. This is true for municipal, industrial and irrigation water supply. The chemicals used in water treatment are also energy intensive to produce.
- Wastewater treatment requires electricity for aeration blowers and for pumping. WWTPs in some climates also require natural gas or other fuels for heating of anaerobic digesters.
- Energy is used for water heating in commercial, industrial and residential applications.
- Energy can be recovered from wastewater in the form of biogas, electricity from biogas, or electricity from incinerable biosolids. Indirectly, energy can be displaced by replacing chemical fertilizers with biosolids.
- Electricity production in thermoelectric plants (nuclear, natural gas, and coal) require water for cooling and for emissions controls.
- Hydroelectricity is produced from water resources and impacts other economic uses and environmental services.
- Production of biofuels may require water for irrigation of energy crops and in conversion of feedstocks to fuels.
- Production of oil and gas may require water for hydraulic fracturing, and often results in a surplus of produced water. Depending on the source and the quality, produced water may require energy for treatment and disposal, or may be treated for beneficial use.

In the future, additional energy may be required for seawater and brackish water desalination and other advanced water treatment. Removal and destruction of contaminants of emerging concern (CECs), including but not limited to per- and polyfluoroalkyl substances (PFAS), could significantly increase the energy required to treat water. Requirements could affect a range of applications including municipal and industrial wastewater treatment, stormwater management, and groundwater cleanup. Additional water may be required for carbon dioxide capture and sequestration from electricity generation and low carbon fuel production. However, a full accounting of energy and water interdependency is beyond the scope of this book.

2.1.2 Differences in scale

The extent to which decarbonization strategies and tactics from the energy system can be adapted to water depends on the similarities, differences, and interdependencies of these two critical infrastructure systems. The lessons learned from the ongoing decarbonization of the energy sector are framed in terms of total system scale, resource substitution, emissions control, quality of service, and sustainability policy. In this chapter, the United States markets for water, energy, and other commodities are used for these framing studies because the US is the largest single country for which well documented energy and water statistics are readily available. Similar lessons will apply to most developed economies, and those lessons can be extended and adapted globally.

As measured by annual use, water is the largest infrastructure/commodity sector in the US by more than a factor of 20. Table 2.1 compares water use to other major energy, agricultural, and material sectors on annualized mass and volume scales. It is important to note that the scale of water infrastructure is ONLY for public water supply (municipal water treatment plants) and that adding wastewater treatment would approximately double that figure. Total water use (including for irrigation, powerplant cooling and other non-municipal uses) is a factor of 10 larger (in the order of 500 000 million metric tons per year)!

From these statistics it is clear that water, as a system, is singular in scale. Society processes water at a flow rate that is orders of magnitude larger than any other commodity, and requires physical infrastructure vastly greater than energy or every other commodity. Decarbonization challenges that scale with flow rate, such as the capital cost of processing systems, will tend to be larger for water than they are for other sectors.

Although the gross material flow through water systems is larger than it is through energy systems, the carbon emissions associated with water and wastewater treatment are smaller than they are for the energy system. In addition, the incremental economic value of each unit of water is significantly smaller than energy or products, reducing the potential revenue available to manage carbon. Chapter 3 of this book introduces a framework for carbon accounting in the urban water cycle. Here we estimate that municipal water and wastewater treatment in the US are responsible for 61 million metric tons (MMT) of CO_2-equivalent greenhouse gas emissions annually. Of that total, 38 MMT (CO_2-e) are associated with methane and nitrous oxide emissions from municipal wastewater treatment and discharge (EPA, 2021). The remaining 23 MMT result from the generation of the ~59 million megawatt hours (MWh)

Table 2.1 Annual flows of widely used material commodities in the US.

Commodity	Mass Flow (million metric tons per year)	Volumetric Flow (million cubic meters per year)	Notes
Water (Dieter *et al.*, 2018)	53 880	53 880	Deliveries from municipal water treatment plants only
Aggregates (USGS, 2021)	2538	1586	Total US consumption of crushed stone, sand, and gravel
Petroleum (EIA, 2021)	895	1133	Total US consumption incl. net imports
Coal (EIA, 2021)	724	804	Total US consumption, excl. exports
Natural gas (EIA, 2021)	607	771 400	Vol. flow calculated at standard temperature and pressure; actual vol. flow is much less because gas lines are pressurized
Corn (USDA, 2021a)	340	479	Total US production
Steel (USGS, 2021)	100	13	Total apparent US consumption including imports which account for ~20%
Wheat (USDA, 2021b)	51	66	Total US production, including exports which account for ~50% of production

of electricity consumed by water and wastewater treatment plants in the US (Greenberg *et al.*, 2017). It is likely that additional GHG emissions are attributable to the water and wastewater treatment sector from onsite natural gas combustion, however no data could be found to quantify this emissions source. Offsite manufacturing of chemicals used for water and wastewater treatment have also been hypothesized to contribute significantly to the sector's life cycle GHG footprint, but estimates of this quantity in the literature vary widely (Kyung *et al.*, 2015; Szulc *et al.*, 2021).

Combustion of fossil fuel across the entire energy sector in the US emits 5300 MMT of CO_2. These two statistics are not directly comparable; the 61 MMT CO_2-e associated with the water sector accounts for CO_2 and other GHGs from scopes 1, 2, and 3 for a specific sector while the 5300 MMT CO_2 in the energy sector accounts for only fossil-fuel derived CO_2. Additionally, scope 2 emissions from water treatment (~23 MMT) are *included* in the 5300 MMT of fossil fuel-derived emissions from the energy sector. However, the vast disparity in scale between these two figures demonstrates that despite managing a far larger quantity of material, the water sector manages a far smaller quantity of carbon. This overlap between water sector GHG emissions and energy sector GHG emissions is a telltale of the Energy-Water Nexus described above.

2.1.3 The carbon-water nexus

Much of the attention on decarbonization focuses on the elimination of carbon dioxide emissions from fossil fuel use in the electric generation, transportation and industrial sectors. The water sector's GHG emissions are a combination of emissions from those sectors (scope 2 emissions), non-CO_2 GHG's from conversion of organic material in wastewater, and the CO_2 product of organic material present in the wastewater itself. Although most of the carbon in wastewater is biogenic in nature (derived recently from atmospheric carbon) it is instructive to consider the total flow of water-borne carbon through wastewater systems. Assuming a chemical oxygen demand (COD) of 350 mg/liter, and that organic matter (CH_2O) represents the bulk of this load, there is approximately 11 mmol carbon per liter of wastewater. Assuming that 32 000 million gallons of wastewater are treated per day in the United States, there are 5.8 MMT per year of carbon passing through wastewater treatment plants with the potential to produce 21.3 MMT of CO_2 emissions from in-plant processes as well as oxidation of biosolids, biogas and remaining BOD/COD in effluents.

Both the energy and water sectors are, in the terminology of the US Department of Homeland Security, 'National Critical Functions' (DHS, 2021), and both move material from the environment to engineered systems, and subsequently back to the environment. However, there are substantial differences. A relatively small amount of energy moves a very large quantity of water, and that water carries with it a small amount of organic carbon. Water resources are acquired from surface or groundwater reservoirs, and water is returned to the surface as impaired or treated water. In the case of energy, diverse resources are drawn from the environment and today's energy systems depend significantly on chemical fossil energy in underground reservoirs of coal, oil and natural gas. Engineered systems separate energy from carbon, delivering services and returning the associated carbon to the environment, most frequently as CO_2 emitted to the atmosphere. The remainder of this chapter will focus on decarbonization trends in the energy system, and how those trends can benefit the decarbonization of the water system through the energy water nexus and through shared technologies, best practices and lessons learned.

2.2 DECARBONIZATION OF THE ENERGY SECTOR

In 2020, the carbon intensity of the energy sector was declining at a rate of approximately 1% per year. Although this pace seems slow, and is certainly not fast enough to reach the emissions targets that climate science indicates are necessary, it represents a substantial change from the prior era. Over the 28-year period from 1977 to 2005, the carbon intensity of energy use barely changed at all, declining from 58.3 to 56.6 million metric tons of CO_2 per exajoule (MMT/EJ), a rate of 0.1% per year. During the 12 years between 2005 and 2017, carbon intensity declined from 56.6 to 49.9 MMT/EJ, a rate of 1% per year (EIA, 2021) (each of these statistics takes the five-year average carbon intensity of energy around the reported year to smooth out noise in the statistics – looking at individual years, it appears that the 2005–2017 trend continued through at least 2019 and was likely accelerating). This ten-fold increase in the pace of decarbonization is due to the following changes in the energy system (listed in order of size of carbon intensity impact):

- a substantial shift from coal to gas in the electricity generation sector;
- substantial increases in electricity generation from wind and solar resources;
- increases in overall vehicle efficiency and the percentage of biofuels consumed in the transportation sector.

In addition to the decrease in the carbon intensity of delivered energy, the energy intensity of the US economy has declined. In real GDP terms (all values quoted in 2012 dollars), the US consumed 13.1 exajoules per trillion dollars (EJ/$T) of economic activity in 1978, 7.1 EJ/$T in 2005 and 5.7 EJ/$T in 2017 (US Bureau of Economic Analysis, 2021). The decline rate of the energy intensity of the economy has been a steady 1.6% per year over that entire time frame. The decreasing energy intensity of the overall economy is due to the following factors:

- structural changes in the economy that favor lower energy intensity commercial activity such as financial and computing/data-driven services over higher energy intensity industrial activity such as iron and steel making;
- improvements in energy efficiency that deliver equivalent economic service for smaller energy inputs such as:
  - improved heavy- and light-duty vehicle fuel efficiency;
  - improved insulation in residential and commercial buildings;
  - efficient devices and appliances such as LED lighting.

Figure 2.1 shows these trends graphically. Despite nearly 300% growth in real GDP (2012 dollars) from ~5.6 $T in 1975 to 19.1 $T in 2019, energy use grew only 40% over that period due to decreased energy intensity, and carbon emissions have begun to decline from their 2005 peak due to both decreased energy and carbon intensity.

There are multiple interrelated factors behind these trends including energy policy that incentivizes sustainable energy use, the cost savings due to energy efficiency in many applications, innovation in energy technology that improves efficiency and reduces emissions, and consumer preference for more sustainable solutions. The remainder of this chapter examines some of these factors and provides examples.

These trends are likely to accelerate into the future. In addition to continued expansion of the lower-emitting and higher efficiency technologies listed above, the following trends and technologies are beginning to roll out at scale in US energy markets. Their impact on overall energy consumption and emissions, while not yet significant, will likely become visible in economy-wide statistics by 2025:

- remote working options (reduced local commuting and long-distance business travel, permanent change initiated by 2020 pandemic);
- growth in online ordering or 'e-commerce' grew steadily over a decade and grew rapidly during the pandemic, dramatically reducing the number of short-distance trips (DOE, 2020);
- electric vehicles (passenger cars and delivery vans);
- high efficiency electric heating (heat pumps).

Further drastic improvements in the energy efficiency and carbon intensity of economic activity are possible with energy technologies that are technically feasible and have been demonstrated at scale, but have not yet achieved performance and/or cost parity with competing technologies. Demand for

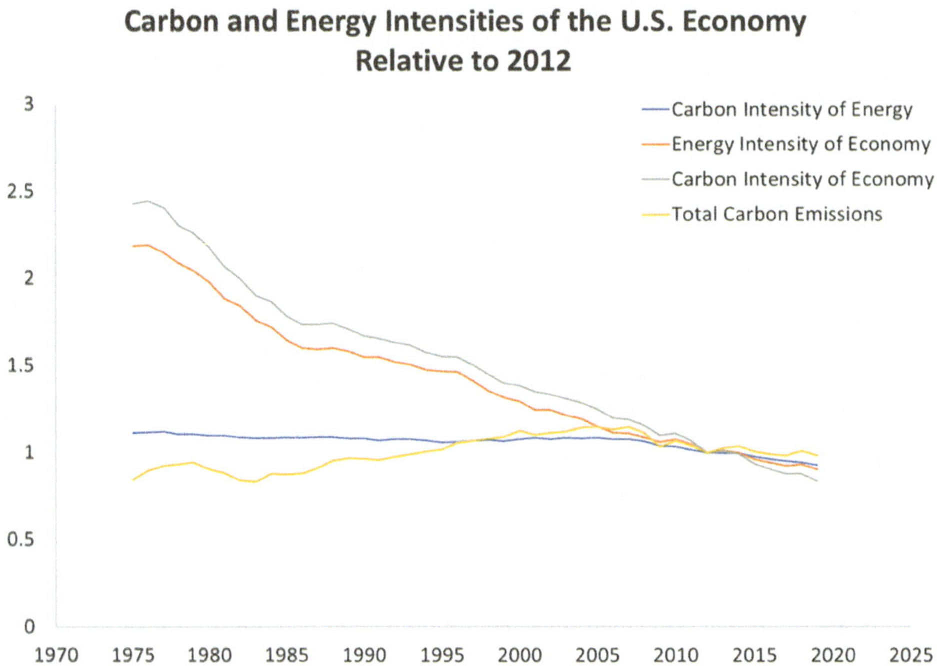

Figure 2.1 Carbon intensity of energy, energy intensity of the economy, carbon intensity of the economy, and economy-wide carbon emissions relative to 2012 for the United States.

these technologies may increase substantially if certain policies are put in place or if a price on carbon emissions is enacted:

- electricity generation with carbon capture and sequestration;
- hydrogen as a transportation fuel, heating fuel, or chemical process input;
- small modular nuclear reactors;
- biomass-derived energy (ethanol, other liquids, biogas, hydrogen or electricity) with carbon capture and sequestration of process emissions.

Europe and some Asian economies have lower energy intensity than the US with similarly advanced economies. While China lags the US and Europe in energy intensity, it is improving far more rapidly. Economic energy intensity is far higher in the developing world, but total and per capita energy use is dwarfed by the advanced economies.

2.3 A FRAMEWORK FOR SUSTAINABILITY FOR ENERGY AND WATER

Energy, water supply, and wastewater management are foundational needs for a modern society. However, unfettered energy and water use are unsustainable due to supply constraints and/or environmental impacts. Here, we introduce a framework, depicted graphically in Figure 2.2, to organize the events, behaviors, and technological advancements that enhance sustainability. This framework applies to both energy and water services. While the framework is roughly hierarchical, with categories listed from least to most effective, there are not strict delineations between each category. Advances in sustainability may be motivated by resource constraints and environmental preservation (left half), and success is achieved through innovation and other 'highly sustainable'

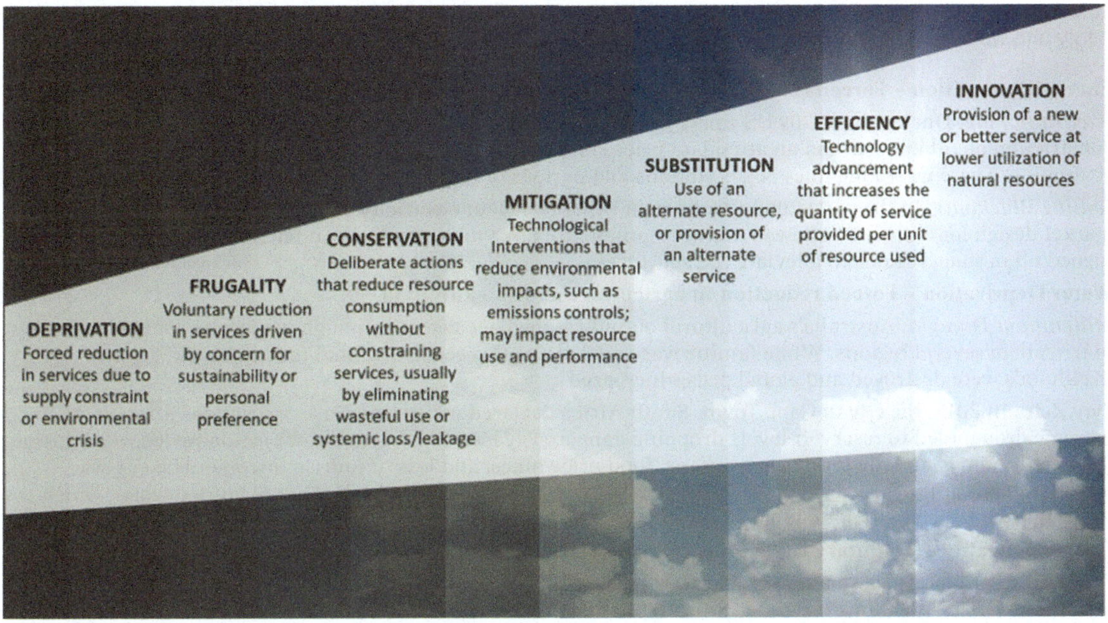

Figure 2.2 A framework to qualitatively assess the drivers of sustainability in energy and water systems, with the least meritorious on the left and most meritorious on the right.

practices (right half). The effectiveness of any specific energy/water intervention depends on both the state of technology and of existing infrastructure. Stressors, behavior changes, technology advancements, and the evolution of energy/water systems do not evolve linearly along this spectrum. The histories of modernization and decarbonization include iterative loops and multi-step hops.

Examples for each of these categories are given in Table 2.2, illustrating the broad applicability of this framework. Lessons learned from both successful and unsuccessful efforts to decarbonize energy may be extended to inform the decarbonization of the water industry.

2.4 THE PACE OF DECARBONIZATION

Frugality and conservation measures reduce emissions in the short term by reducing demand for services associated with energy and water use. Over the long term, however, the demand continues to grow for the services that energy and water provide. Therefore, emissions from these sectors are controlled by the efficiency and emissions of the capital equipment that transforms resources into services. The pace of decarbonization is almost entirely determined by the rate of capital stock turnover. Emissions associated with a piece of equipment in an energy or water system will persist throughout the useful life of that equipment. Capital turnover is the expected time for replacement of energy related devices, facilities, and infrastructure. This time period depends on the functional life of the systems and the relative value of potential replacements. As illustrated above, the replacement value can be increased by innovation (better service), efficiency (lower energy consumption), substitution (lower costs of alternative inputs), and mitigation (better environmental footprint). Technology developments and policy incentives can accelerate decarbonization by increasing the value of replacement, but the capital cost of a system and its anticipated remaining useful lifetime have major, if not dominant, impacts on the overall pace.

Table 2.2 Historical examples from each part of the sustainability framework that have alleviated the impacts of energy and water use, demonstrating the adverse consequences of deprivation and benefits of innovation.

Energy Deprivation – Forced reduction in energy services and quality of life
Fuel Rationing: During the 1970s US Energy Crisis, international political tensions caused a sudden constriction on oil imports and an attendant price spike for the fuels used in transportation and electricity generation. There were long lines at gas stations and periods of fuel unavailability
Rolling Blackouts: In the early 2000s, California experienced an electricity crisis. Flawed regulations and market design for electricity allowed market manipulators to constrict supply and drive up prices. Power was turned off in some regions to alleviate the shortages
Water Deprivation – Forced reduction in agriculture and/or sanitation
Millennium Drought: Australia's agricultural output crashed because not enough water was available to irrigate in several regions. While famine was avoided due to a resilient global food network, farmers livelihoods were destroyed and global prices increased
Day Zero: In 2018, the city of Cape Town, South Africa declared major water use restrictions after an extreme drought led to reservoir levels dropping dangerously low. This severe water rationing led to job losses (especially among already low wage workers), food price hikes, and loss of tourism income. The city was forced to plan for the forced shutdown of municipal water supplies and distribution of bottled water, which would have resulted in major quality-of-life disruptions for all residents
Energy Frugality – Voluntary reductions in quality of life associated with energy services
Compact car: Choosing a smaller vehicle affords the purchaser less passenger room and (usually) fewer amenities. These reductions in service are compensated by lower fuel expenses, and are accompanied by lower emissions and other environmental impacts
Thermostat settings: Lower setpoint temperature (in winter) trades lower fuel expenses and carbon emissions for reduced comfort
Water Frugality – Voluntary reductions in the quality of life associated with abundant water
Landscape irrigation: Letting a lawn/field go brown by reducing/foregoing irrigation during water shortages lowers water bills and imparts a sense of 'doing one's part.' However, it impacts residential aesthetics and outdoor comfort. Lower water use reduces the energy and carbon emissions associated with treating irrigation water
Let it mellow: Choosing to forego flushing after urination reduces water use, thereby reducing energy and associated emissions from water supply and treatment. However, it incurs odor and may reduce the appearance of appropriate sanitation
Energy Conservation – Deliberate action to reduce energy waste
Turning off lights: Motion sensors, timers, and/or constant vigilance that disables lighting when not in use trades lower energy use and associated emissions for a small investment in controls and/or minor inconvenience
Engine start/stop: Systems that automatically turn off vehicle engines trade the benefit of fuel savings and reduced idling pollution for higher-cost starting equipment and emissions controls
Water Conservation – Deliberate action to reduce water waste
Fixing leaks: Leaks in underground infrastructure can be invisible, and even small leaks such as dripping faucets can substantially increase water consumption in a single residence. Remediation requires vigilance and (sometimes) costly intervention, and results in reduced water costs for the consumer as well as energy/GHG savings at the point of treatment
Low-flow fixtures and appliances: Toilets, faucets, dishwashers and other appliances can be designed to deliver the same sanitation benefits while reducing the amount of water bypassed during use. This reduces overall water consumption and energy for supply and wastewater treatment. Some water conserving equipment provides identical services to the consumer, while others deliver reduced comfort or convenience
Energy Mitigation – Costs incurred to reduce the environmental impact of energy use
Carbon capture and sequestration: Carbon dioxide generated from fossil energy use can be separated at the exhaust stack, pressurized, and re-injected into the subsurface. This process requires costly equipment and reduces overall energy efficiency, but it can mitigate 90% of greenhouse gas emissions at some facilities

(*continued*)

Table 2.2 Historical examples from each part of the sustainability framework that have alleviated the impacts of energy and water use, demonstrating the adverse consequences of deprivation and benefits of innovation (*Continued*).

Landfill gas recovery: Interception and recovery of methane from organic decay in landfills avoids high GWP emissions. Conversion of landfill gas to electricity has a small side benefit of avoiding some fossil fuel use
Water Mitigation – Investments to reduce the emissions associated with wastewater treatment
Anaerobic digestion: Organic material is digested in a bio-reactor, producing streams of biogas and less energetic sludge. The cost of installing and operating AD results in a useful energy product which may offset fossil fuel use and lower environmental impacts from organic discharge
Energy Substitution – Use of an alternate resource with potential cost or reliability impacts
Cleaner fuels: Natural gas replaces coal in electricity generation, resulting in lower CO_2 emissions per unit electricity. When the price of natural gas per unit energy became comparable to coal, there was little reason not to switch
Renewable electricity: Electricity from solar panels and wind turbines can displace fossil-fired generation. Solar energy trades higher capital cost and inherent intermittency for zero fuel cost and no carbon emissions
Water Substitution – Alternate resource or technology which may drive system reconfiguration
Non-potable reuse: Purple pipe systems deliver tertiary-treated wastewater to irrigation and some industrial/cooling applications. This reduces the demand for freshwater supply, and may offset the energy used for pumping and treatment of potable water
Ultraviolet disinfection: UV light can be substituted for the chlorine that is used to kill pathogens in water supply or recycled wastewater. UV does not require chemical delivery or dosing equipment and avoids the creation of disinfection byproducts. However, UV reactors incur significant upfront cost and ongoing energy costs.
Energy Efficiency – Technological increase in services provided from equivalent resources
Aerodynamics: Refinements in vehicle shape reduce drag, enabling cars and trucks to go substantially farther on the same amount of fuel with the same weight and volume of passenger/cargo capacity
Heat recuperation: Power generation and many industrial processes transfer thermal energy from exhaust to intake. This process requires costly heat exchange equipment and results in substantial energy savings and therefore emissions reduction
Variable speed drives: Novel electronics enable the motors that drive compressors and pumps to operate at lower speeds (and therefore power) without loss in efficiency. This saves substantial energy during periods when aeration or pumping needs are low
Water Efficiency – Increase in the benefits per unit water delivered
Drip irrigation: Replacing broadcast with drip irrigation drastically reduces the water lost to evaporation and percolation, thereby reducing pumping and treatment requirements (and associated energy and emissions) for water supply. However, drip systems are more costly to install and maintain
Energy Innovation – Delivers a better energy service for fewer resources consumed
LED lighting: New diode and phosphor materials enable drastically lower energy use at the same (or better) quality and intensity of light, with longer life and lower heat generation
Hybrid and electric vehicles: Higher energy density and more durable batteries enable regenerative braking and electric fueling, thereby increasing efficiency and reducing emissions. Electric cars are quieter and eliminate local pollution. They accelerate and handle better than comparable conventional vehicles, and can sometimes be fueled at home
Water Innovation – Delivers better treatment for less energy/material input
Membrane aerobic bioreactors: New materials and tube configurations enable oxygen delivery for organic deconstruction in wastewater at much lower pumping energy than traditional aeration, thereby reducing emissions

Water and energy investments exhibit a wide range of capital turnover rates. Capital turnover for both water and energy equipment tends to be fastest in the residential and commercial end-use sectors. Turnover of transportation equipment is slower. The large capital intensity of equipment in the industrial and utility sectors tends to drive the slowest turnover rates. Energy and water distribution and collection infrastructure is also designed for long service life and therefore very slow to change.

2.4.1 Residential and commercial equipment

Capital turnover in the residential and commercial sectors falls into three general categories. With 'devices' such as lightbulbs, electronics, and small appliances, replacement timelines are on the order of five years. They may be upgraded on the basis of consumer preference. Consumers often choose devices with state-of-the-art efficiency at the time of purchase. Energy- and water-consuming 'major appliances' such as furnaces, water heaters, air conditioners, and refrigerators have service lives of approximately 20 years. They are typically replaced upon failure. There have been federal and state-level incentives to improve efficiency. These incentives are effective in influencing consumers to choose higher efficiency devices when replacement is needed, but they only accelerate the decision to replace inefficient devices before end-of-life among affluent consumer groups. The 'dwelling' itself is a family's largest expense (as rent or a capital purchase). Housing capital stock turnover is typically measured in lifetimes and is difficult to assess for decarbonization. Major changes to electrical, fuel, and plumbing systems are seldom undertaken with sustainability as the primary motivation. An exception to this is the addition of solar energy, which is becoming more common as prices have dropped and innovative financing models have become widespread.

2.4.2 Transportation equipment

After housing, the largest capital investments for most Americans are personal vehicles. Vehicle energy use and carbon emissions are subjected to several sensitivities. Consumers typically make decisions based on current market conditions, that is the price of fuel at the pump, and not on total cost of ownership. Therefore, when fuel prices are higher, consumers tend to purchase more efficient, and therefore lower carbon-emitting vehicles, and when the price of fuel is low, consumers tend to purchase significantly less efficient vehicles. As vehicle manufacturing technology has improved, vehicle service life has extended, and is approaching 15 years. From a life cycle consideration, the long life of the vehicle avoids energy consumption and carbon emissions from the manufacturing process. It also retards widescale deployment of more efficient technologies.

Electrification of transportation is projected to have one of the largest impacts on economy-wide carbon emissions. With a low-carbon power generation, battery electric vehicles (EVs) offer a strong pathway to decarbonization. However, the impact of more durable conventional vehicles being sold today is that their extended service lives will contribute to long-term carbon emissions. Several vehicle manufacturers have announced plans to only manufacture EVs by 2035. With the capital turnover of about 15 years, this suggests that we will still have emissions from internal combustion engines out to 2050. This is the time frame that most advanced economies are targeting for net carbon zero. Therefore, there is limited room for delays in electrification.

2.4.3 Utility equipment

As with water, capital turnover at the energy utility scale can be very slow. As an extreme example, the BP petroleum refinery in Whiting Indiana was originally built in 1889 by Standard Oil and is still the largest petroleum refinery in the US. Most relevant to decarbonization is capital turnover in the power sector. Utility scale (100's of MW to GW) thermoelectric power plants served as the base for most of the power sector. Coal, nuclear, and more recently natural gas plants, are depreciated over a decade or more but continued to be used for 50 years. Depreciated capital offers operational advantage to older plants. With a decline in coal on a global scale, driven by societies demand for improvements in air quality, coal power has been declining well before drives for decarbonization became prevalent. The US has retired almost half of its coal power capacity over the past decade, declining from a more than 60% of total capacity to about 20%. Some utilities are looking to leverage coal power plant infrastructure and retrofit coal plants with cleaner energy sources. As with coal, operational nuclear power plants have almost all completed depreciated. With carbon-free emissions, there are strong incentives to maintain the nuclear fleet. The challenge is to remain profitable in a market where nuclear plants operate with constant output while demand and pricing are dynamic due to growth in wind and solar generation. There has not been a new plant commissioned since the partial meltdown of the Three Mile Island plant in 1979. There is one nuclear power facility under construction in Georgia.

In comparison to the 50+ year capital turnover for thermoelectric power plants, renewable plants tend to be both more distributed and have higher capital turnover rates. Wind turbines and solar photovoltaic (PV) facilities are projected to have a lifetime of about 20 years. This is based on facilities commissioned in the 1970s and 1980s that reached the end of their useful life in the 1990s and early 2000s. New wind turbines have nameplate capacities of 1–3 MW (rather than 10 s to 100 s of MW for gas/thermal turbines), and new wind farms have total capacities of 10 s to 100 s of MW. Individual solar panels have nameplate capacities in the 100 s of watts, making solar plant design and installation extremely modular. With the ability to add capacity incrementally, wind and solar generation have been growing steadily, and this trend is expected to continue. In comparison to wind and solar, capital turnover in hydropower can be extremely long. Century-old hydropower plants are still in operation. Dam-based hydropower plants can significantly disrupt wildlife, for example fish spawning. Recent investments in hydropower have replaced dams with 'run of the river' systems to address society demand and environmental regulations.

2.4.4 Integration

Capital turnover in the energy sector may present some unique challenges. For example, decisions on vehicle electrification are expected to have a strong impact on both liquid fuels production and power generation. A large increase in electricity demand for vehicles may trigger a new wave of capital expenditures in the electric sector, and/or a major change in the operation of existing generation and transmission assets. Similarly, a large drop in liquid fuel consumption will cause significant disruptions to gasoline and ethanol markets (see section 2.5.3). There are not similar 'fuel switching' capital replacement options for water consumers.

2.5 CASE STUDIES

2.5.1 Energy efficient lighting

The penetration of energy efficient lighting into the market in the years between 2010 and 2020 was an enormous success for new technology adoption. In the space of approximately ten years, the energy intensity of lighting in the residential sector dropped by 75–88% (a ~5× increase in energy efficiency), saving approximately 500 petajoules of energy per year in the United States. With an average carbon intensity of 450 gCO_2/kWh, this change resulted in an emissions reduction from the electricity sector of 62.5 million tons of CO_2 per year. Light emitting diodes (LEDs) replaced incandescent lights not only because they are more energy efficient, but because they are longer lasting (requiring less maintenance by the user and saving money on new bulbs over the long term) and because they provide a better lighting service with a choice of 'color temperatures' that appeal to many different consumers.

It had long been known that incandescent light bulbs were extremely inefficient. Approximately 3% of the electricity consumed by an incandescent bulb is radiated as visible light. The remainder is emitted as heat. This fact indicates that the same service could be accomplished with far less electricity if a new technology were used. Additionally, the excess heat given off by incandescent bulbs increases the load on air conditioning systems in warm climates, further increasing energy demand.

The transition to energy efficient lighting was not without challenges though. Compact fluorescent lamps (CFLs), an earlier generation of energy efficient lighting technology, failed to attain consumer acceptance. CFLs were nearly as efficient as LEDs, saving 70–80% of lighting electricity over their incandescent predecessors. However, CFLs were disliked by consumers because the quality of the light they produced was inferior. CFL light had a high color temperature (bluish tint) and many users perceived a flickering nature to it. CFLs were advertised as having much longer lives than incandescents, but they burned out earlier than predicted. The failure of CFLs in the marketplace is proof that consumers may be unwilling to trade quality of service for energy savings, even if there is a comparable quantity of service and modest cost savings over the long term. Policy initiatives that supported the transition to more efficient lighting (efficiency standards and incandescent 'bulb bans') were met with fierce opposition when the only viable alternative to incandescent bulbs was CFL.

Government research institutions and private industry committed significant resources to developing LED technology. Some of these investments were based on evidence that LED would ultimately be a better technology. Some were responsive to consumer demand created by the policy incentives described above. Today, it is nearly unthinkable to purchase incandescent lighting in the residential sector for anything but the most niche applications. Manufacturing know-how has advanced so that LED bulbs can be produced to meet the demands of almost any application and form factor.

2.5.2 Electric vehicles

Adoption of electric vehicles (EVs) represents a sea change in transportation that is starting to transform societal energy use and carbon emissions. Battery technology, which had been stagnant for decades, began to change dramatically in the 1970s, first with discovery science and then with scaled-up manufacturing. Nickel metal hydride chemistry was quickly surpassed by lithium-ion technology in the early 2000s. Improving battery technology has impacted a broad range of market sectors. Compared to other battery chemistries, lithium-ion batteries offer flexibility in recharging, much higher energy density (energy per unit mass or unit volume), and much higher power density (higher current at the same voltage). Li-ion technology transformed the small electronics sector, ironically creating an increase in energy demand. While limited range vehicles such as golf carts could operate with traditional rechargeable lead-acid batteries as the primary energy source, longer range road vehicles were beyond the range of available at existing energy capacity. Li-ion offered the potential for long ranges in vehicles suitable for the roadway.

While some early vehicles in the 19th century were electric, they faded from favor in comparison to the significantly higher power and energy density of internal combustion engines (ICEs) of that era. At the dawn of the 21st century, the first broadly commercialized battery technology for 'electrifying' mobility were launched using hybrid vehicles such as the Toyota Prius. The Prius employed both a traditional ICE operating on liquid fuels and an electric motor powered by rechargeable batteries. The batteries were recharged by recovering energy using regenerative braking and the vehicles enjoyed ~50% increase in fuel mileage. Energy storage, while not used as the primary energy supply, was able to overcome the weaknesses of conventional powertrain design that led to substantial inefficiency. Thus, electrification (batteries, motors, and drivetrain-capable power electronics) gained a foothold in the automotive industry.

Subsequently, the rise in petroleum prices in the mid 2000s sparked entrepreneurial interest in all-electric vehicles. While the battery in a hybrid vehicle typically offers less than a 20-mile range, a battery EV requires a minimum of a 100-mile range, and preferably greater than 300 miles. This required multiple innovations in battery chemistry, electrode design, and cell pack assembly. Research from academic and research laboratories discovered new chemistries and designs, and large chemical companies took a strong interest in developing the manufacturing technologies to deploy them. Within a few years, both Tesla and GM, (a start-up and a global mega-corporation respectively), as well as other market entrants brought light-duty EVs to market. In 2021, light-duty EVs captured about 2% of the market in the US. In northern Europe, high fuel prices have driven EV sales to greater than 50%

of the new vehicle market. The largest global market for light-duty vehicles is China, and China has the largest EV fleet. While the upfront costs for EVs are higher than ICEs, the total cost of ownership when considering the cost of fuel, repairs, and vehicle life make EVs lower in cost than ICEs. At current US energy prices and typical vehicle energy efficiencies, the fuel cost for electric vehicles is much lower than for gasoline-powered cars. In comparison to ICEs, EVs have fewer moving parts, generate less heat, and do not need to replace lubricating oils, cooling fluids, and brake pads as often. Therefore, repairs are less frequent, and except for the replacement of the battery (~10 years), vehicle life is significantly longer, and maintenance is significantly cheaper.

However, as markets expand for light-duty EVs, it is becoming clear that the access to vehicle charging will be a limiting factor on widespread EV adoption. With upfront costs of EVs higher than ICE vehicles, most early adopters have been affluent buyers with ready access to overnight charging in private garages. Less affluent drivers who live in urban and suburban rental units will not have the same opportunity. Similarly, public charging infrastructure is being deployed in urban areas and along high-use transportation corridors, so rural users are disadvantaged. Finally, taxi and delivery drivers will have substantially different charging needs than the owners of vehicles whose use is purely personal.

There are important lessons for the water industry in the adoption of electric vehicles. Transportation is a contributor to carbon emissions in the US and light-duty EVs account for greater than 60% of liquid fuel use. The world cannot achieve any meaningful decarbonization goals without transforming the light-duty fleet. Note that electrifying the fleet will only achieve the decarbonization targets if the vehicles are charged with carbon-free electricity. Similarly, electrifying energy input or unit operations in the water and wastewater sectors will only be effective if the grid is decarbonized. Furthermore, success requires investments in both fundamental science and engineering as well as underlying infrastructure. Nascent science and technologies can grow into major business opportunities. For example, Tesla has become the world's most valuable vehicle manufacturer since before the pandemic. However, certain performance targets must be met before a low-carbon technology will be adopted at scale. In the case of electric vehicles, the performance target was the energy density of the battery, and the market needed to wait for lithium-ion chemistry to be sufficiently advanced (reliable, manufacturable) to be adopted. The water industry must identify performance targets for decarbonized systems and seek investment in technologies that can reach those targets.

2.5.3 Cellulosic biomass

Engines that run on agriculturally-derived fuels (alcohols or converted vegetable oils) have existed almost as long as engines that run on fossil fuels. However, petroleum fuels far out-perform biofuels on a cost and energy-return basis in most cases. Despite the interest in biofuels generated by the oil crises of the 1970s, the markets for these fuels remained very small for decades.

With rising petroleum prices in the mid 2000s, the US passed first the Energy Policy Act (EPACT) of 2005 and the Energy Independence Security Act (EISA) of 2007 (EPA, 2007). These laws were intended to ensure a reliable domestic fuel supply and to simultaneously create economic opportunities for farms and rural regions. EPACT set national blend volume mandates for ethanol. Due to the benefits, fuel manufacturers readily exceeded the ethanol blend mandates. EISA created more aggressive blend requirements, and for the first time mandated life cycle-based GHGs emissions reductions. Life cycle analysis (LCA) indicates that corn starch ethanol, despite being biogenic in nature, reduces GHGs by only about 20% because of the extensive fossil energy requirements for farming and process heat. Cellulosic ethanol has the potential for much lower life cycle GHG emissions because it uses more of the plant material (and thereby reduces total acreage farmed per ton of feedstock), and because it is designed to use biomass for process energy. EPACT and EISA created a pathway for starch ethanol as an early market entrant, and with the expectation that cellulosic ethanol would dominate production and plateau in 2021. EISA 2007 generated significant interest from venture investors, entrepreneurs, and scientists to focus on cellulosic research.

With the early mandates for corn starch ethanol, investors were incentivized to increase the size of conventional biorefineries and production outpaced the targeted EISA volumes. Within a few years, corn ethanol utilized 40% of corn crop production, largely achieving one of the original goals of EPACT in supporting rural economic development. Ethanol rapidly achieved ~10% volumetric blend of the gasoline supply, extending the liquid fuel supply as the mandates targeted.

With cellulosic biofuels, progress was slower. Originally, the limiting technical factor was considered enzymes to breakdown recalcitrant cellulosic into fermentable sugars. Cellulose is a structural polymer composed of sugars monomers that are difficult to depolymerize. In comparison starch is a nutrient source composed of readily digestible sugar polymers. As the science of cellulosic enzymes advanced, other technical challenges were identified in the cellulosic biofuel process. When pioneer cellulosic biorefineries were constructed, initial estimates were that they would have about twice the capital cost per unit of product volume. With lower overall productivity, cellulosic biorefineries capital costs grew to five- to ten-fold in comparison to mature starch ethanol biorefineries. This resulted in commercialization delays and unique challenges to the cellulosic industry. Fourteen years after EISA created mandates and incentives for cellulosic biofuels, the industry has yet to substantially impact decarbonization.

Two distinct challenges rapidly developed for the cellulosic and overall biofuel markets. The first challenge is that the market is structurally limited in size. As corn starch ethanol production grew rapidly, the US soon produced enough fuel to achieve 10% volume of the entire gasoline market. At that time, most vehicles and most fuel infrastructure were limited to a 10% ethanol (E10) blend due to materials compatibility. The conventional technology, which is relatively ineffective at decarbonization, had exceeded the ability of the market to consume it. Flexible fuel vehicles (FFVs), which are capable of using fuel with up to 85% ethanol fuel, were proposed as a solution. The manufacturing cost differential is only about $100. The vehicle manufacturers received credits for the vehicles as if the vehicle always used 85% ethanol (E85) fuel to meet fleet-wide corporate average fuel economy (CAFE) standards. The fuels market did not have any incentive to market E85 fuels so FFVs continued to operate on conventional E10 fuel. Therefore, FFVs resulted in only incidental increase in ethanol usage, and therefore minimal impact on GHG reductions. Sixteen years after EPACT, corn ethanol accounts for about 10% of the gasoline market and each gallon reduces GHGs by about 20%. Ethanol, therefore, results in about ~2% reduction in GHGs. EISA 2007 and cellulosic biofuels have had essentially no additional impact on GHG emissions.

The second challenge in the biofuels market, and notably the cellulosic biofuel, is a significant warning for the water sector. Realizing that cellulosic production was a nascent industry in 2007, EISA created a regulatory requirement that the US EPA monitor cellulosic biofuel production capacity on an annual basis. The EISA mandate for cellulosic fuels is adjusted annually to avoid fuel blenders being mandated to use cellulosic biofuels that do not exist. Since the first blending requirements, manufacturing capacity has lagged blend mandates, so EPA adjusted the volumetric requirements. A cellulosic biorefinery is a complex operation and requires several years to construct and deploy. One of the authors of this chapter interviewed project investment banks and described the EISA mandates, EPA regulatory role, and the time and cost to build (Blazy *et al.*, 2015). It was uniformly considered a poor investment decision. Therefore, few cellulosic biorefinery projects have even been launched, and there is little-to-no success in the industry. The water sector should learn that mandates without consideration of markets, economics, fundamental science, state of technology, and the full scope of the mandates could lead to failed investments and little progress in achieving decarbonization goals.

2.5.4 Wind and solar

One of the true success stories in decarbonization of the energy sector is the substantial growth in wind and solar power. While the technologies are quite distinct, we assess their combined impact here. Solar has the potential to meet all of society's energy demand (Hermann, 2006), and has been considered as an ultimate solution to decarbonization. While wind is more limited in terms of total potential it has exhibited faster growth than solar. The advantages of both wind and solar is that they release no carbon emissions and have no fuel requirements. The need for fuel creates supply chain risks and also adds fuel price volatility risk to the total cost. Together, renewables are, at the time this book is published, the second largest generator of power in the US after natural gas, catching up to nuclear and surpassing coal.

The growth in wind and solar demonstrate a technology 'learning curve' that water decarbonization may emulate.

Driven by incentives including both investment tax credits (ITCs) and production tax credits (PTCs), wind capacity has exhibited the largest overall increase in capacity of any type of generation. This rapid build-out has catalyzed 'learning-by-doing,' and the modular nature of wind power allows for continuous innovation in the design, manufacturing, and construction of turbines. Focused R&D in wind has resulted in only incremental improvements, but for wind technology, incremental improvement delivers outsized gains in performance. For example, the generation capacity of a turbine scales as the square of the length of its blades. Therefore, small increases in blade length enabled by novel designs and materials have resulted in a non-linear increase in turbine capacity. Similarly, taller towers enable turbines to access more reliable wind resources. Higher reliability translates to a more valuable electricity resource, in addition to the bulk increase in kWh generated. Ultimately, these improvements increase land use efficiency.

There has been enormous research investment in new PV solar materials, however, no new materials have been deployed at scale. Rather, the dramatic drop in prices (and commensurate rapid build-out of PV-based generation) has been driven largely by reductions in manufacturing costs of conventional solar materials (polysilicon, and to a lesser extent cadmium telluride) and installation costs. China has driven the reduction in manufacturing costs. The ITC accelerated the domestic market for installation, and again, learning-by-doing drove down installation costs. Each new PV installation enabled incremental innovations in racking, interconnection, and construction logistics for ground-mount and roof-mount systems. As this book is published, solar PV offers the lowest cost of power costs in sunny regions such as the US southwest. However, because generation peaks during the middle of the day and demand peaks at other times, the value of additional solar installation is beginning to decline in areas with substantial solar penetration (Bolinger *et al.*, 2021).

The challenge to both wind and solar is intermittency. The ultimate solution is to link intermittent renewable power production to energy storage. Energy storage includes batteries, supercapacitors, pumped hydropower, other mechanical systems, or even thermal systems. The value of grid-scale energy storage is less than battery electric vehicles, so grid storage is learning and adapting from EVs.

Renewable power generation provides an important template for decarbonizing the water sector. Significant advancements in technology were not required for transition in the market. Rather, policy incentives drove the economics enough to foster capacity expansion. Increased installations drove manufacturing and infrastructure support down a learning curve to further incentivize deployment. For several years, incentives made renewable power the lowest cost pathway to increase capacity. As the manufacturing and installation processes matured, the ITCs were no longer required. As capacity grew to where it was disrupting grid resilience, storage and other mechanisms are developing as solutions. Incentives for storage and now driving storage capacity growth. It is largely expected that renewable power plus grid-scale storage will offer a cost-competitive and reliable carbon-free power sector.

ACKNOWLEDGEMENTS

This work was supported by the US Department of Energy through contract DE-AC07-05ID14517 (Idaho National Laboratory). This work was performed under the auspices of the US Department of Energy by Lawrence Livermore National Laboratory under Contract DE-AC52-07NA27344. This document was prepared as an account of work sponsored by an agency of the United States government. Neither the United States government nor Lawrence Livermore National Security, LLC, nor any of their employees makes any warranty, expressed or implied, or assumes any legal liability or responsibility for the accuracy, completeness, or usefulness of any information, apparatus, product, or process disclosed, or represents that its use would not infringe privately owned rights. Reference herein to any specific commercial product, process, or service by trade name, trademark, manufacturer, or otherwise does not necessarily constitute or imply its endorsement, recommendation, or favoring by the United States government or Lawrence Livermore National Security, LLC. The views and opinions of authors expressed herein do not necessarily state or reflect those of the United States government or Lawrence Livermore National Security, LLC, and shall not be used for advertising or product endorsement purposes.

REFERENCES

Blazy D., Pearlson M. N., Miller B. and Bartlett R. E. (2015). A Monte Carlo-Based Methodology for Valuing Refineries Producing Aviation Biofuel, Chapter 15. Royal Society of Chemistry, Cambridge UK, pp. 336–351. https://doi.org/10.1039/9781782622444-00336

Bolinger M., Seel J., Warner C. and Robson D. (2021). Utility Scale Solar 2021 Technical Brief. Lawrence Berkeley National Lab, Berkeley, CA. https://emp.lbl.gov/sites/default/files/utility-scale_solar_2021_technical_brief.pdf (accessed 2 November 2021)

DHS (2021). National Critical Functions. U.S. Department of Homeland Security, Washington, DC. https://www.dhs.gov/cisa/national-critical-functions (accessed 30 October 2021)

Dieter C. A., Maupin M. A., Caldwell R. R., Harris M. A., Ivahnenko T. I., Lovelace J. K., Barber N. L. and Linsey K. S. (2018). Estimated use of Water in the United States in 2015. In *USGS* (Circular 1441). U.S. Geological Survey, Reston VA. https://doi.org/10.3133/cir1441

DOE (2014). The Water-Energy Nexus: Challenges and Opportunities. U.S. Department of Energy, Washington DC. https://www.energy.gov/sites/prod/files/2014/07/f17/Water Energy Nexus Executive Summary July 2014.pdf (accessed 20 April 2020)

DOE (2020). Moving Goods in A SMART Mobility System. U.S. Department of Energy, Washington DC. https://www.energy.gov/eere/vehicles/downloads/smart-webinar-6-moving-goods-smart-mobility-system (accessed 30 October 2021)

EIA (2021). Monthly Energy Review. U.S. Department of Energy, Washington DC. http://www.eia.gov/totalenergy/data/monthly no (accessed 24 November 2020)

EPA (2007). Energy Independence and Security Act. Public Law 110-140. U.S. Environmetnal Protection Agency, Washington DC. https://www.epa.gov/laws-regulations/summary-energy-independence-and-security-act (accessed 11 January 2022)

EPA (2021). Inventory of U.S. Greenhouse Gas Emissions and Sinks 1990–2019 (No. EPA430-R-21-005). U.S. Environmental Protection Agency, Washington DC. https://www.epa.gov/ghgemissions/inventory-us-greenhouse-gas-emissions-and-sinks-1990-2019 (accessed 30 October 2021)

EPRI (2013). Electricity Use and Management in the Municipal Water Supply and Wastewater Industries (No. 3002001433). Electric Power Research Institute, Palo Alto CA. https://publicdownload.epri.com/PublicDownload.svc/product=000000003002001433/type=Product (accessed 14 January 2020)

Gleick P. H. (1994). Water and energy. *Annual Review of Energy and the Environment*, **19**(1), 267–299. https://doi.org/10.1146/annurev.eg.19.110194.001411

Greenberg H. R., Simon A. J., Singer S. L. and Shuster E. P. (2017). Development of Energy-Water Nexus State-Level Hybrid Sankey Diagrams for 2010. In *LLNL* (LLNL-TR-669059). Lawrence Livermore National Lab, Livermore CA. https://flowcharts.llnl.gov/report (accessed 7 January 2020)

Grubert E. and Sanders K. T. (2018). Water use in the United States energy system: A national assessment and unit process inventory of water consumption and withdrawals (research article). *Environmental Science and Technology*, **52**(11), 6695–6703. https://doi.org/10.1021/acs.est.8b00139

Hermann W. (2006). Quantifying global exergy resources. *Energy*, **31**(12), 1685–1702. https://doi.org/10.1016/j.energy.2005.09.006

Kyung D., Kim M., Chang J. and Lee W. (2015). Estimation of greenhouse gas emissions from a hybrid wastewater treatment plant. *Journal of Cleaner Production*, **95**, 117–123. https://doi.org/10.1016/j.jclepro.2015.02.032

Roy M., Edwards M. R. and Trancik J. E. (2015). Methane mitigation timelines to inform energy technology evaluation. *Environmental Research Letters*, **10**(11), 114024. https://doi.org/10.1088/1748-9326/10/11/114024

Szulc P., Kasprzak J., Dymaczewski Z. and Kurczewski P. (2021). Life cycle assessment of municipal wastewater treatment processes regarding energy production from the sludge line. *Energies*, **14**(2), 356. https://doi.org/10.3390/en14020356

US Bureau of Economic Analysis (2021). National Income and Product Accounts, Table 1.1.4 Price Indexes for Gross Domestic Product and Table 1.1.5. Gross Domestic Product. U.S. Bureau of Economic Analysis, Suitland MD. https://apps.bea.gov/iTable/index_nipa.cfm (accessed 30 October 2021)

USDA (2021a). CORN Quickstats: Acreage, Yield, and Price. National Agricultural Statistics Service. U.S. Department of Agriculture, Washington DC. https://www.nass.usda.gov/Quick_Stats/Lite/result.php?B565104E-D426-36E7-9481-03007890E2F2 (accessed 8 October 2021)

USDA (2021b). Wheat Sector at A Glance. USDA Economic Research Service. U.S. Department of Agriculture, Washington DC. https://www.ers.usda.gov/topics/crops/wheat/wheat-sector-at-a-glance/ (accessed 8 October 2021)

USGS (2021). Mineral Commodity Summaries 2021. U.S. Geological Survey, Reston, VA. https://doi.org/10.3133/mcs2021

doi: 10.2166/9781789061796

Chapter 3

Greenhouse gases in the urban water cycle

Andrew Shaw[1]*, Kubeshnee Chetty[2], Tak Fan Chan[2], Elena Lindsey[2], Anjana Kadava[1] and Ben Stevenson[3]

[1]Black & Veatch Corporation, 920 Memorial City Way, Suite 600, Houston, TX 77024, USA
[2]Binnies, 60 High Street, Redhill, Surrey, RH1 1SH, UK
[3]Black & Veatch Ltd, Queens House, 19–29 St. Vincent Place, Glasgow, G1 2DT, UK
*Correspondence: shawar@bv.com

3.1 INTRODUCTION

This chapter gives an overview of greenhouse gas (GHG) emissions in the context of the urban water cycle. Starting with an overview of the urban water cycle and a definition of different equivalent carbon emission scopes, the chapter then gives a description of the major GHGs in each part of the urban water cycle, in order to identify opportunities for decarbonization. In the latter sections of the chapter a framework is presented for carrying out a carbon footprint assessment along with an overview of available and relevant protocols and methods for assessing GHG emissions.

3.1.1 Overview of the urban water cycle

The urban water cycle can be defined as the cycle containing processes to provide potable drinking water to society (Bakhshi, 2009). This also includes the removal and reclamation of wastewater and sewage, and redirection of stormwater as a natural resource. It provides a vital balance between potable water demand and natural resource provision. There are seven key stages to the urban water cycle: abstraction, treatment, distribution, use, wastewater collection, wastewater treatment, and discharge. These are shown in Figure 3.1. The following section gives a brief description of each of these stages with an initial indication of the energy used for each, which can be an indicator of the overall decarbonisation potential of that stage.

Abstraction takes place from surface water bodies, such as rivers, or from groundwater sources. Energy is required in the extraction processes of raw water from its source. Groundwater is typically of a naturally higher quality than other sources and therefore requires less energy input for the treatment process. However, groundwater extraction requires approximately 30% more electricity per unit than extraction from a surface water source (Appelbaum, 2002). Water is stored by dams and reservoirs where necessary to ensure an adequate supply for treatment. Water can also be abstracted from brackish or saline sources to be treated at desalination plants.

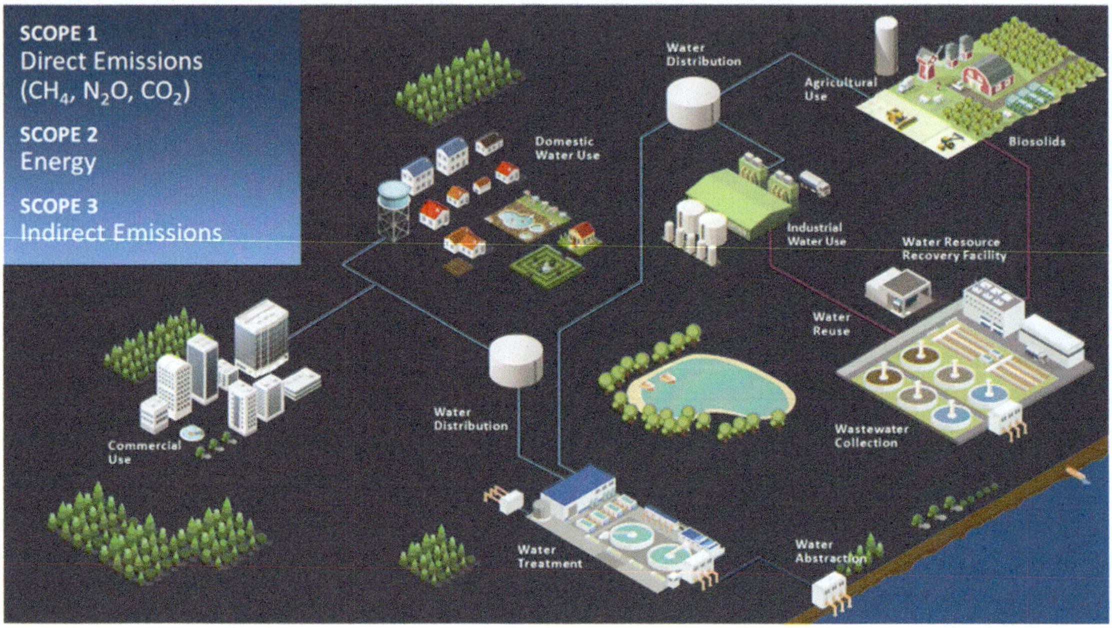

Figure 3.1 The urban water cycle.

After abstraction, water is treated to potable standards, the quality of which is dependent on national or regional requirements. There are various treatment technologies in use today, selected based on site specific conditions and on the treatment standards required. The advancement of technological processes does not directly correlate to higher emissions, as modern processes seek to reduce emissions. However, Bakhshi (2009) posits that energy use, and therefore emission release, will increase as various water acts (e.g. the Safe Drinking Water Act) demand higher quality water.

Whether using distributed or centralized treatment plants, storage facilities, pumps, and pipes are required to distribute the clean water to the end user. Although some systems rely solely on gravity, most require pumping. Pumping maintains the pressure and movement of water to not only relocate the water but to minimise corrosion and contamination of the pipe network (Klein, 2005). This can often be an energy intensive process to achieve the required system pressure throughout the distribution network. Due to the age of many water networks, leaks and issues with infrastructure contribute to increased energy use within the associated systems. Moreover, energy use is primarily driven by increased demand and urban growth.

Water is consumed by homes, businesses and industries, with further energy use within them: treatment, circulation, heating, cooling. Surface water run-off can be considered in the 'use' category although its source – precipitation – is from the natural water cycle. Surface water runoff and stormwater management thus become processes of the urban water cycle. Wastewater from a home or industry that does not contain fecal contamination is called grey water. Grey water can be cleaned and used onsite, for example for cooling processes.

Wastewater collection systems often use gravity to move wastewater to a facility to be treated typically by positioning the wastewater treatment facility 'downstream' of the urban area. However, urban areas that are large or very flat will require pumping of wastewater which significantly increases their energy use.

As the incoming water is 'dirty', wastewater requires more energy to treat than fresh water. The type of waste treatment depends on the final discharge or reuse point and therefore the energy and greenhouse gas emissions vary widely. Wastewater originates from a variety of sources, varying in contamination from industrial processes to surface water runoff. Therefore, the treatment of wastewater requires more consideration than water treatment. Wastewater can be cleaned sufficiently for several reuse applications, including advanced wastewater treatment to produce potable water without returning water to its natural source, that is potable water reuse.

Water is discharged into surface water bodies for integration back into the natural water cycle or for reuse by the urban water cycle.

3.1.2 Definition of scope 1, 2 and 3 emissions

Emissions can be categorised as Scope 1, 2, or 3. The Intergovernmental Panel on Climate Change (IPCC) has well-defined definitions of the three scopes (IPCC, 2014a):

- **Scope 1** represents direct GHG emissions, from sources owned by the reporting entity. GHG emissions arising from sources controlled or managed by the reporting entity can also be classified as Scope 1.
- **Scope 2** emissions cover those that arise indirectly from the production of energy that has been purchased by the reporting entity. Emissions are described as indirect if they are a consequence of activities from a specific region, sector, company, or other distinct boundary. The source of scope 2 emissions is generally the production of electricity, heat, or steam.
- **Scope 3** emissions represent all other indirect emissions. These may include emissions associated with material production, including extraction of raw materials and the production processes of purchased materials, fuels, or services (Hertwich & Wood, 2018). Generally, any outsourced activity is classed as Scope 3, such as waste disposal, hire cars, or contracted maintenance. Scope 3 emissions can be difficult to globally quantify as a scope 3 emission for one entity may also be categorised as a scope 1 or 2 emission by an alternate organisation (Ghaemi & Smith, 2020), thus leading to double-counting of the emissions.

3.1.3 Water footprint and carbon footprint

Water footprint is a concept similar to carbon footprint that attempts to account for water used in human activities. ISO 14046 is a standard approach for conducting a water footprint. In their research, 'Water Footprint: A New Concept for Sustainable Water Utilities' (WRF, 2014), the Water Research Foundation investigated the use of water footprint for water utilities. The intersection of carbon and water footprints is a natural discussion with the developers of water footprinting methods.

3.2 GREENHOUSE GASSES IN THE WATER CYCLE

The GHG emissions in the water cycle need to fully account for emissions arising from all stages of the water cycle, from abstraction through to water treatment, consumer use, disposal, sewer network emissions, waste treatment and final discharge. Based on 2014 data shown in Figure 3.2, the waste sector (e.g. landfills, wastewater treatment) contribute towards approximately 3% of global anthropogenic GHG emissions. The water supply and treatment process only accounts for 11% of the GHG emissions from the water cycle with the majority (89%) arising from carbon emissions associated with domestic water use and disposal 'in homes' which also includes the energy used in the heating of the water (Rissman *et al.*, 2020). Guohua *et al.* (2019) investigated the impact of the urban water cycle in Beijing, China (Figure 3.3). They found that the water cycle accounted for 33% of the total urban energy use, however 90% of this was associated with the end use, such as heating of water in households. They confirmed that their findings for energy intensity (kWh/m^3) were consistent with other studies, showing that the use phase for water is most significant.

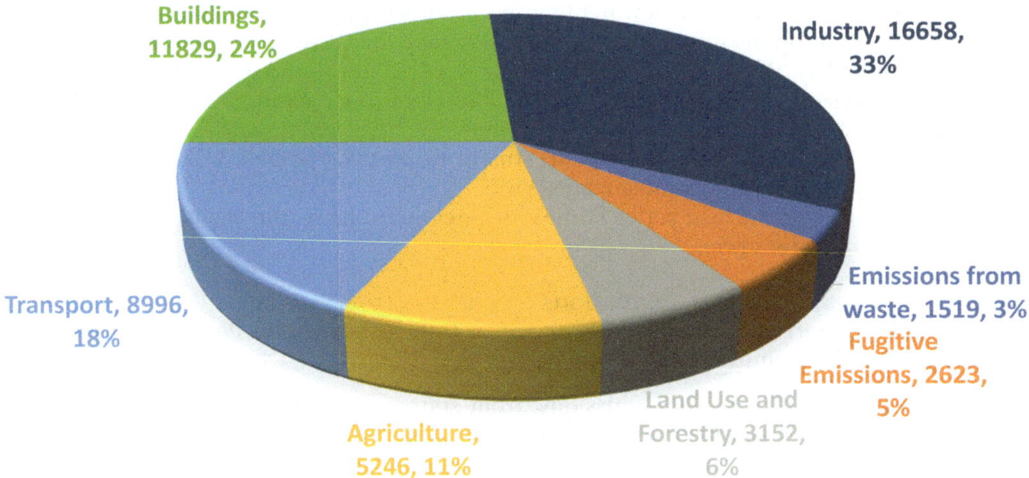

Figure 3.2 Global GHG emissions by sector in 2014, Mt CO_2e (based on Rissman *et al.*, 2020).

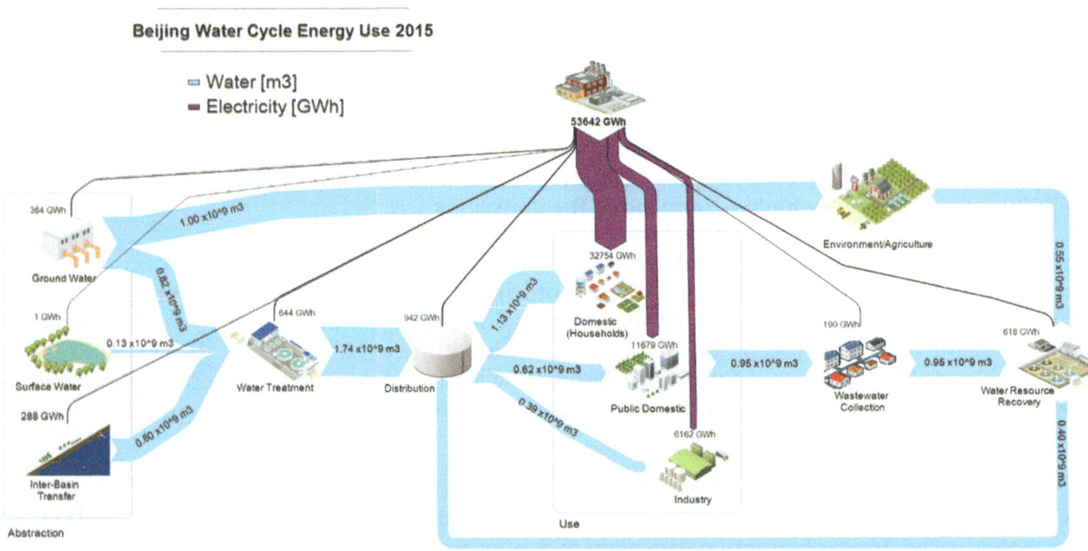

Figure 3.3 Sankey diagram of energy use in the whole water cycle process in Beijing in 2015 (adapted from He, 2019).

This chapter will provide an overview of scope 1, 2 and 3 emissions associated with abstraction, water and wastewater treatment and biosolids management process. To truly achieve net zero carbon emissions, policy makers will need to switch focus from supply and treatment to the consumer side of the equation and target interventions at either reducing or making consumption more efficient, thereby reducing the GHG emissions from the whole water life cycle and not just on conveyancing and treatment.

The water industry in the UK reported to have achieved 45% emission reduction of 2.4 $MtCO_2e$ between 2011 and 2018 and consumed 6.8 TWh of electricity for a population of approximately 67 million. Most of the emissions are CO_2 associated with the electricity, which is approximately 2% of the electricity consumption in the UK to pump water to customer and wastewater treatment. The industry has pledged to reach net zero carbon emissions by 2030 (Water UK, 2020).

3.2.1 Scope 1 – direct emissions – from own and controlled sources

3.2.1.1 Design and construction of new assets

Embedded carbon emissions are generally associated with upsizing or installing of new assets on a treatment works. The broad assumption is that new schemes are implemented to meet rising demand rather than replacing existing assets. Hence with increase in demand comes increase in embedded carbon emissions as treatment capacity is built to align with growth in demand.

Water companies can reduce increases in demand by investing in 'smart devices' that help consumers monitor and track consumption such as: water saving devices for toilets, showers and baths; water meters; water efficient domestic appliances; rainwater collection systems; grey-water recycling (i.e. water from showers, baths and sinks used for toilet flushing) and water mains leakage reduction.

3.2.1.2 Water and wastewater collection systems

Water collection and abstraction: The majority of operational emissions associated with storage reservoirs; regional water grids via transfer pipelines resulting from operational activities and so on. are normally classified as scope 2 emissions as it involves some degree of pumping utilizing energy from the grid. Other emissions directly related to water collection systems are emissions resulting from vehicle movements on site as well as emissions associated to directly delivered maintenance activities on site.

Wastewater collection: Methane, nitrous oxide and hydrogen sulphide can be produced in sewers biologically in the absence of oxygen. Methane CH_4 and N_2O emissions contribute up to 3% of the sewer life cycle carbon footprint (Eijo-Río *et al.*, 2015). The carbon emissions arising from sewer systems are often not included in carbon calculations for treatment works which results in a gap in the whole cycle carbon assessment of water cycle. The paper by Zawartka *et al.* (2020), concluded similar findings for transporting of wastewater to a wastewater treatment plant, as well as the gas emissions for collecting and treating wastewater. Moving forward, a focus on waste collection systems will be required to truly achieve et zero emissions.

3.2.1.3 Water and wastewater treatment and sludge management

Water treatment: The nature of water treatment processes generally emits insignificant amount of GHG during the production of potable water.

Wastewater treatment: The GHG emissions from a wastewater treatment plant is normally through the production of carbon dioxide (CO_2), methane (CH_4), and nitrous oxide (N_2O) arising from the biological wastewater treatment process. The calculations presented in the European Commission's report regarding emissions from wastewater sewage systems and treatment processes accounted for 9 and 3% of the total world CH_4 and N_2O emissions (Zawakati, 2020), whereas in comparison a quarter of the emissions reported in the UK as per the Carbon Accounting Workbook (CAW) were attributed to methane and N_2O arising out of treating sewage and recycling wastewater from 28 million homes.

CO_2 emissions are generated from both the biological treatment process as well as on site electricity consumption. The organic carbon of wastewater is either incorporated into biomass or oxidized to CO_2 (Campos *et al.*, 2016). Driving energy efficiency through the operation of the treatment works will assist in the overall reduction of CO_2 emissions from site, thereby contributing to a reduction in treatment costs by enhancing energy savings whilst simultaneously reducing the environmental impact of the operational activities.

N_2O emissions are generated by nitrification and denitrification processes used to remove nitrogenous compounds from wastewater. It is produced most commonly in the activated sludge process where nitrifying bacteria produce N_2O under aerobic or anoxic conditions. In anoxic conditions both ammonia and nitrite-oxidizing bacteria produce N_2O, while only ammonia-oxidizing bacteria produce N_2O in aerobic conditions. A small percentage of N_2O is also generated from onsite grit and sludge storage tanks.

Nitrous oxide (N_2O) and methane (CH_4) are potent GHG gases with global warming potentials (expressed in terms of CO_2-equivalents) of 265 and 28, respectively. When emitted to the atmosphere, they significantly contribute to climate change. Research by Oshita *et al.* (2014), show that most of the methane emissions from WWTPs are closely related to processes involved in the sludge line. CH_4 emissions mainly arise from the anaerobic digestion process, its associated process units such as primary sludge thickener, the centrifuge, the exhaust gas of the cogeneration plant, the buffer tank for the digested sludge, and the storage tank for the dewatered sludge. Daelman *et al.* (2012) found that around 1% of incoming chemical oxygen demand (COD) to waste water treatment works was emitted as methane and that the sludge process units contribute to 72% of the methane emissions from a waste water treatment works.

Due to the apparent inconsistency between the CAW methodology for calculating process emission and current GHG emission accounting practice, there is an uncertainty around the scale of process emissions in the UK. This is partly due to an update to the CAW methodology which can increase scope 1 emissions by 0.2 $MtCO_2e/y$ (~8% of UK water sector's net emission) and an associated increase in baseline process emissions by ~30%.

Fundamentally, the core focus for reduction of operational carbon emissions generally relates to the modification or optimization of the operational conditions of the treatment plant as this is the most economical and cost-efficient method, however this is not always possible due to the operational limitations of the installed units. Other ways to mitigate GHG emissions also include treatment of gaseous streams or the installation of new processes to remove both organic matter and pollutants.

Sludge management: Advanced digestion treatment methods are used to process sludge generated onsite as well as imported from other sites to enhance the biogas (CH_4) yield from the sludge treatment process. The biogas can be used in onsite combined heat and power engines (CHP) to generate renewable energy. Other uses for biogas also include gas to grid applications.

The stabilized and dewatered biosolids is a valuable product as fertilizer and soil conditioner for agricultural use. There are biosolids assurance schemes, code of practices and regulations in different countries to ensure its safe recycling.

3.2.2 Scope 2 – GHGs from energy use

The water industry in the UK consumes approximately 3% of the total electricity generated in the country for pumping, water treatment and wastewater treatment (Nair *et al.*, 2014). Electricity is needed throughout the water cycle for potable water and wastewater operations, which include water abstraction, treatment and distribution followed by sewage collection and wastewater treatment and where applicable water reuse.

3.2.2.1 Pumping

The energy required for abstraction depends on the source of water. Pumping is often required to lift groundwater from the water table, while energy for abstracting surface water depends on the distance and profile between the source and the treatment facility.

After the water has been used by the consumers the used water or wastewater is collected and, depending on the local ground profile and the relatively location of the wastewater treatment facility, pumped through a network or sewer for treatment to remove the pollutants. The treated wastewater is often discharged into the local receiving water by gravity and pumped discharge is not often required for most works.

3.2.2.2 Water treatment process

Water treatment plants are designed for gravity flow where possible unless inter-stage pumping is necessary due to site profile or hydraulic requirement for the treatment processes. The treatment processes chemical dosing, mixing and filtration systems consume energy. Advanced processes such as membrane filtration, oxidation, disinfection by UV and ozonation are more energy intensive. Desalination is highly energy intensive even with the introduction of energy efficient reverse osmosis membrane and the move from the desalination of sea water to brackish or treated wastewater.

A significant amount of energy is required to distribute treated water into the potable network. However, reportedly 20% of the water put into supply in England is lost through leakage compared with approximately 3 and 5% reported in Germany and Singapore respectively (C40 Cities, 2021).

3.2.2.3 Wastewater treatment process

Wastewater treatment plants are often designed for gravity flow where possible, which also provides process security. Where necessary, terminal pumping stations would be designed to deliver the wastewater flow to the inlet of the treatment works. Inter-stage pumping is avoided where possible to reduce energy requirement. Energy use in wastewater treatment is determined by the treated flow, pollutant load, final effluent quality, the types of treatment process employed, level of monitoring and automation and experience of the operations staff.

The most significant amount of energy is consumed by secondary/biological treatment processes (aeration, mixed liquor return and flow recirculation), conventional and especially advanced anaerobic digestion (heating). Sludge pumping, aerobic sludge digestion, sludge processing equipment for sludge dewatering, and in particular drying, are energy intensive processes. Where advanced treatments are required such as membrane bioreactor, oxidation, disinfection, there will be a substantial increase in energy demand. Ancillary processes such as chemical dosing and mixing consume a reduced amount of energy. Selection of energy efficient process equipment and, more recently, the application of a real time control system have the benefit of reducing overall energy consumption in a treatment plant.

A wide range of carbon emissions is reported for wastewater treatment from 0.057 to 0.28 kg CO_2 produced when electricity is used to treat 1 m^3 domestic sewage (Gu *et al.*, 2016).

3.2.2.4 Scope 2 – energy generation

Currently indigenous and imported sludges are anaerobically digested to produce biogas in larger sludge treatment centres. The biogas can then be collected and used to power the combined heat and power (CHP) plant to provide both hot water/steam and electrical energy to be used within the works or exported to the grid. Advanced processes such as enzymic and thermal pre-treatment of the sludge before digestion enhanced biogas production. Although additional energy is required to operate these advanced processes, the energy, as both heat and power, is generated by the additional biogas produced and thus provides a net reduction in GHG emission. In recent years, biogas to grid or for fleet usage is becoming more widely practiced.

The co-digestion of imported high strength waste, food waste and fat oil and grease, where regulations permit, can maximize the output from existing facilities as well as divert GHG production if landfilled or treated elsewhere.

Many water utilities around the world have installed solar photovoltaics (PVs) and wind turbines on-site as well as generating biogas from biosolids at wastewater treatment plants. Two of the largest water and wastewater utilities in the UK, Thames Water and United Utilities, self-generated 24% of their electricity needs in 2019–2020 (Thames Water, 2020; United Utilities, 2020) and this is expected to increase further in the future.

3.2.3 Scope 3 – indirect emissions from other activities

Scope 3 emissions for the water treatment cycle usually accounts for all sources not within the water treatment cycle scope 1 and scope 2 boundaries. These include all emissions arising from indirect activities both upstream and downstream of the water treatment and supply value chain. Areas that generally apply to scope 3 emissions are as follows:

- purchased goods & services;
- fuel and energy related activities;
- transportation and distribution (both upstream and downstream);
- treatment of waste generated;
- business travel – using indirect sources such as planes, trains and so on.;
- employee commuting;
- lease/hired equipment;
- use of sold products;
- end of life treatment of sold products.

With emphasis now shifting towards full accountability and net zero emissions, more organisations are reaching into their value chain to understand the full GHG impact of the operation. Although the accounting and understanding of scope 3 emissions are not in water organisation's direct control, this will present an opportunity for the reduction of overall GHG emissions, as organisations can influence its suppliers or streamline procurement by contracting with vendors who fully account for their GHG emissions.

3.2.4 Carbon sequestration and mitigation

When all avenues for reduction of operational GHG emissions arising out of the water treatment cycle have been completed, a further opportunity to shift the carbon balance is to consider carbon sequestration (off setting) to achieve net zero emissions. Carbon sequestration is a process of capturing and storing atmospheric CO_2 which can be achieved using geologic or biologic methods. Biological carbon sequestration is most commonly used and refers to the storage of atmospheric carbon in forestation, soils, aquatic environments, natural vegetation and other wetlands. Carbon sequestration requires large geographical land for application and those that are limited with land generally purchase green carbon credits. The land-application of biosolids from a wastewater treatment plant can be used to increase carbon sequestration and provide a credit for the utility.

3.3 PROTOCOLS

Protocols consist of a set of standardized frameworks to estimate the GHG emissions from a process or an activity. Protocols typically use a set of emissions factors (EF) that relate a task or a process to the amount of GHG emitted by a similar standard process. There are specific protocols that are applicable to a certain industry in a specific region, so selection depends on location and purpose of the project.

A protocol also provides the guidelines to define goal and boundary conditions for a project and categorizes the activities into various scopes. These protocols are listed in the sections below.

3.3.1 International protocols

3.3.1.1 IPCC

Originally developed in 1988, IPCC provides a set of guidelines that is developed based on the latest science and understanding of the GHG emissions. The emission inventory is calculated by taking the activity data and multiplying it by the emission factors laid out in the protocol. This protocol is the basis for the majority of regional and specialized protocols.

As IPCC is a panel of scientists, it only produces a set of reports on various methodologies and is not directly involved in active research about the anthropogenic GHG emissions. The panel produces a revised report/addendum, called the Assessment Report (AR), every 5–7 years that summarises the latest science on GHG emissions. AR5 (fifth Assessment Report) published in 2014 is the latest available update on IPCC protocol.

The various updates to the original guidelines are presented Table 3.1 below.

Table 3.1 Various updates to original guidelines.

Version Number	Description	Year Developed
FAR	First Assessment Report	1990
SAR	Second Assessment Report	1995
TAR	Third Assessment Report	2001
AR4	Fourth Assessment Report	2007
AR5	Fifth Assessment Report	2014
AR6	Sixth Assessment Report	2022

3.3.1.2 World resources institute (WRI)

The defining feature of the GHG Protocol developed by WRI over other protocols is that it uses a holistic approach to provide sustainable solutions to various organizations. This protocol focuses on the following areas:

- climate;
- energy;
- food;
- water;
- forests;
- sustainable cities;
- ocean.

The WRI protocol is typically preferred when the GHG reporting agency is a business or an organisation. The global warming potential (GWP) of various greenhouse gases that compares their strength to the standard CO_2 gas is adopted from the fifth assessment report (AR5) produced by IPCC.

3.3.2 Regional protocols

The regional protocols focus on modifying or refining the emissions factors laid out in IPCC to fit the specified regional government or businesses. The following protocols include guidance that is relevant to water utilities in a local context, using the IPCC assessment reports as their basis.

3.3.2.1 United Kingdom – UKWIR

The United Kingdom Water Industry Research (UKWIR) together with the Water Research Centre (WRc) developed a set of guidelines based on IPCC that is specific to the water industry. The original protocol that was published in 2004 was refined and currently version 13 is in use. The primary focus of this protocol is:

- water treatment processes;
- wastewater treatment processes;
- sludge treatment and disposal;
- sludge disposal to land and landfills.

3.3.2.2 United States – LGOP

Local government operating protocol, or LGOP, was developed in association with California local government and provides reporting guidelines to local government officials, wastewater and potable water treatment facilities, planners and stakeholders. LGOP has been officially accepted as the standard reporting tool for local agencies within the United States. Two versions of LGOP have been published to date:

(1) Version 1 – Originally adopted in 2008
(2) Version 1.1 – Revised in May 2009

3.3.2.3 Germany – ECAM tool

Developed in association with water and wastewater companies for climate mitigation (WaCCLiM), and Catalan Institute of Water Research (ICRA), the Energy Performance and Carbon Assessment and Monitoring or ECAM tool helps with GHG calculation and reporting. This tool is based on the IPCC guidelines and uses GWP based on AR5.

3.3.2.4 Australia – NGER system

The NGER (National Greenhouse and Energy Reporting) Act provides a tool and set of guidelines for the registered organisations in Australia to report their annual GHG emissions. It was designed in 2007–2008 by Australia' s Department of Environment and Department of Climate Change. This system includes guidelines for GHG measuring, reporting and verification. The global warming potential (GWP) of various greenhouse gases that compares their strength to the standard CO_2 gas is adopted from the fourth assessment report (AR4) produced by IPCC.

Table 3.2 provides the different versions of the protocol.

3.3.2.5 CCME – Canadian council of ministers of the environment

Originally developed in 2009 by Canada's Ministry of Environment, the CCME primarily focused on emissions from biosolids management. The BEAM tool developed as a part of this protocol provides the emissions associated with the following treatment processes:

- sludge stabilization;
- drying;
- thickening and dewatering processes;
- combustion;
- land application and composting.

3.3.2.6 Summary of regional protocols

Table 3.3 compares and summarizes the regional GHG protocols.

Table 3.2 Different versions of protocol.

Version Name	Update to the Original Version[1]
Amendment 2013–14	
Amendment 2014–15	Revisions to scope 2 emissions factor and emissions from landfills; fevised method to estimate emissions from sludge lagoons
Amendment 2015–16	Updates to methods for municipal waste emission calculation
Amendment 2016–17	Refinement to scope 2 emissions factor and additional method to estimate fugitive emissions from geological formations
Amendment 2017–18	Updates to method for scope 2 calculations
Amendment 2018–19	Refinement to scope 2 emissions factor; Language about reporting requirements for wastewater utilities
Amendment 2019–20	Revisions to scope 2 emissions factor

Note 1: Important updates only.

Table 3.3 Comparison and summary of regional GHG protocols.

Regional Protocol	Acronym	Region of Primary Focus	GWP Source	Link to the Protocol
Local Government Operating Protocol	LGOP	United States		https://ww2.arb.ca.gov/local-government-operations-protocol-greenhouse-gas-assessments
European Commissions	EC/EMAS	European Union	IPCC AR5	https://ec.europa.eu/environment/emas/index_en.htm
Water and Wastewater Companies for Climate Mitigation	ECAM	Europe	IPCC AR5	https://wacclim.org/ecam/
National Greenhouse and Energy Reporting	NGER	Australia	IPCC AR4	http://www.cleanenergyregulator.gov.au/NGER
Canadian Council of Ministers of Environment	CCME/BEAM	Canada	IPCC AR4	https://www.ccme.ca/en/about/index.html

3.4 METHODS OF GHG QUANTIFICATION

3.4.1 Emission factors

The most widely adopted method to quantify GHG emissions is through the use of emission factors (EF). EFs are a measure of the GHG intensity of a particular activity, process or material. The application of EFs requires the collection of activity data, such as electricity consumption or fuel use, and applying a relevant EF, usually given in terms of tCO_2e. This approach removes the requirement for an organisation to deploy a measurement campaign or to actively measure emissions, thus reducing the resource requirements. Published EFs can be obtained through specific databases, such as the IPCCs 'Emission Factor Database', or from relevant literature. Calculated or measured EFs can be used to reflect a more representative emission for a specific process or material.

Given that the significant proportion of the water industry's emissions come from energy and transport, the use of EFs is an effective approach. The collection of activity data is far simpler and cost effective than measuring emissions. Much of the data is already widely available to organisations through electricity use, fuel purchases and vehicle movements. This extends to the quantification of embodied carbon, whereby EFs for materials are widely available. For example, the University of Bath has developed an open-access database to collate and report per-reviewed energy and carbon values of various materials (Hammond & Jones, 2008).

Global average EF data is currently readily available for a range of good or services, and EFs at a national level are becoming increasingly reported allowing for better representative data (Ercin & Hoekstra, 2012). EFs are often regularly reviewed and updated to reflect the most accurate emission intensity. This is particularly useful when considering emissions sources such as electricity consumption, whereby the source of electricity can change seasonally and annually, particularly as distribution and transmission systems decarbonize.

In the UK, water companies in England and Wales are obligated to report annual (operational) GHG emissions via the CAW. This approach uses published emission factors and requires companies to submit operational activity data, thus allowing for a simplified and standardised method of estimating operational GHG emission.

The simplicity of using emission factors has its limitations. Site specific conditions are not factored in, and therefore any use of generalized emission factors can lead to over- or under-estimates on smaller scales. For example, N_2O emissions associated with WWTPs are highly dependable on operating conditions and therefore will vary both spatially and temporally (Law *et al.*, 2012; Parravicini *et al.*, 2016). The IPCC have recently acknowledged that N_2O emissions were likely being underestimated under the 2006 guidelines and provided revised emission factors and quantification methodologies. The revised EF is a result of further appreciation for emissions during the treatment process and disposal of effluent to aquatic environments (IPCC, 2019). Wallace *et al.* (2020) reported that this revision led to a 40-fold increase in calculated N_2O emissions for a WWTP in Christchurch, New Zealand, which substantially increased the plant's overall carbon footprint, thus highlighting the sensitivity in GHG reporting to the accuracy and representativeness of EFs.

3.4.2 Direct measurement

Nitrous oxide can be directly measured from wastewater treatment processes either in the gas- or liquid-phase, and measurement campaigns have been deployed at both laboratory and plant scales. Closed chambers can be deployed to capture emitted N_2O gas from sludge tanks. Sampling can either be carried out 'off-line' with the repeated extraction of gas samples extracted, or 'on-line' with continuous monitoring. Given the emission flux of N_2O is dynamic in nature, the on-line approach is likely to garner more accurate quantification results. The gas analysis can be carried out using a variety of analysers or spectrometry approaches, but the use of infrared analysers is reported to be the preferred approach given the broad N_2O measurement range (Law *et al.*, 2012). Chamber methods can also be used to measure the release of CH_4 from treatment units (Hwang *et al.*, 2016).

Liquid-phase N_2O measurements can be carried out through the deployment of microsensors in sludge treatment lines. Similar to that of gas-phase measurements, continuous monitoring is likely to capture dynamic fluxes in dissolved N_2O concentrations (Baresel *et al.*, 2016). Mathematical models are required to convert the dissolved N_2O measurements into emission rates. Whilst this approach has previously been considered more appropriate for improving the understanding of emission processes (Law *et al.*, 2009), more recent studies have shown agreement between gas- and liquid-phase measurements (Baresel *et al.*, 2016).

The sources and sinks of methane are poorly understood and there is a growing international effort to better understand current emission rates. The quantification of methane for emission inventories is primarily achieved through the use of emission factors, however, there is a growing body of research assessing the accuracy and reliability of using satellites or aircraft to quantify methane emissions at both local and regional scales. Satellite data can be used to construct models to infer methane emissions (e.g., Jacob *et al.*, 2016; Turner *et al.*, 2015). Meanwhile, methane can either be sampled (e.g., Schwietzke *et al.*, 2017) or inferred using lidar (e.g., Riris *et al.*, 2012) from aircrafts. However, much of the focus of published research is on regional methane emissions, and any studies on identifying individual sources tends to centre on releases from natural sources, such as wetlands, or large anthropogenic emissions sources such as natural gas plants. As such, there appears to be no studies dedicated to using remote sensing techniques to specifically quantify methane emissions associated from wastewater treatment plants.

3.4.3 Models

Developing mathematical models to aid in the quantification of GHG emissions from wastewater plants offers a solution to the use of generalised emission factors or resource-intensive measurement campaigns. Modelling aims to identify and help better understand the various complex relationships along the treatment process pathway, thus paving the way for solutions that can ultimately reduce the carbon footprint (Mannina *et al.*, 2016). The boundaries of a model can vary from the analysis of a single process or component within a wastewater treatment plant, up to plant-wide approaches that consider all processes and the interactions between them (Mannina *et al.*, 2016).

Modelling approaches to estimating GHG emissions from wastewater treatment plants can be divided into categories based on complexity. Simple comprehensive process models have been shown to be effective in helping identify factors that influence GHG emissions, but their simplicity requires assumptions that ultimately impact the accuracy of results (Mannina *et al.*, 2016). Furthermore, they are restricted to steady-state analysis, which limits the effectiveness in quantifying N_2O emissions due to their dynamic nature (Corominas *et al.*, 2012).

Dynamic process-based models have been shown to be more effective at capturing the variability in GHG emissions, largely a result of the dynamic nature of N_2O emissions. System configuration, operating settings and atmospheric conditions influence the release of GHGs, all of which are difficult to account for in simple models (Corominas *et al.*, 2012). Dynamic process-based models are complex and require both high computational power and large amounts of data for calibration. However, when deployed at plant-wide scales, they offer the potential to improve the description and quantification in GHG emissions. Current knowledge of N_2O formation is a clear limiting factor in the accuracy of such models (Corominas *et al.*, 2012; Mannina *et al.*, 2016). The direct measurement and monitoring of N_2O emissions, alongside the collection of various other parameters such as pH and temperature, via sites' SCADA systems, can be used to develop and improve models to better understand dynamic relationships within wastewater treatment processes that ultimately influence emission rates (Baresel *et al.*, 2016; Law *et al.*, 2009).

Wallace *et al.* (2020) applied a mechanistic model of a WWTP to determine the N_2O emission rate. The calculated value was 25% lower than the N_2O emission factor reported by the IPCC.

3.4.4 Quantification method selection

Current protocols commonly use emission factors to develop GHG estimates. This is a reasonable approach for scope 2 emissions, however, for scope 1 (direct) emissions it can lead to poor estimates depending on the assumptions. It is self-evident that the preferred method for scope 1 emissions is direct measurement, wherever possible. Where this is not possible, modelling can be used to 'fill-in' data gaps and assist in the understanding of the mechanisms and influencing factors that can reduce emissions.

3.5 A FRAMEWORK FOR CARBON FOOTPRINT ANALYSIS

As described in the previous sections there are several protocols and methodologies for developing carbon footprints. It is helpful to take a broad and systematic approach to carrying this out. In their paper Pagilla *et al.* (2009) developed a framework for establishing a carbon footprint for water resource recovery facilities that can also be applied to other parts of the urban water cycle. Figure 3.4 is a flow chart of the framework which is consistent with available protocols.

The steps in the framework are listed here, with a brief description for each.

(1) *Determine objectives* – An important step in any project, study or assessment is to clearly define the objectives, based on the purpose. For example, if the carbon footprint is being developed simply to look at ways to reduce emissions for a single facility, then any one of several protocols and methods may be applicable. However, if the carbon footprint is being used as part of a broader city-wide or community assessment, or for carbon trading, then the protocol and methods must be consistent with them.
(2) *Select protocol* – Based on the purpose and geographical location, a protocol can be selected.
(3) *Define boundaries* – In order to avoid double-counting or omitting significant emissions, it is important to define boundaries clearly. This is often most easily done with a simple process schematic showing what is included and what is excluded from the system being assessed

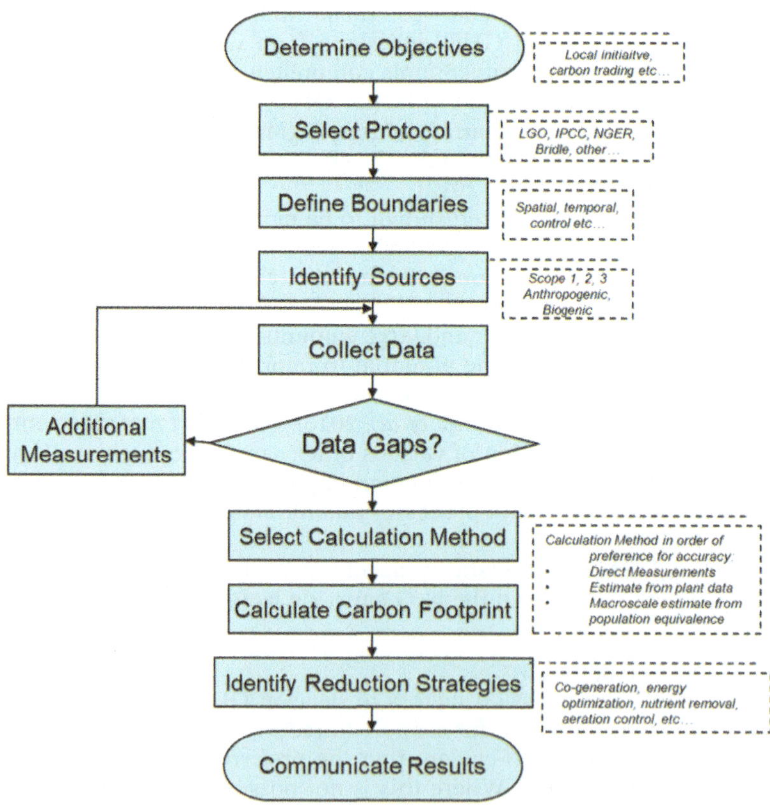

Figure 3.4 Decision framework for carbon footprint analysis (from Pagilla, 2009).

(4) *Identify carbon sources and sinks* – Having defined the boundaries, carbon sources and sinks can be identified, using approaches that are acceptable for the selected protocol.
(5) *Collect and assess data* – Often the most labour-intensive and costly part of the project is collecting and assessing the data needed to carry out the carbon footprint calculation. Data gaps may be identified requiring additional measurements to be made, or alternative data to be gathered to provide a different method for estimating emissions (e.g. developing a process model if no direct measurements are available).
6. *Select method and calculate carbon footprint* – Once all available data is gathered, the carbon footprint can be calculated using appropriate methods. In general, the preferred method is to use direct measurements if possible, followed by estimated emissions based on other facility data. If neither are available, then macro-scale estimates can be made based on population equivalents.
7. *Identify reduction strategies* – Carbon reduction may or may not be an explicit objective of the carbon footprint development, but regardless, having completed the calculation, this is the ideal point at which to identify potential hot spots and opportunities for reducing them.
8. *Communicate the results* – Displaying results in a clear and meaningful way is an important last step in developing the carbon footprint. In many instances the final audience for the results may not be experts in carbon emissions and so it is important that terminology, assumptions and the significance of the results are communicated well.

3.5.1 A roadmap to reducing carbon footprint in the water cycle

Step 7 of the framework is to identify reduction strategies. Throughout the complexity of the urban water cycle, there are many opportunities for reducing carbon emissions. The following is a list of holistic concepts that can be applied to reducing the carbon footprint of urban water cycles, providing a road map for improvement.

3.5.1.1 Conserve water

In considering the pathway of water through the urban water cycle (Figure 3.1), the main driver for the quantities of water that have to be abstracted, treated and conveyed, is the demand for water by the users (domestic, commercial, industrial). Similarly, the quantity of wastewater that has to be collected and treated is also governed by water use. If the water users can conserve water, this has a knock-on effect in reducing the energy needed throughout the water cycle and hence reduces scope 2 carbon emissions. Adding to this the considerable energy demand within homes and businesses to heat and use water, it is obvious that water conservation (particularly in reducing the use of hot water, or operating the hot water systems at lower temperatures) has a direct impact on the water cycle.

The one caveat to the generally positive effect of water conservation is that wastewater strength will increase (less dilution) which can increase the production of methane in sewer systems. This requires further investigation and consideration.

3.5.1.2 Reduce water loss (distribution) and infiltration (collection)

In a similar vein to water conservation at the user level, any water that is abstracted and treated, but then lost, represents an inefficiency not only in the water production but also the energy used to treat it and convey it. On the wastewater collection side of the water cycle, any inflow or infiltration (I&I) into the sewer system adds flows (and in some cases loads) that must be conveyed and treated. These add to the overall carbon footprint of the system.

3.5.1.3 Maximize energy generation

As noted in section 3.2.2.4 on scope 2 energy generation, wastewater treatment facilities have the potential to generate energy by converting the organics in the wastewater to fuel, heat or electricity.

Many water and wastewater facilities also have land (e.g. for buffers) on which solar or wind turbines can be installed for renewable energy generation.

3.5.1.4 Be energy efficient

Energy efficient processes and equipment can be selected for pumps, aeration, and solids processing. Efficient hydraulic design and avoiding 'double pumping' will reduce energy and so attention to plant hydraulics is important in reducing carbon footprint. In some facilities the focus is just on the largest single energy use (typically the blower for a wastewater treatment system and large pumps for both water and wastewater treatment), but attention should be paid to the multiple smaller energy uses on treatment facilities (e.g. mixers and centrifuges) as these can add up to be as significant as the larger energy users.

Another consideration in selecting equipment is to avoid oversizing, for example by installing multiple smaller units, so that they can operate closer to their design point and be more efficient overall. This will entail a higher capital cost but lower operating cost and often lower overall life cycle cost, as well as lower carbon footprint.

3.5.1.5 Maintain equipment

Installing energy-efficient equipment is only effective if the equipment is properly maintained. An example of this is in the selection of high-efficiency air diffusers for wastewater treatment. If they are not cleaned regularly, they become fouled, their efficiency drops significantly, and the carbon footprint will increase. Storing of sludge in tanks for extended periods or neglecting to clean our tanks can result in methane emissions. Good housekeeping, good maintenance, and operating facilities as they were intended to be are keys to ensuring a plant operates well which will keep direct and indirect carbon emissions lower.

REFERENCES

Appelbaum B. (2002). Water and sustainability (vol. 4). US electricity for consumption for water supply and treatment systems. *Urban Water*, **4**, 153–161, https://doi.org/10.1016/S1462-0758(02)00014-6

Bakhshi A. A. (2009). Carbon Footprint Estimation of Municipal Water Cycle. George Mason University. Mason Archival Repository. Retrieved from Mason Archival Repository: http://mars.gmu.edu/xmlui/handle/1920/5659 (accessed 25 February 2022)

Baresel C., Andersson S., Yang J. and Andersen M. H. (2016). Comparison of nitrous oxide (N_2O) emissions calculations at a Swedish wastewater treatment plant based on water concentrations versus off-gas concentrations. *Advances in Climate Change Research*, **7**(3), 185–191, https://doi.org/10.1016/j.accre.2016.09.001

C40 Cities (2021). The C40 Cities Website. Retrieved from https://www.c40.org/ (accessed 25 February 2022)

Campos J. L., Valenzuela-Heredia D., Pedrouso A., Val del Rio A., Belmonte M. and Mosquera-Corral A. (2016). Greenhouse gases emissions from wastewater treatment plants: minimization, treatment and prevention. *Journal of Chemistry*, **12**, 1–12.

Corominas L., Flores-Alsina X., Snip L. and Vanrolleghem P. A. (2012). Comparison of different modeling approaches to better evaluate greenhouse gas emissions from whole wastewater treatment plants. *Biotechnology and Bioengineering*, **109**, 2854–2863.

Daelman M. R., Voorthuizen E. M., Dongen U. G., Volcke E. I. and Loosdrecht M. C. (2012). Methane emission during municipal wastewater treatment. *Water Research*, **46**(11), 3657–3670, https://doi.org/10.1016/j.watres.2012.04.024

Eijo-Río E., Petit-Boix A., Villalba G., Suarez-Ojeda M. A., Amores M. J., Aldea X., Rieradevall J. and Gabarell X. (2015). Municipal sewer networks as sources of nitrous oxide, methan and hydrogen sulphide emissions: A review and case studies. *Journal of Environmental Chemical Engineering*, **3**(3), 2084–2094, https://doi.org/10.1016/j.jece.2015.07.006

Ercin E. and Hoekstra A. Y. (2012). Carbon and Water Footprints: Concepts, METHODOLOGIes and Policy Responses (World Water Assesment Programme; No. 4). United Nations Educational, Scientific and Cultural Organization (UNESCO), 7, place de Fontenoy, 75352 Paris 07 SP, France.

Ghaemi Z. and Smith A. D. (2020). A review on the quantification of life cycle greenhouse gas emissions at urban scale. *Journal of Cleaner Production*, **252**, 119634, https://doi.org/10.1016/j.jclepro.2019.119634 (accessed 25 February 2022)

Gu Y., Dong Y.-N., Wang H., Keller A., Xu J., Chiramba T. and Li F. (2016). Quantification of the water, energy and carbon footprints of wastewater treatment plants in China considering a water-energy nexus perspective. *Ecological Indicators*, **60**, 402–409, https://doi.org/10.1016/j.ecolind.2015.07.012

Guohua H., Yong Z., Jianhua W., Yongnan Z., Shan J., Haihong L. and Qingming W. (2019). The effects of urban water cycle on energy consumption in Beijing, China. *Journal of Geographical Sciences*, **29**(6), 959–970, https://doi.org/10.1007/s11442-019-1639-5

Hammond G. P. and Jones C. I. (2008). Embodied energy and carbon in construction materials. *Proceedings of the Institution of Civil Engineers – Energy*, **161**(2), 87–98, https://doi.org/10.1680/ener.2008.161.2.87

Hertwich G. E. and Wood R. (2018). The growing importance of scope 3 greenhouse gas emissions from industry. *Environmental Research Letters*, **13**(10), 104013, https://doi.org/10.1088/1748-9326/aae19a

Hwang K.-L., Bang C.-H. and Zoh K. D. (2016). Characteristics of methane and nitrous oxide emissions from the wastewater treatment plant. *Bioresource Technology*, **214**, 881–884, https://doi.org/10.1016/j.biortech.2016.05.047

IPCC (2014a). Annex I: glossary. In: IPCC, Climate Change 2014: Synthesis Report, K. J. Mach, S. Planton and C. von Stechow (eds.), IPCC, Geneva, pp. 117–130.

IPCC (2019). Refinement to the 2006 IPCC Guidelines for National Greenhouse Gas Inventories, volume 5, Waste. Available at: https://www.ipcc-nggip.iges.or.jp/public/2019rf/vol5.html (accessed 25 February 2022)

Jacob D. J., Turner A. J., Maasakkers J. D., Sheng J., Sun K., Liu X., Chance K., Aben I., McKeever J. and Frankenberg C. (2016). Satellite observations of atmospheric methane and their value for quantifying methane emissions. *Atmospheric Chemistry and Physics*, **16**, 14371–14396. https://doi.org/10.5194/acp-16-14371-2016

Klein G. (2005). California's Water – Energy Relationship. California Energy Commission, CA, USA.

Law Y., Ni B.-J., Lant P. and Yuan Z. (2012). N_2O production rate of an enriched ammonia-oxidising bacteria culture exponentially correlates to its ammonia oxidation rate. *Water Research*, **46**(10), 3409–3419, https://doi.org/10.1016/j.watres.2012.03.043

Mannina G., Ekama G., Caniani D., Cosenza A., Esposito G., Gori R., Garrido-Baserba M., Rosso D. and Olsson G. (2016). Greenhouse gases from wastewater treatment – A review of modelling tools. *Science of the Total Environment*, **551–552**, 254–270, https://doi.org/10.1016/j.scitotenv.2016.01.163

Nair S., George B., Malano H. M., Arora M. and Nawarathna B. (2014). Water-energy-greenhous gas nexus of urban water systems: review of concepts, state-of-art and methods. *Resources, Conservation and Recycling*, **89**, 1–10, https://doi.org/10.1016/j.resconrec.2014.05.007

Oshita K., Okumura T., Takaoka M., Fujimori T., Appels L. and Dewil R. (2014). Methane and nitrous oxide emissions following anaerobic digestion of sludge in Japanese sewage treatment facilities. *Bioresource Technology*, **171**, 175–181, https://doi.org/10.1016/j.biortech.2014.08.081

Pagilla K., Shaw A., Kunetz T. and Schiltz M. (2009). A systematic approach to establishing carbon footprints for wastewater treatment plants. *Proceedings of the Water Environment Federation*, **2009**(10), 5399–5409. https://doi.org/10.2175/193864709793952350

Parravicini V., Svardal K. and Krampe J. (2016). Greenhouse gas emissions from wastewater treatment plants. *Energy Procedia*, **97**, 246–253, https://doi.org/10.1016/j.egypro.2016.10.067

Riris H., Numata K., Li S., Wu S., Ramanathan A., Dawsey M., Mao J., Kawa R. and Abshire J. B. (2012). Airborne measurements of atmospheric methane column abundance using a pulsed integrated-path differential absorption lidar. *Applied Optics*, **51**, 8296–8305, https://doi.org/10.1364/AO.51.008296

Rissman J., Bataille C., Masanet E., Aden N., Morrow W. R., Zhou N. and Williams E. D. (2020). Technologies and policies to decarbonize global industry: review and assessment of mitigation drivers through 2070. *Applied Energy*, **266**, 114848, https://doi.org/10.1016/j.apenergy.2020.114848

Schwietzke S., Pétron G., Conley S., Pickering C., Mielke-Maday I., Dlugokencky E. J., Tans P. P., Vaughn T., Bell C., Zimmerle D., Wolter S., King C. W., White A. B., Coleman T., Bianco L. and Schnell R. C. (2017). Improved mechanistic understanding of natural gas methane emissions from spatially resolved aircraft measurements. *Environmental Science & Technology*, **51**(12), 7286–7294, https://doi.org/10.1021/acs.est.7b01810

Thames Water (2020). Combined Report – Annual Report, Annual Performance Report and Sustainability Report 2019/20. Available at: https://www.thameswater.co.uk/media-library/home/about-us/investors/our-results/current-reports/annual-report-2019-20.pdf (accessed 21 February 2021)

Turner A. J., Jacob D. J., Wecht K. J., Maasakkers J. D., Lundgren E., Andrews A. E., Biraud S. C., Boesch H., Bowman K. W., Deutscher N. M., Dubey M. K., Griffith D. W. T., Hase F., Kuze A., Notholt J., Ohyama H., Parker R., Payne V. H., Sussmann R., Sweeney C., Velazco V. A., Warneke T., Wennberg P. O. and Wunch D. (2015). Estimating global and North American methane emissions with high spatial resolution using GOSAT satellite data. *Atmospheric Chemistry and Physics*, **15**, 7049–7069. https://doi.org/10.5194/acp-15-7049-2015

United Utilities (2020). Annual Report and Financial Statements for the Year Ended 31 March 2020. Available at: https://www.unitedutilities.com/corporate/investors/results-and-presentations/annual-reports/ (accessed 21 February 2021)

Wallace E., Bouman R., Bourke M. and Benter-Lynch M. (2020). IPCC emissions factors – Challenging a fortyfold jump in nitrous oxide. Proceedings of the Water New Zealand Conference, 17–19 November 2020, Hamilton, New Zealand.

Water UK (2020). Net Zero 2030 Roadmap. Retrieved from https://www.water.org.uk/routemap2030/wp-content/uploads/2020/11/Water-UK-Net-Zero-2030-Routemap.pdf (accessed 25 February 2022)

WRF (2014). Water Footprint: A New Concept for Sustainable Water Utilities. Water Research Foundation Project #4378, Completed 2014. Available at: https://www.waterrf.org/research/projects/water-footprint-new-concept-sustainable-water-utilities (accessed 25 February 2022)

Zawartka P., Burchart-Korol D. and Blaut A. (2020, April 2). Model of Carbon Footprint Assessment for the Life Cycle of the System of Wastewater Collection, Transport and Treatment. Retrieved from Scientific Reports: https://www.nature.com/articles/s41598-020-62798-y (accessed 25 February 2022)

doi: 10.2166/9781789061796

Chapter 4

Operational optimization and control strategies for decarbonization in WRRFs

*Krishna R. Pagilla**

Department of Civil and Environmental Engineering, University of Nevada, Reno, NV, USA
**Correspondence: pagilla@unr.edu*

4.1 INTRODUCTION

Natural water resources are stressed due to population growth and reduced and/or uneven precipitation induced by climate change impacts. Consequently, more extreme measures are taken to meet the water demands including energy intensive water extraction, conveyance, and treatment systems. Likewise, the wastewater generated requires energy to collect and treat to make it suitable for reuse and environmental discharge. As water and other resources are extracted from wastewater for reuse, there are costs in terms of energy usage and GHG emissions from these operations. No other part of the water infrastructure is more energy intensive per unit of water handled than water reclamation from wastewater, and it is the sector that has the highest potential to decarbonize or become carbon neutral/positive. The wastewater itself is a potential source of energy for recovery. The embedded energy in wastewater includes heat, organic and inorganic reduced species that constitute chemical energy, and potential hydraulic energy in certain situations. The case for terming wastewater as a reNEWable resource and wastewater treatment plants as water resource recovery facilities (WRRFs) stems from the fact that nutrients (nitrogen and phosphorus), energy, and water can be recovered while reducing the carbon footprint of the facility and its operations. After a WRRF has been built and commissioned, the operational aspects and capacity utilization of the facility are mainly responsible for the carbon footprint for the entire operational life. Hence, the focus of this chapter is to explore key opportunities for decarbonization through optimization at process level and at whole facility level. The aspects addressed in this chapter do not consider various process options/substitutions requiring capital infrastructure changes for water reclamation and sludge treatment and management. The emphasis is on how a conventional (solid–liquid separation, secondary treatment for carbon removal/recovery and nitrification only) or advanced WRRF performing C, N, and P removal/recovery processes can be optimized for decarbonization. Further strategies to reduce chemical use, capacity utilization, and energy recovery by operational and process optimization strategies are discussed here.

With regard to energy demand level and GHG emissions, it is known that the carbon footprint is nearly identical to the energy footprint in most WRRFs. Figure 4.1 shows an example of an approximately 38 000 m^3/day activated sludge based conventional WRRF in the US that treats predominantly domestic wastewater and discharges the treated effluent to the environment (Pagilla *et al.*, 2009). It can be

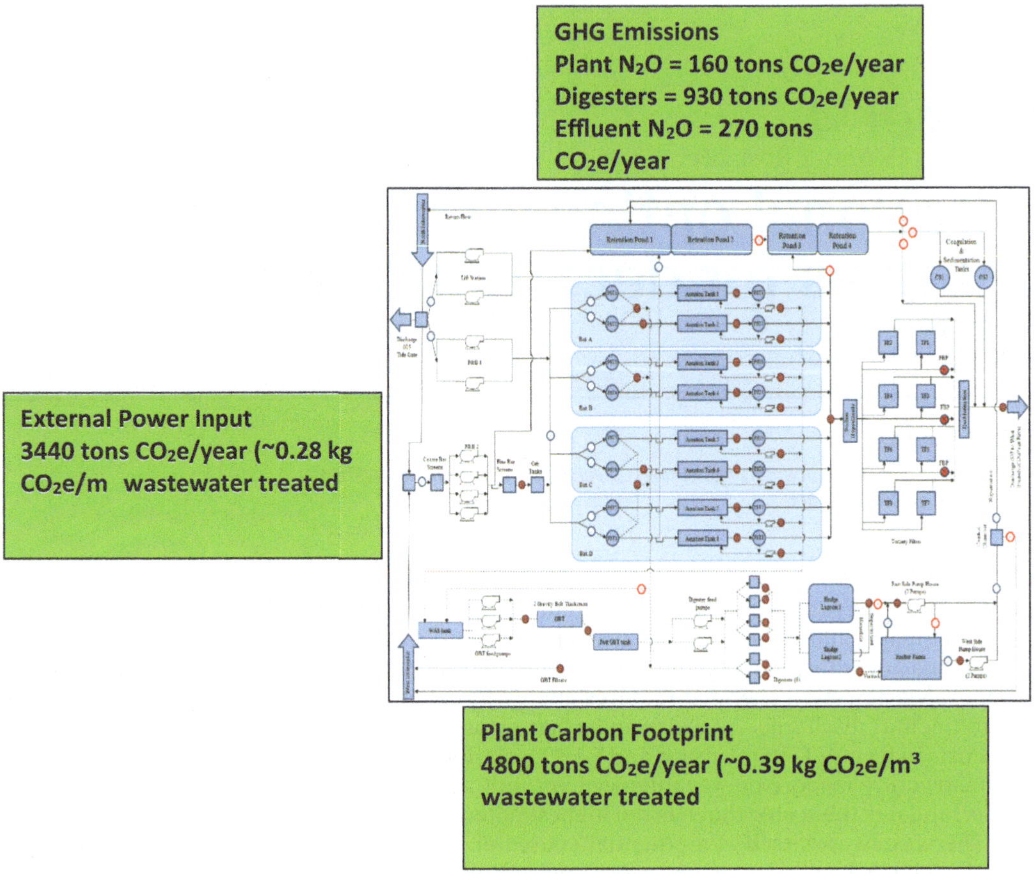

Figure 4.1 Carbon footprint of a WRRF and the external site power for operating the processes (adapted from Pagilla *et al.* 2009).

seen that the external site electrical power used in terms of carbon emissions equivalent is nearly 80% of the GHG emissions from the facility, while the rest comprises of emissions from the facility, excluding biogenic CO_2. The carbon footprint and hence the energy footprint can be minimized by energy and chemical inputs to the treatment processes, and by maximizing energy generation from the wastewater by carbon and heat capture. Figure 4.2(a) shows a simplistic rendition of the processes and inputs/outputs in a WRRF and Figure 4.2(b) shows the key energy inputs and outputs in a WRRF. At the fundamental level, the energy present in the wastewater and its extraction in usable form create opportunities to reduce carbon footprint or decarbonize the WRRFs. The energy inputs for treatment are costs that make-up the carbon footprints. Maximizing the opportunities and minimizing the inputs through optimization is the key to decarbonization in the WRRF.

4.2 OPTIMIZATION STRATEGIES AT THE PROCESS/OPERATION LEVEL

The opportunities for decarbonization can be considered at pumping, preliminary, primary, secondary, advanced, and sludge treatment level within a WRRF. The energy intensity of treatment in each of

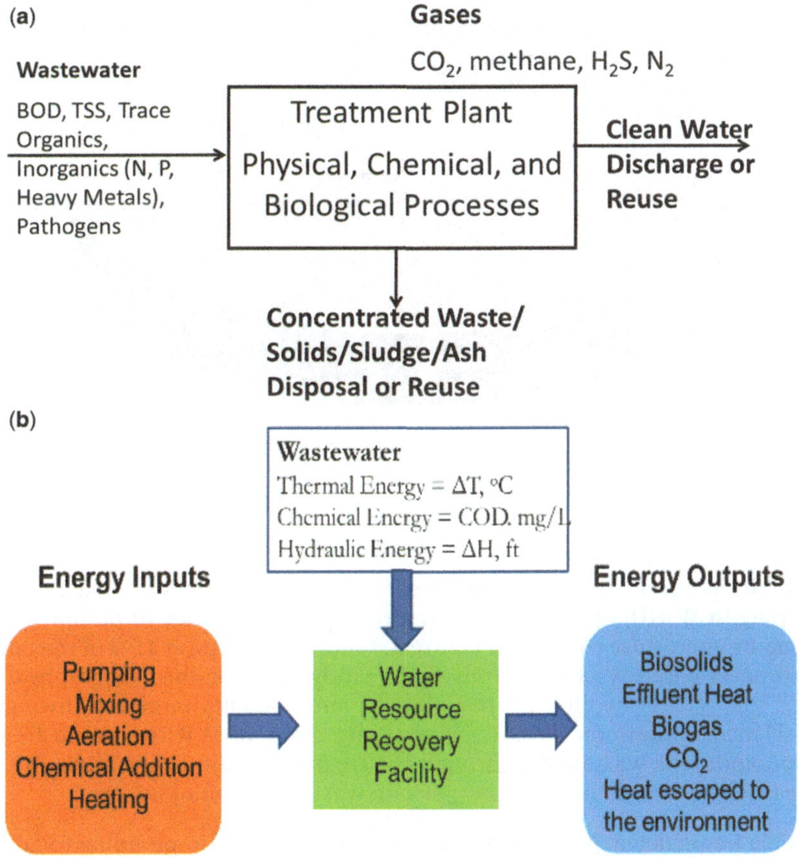

Figure 4.2 (a) Inputs and outputs of a conventional WRRF; (b) Energy inputs and outputs of a conventional WRRF.

these process/unit levels vary from facility to facility depending on the existing infrastructure or treatment method employed. As facilities employ more advanced treatment methods to recover water for reuse and nutrients from wastewater and sludge, the overall WRRF becomes energy intensive and increases its carbon footprint due to more energy intensive processes employed. There can be offsets due to the embedded carbon value of the recovered resources. The typical operations up to secondary treatment and the corresponding energy needs are well established. Another way to represent energy use in a WRRF is by function or use type for existing WRRFs with typical configuration. The typical energy use distribution based on numerous data sources in the literature in an activated sludge process-based WRRF with sludge treatment is shown in Figure 4.3. The typical standard deviation value for each category is approximately 20% of the respective percent energy use. The main process areas of concern include aeration for the activated sludge process, pumping of wastewater and sludge, sludge treatment and processing, and others (buildings and lighting, odor control, disinfection, etc.). It can be seen that the best opportunities for energy footprint reduction and decarbonization lie in two or three main categories including aeration for process needs. At the process level, the energy use and carbon footprint predominantly lie in wastewater pumping, secondary treatment, and sludge treatment and are discussed further.

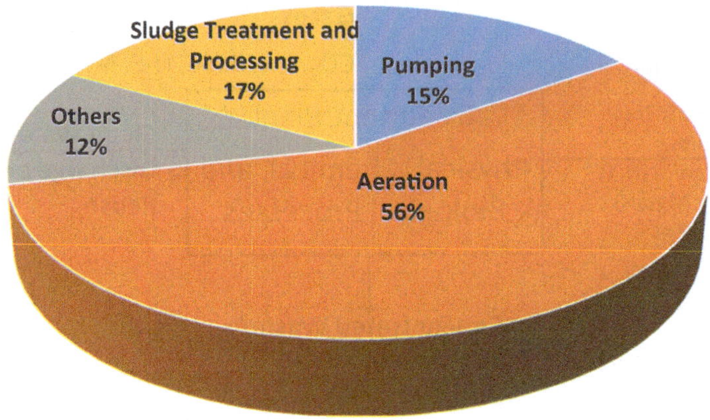

Figure 4.3 Typical energy use distribution in activated sludge process-based WRRFs.

4.2.1 Wastewater pumping

Wastewater pumping, from the headworks of the WRRF to effluent for advanced reclamation or environmental discharge, contributes to a significant portion of the energy use in a WRRF. Hence, minimizing pumping in WRRF operations and improving the efficiency of pumping are two critical pathways for reducing energy use. Although Figure 4.3 shows pumping at 15% of the energy used in an activated sludge process-based WRRF, it varies significantly based on the topography of the collection system as well as the WRRF layout, and the need for intermediate pumping through the facility's treatment train. Hence, a common basis for comparing different WRRFs would be to be consider energy used for pumping the wastewater through the treatment train to the fence line of the WRRF. The key strategies identified in reducing energy use in wastewater pumping are:

- use of variable frequency drives (VFDs) and/or smaller pumps in pumping to adapt to changing flow rates received by the facility;
- pump maintenance to keep the performance of the pumps at optimum levels;
- pumping control including flow management through in-line equalization and/or storage;
- data driven strategies to optimize pumping for economic and energy benefits.

Use of VFDs in influent pumping instead of fixed speed pumps is a common strategy to reduce energy use due to variable influent flows over a day, week, and month/season/year. This is particularly true in facilities that expect wide variations in influent flow due to significant inflow and infiltration of stormwater. Although VFDs have the ability to reduce energy consumption, excessive pump speed control without considering the pump characteristics can be counter-productive in terms of energy use (Kato *et al.*, 2019). A number of factors including static head, operational range of pump rotational speed, and number of pumps in service influence the overall energy consumption (WEF, 2009). Hence, power consumption analysis of plant specific pumping systems and optimizing for energy used per unit discharge (specific energy) should be the overarching strategy to reduce energy/carbon footprint of pumping. Over 30% improvement in energy efficiency was achieved in WRRFs employing such analysis in plants using VFDs for influent pumping (Kato *et al.*, 2019). Similar outcomes can be expected for other functions such as recycles and sludge pumping.

Raw or screened wastewater pumping involves fluids that have high debris content and grit, thereby creating rapid degradation of pump performance in terms of specific energy. Both pumps and pipeline maintenance are critical to maintain pumping efficiency in WRRFs. A pump performance monitoring strategy such as continuous pressure measurements, periodic cleaning, and maintenance through pigging or short duration high-speed pumping are needed during ongoing operations to maintain high efficiency (Larsen *et al.*, 2016).

Since wastewater pumping systems are designed for a maximum flow rate received by the facility, there is considerable deviation from peak performance (specific energy) of the pumping system at average and minimum flow conditions. Although flow management to create wastewater storage or in-line equalization in the sewer network is feasible and could be energy efficient for wastewater pumping, the unintended consequences of pipeline degradation due to deposition of sediments, fats/oils/greases (FOG), and odor issues need to be considered. Integrated control strategies which simultaneously control both a WRRF and sewer system are necessary without structural changes under dry weather and wet weather flow conditions (Kroll *et al.*, 2018). This results not only in energy savings, but also improves the effluent quality of WRRFs.

Overall, influent wastewater pumping has tremendous potential for reducing specific energy and hence, decarbonization of WRRFs if pumping systems have the control and sensing equipment needed. Among numerous examples, data driven strategies and control of wastewater pumping using fuzzy logic, data-mining, and bench-marking with the best cases seem to have much potential to reduce specific energy and improved energy savings (Torregrossa *et al.*, 2017). These strategies need more full-scale experience in multiple facilities before they become routine in the broader sector of water reclamation and sludge treatment. Benchmarking with other facilities of similar size and capacity utilization in terms of specific energy for pumping would reveal energy use reduction and hence decarbonization potentials.

4.2.2 Secondary treatment

Secondary treatment in a WRRF uses maximum energy due to the need for aeration in BOD and ammonia oxidation. Furthermore, secondary treatment produces biogenic CO_2 and N_2O emissions during carbon and nitrogen removal from wastewater. Extensive literature is available on methods to assess the carbon footprint of secondary treatment and strategies to reduce it. Modeling tools to estimate carbon emissions from secondary treatment are based on popular process modeling tools (Flores-Alsina *et al.*, 2011; Mannina *et al.*, 2016). The published literature includes numerous case studies to critical reviews. The strategies which involve novel and emerging technologies for carbon capture and utilization during wastewater treatment hold promise but are not widely implemented in existing WRRFs. They include electrochemical or bio-electrochemical methods, wetlands to produce biomass, and algae cultivation (Lu *et al.*, 2018). The key feasible strategies to reduce the carbon footprint of secondary treatment to remove/recover organic carbon, nitrogen, and/or phosphorus from wastewater include the following:

- alternative carbon capture strategies in secondary treatment;
- maintenance of physical aeration systems for peak performance;
- real time dissolved oxygen monitoring and control of aeration;
- real time nitrogen species monitoring and control including N_2O emissions.

Secondary treatment by biological processes, such as the activated sludge process, includes carbon capture (as biomass or storage products), carbon removal by oxidation with oxygen (aerobic) or nitrate (anoxic), and ammonia oxidation with oxygen. Other biological processes such as granular processes, membrane bioreactor systems, and fixed film systems also work on similar principles. Anaerobic wastewater treatment, which is not common, converts a part of the organic matter into biogas, a different form of carbon capture from wastewater. Carbon capture as biomass by aerobic, anoxic, and anaerobic process in secondary treatment is based on the biomass yield per unit of carbon utilized. The biomass yield is a function of the mean cell residence time (MCRT) of the process. In principle, most conventional WRRFs set the MCRT based on the treatment capacity available since longer MCRTs require a greater inventory of biomass in the main aeration tank. Short MCRT (high rate) in the activated sludge process can increase the biomass yield by capturing the influent organic carbon instead of oxidizing it, but could result in challenging conditions for subsequent denitrification and also for sludge handling. The use of a so-called high rate activated sludge system (HRAS) preceding a conventional activated sludge process or contact stabilization (CS) process has been found to be energy efficient in secondary treatment (Rahman *et al.*, 2019). Furthermore, adding chemically enhanced primary treatment to the HRAS or CS process would recover 200% more carbon from wastewater. As novel process technologies such as Anammox become common practice for nitrogen removal, the potential for HRAS process with low MCRT ($\sim$0.3 days) can yield good BOD removal and thereby allow for energy optimized WRRFs (De Graff *et al.*, 2016). Good design of the high rate process for optimum sludge production and processing of sludge for energy production is a critical aspect of successful carbon capture above conventional levels. Operational stability and reliability can then reduce the carbon footprint of secondary treatment by carbon capture rather than carbon oxidation.

Aeration is the single largest source of energy use in most WRRFs and fine bubble diffused aeration is the predominant type used in conventional activated sludge process-based WRRFs. Alternative diffusers and control systems taking advantage of the newer developments in membranes and online sensing of aeration systems have been developed to make aeration energy efficient, but diffuser fouling and air distribution to optimize energy use are critical factors in on-going energy optimization during operations of WRRFs. A combination of diffuser scaling and backpressure can significantly increase energy use and reduce the performance of diffused aeration systems. Maintenance and upkeep of the aeration systems including blowers, piping, diffusers, and online sensors will enable optimum air to be delivered for process needs and maintain energy efficiency. However, the frequency of cleaning needed is site specific and off-gas measurements to determine oxygen transfer efficiency are critical for each plant based on its wastewater composition and process operations (Leu *et al.*, 2009). The key factors of interest with regard to aeration systems evaluation and maintenance for energy efficiency have been widely discussed in the literature (Aviles *et al.*, 2020; Drewnowski *et al.*, 2019) and can be summarized as follows:

- blower maintenance, sequencing, and optimization;
- minimizing pressure losses through headers and distributor piping;
- diffuser fouling and cleaning;
- maintenance of on-line dissolved oxygen, ammonia, total organic carbon sensors that control air supply;
- replacement of non-functioning diffusers.

The application of real-time control of aeration systems to balance between air supply and demand as a function of loadings and variations (diurnal, seasonal, annual, wet weather vs. dry weather, etc.) and process conditions (temperature, treatment limits, etc.) could be very successful in achieving aeration costs reduction and carbon footprint reduction in activated sludge process secondary treatment. This approach has been in place for a few large facilities which were focused on operational efforts to reduce energy costs during aeration for wastewater treatment (Leu *et al.*, 2009). However, successful full-scale case studies with real time monitoring and control of aeration systems and demonstrated evidence of GHG emissions reductions are scant in the literature. The future potential lies in integrating aeration systems real time monitoring and control strategy with plant process control systems at plant scale for overall decarbonization.

Other potential opportunities to reduce aeration energy use in the activated sludge process include operational densification of the activated sludge and low DO operations (Arnaldos & Pagilla, 2014; Jassby *et al.*, 2014). It is common knowledge that bulking due to excessive filamentous bacteria presence in activated sludge makes it less dense and reduces the treatment capacity of the activated sludge process per unit reactor volume. Jassby *et al.* (2014) showed that the higher the filamentous bacteria content, the lower is the density of activated sludge, and hence lower is the settleability. High filamentous bacteria in activated sludge in effect disallows high biomass concentrations in the aeration tank because of poor settleability in the secondary clarifiers. Elimination of settling problems in the activated sludge process through various proven methods including use of biomass selectors, feeding strategies for optimum food-to-microorganisms ratio, aeration control, and chemical control of bulking, are imperative to reduce energy use in the activated sludge process (Jenkins *et al.*, 2003). Arnaldos and Pagilla (2014) demonstrated that the activated sludge process can be operated at low DO by acclimating the biomass over an extended period of time and, thereby reduce 20% aeration energy use and 20% improvement in oxygen mass transfer efficiency. This strategy, although employed in some full-scale facilities based on empirical reports, is not common practice in most full-scale facilities.

Monitoring of alternative end points of aeration systems such as N species, pH, and ORP are also possible in full-scale decarbonization. Combined monitoring of ORP, pH, and DO using sensors and managing aeration have been demonstrated for a long period of time (Paul *et al.*, 1998). The strategy has been demonstrated for aeration control even under varying loading conditions in the activated sludge process. Similarly, real-time control of biological nitrogen removal to achieve process performance and aeration efficiency has also been successfully demonstrated (Zanetti *et al.*, 2012). In fact, aeration control of the activated sludge process using ammonia as the controlling variable can not only provide energy use reduction, but also improve N removal and biological P removal performance. The ammonia-based aeration control using a data-driven modeling approach was found to be more effective than the DO-based control (Newhart *et al.*, 2020). As with real-time DO control for aeration, the N-based control systems in full-scale facilities are scant at the present. Therefore, real-time monitoring and control using sensors and parameters combined with data-driven modeling of full-scale facilities provides a great opportunity to find decarbonization strategies in aeration systems.

As can be seen from Figure 4.1, N_2O emissions from biological processes during nitrification-denitrification constitute a significant portion of the overall direct GHG emissions (scope 1) in WRRFs. The facility shown in Figure 4.1, is a nitrification facility without a denitrification requirement. It has been clearly demonstrated that N_2O emissions occur from both nitrification and denitrification steps in N removal (Rassamee *et al.*, 2011). This is particularly the case when DO levels are controlled in response to ammonia levels in wastewater causing either/and/or incomplete nitrification and denitrification evidenced by high nitrite levels. In a survey of 12 WRRFs in the US, it was found that there is a high degree of variability in N_2O emissions from biological N removal plants due to process and operational variations (Ahn *et al.*, 2010). The N_2O emission factors determined from these facilities ranged over two orders of magnitude. Considering only biological N removal plants with an estimated emission factor of 7.0 g N_2O/PE/year translates to a flow-based emission factor of approximately 51 mg/m^3 with US average of 378 L wastewater/PE. Overall, it can be seen from the example in Figure 4.1 that up to 10% of the facility's carbon footprint is due to N_2O emissions in a nitrifying activated sludge plant. Therefore, the potential to reduce those emissions lies in the secondary treatment due to the N_2O production there. Efforts should focus on complete nitrification and denitrification in BNR processes while careful attention should be paid to nitrite formation in short cut N pathways processes which is the main factor in N_2O production.

4.2.3 Sludge treatment

The most common configuration of sludge treatment in a WRRF includes pre-thickening of primary and/or waste activated sludge, stabilization by aerobic or anaerobic digestion, dewatering of the stabilized sludge for further processing and reuse/disposal. Further processing methods such as composting, incineration, thermal drying, and other end-use technologies are not considered here. For the purpose of this discussion, anaerobic digestion is the sludge stabilization method since it is practiced by more facilities than aerobic digestion which needs aeration with a large carbon footprint. A conceptual representation of the direct and indirect GHG emissions from wastewater sludge treatment in a WRRF can be seen in Figure 4.4. It identifies the potential opportunities for emissions reductions, although the magnitude of reduction is plant specific. The GHG equivalent inputs to sludge treatment include electrical energy, heat energy for temperature control of sludge in

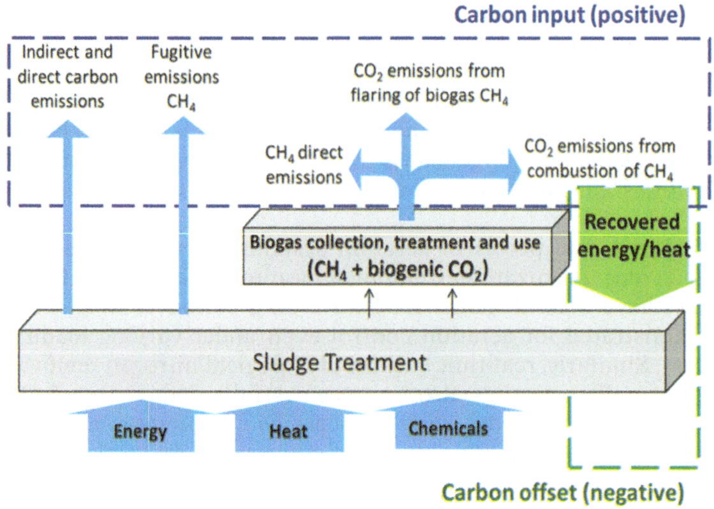

Figure 4.4 Conceptual diagram to identify GHG emissions reductions in sludge treatment.

anaerobic digestion, and chemicals used in thickening and dewatering. An additional operation that is integrated with sludge treatment is energy generation and use (in the form of heat and electricity) through biogas collection, conditioning or treatment. The recovered energy and heat serve as carbon offsets in the overall sludge treatment system.

The key opportunities for decarbonization through sludge treatment in a WRRF are:

- improved operations to generate more biogas for energy recovery;
- reduction in fugitive emissions of methane and flaring of unused biogas;
- reduction of chemical use in sludge concentration steps.

Anaerobic digestion of wastewater sludge is the energy producing operation in a WRRF and has significant potential to decarbonize a WRRF. It is possible to reduce the energy demand of the anaerobic digestion process itself by optimizing sludge temperature and mixing, both of which require energy inputs. Biogas production in anaerobic digestion is dependent on temperature, and often higher operating temperatures have a positive effect in both mesophilic or thermophilic types. Temperature optimization can lead to higher biogas production while balancing the energy needs for sludge heating. Effects of temperature on mesophilic anaerobic digestion of sludge from 32 to 37.5°C showed that there is no significant difference in biogas production when the temperature was reduced from 37.5 to 35°C; however, a further decrease in digestion temperature led to a biogas production decrease (Andersson *et al.*, 2020). Therefore, a good temperature sensing system in feed sludge and digester contents, and a feedback control operations strategy to maintain the optimum temperature under varying ambient conditions and feed sludge temperature conditions, is critical to reduce energy demand and increase biogas production.

Mixing is another important aspect of anaerobic digestion that not only influences energy use but also process performance and operating issues. Uniform substrate conditions in the digester are desired, but excessive mixing leads to foaming issues which can impact operations and hence biogas production (Pagilla *et al.*, 1997). It was later demonstrated that in most high-rate anaerobic digesters, excessive mixing is highly likely and can cause foaming issues which impact digestion and biogas production (Subramanian & Pagilla, 2015a; Subramanian *et al.*, 2015b). The natural mixing due to biogas production and sludge recirculation through the sludge heating loop are sufficient to maintain homogeneity in the digester and process performance.

Extensive literature exists on operational strategies to increase biogas production in anaerobic digestion. Balancing of any single or multiple operational strategies considering the energy or chemical demand while maximizing the biogas production is the overarching goal in WRRFs. The main factors influencing biogas production are feed sludge quality, feeding patterns and rates, operating temperature, pH/alkalinity in the digester, mixing intensity, and operating retention time. It is well known that anaerobic digestion process stability and optimization is dependent on these operational parameters (Wu *et al.*, 2021). A summary of example cases of process instability due to operational aspects and countermeasures are reviewed by Wu *et al.* (2021). It is important for the operational parameters to be controlled within the optimum range to maintain anaerobic digestion performance and maximize biogas production for energy production in a WRRF. Table 4.1 presents the optimum operational parameter values and operational strategies to maintain the optimum conditions. Some examples of literature sources that can provide more information or detailed operational strategies for control of anaerobic digestion to enhance biogas production in WRRFs are also shown in Table 4.1. These strategies do not include feedstock augmentation with non-municipal sludge carbon sources. Addition of supplementary carbon sources, particularly from organic food wastes, is well documented in the literature and hence is not a focus here.

Although fugitive methane emissions can be caused at multiple locations in a WRRF, the predominant source appears to be in the anaerobic digestion of sludge due to high biogas production levels due to intentional methanogenesis. Other significant sources can be from sludge storage tanks

Table 4.1 Operational parameters for optimization of anaerobic digestion for biogas production in a WRRF.

Process Parameter	Optimum Values or Conditions	Operating Strategies	Example Reference
Temperature	35–37°C (mesophilic)	Feed sludge and digester contents online temperature sensing and control	Andersson *et al.* (2020)
pH	6.6–7.6	Ensure sufficient feed sludge alkalinity for buffering; control loading rates to maintain stable pH	Rajagopal *et al.* (2013)
Alkalinity	1000–5000 mg/L as $CaCO_3$	Ensure sufficient feed sludge alkalinity; minimize variations in organic loading rates	Metcalf and Eddy, Inc. (2014)
Feed quality	No toxicity, digestable substrates, good C:N ratio	Increase feed sludge digestibility; minimize inhibitors and toxicants if present; avoid constituents that can cause operating problems such as surfactants, inerts, high ammonia, and so on	Wu *et al.* (2021)
Feeding rates	Below maximum volumetric and mass loading rates allowed; minimize organic loading rate variations	Steady state loading, practice loading rates below maximum volatile solids loading rate	Dalmau *et al.* (2010)
Feeding patterns	Continuous or semi-continuous	Feed uniformly in semi-continuous or continuous mode to minimize rapid gas production and digester conditions changes	Dalmau *et al.* (2010)
Mixing intensity	Site specific; low to no mixing in high rate digesters; bottom-up mixing mode	Minimize mixing to eliminate foaming problems; more mixing is needed at the bottom of the digester than the top	Subramanian and Pagilla (2015a)
Sludge retention time	15–25 days for high rate digesters	Avoid short SRT AD to minimize loading rate effects	Metcalf and Eddy, Inc. (2014)

Table 4.2 Chemicals used, purpose, and GHG emission factors in a 113 000 m^3/day flow rate advanced water reclamation facility in the US.

Chemical	Emission Factors kg CO_2e/unit[a]	Purpose	Quantity	Indirect GHG Emissions, kg CO_2e/day
Sodium hypochlorite	0.92	Effluent disinfection	9100 L/day @12.5% by weight	8372
Sodium bisulfite	0.44	Effluent dechlorination	400 L/day @39% by weight	176
Polymers	2.2	Sludge thickening and dewatering	1000 L/day @42% by weight	2200
Sulfuric acid	0.14	pH control	1200 L/day @50% by weight	168
Aluminum sulfate	0.50	P control and coagulant	2000 L/day @28% by weight	1000
Methanol	1.47	Denitrification	10 580 L/day @99.9% by weight	15 553
All Chemicals				27 469

[a]Determined from various literature sources.

or fields due to residual incidental or residual methane production, and raw wastewater headworks subjected to anaerobic conditions. The key sources of methane from anaerobic digestion include the digested sludge, digester floating cover annular space, sludge buffer tanks, and leaks from gas handling systems. Over 70% of total methane emissions from a WRRF are due to emissions from the anaerobic digestion complex (Daelman *et al.*, 2012). Therefore, operational strategies such as prevention of biogas leaks, biogas collection from digested sludge storage and buffer tanks, and returning sludge liquor recycles to the activated sludge process to oxidize dissolved methane are likely to reduce fugitive emission of methane in a WRRF.

Small WRRFs with low biogas production or biogas production in excess of heating needs often tend to flare the biogas for disposal. As a WRRF size decreases, the amount of biogas produced from AD that is flared instead of utilized increases (Shen *et al.*, 2015). The economics of biogas cleanup for utilization has been cited as a major barrier in small WRRFs due to the low cost of natural gas in recent years.

Another operational strategy that can make significant progress in decarbonization of WRRFs is the reduction in the use of chemicals for sludge treatment. The main chemicals used in sludge treatment are polymers and inorganic coagulants for sludge concentration and dewatering. A case study of chemical uses in a WRRF and the contribution of polymer use for sludge thickening and dewatering is nearly 10% of the GHG emissions equivalent due to all chemicals used in the WRRF (Table 4.2) and is further discussed in the following section.

4.3 OPERATIONAL STRATEGIES AT THE WHOLE FACILITY LEVEL

At the whole facility level in a WRRF, numerous operational strategies can be employed to achieve decarbonization in addition to individual processes or unit level strategies. Broadly, the most attractive decarbonization approaches or pathways can be classified into the following:

- plant level benchmarking with other facilities to identify opportunities;
- capacity utilization methods including peak flow and load management, base load operational approach, and parallel treatment trains out-of-service or in-service;
- optimized use or reduction of chemicals and additives in plant operations.

The use of benchmarking to determine relative energy use per unit wastewater treated or per capita, and other normalized metrics, is well discussed in the literature (Longo *et al.*, 2016). Benchmarking itself does not improve energy use efficiency in WRRFs, but reveals factors effecting high or low energy use in a particular plant such as plant size, wastewater strength, flow rate and its variations, capacity utilization, reclaimed water quality, and other regulatory and plant requirements. Benchmarking also shows how a plant is performing relative to other WRRFs of similar type based on key performance indicators (KPIs) (Longo *et al.*, 2016) at the whole facility level and at individual process/operation level. The energy analysis methods should include other KPIs such as kWh/PE, kWh/kg COD removed, kWh/kg N removed, and so on, to show nuances in specific energy in different facilities corresponding to different inputs and outputs and functions. For example, using energy benchmarking including chemical use, Belloir *et al.* (2014) showed that two facilities that have similar treatment processes had starkly different energy use per unit volume of wastewater treated due to process and operational differences. Similar analysis can be carried out for a plant of interest, and then target process level or plant level strategies for decarbonization during operations.

When KPIs are estimated in terms of energy use in a WRRF, a key consideration of importance is the operational capacity of the facility versus its design capacity. For example, two facilities with similar treatment processes and the same designed capacity may have different specific energy use because of variations in capacity utilization in terms of hydraulic and mass loading rates. This also reflects in operational costs and GHG emissions per unit flow rate treated. Although larger facilities have the ability to take parallel units out-of-service during low flow or load conditions, this is not always practiced because other considerations. The key considerations include availability of labor and operational ease with which the units can be taken out of service or put back into service. Any facility that has a design capacity well in excess of the operating capacity should conduct an analysis of operational strategies impacting specific energy use, operating costs, and GHG emissions. A data-driven approach based on time-variant flows and loads information to analyze capacity utilization and its impact on energy utilization in WRRFs is needed for operational decision making (Torregrossa *et al.*, 2019). A possible strategy to overcome capacity under-utilization are operating a base flow or load facility with partial capacity utilization while keeping the rest of the capacity on standby mode for transient flow/load treatment. This strategy requires that the facility has parallel trains which can be easily isolated and kept in standby mode. Even small WRRFs which do not have parallel units have high specific energy use due to under-utilization of the design capacity (Foladori *et al.*, 2015), suggesting that novel operational and control strategies are needed for them. Certain plants have the ability to store the influent wastewater either in reservoirs or in the sewer system to manage short term peak flows and loads.

Chemicals used in wastewater treatment and water reclamation contribute to the indirect emissions (scope 2) of GHGs from WRRFs. Therefore, operational efforts and strategies to minimize chemical use during facility operations are critical for overall decarbonization of the WRRF. The chemicals used in WRRFs at the plant scale which are significant and used in most advanced water reclamation facilities are disinfection chemicals, coagulants and flocculants, pH control, precipitation chemicals, and carbon augmentation chemicals. Furthermore, if facilities employ more advanced treatment processes for P recovery, softening, biogas recovery, and water reuse, additional types or quantities of multipurpose chemicals and additives (hydrogen peroxide, ozone, iron materials) may also be used in a WRRF. The relative carbon footprint or indirect emissions per unit quantity is a function of supply chain and source of the chemicals in a specific WRRF. A most recent case study to determine the comprehensive carbon footprint of WRRFs in the Baltic Sea region is a good example of estimating decarbonization potentials in WRRFs (Maktabifard *et al.*, 2022). Table 4.2 shows typical chemicals and estimated indirect emissions in a 113 000 m^3/day operating flow, advanced water reclamation facility in Reno, NV, USA. The facility includes advanced treatment for N and P control and chlorine-based disinfection (Lacroix *et al.*, 2020).

Table 4.3 Variability in energy footprint due to capacity utilization, level of treatment, and chemicals used in five BNR facilities in the US.

Process Type, Parameter	5-Stage Bardenpho	3-Stage Westbank	Phased Oxidation Ditch	3-Stage BNR	Activated Sludge Process and Denitrification Filter
Design/actual flow, m^3/d	37 800/20 800	39 000/32 000	45 400/26 500	113 400/87 000	26 500/15 500
Effluent TN, mg N/L	2.2	4.6	3.5	1.7	2.1
Effluent P, mg P/L	0.2 (w/alum)	0.2	0.4	0.47 (w/alum)	0.27 (w/alum)
Methanol	No	No	No	Yes	Yes
Energy footprint, KWH/m^3	0.55	0.55	0.18	1.01	0.84

The overall indirect emissions (scope 2) due to chemical use in this facility is about 0.24 kg CO_2e/m^3 of wastewater treated. The largest contribution to the carbon footprint of this facility is the use of methanol for tertiary denitrification which is equivalent to 0.14 kg CO_2e/m^3 of wastewater treated. Therefore, the largest decarbonization potential in this facility can be realized by either finding internal carbon sources to replace methanol or optimization of N removal in the overall facility to minimize methanol use for denitrification. A comparison of full-scale data collected showed that biological nutrient removal facilities that practice chemical addition such as alum and methanol for denitrification and P removal can double the energy/carbon footprint of wastewater treatment compared to the facilities that do not. Table 4.3 shows the role of capacity utilization and chemical use on the energy footprint of wastewater per unit volume of wastewater treated.

A case study which investigated an alternative sludge treatment processing strategy which led to whole plant chemical use reduction and operating cost reduction was described by Mentzer *et al.* (2021). The major goal of this study was to enhance the dewaterability of final sludge being sent to a landfill for disposal. The facility investigated bypass of thickened WAS from anaerobic digestion and combined dewatering of digested primary sludge and un-digested thickened WAS. This operational strategy not only reduced the overall polymer use in sludge dewatering, but either eliminated or considerably reduced the use of other chemicals such as $Mg(OH)_2$ (for struvite recovery), sulfuric acid for pH adjustment, and methanol due to lower N in the recycles, and others. In fact, the outcomes were dramatic in terms of enhanced anaerobic digestion capacity, eliminating the need to treat dewatering centrate, and reduced operational problems such as struvite scaling, gas conditioning media fouling, and so on.

4.4 PATHWAYS FOR DECARBONIZATION AND FUTURE PERSPECTIVES

The role of WRRF operations and process optimization is critical for not only meeting the water quality goals of the facility, but also in achieving resource recovery and sustainability goals or outcomes. For an existing facility, the carbon footprint lies in operational activities and hence the decarbonization potentials. Although the preceding discussion was focused on proven methods or feasible options for decarbonization in a WRRF, future potentials lie in both operation and technology selection for upgrades and newer designs. Planning and design of future upgrades of the facilities with sustainability goals in mind including potentials for decarbonization is critical. At the same time, the early wins to set the path for future decarbonization pursuits lie in current operations. Major areas of operations which have maximum potentials for decarbonization in WRRFs and their associated sewer systems are as follows:

(a) Minimize dilution of wastewater and increase the energy density of wastewater through reduced water use and inflow/infiltration into sewers.
(b) Source control of trace pollutants that have low concentration impact thresholds and are energy intensive to treat through advanced treatment methods. These include pharmaceuticals, forever chemicals, and anthropogenic nanomaterials.
(c) Source control of salinity to minimize energy intensive extraction of reclaimed water for reuse.
(d) Effective flow management to minimize pumping and hydraulic overloads to the facility.
(e) Carbon capture and P recovery instead of energy intensive and chemical intensive biological and chemical processes.
(f) Anaerobic treatment systems with complete methane capture for carbon recovery.
(g) Efficient aeration systems (if cannot be replaced with anaerobic treatment options) that supply oxygen to meet actual metabolic demands of the process instead of open tank aeration providing less than 20% oxygen transfer efficiency.
(h) Densification of WRRF operations through process optimization and technology selection.
(i) Online sensing and feedback/feed-forward control of processes to achieve treatment goals, energy conservation, and minimize emissions.
(j) Minimize nitrous oxide and methane emissions through operational control of processes.
(k) Minimize and eliminate chemicals use, particularly those with high life cycle GHG emissions in WRRF operations. Use of external carbon sources such as methanol and coagulants/flocculants/polymers present immediate decarbonization potential in existing WRRFs.
(l) Novel in-plant modifications to concentrate and dewater sludge better for further processing or reuse.
(m) Enhance biogas production from anaerobic digestion through operational and process modifications and optimizations.
(n) Optimizing of auxiliary facilities such as odor control, gas cleaning, and in-plant transportation.

Although the above is an extensive list of potentials and opportunities, they can be prioritized and addressed for the overall decarbonization of the WRRFs by careful carbon footprinting of the facility and its operations. The future lies in more WRRFs conducting operational carbon footprint determinations at least at scope 1 level so that more full-scale data can be developed to carry out benchmarking of WRRFs with similar process trains, capacity utilization, resource quality requirements, and other variables of interest. This allows the water sector to develop KPIs that can be compared across various WRRFs to set goals for decarbonization in each facility. The heterogeneity of the carbon footprints in drinking water facilities and WRRFs is large and hence, best practices from more full-scale facilities can be implemented at others with higher carbon footprint KPIs. For example, the GHG emissions from energy consumption by drinking water and wastewater treatment facilities in the US were in the range of <0.1–0.8 and <0.1–0.65 kg CO_2e/m^3, respectively (Zib III *et al.*, 2021). Such aggregate values do not reflect ground reality of the heterogeneity among facilities. The ability to embark on large-scale water sector decarbonization is dependent on the availability of this full-scale data and practical strategies to decarbonize at process and whole plant level.

The broader and significant opportunity and positive impact to decarbonize in WRRFs and the water sector as a whole in existing facilities is not possible without the education and training of facility personnel and staff to understand and implement feasible decarbonization strategies. The goal should be to clearly demonstrate the operational costs savings, enhanced treatment performance and efficiency, and how they can contribute to addressing the climate change effects by decarbonization in their respective facility.

REFERENCES

Ahn J. H., Kim S., Park H., Katehis D., Pagilla K. and Chandran K. (2010). Spatial and temporal variability in atmospheric nitrous oxide generation and emission from full-scale biological nitrogen removal and non-BNR processes. *Water Environment Research*, **82**(12), 2362–2372.

Andersson J., Helander-Claesson J. and Olsson J. (2020). Study of reduced process temperature for energy optimization in mesophilic digestion: a lab to full-scale study. *Applied Energy*, **271**, 115108. https://doi.org/10.1016/j.apenergy.2020.115108

Arnaldos M. and Pagilla K. R. (2014). Implementation of a demand-side approach to reduce aeration requirements of activated sludge systems: directed acclimation of biomass and its effect at the process level. *Water Research*, **62**, 147–155. https://doi.org/10.1016/j.watres.2014.05.040

Aviles A. B. L., Velazquez F. D. C. and del Riquelme M. L. P. (2020). Integral control system for the aeration stage in the biological process of activated sludge and the membrane biological reactor. *Sensors*, **20**, 4342. https://doi.org/10.3390/s20154342

Belloir C., Stanford C. and Soares A. (2014). Energy benchmarking in wastewater treatment plants: the importance of site operation and layout. *Environmental Technology*, **36**, 260–269. https://doi.org/10.1080/09593330.2014.951403

Daelman M. R. J., van Voorthuizen E. M., van Dongen U. G. J. M., Volcke E. I. P. and van Loosdrecht M. C. M. (2012). Methane emission during municipal wastewater treatment. *Water Research*, **46**, 3657–3670. https://doi.org/10.1016/j.watres.2012.04.024

Dalmau J., Rodriguez-Roda I., Comas Q., Steyer J. P. and Pagilla K. R. (2010). Model development and simulation for predicting risk of foaming in anaerobic digestion systems. *Bioresource Technology*, **101**, 4306–4315. https://doi.org/10.1016/j.biortech.2010.01.056

De Graff M. S., van den Brand T. P. H., Roest K., Zandvoort M. H., Duin O. and van Loosdrecht M. C. M. (2016). Full-scale highly-loaded wastewater treatment processes (A-stage) to increase energy production from wastewater: performance and design guidelines, *Environmental Engineering Science*, **33**, 571–577. https://doi.org/10.1089/ees.2016.0022

Drewnowski J., Remiszewska-Skwarek A., Duda S. and Lagod G. (2019). Aeration process in bioreactors as the main energy consumer in a wastewater treatment plant – review of solutions and methods of process optimization. *Processes*, **7**, 311. https://doi.org/10.3390/pr7050311

Flores-Alsina X., Corominas L., Snip L. and Vanrolleghem P. A. (2011). Including greenhouse gas emissions during benchmarking of wastewater treatment plant control strategies. *Water Research*, **45**, 4700–4710. https://doi.org/10.1016/j.watres.2011.04.040

Jassby D., Xiao Y. and Schuler A. J. (2014). Biomass density and filament length synergistically affect activated sludge settling: systematic quantification and modeling. *Water Research*, **48**, 457–465. https://doi.org/10.1016/j.watres.2013.10.003

Jenkins D., Richard M. G. and Daigger G. T. (2003). Manual on the Causes and Control of Activated Sludge Bulking, Foaming, and Other Solids Separation Problems, 3rd edn, Taylor and Francis, New York, NY.

Kato H., Fujimoto H. and Yamashina K. (2019). Operational improvement of main pumps for energy-saving in wastewater treatment plants. *Water*, **11**, 2438–2450. https://doi.org/10.3390/w11122438

Kroll S., Fenu A., Wambecq T., Weemases M., Impe J. V. and Willems P. (2018). Energy optimization of the urban drainage system by integrated real-time control during wet and dry weather conditions. *Urban Water Journal*, **15**, 362–370. https://doi.org/10.1080/1573062X.2018.1480726

Lacroix A., Mentzer C. and Pagilla K. R. (2020). Full-scale N removal from centrate using a sidestream process with a mainstream carbon source. *Water Environment Research*, **92**, 1922–1934. https://doi.org/10.1002/wer.1345

Larsen T., Arensman M. and Nerup-Jensen O. (2016). Including pressure measurements in supervision of energy efficiency of wastewater pump systems. *Journal of Hydraulic Engineering*, **142**(2), 04015048.

Leu S., Rosso D., Larson L. E. and Stenstrom M. (2009). Real-time aeration efficiency monitoring in the activated sludge process and methods to reduce energy consumption and operating costs. *Water Environment Research*, **81**, 2471–2481. https://doi.org/10.2175/106143009X425906

Longo S., d'Antoni B. M., Bongards M., Chaparro A., Cronrath A., Fatone F., Lema J., Mauricio-Iglesias M., Soares A. and Hospido A. (2016). Monitoring and diagnosis of energy consumption in wastewater treatment plant – A state of the art and proposals for improvement. *Applied Energy*, **179**, 1251–1268. https://doi.org/10.1016/j.apenergy.2016.07.043

Lu L., Guest J., Peters C. A., Zhu X., Rau G. H. and Ren Z. J. (2018). Wastewater treatment for carbon capture and utilization. *Nature Sustainability*, **1**, 750–758. https://doi.org/10.1038/s41893-018-0187-9

Maktabifard M., Awaitey A., Merta E., Haimi H., Zaborowska E., Mikola A. and Makinia J. (2022). Comprehensive evaluation of the carbon footprint components of wastewater treatment plants located in the Baltic Sea region. *STOTEN*, **806**, 150436.

Mannina G., Ekama G., Caniani D., Cosenza A., Esposito G., Gori R., Garrido-Baserba M., Rosso D. and Olsson G. (2016). Greenhouse gases from wastewater treatment – a review of modelling tools. *STOTEN*, **551–552**, 254–270.

Mentzer C., Drinkwater M. and Pagilla K. R. (2021). Investigation of direct waste activated sludge dewatering benefits and costs in a water resource recovery facility. *Water Environment Research*, **93**, 1–13. https://doi.org/10.1002/wer.1651

Metcalf & Eddy, Inc. (2014). In: 2014 Wastewater Engineering: Treatment and Resource Recovery, G. Tchobanoglous, H. D. Stensel, R. Rsuchihashi and F. Burton (eds), 5th edn, McGraw Hill, New York, NY, USA.

Newhart K. B., Marks C. A., Rauch-Williams T., Cath T. Y. and Hering A. (2020). Hybrid statistical-machine learning ammonia forecasting in continuous activated sludge treatment for improved process control. *Journal of Water Process Engineering*, **37**, 101389. https://doi.org/10.1016/j.jwpe.2020.101389

Pagilla K. R., Craney K. C. and Kido W. H. (1997). Causes and effects of foaming in anaerobic sludge digesters. *Water Science and Technology*, **36**, 463–470. https://doi.org/10.2166/wst.1997.0624

Pagilla K., Shaw A. R., Kunetz T. and Schiltz M. (2009). A Systematic Approach to Establishing Carbon Footprints for Wastewater Treatment Plants. Proceedings of WEFTEC. Water Environment Federation, Alexandria, VA, USA.

Paul E., Plisson-Saune S., Mauret M. and Cantet J. (1998). Process state evaluation of alternating oxic-anoxic activated sludge using ORP, pH, and DO. *Water Science and Technology*, **38**, 299–306. https://doi.org/10.2166/wst.1998.0224

Rahman A., De Clippeleir H., Thomas W., Jimenez J., Wett B., Al-Omari A., Murthy S., Riffat R. and Bott C. (2019). A stage and high rate contact-stabilization performance comparison for carbon and nutrient redirection from high strength municipal wastewater. *Chemical Engineering Journal*, **357**, 737–749. https://doi.org/10.1016/j.cej.2018.09.206

Rajagopal R., Masse D. I. and Singh G. (2013). A critical review of inhibition of anaerobic digestion process by excess ammonia. *Bioresource Technology*, **143**, 632–641. https://doi.org/10.1016/j.biortech.2013.06.030

Rassamee V., Sattayatewa C., Pagilla K. and Chandran K. (2011). Effect of oxic and anoxic conditions on nitrous oxide emissions from nitrification and denitrification processes. *Biotechnology Bioengineering*, **108**(9), 2036–2045.

Shen Y., Linville J. L., Urgun-Demirtas M., Mintz M. M. and Snyder S. W. (2015). An overview of biogas production and utilization at full-scale wastewater treatment plants (WWTPs) in the United States: Challenges and opportunities towards energy-neutral WWTPs. *Renewable & Sustainable Energy Review*, **50**, 346–362.

Subramanian B. and Pagilla K. R. (2015a). Mechanisms of foam formation in anaerobic digesters. *Colloids and Surfaces B: Biointerfaces*, **126**, 621–630. https://doi.org/10.1016/j.colsurfb.2014.11.032

Subramanian B., Miot A., Jones B., Klibert C. and Pagilla K. R. (2015b). A full scale study of mixing and foaming in egg-shaped digesters, *Bioresource Technology*, **192**, 461–470. https://doi.org/10.1016/j.biortech.2015.06.023

Torregrossa D., Hansen H., Hernandez-Sancho F., Cornelissen A., Schutz G. and Leopold U. (2017). A data-driven methodology to support pump performance analysis and energy efficiency optimization in waste water treatment plants. *Applied Energy*, **208**, 1430–1440. https://doi.org/10.1016/j.apenergy.2017.09.012

Torregrossa D., Castellet-Viciano L. and Hernandez-Sancho F. (2019). A data analysis approach to evaluate the impact of the capacity utilization on the energy consumption of wastewater treatment plants. *Sustainable Cities and Society*, **45**, 307–313. https://doi.org/10.1016/j.scs.2018.11.036

WEF (2009). Energy Conservation in Water and Wastewater Facilities. WEF Manual of Practice, Alexandria, VA, USA.

Wu D., Peng X., Li L., Yang P., Peng Y., Liu H. and Wang X. (2021). Commercial biogas plants: review on operational parameters and guide for performance optimization. *Fuel*, **303**, 121282. https://doi.org/10.1016/j.fuel.2021.121282

Zanetti L., Frison N., Nota E., Tomizioli M., Bolzonella D. and Fatone F. (2012). Progress in real-time control applied to biological nitrogen removal from wastewater – a short-review. *Desalination*, **286**, 1–7. https://doi.org/10.1016/j.desal.2011.11.056

Zib III L., Byrne D. M., Marston L. T. and Chini C. M. (2021). Operational carbon footprint of the U.S. water and wastewater sector's energy consumption. *Journal of Cleaner Production*, **321**, 128815. https://doi.org/10.1016/j.jclepro.2021.128815

doi: 10.2166/9781789061796

Chapter 5

Energy and resource recovery using the anaerobic digestion platform

Prathap Parameswaran[1], Jessica A. Deaver[2], Sudeep C. Popat[2], Vikas Khanna[3], Madison Kratzer[3] and Mel Harclerode[4]*

[1]Kansas State University, Manhattan, KS 66506, USA
[2]Clemson University, Clemson, SC, USA
[3]University of Pittsburgh, Pittsburgh, PA, USA
[4]CDM Smith Inc., NJ, USA
**Correspondence: prathapp@ksu.edu*

5.1 CURRENT STATE OF THE ART FOR ANAEROBIC DIGESTION IN MUNICIPAL WASTEWATER RESOURCE RECOVERY FACILITIES

Anaerobic digestion (AD) is a promising Environmental Biotechnology platform integrated into municipal wastewater treatment infrastructure for sludge treatment in most municipalities in the US and across the world. AD provides obvious benefits such as energy utilization from the produced biogas (as heat or electric power), valuable nutrient capture options from the centrate in the form of struvite or other fertilizer products, reduction of the sludge quantity to be disposed, and enhanced biosolids quality enabling its land application. However, its widespread adoption in municipal wastewater facilities is impeded by the susceptibility for process instability and failure, a need for trained personnel for process optimization, odor issues, little to no monetary returns from the produced biogas or biosolids due to absence of carbon credits or energy subsidies.

Despite these challenges, AD has emerged as a clear technology platform of choice to achieve energy and carbon neutrality in medium to large-scale municipal wastewater resource reclamation facilities (WRRFs), especially for wastewater flows greater than 190 million liters per day. AD is a key unit operation in WRRFs with the ability to recover internal energy locked in the sludge, creating a favorable impact for the energy and carbon footprint for the facility, as shown from an example energy balance in Figure 5.1.

Energy neutral wastewater treatment with sidestream AD and sustainable internal carbon utilization has been demonstrated in several WRRFs, as shown in Table 5.1. It is to be noted that

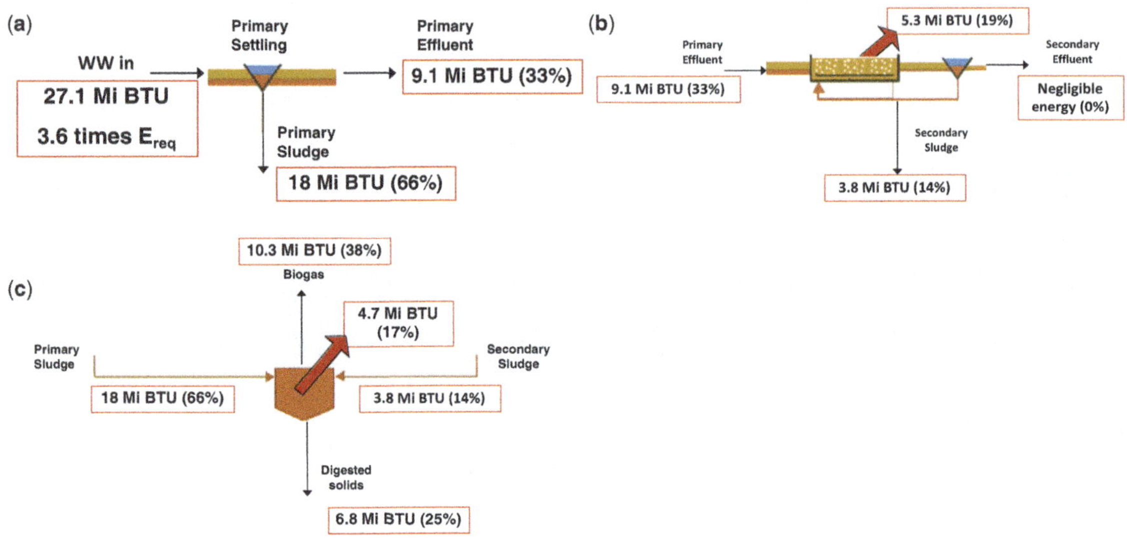

Figure 5.1 Energy balance analysis based on actual data collected from a WRRF and adapted from Shizas and Bagley (2004). The anaerobic digestion step is the main source of internal energy recovery from the sludge in a WRRF, as shown in these figures. It is to be noted that the 17% of the energy requirement is for both mixing and heating requirements of the digester, which can be optimized further to enhance the net recovery and sequestration of the valuable energy and carbon. (a) Primary sedimentation basin – energy balance. (b) Aeration basin – energy balance. (c) Anaerobicdigester – energy balance.

this list is limited to municipal WRRFs in the United States which accept external waste from outside the facility to perform anaerobic co-digestion, which puts them on a positive trajectory to achieve energy neutrality and subsequent beneficial recovery of nutrients and end use of the produced biogas.

The objective of this chapter is to review and summarize current knowledge about sludge management and provide a basis for future practices. The chapter focuses on how current management practices help decarbonization, the role of sludge management strategies in achieving the decarbonization targets of utilities, and how challenges (e.g., emerging contaminants, odors, public scrutiny and upset) in sludge management can be addressed in meeting such targets. It should be noted that residual digestate after AD of sludge also needs additional treatment and disposal. New emerging concepts, namely Water-Energy Nexus, Circular Economy, and Nutrient Trading, are important vehicles for decarbonization in shaping the future sludge management practices. These concepts significantly help to reduce the financial burdens of sludge management on societies and overcome ecological issues and resource scarcity. New technologies and approaches need to be developed to extract energy and nutrients from sludge and improve process and energy efficiency. Recovered energy and nutrients help utilities become a source of revenue generation, overcoming their reputation as pollution mitigation entities. In turn, they will become entities contributing to reduction in carbon emissions and achieve decarbonization of the water sector. Renewable energy production and resource recovery are presented as areas of sludge management to close the linearity of waste production and implement a circular economy of waste management. The chapter closes with a section on implementation of decarbonization at the utilities and future strategies and pathways.

Table 5.1 A sample list of municipal wastewater resource recovery facilities in the United States (with average flow capacity >190 million liters per day) that have demonstrated energy neutral or energy positive operation with carbon diversion to valuable products for internal or external reuse.

Facility	Average Capacity, MLD	Biogas Utilized – HVAC/ Machinery	Sidestream Carbon Capture/Diversion *Pilot Study	Nutrient Products Sequestration (yes/no, type)
East Bay Municipal Utility District (EBMUD), San Francisco, USA	238	Energy positive operation	Yes, for EBPR through sidestream fermentation*	No
DC Water, USA	1450	10 MW (*67% energy neutrality)	Yes, for EBPR through sidestream fermentation*	N & P for silviculture
Des Moines Metro Wastewater Reclamation Authority, IA	227	14 000 MWh/year	NA	Biosolids for land application
Buffalo Sewer Authority, Buffalo, NY	465	Yes	NA	Biosolids for land application
Hyperion Treatment Plant, Los Angeles, CA	1135	Yes	NA	Class A biosolids for fertilizer
Sacramento Regional County Sanitation District, Sacramento, CA	567	Yes	NA	Biosolids for land application
Stickney WRRF, Stickney, IL	2840	Yes	Primary effluent fermentation for EBPR	Biosolids; struvite P recovery with Ostara®
Southeast Water Pollution Control Plant, San Francisco, CA	212	Yes	NA	Biosolids for land application
Bergen City Utility, Little Ferry, NJ	284	Yes 2.69 MW for internal uses	NA	Biosolids for land application
Essex & Union City Joint meeting STP, Elizabeth, NJ	234	Yes	NA	Soil conditioner or composting
Bird Island, WWTP, Buffalo, NY	473	Yes 1–1.2 MW from biogas + incinerator offgas	NA	Biosolids for land application
Coney Island WWTP, Brooklyn, NY	370	Yes for digester heating	NA	Biosolids for land application
Metropolitan Syracuse WWTP, Syracuse, NY	246	Yes combined heat and power	NA	Biosolids for land application
NCSD #3, Wantagh, NY	246	Yes	NA	Biosolids for land application
NCSD # 2, East Rockaway, NY	208	Yes combined heat and power	NA	Biosolids for land application
Joint Water Pollution Control Plant (JWPCP), Carson, CA	1135	Yes		Biosolids for land application or composting
Point Loma Wastewater treatment plant, San Diego, CA	681	Yes6.4 MW through cogeneration		
San José-Santa Clara Regional Wastewater Facility, CA	416	Yes	NA	NA
Central Wastewater Treatment Plant, Nashville, TN	378	Yes	NA	Biosolids for land application

This information was screened from the database presented in www.resourcerecoverydata.org and was subject to further screening criteria such as published information available about nutrient products sequestration. Detailed information foused mainly on energy self sufficiency for WRRFs with AD in the US, Canada, and Europe can be found in Shen *et al.* (2015).

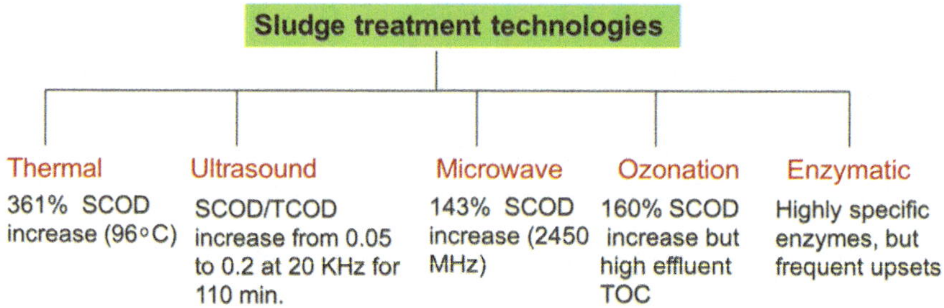

Figure 5.2 A summary of key sludge pretreatment mechanisms under each of which there are patented and/or commercial platforms available and being used in municipal, industrial, and agricultural wastewater installations worldwide. The key findings summarized here are from Kim *et al.* (2003), Khanal *et al.* (2007), Rittmann *et al.* (2008) and Burger and Parker (2013).

5.2 NEED FOR SLUDGE PRETREATMENT TO ENHANCE VIABILITY OF ANAEROBIC DIGESTION

Sludge pretreatment prior to anaerobic digestion enhances the overall rate of anaerobic energy conversion. There are a variety of pretreatment technologies that have been evaluated within the context of municipal, agricultural, and industrial wastewater treatment, as shown in Figure 5.2. Heat, chemical, mechanical, and electrical methods are the most popular for sludge treatment. The advantages of sludge pretreatment include: (1) increase of the surface area of solid particles and thus increase of solubilization by enzymatic hydrolysis; (2) improved biogas production; and (3) reduction of volatile solids (VS). Final concentrations of methane, VS, soluble COD (SCOD) as compared to input solids are indicators of pretreatment performance and AD operation. Thermal pretreatment typically reaches over 100°C (Chauzy *et al.*, 2007; Eskicioglu *et al.*, 2006; Haug *et al.*, 1978; Kim *et al.*, 2003; Pickworth *et al.*, 2005). Acid or alkaline chemicals as well as strong oxidants (e.g., ozone and hydrogen peroxide) have been used for chemical pretreatment (Haug *et al.*, 1978; Kim *et al.*, 2003, 2007; Li *et al.*, 2008). Ultrasonication and microwaves are commonly applied as mechanical treatments (Khanal *et al.*, 2007; Kim *et al.*, 2003; Nickel & Neis, 2007; Wolff *et al.*, 2007). Pulsed-electric-field (PEF) as an electrical method of pretreatment is applied for sludge treatment (Lee *et al.*, 2010; Rittmann *et al.*, 2008; Salerno *et al.*, 2009). Also, various combinations of these pretreatment technologies have been studied (Kim *et al.*, 2003; Ki *et al.*, 2015; Vlyssides & Karlis, 2004). However, further optimization and economic analysis is needed. Many publications show significant improvement in AD performance with several pretreatment technologies (Carlsson *et al.*, 2012; Carrère *et al.*, 2010; Rittmann *et al.*, 2008). Specifically, thermal pretreatment processes like CAMBI™ and EXELYS™ have already been implemented around the world in full-scale wastewater treatment plants to improve AD process performance (Burger & Parker, 2013; Carrère *et al.*, 2010; Gonzalez *et al.*, 2018). Thermal pretreatment processes generally increase WAS temperatures to 90–190°C under pressure, resulting in increased cell lysis and chemical oxygen demand (COD) solubilization (Kim *et al.*, 2003).

However, these methods have not been widely adopted in full-scale operations because the net benefits have not been proven (Rittmann *et al.*, 2008). Investment in and installation of new units as well as the addition of extra energy and/or chemicals present serious operating problems due to toxic by-products, odors, corrosion, or maintenance and have retarded scaling-up to full capacity and commercialization.

Unintended negative consequences from thermal pretreatment such as recalcitrant dissolved organic nitrogen and its impact on the digestate; complexities arising from pretreatment of mixed waste streams, for instance, conductivity of thickened mixed sludge renders pulsed electric field pretreatment infeasible compared to pretreating thickened waste activated sludge (Lee *et al.*, 2010; Zhang *et al.*, 2020).

5.3 DIVERSIFYING PORTFOLIO OF ANAEROBIC DIGESTION AT MUNICIPAL WASTEWATER FACILITIES – THE ADVENT OF ANAEROBIC CO-DIGESTION

The advent of anaerobic co-digestion (ACoD) enabled diversification of anaerobic digestion (AD) profiles at municipal wastewater treatment facilities. ACoD is the simultaneous digestion of two or more substrates. Early study of co-digestion arose to advance digestion of the organic fraction of municipal solid waste; addition of sewage sludge was suggested to increase biogas production by improving environmental conditions within the digester (Cecchi *et al.*, 1988). Further development led to application of a range of co-substrates to overcome limitations of mono-substrate digestion and to increase the economic feasibility of AD (Mata-Alvarez *et al.*, 2000, 2014). Implementation of ACoD at wastewater treatment facilities has enabled improved biogas yields, greater solids destruction, better buffering capacity, enhanced biosolids quality, and dilution of toxic or inhibitory compounds such as heavy metals, ammonia, and sodium (Hagos *et al.*, 2017). These benefits increase the economic viability of AD and contribute towards the decarbonization of wastewater utilities.

5.3.1 Theoretical basis/substrates used

Theoretically, mixing two or more substrates at ideal ratios yields more agreeable operating conditions to ultimately improve biogas volume and percent methane. Therefore, selection of a compatible co-substrate is vital to provide the necessary balance of nutrients, moisture, and physiochemical operating conditions and to increase microbial community diversity. A low C:N ratio, high ammonia and alkalinity, and abundant macro- and micro-nutrients characterize municipal sewage sludge (Tyagi *et al.*, 2018). Common co-substrates complementary to municipal sewage sludge include food waste (FW), grease trap waste (GTW)/fats, oils, and grease (FOG), and the organic fraction of municipal solid waste (OFMSW) (Grosser & Neczaj, 2016; Tandukar & Pavlostathis, 2015; Tyagi *et al.*, 2018; Yang *et al.*, 2019). These substrates raise the C:N ratio, dilute high ammonia and alkalinity, and are low in nutrients required for growth of anaerobic digestion microorganisms. Consequently, co-digestion of municipal sewage sludge with FW, GTW, FOG, or OFMSW may improve the overall performance of a digester previously only fed municipal sewage sludge. Co-digestion of municipal sludge with some agricultural and industrial co-substrates have also been explored and are promising co-substrates for more robust bio-methane production (Mata-Alvarez *et al.*, 2014; Yang *et al.*, 2019).

5.3.2 Challenges of ACoD

Anaerobic co-digestion succeeds when operational parameters, including correct nutrient balances, organic loading rate, HRT/SRT, and dilution of toxic compounds, are optimized and economic considerations, such as operating costs, storage and handling of digester substrate, and transportation costs for co-substrates, are met (Tandukar & Pavlostathis, 2015). However, there are challenges in meeting these operational and economic demands that prevent co-digesters from attaining maximum performance goals. Hydrolysis rates of complex particulate matter are one potential bottleneck, and they are often considered the primary rate limiting step of AD (Pavlostathis & Giraldo-Gomez, 1991). Therefore, hydrolysis rates are vital for determining the speed of digestion and determining waste biodegradability. When assessing substrate compatibility, it is necessary to pair fast hydrolyzing substrates with slow hydrolyzing substrates to avoid bottlenecks (Hagos *et al.*, 2017). Additionally, there are waste-specific bottlenecks that disadvantage co-digestion. Potential co-substrates with elevated protein concentrations, and therefore ammonia may be toxic to the AD microbial consortium, in particular to methanogens (Amha *et al.*, 2017). Wastes rich in lipids and long chain fatty acids, such as GTW and FOG, also exhibit toxic properties that lead to digester failure rather than performance enhancement (Long *et al.*, 2012).

The economic viability of ACoD derives from its ability to use a single reactor for degradation of multiple substrates and improved biogas production for energy generation. Ensuring reduction in the transportation distances and storage costs of co-substrates is also critical to enhance economic benefits (Tandukar & Pavlostathis, 2015). Several studies have demonstrated the economic advantage of successful co-digestion. Krupp *et al.* (2005) investigated the ecological and economic impacts of sludge co-digestion with OFMSW in oversized, full-scale digesters. Life cycle assessments revealed that compared to composting and mono-substrate digestion, ACoD was more beneficial in the climate change category, and when applied at a larger plant its economic benefit was best. Pavan *et al.* (2007) and Righi *et al.* (2013) also demonstrated under specific conditions at smaller, full-scale treatment plants in rural areas, that co-digestion can be a beneficial tool for improving the economic balances of WWTP. Important considerations include size of the digesters, volume of co-substrates for treatment, and reduced transportation and storage time.

5.3.3 Current research on ACoD

ACoD continues to advance via research on the co-digestion process, the refinement of downstream processes, the exploration of nutrient recovery from ACoD, and the models used to predict function when applying waste streams of variable characteristics. Studies on the ACoD process seek to refine characterization of substrates, define ideal mixing ratios for co-substrates, and optimize operational parameters, such as organic loading rate. Downstream process improvements focus on enhancing biogas quality, in particular increasing the methane proportion, improving digestate dewaterability, and reducing odor emissions from biosolids (Xie *et al.*, 2018). Integrated technologies from the recovery of carbon, nitrogen, and phosphorous may also add to the benefits of ACoD, therefore enhancing its economic value as well. Finally, efforts to refine ADM1 to apply it to the co-digestion process are underway. Improving the ADM1 model for co-digestion would help to predict the performance of a digester when adding a new co-substrate or changing operational parameter (Hagos *et al.*, 2017).

5.4 ENHANCING THE VALUE OF THE PRODUCED BIOGAS THROUGH CO-GENERATION AND FURTHER PURIFICATION TO NATURAL GAS FOR PIPELINE DELIVERY

Carbon capture from anaerobic digestion yields a valuable gaseous product, namely biogas, which typically consists of 50% methane: 50% CO_2 along with other gases and impurities such as H_2S, ammonia, mercaptans, siloxanes and other minor ingredients. This presents a valuable opportunity to sequester the gaseous product and utilize it for combined heat and power production, direct electricity generation via gas turbines, biogas upgrading to natural gas quality for pipeline delivery, or biogas use in the transportation sector. Despite the obvious economic, energetic, and environmental benefits of carbon capture, roughly only a third of the municipal wastewater facilities across the United States practice some form of methane value recovery, while it is roughly two times more prevalently practiced across the European Union and some other parts of the world (Scarlat *et al.*, 2018; Shen *et al.*, 2015). Each of these scenarios are discussed briefly.

5.4.1 Biomethane for combined heat and power production

This option involves the combustion of the biogas in a boiler with or without other fuel sources to produce electricity as well as heat that can be used for such applications as space heating and other temperature-based operations. While this is the most practiced option for wastewater facilities that do practice energy recovery, some of the challenges with this approach include significant loss of process efficiency due to process limitations, suitability limited to temperate regions or during colder seasons, and corrosion and equipment damage due to processing of the biogas without cleaning.

5.4.2 Biomethane for electricity generation

Direct generation of electricity from the produced biogas is possible by combustion of the biogas in a gas turbine utilizing the Rankine cycle. However, significant bottlenecks exist with this option, mainly corrosion of the turbine blades by biogas impurities, especially siloxanes. Targeted research such as novel gas scrubbing, biological gas treatment, and sludge pretreatment have focused on the removal of siloxanes either from the sludges or from the produced biogas prior to being sent to the gas turbines (Dewil *et al.*, 2006; Lee & Rittmann, 2016; Popat & Deshusses, 2008).

5.4.3 Biomethane for upgrading and pipeline delivery

An option gaining popularity for biogas end use is to purify it using advanced separation techniques such as pressure swing adsorption (PSA), which removes CO_2 and other impurities to produce >90% purity CH_4, which can be injected into natural gas pipelines for offsite energy generation. Key bottlenecks include the high capital costs for the PSA addition and the low economic value for the produced natural gas, especially in North America due to the lower prevailing price for natural gas and lack of sufficient subsidies to incentivize production and purification.

5.4.4 Biomethane for transportation

Biomethane can be used in existing liquified natural gas (LNG) and compressed natural gas (CNG) refueling infrastructure as well as public transport infrastructures. Purification of the biogas to natural gas standard is a pre-requisite for transportation fuel purposes (Augelletti *et al.*, 2017; Kim *et al.*, 2015) Another emerging alternative is to use the purified biomethane as an electron donor in chemical fuel cells to produce electric power, albeit there are several bottlenecks to enhance its efficiency, such as catalyst poisoning due to biogas impurities (Alves *et al.*, 2013; Lanzini & Leone, 2010).

5.4.5 Biogas to valuable chemicals

Volatile fatty acids (VFAs), produced as byproducts of anaerobic digestion, are high value building block chemicals. A technique for capture and reuse of AD byproducts is through selective mixed culture fermentation which allows for the elongation of acetate and other short chained carbonic acid based compounds. These medium chain (VFAs) are high value and more easily extracted from the product than acetate and other short chained VFAs (Steinbusch *et al.*, 2011). Preventing conditions for methanogenesis in a steady state reactor remains a challenge for mixed culture fermentation (Agler *et al.*, 2012; Steinbusch *et al.*, 2011).

5.5 ALTERING THE AD PLATFORM FOR HIGHER ORGANIC CARBON PRODUCT CAPTURE COUPLED WITH WATER REUSE AND NUTRIENT RECOVERY

Strong drivers such as global climate change, changing economic landscape, and increased demand for chemicals and sustainably derived plastics could soon transform carbon capture from wastewater treatment facilities to the modern wastewater resource recovery facility. Short- and medium-chain carboxylic acids are an essential intermediate in the anaerobic food web that leads to methanogenesis, which when controlled can accumulate to high concentrations in the bioreactor. Several strategies have been proposed to manage arrested methanogenesis in an anaerobic digestion platform to facilitate the hydrolysis and acidogenesis products to accumulate and be recovered through subsequent separation techniques. Recent research has indicated that a bio-electrochemically assisted anaerobic digester may not only serve to enhance the overall hydrolysis and fermentation rates, but also promote greater accumulation of the higher organic acids by selective consumption of acetate by the electroactive bacteria through thermodynamic and kinetic benefits. The H_2 rich environment in the AnMBR should not only arrest methanogenesis but also facilitate secondary fermentation reactions leading to higher order VFA synthesis (Bhatt *et al.*, 2020; De Vrieze *et al.*, 2018; Jiang *et al.*, 2018).

Separation of VFAs from fermented broths is challenging due to low VFA concentrations in ion-rich solutions. Consequently, separation capacity and selectivity with traditional solvents and adsorbents are both compromised. In the recent extraction literature, ionic liquids (ILs) have been reported for extraction of VFAs, and some have been noticed to be better than conventional solvents in terms of extraction efficiency. ILs exist as molten salts at ambient temperature and consist entirely of ions, usually a charge-stabilized organic cation and an inorganic or organic anion. A study concluded that phosphonium-based ILs are better extractants than the traditional organic solvents for recovery of short chain organic acids from aqueous dilute solutions. They succeeded to obtain higher distribution coefficients as compared to most traditional solvents by using phosphonium-based ionic liquids for the extraction of low concentrated lactic acid solutions, and obtained an extraction efficiency of 98.4%

with a two-step extraction (Liang *et al.*, 2017; Oliveira *et al.*, 2012). IL-mediated esterification for reactive extraction of low-value VFA from dilute aqueous streams was also reported. Distillation or evaporation will be needed as the final VFA purification step. Membrane based non-reactive or reactive separations from integrated anaerobic digesters is an equally effective separation platform that has been receiving increased attention (Zhu *et al.*, 2021).

5.6 ENERGY MANAGEMENT IN ANAEROBIC DIGESTION FOR OVERALL ENERGY NEUTRALITY OR ENERGY POSITIVE TREATMENT – THE CASE FOR DIRECT ANAEROBIC TREATMENT THROUGH AnMBRs

Secondary treatment of sludges via the anaerobic digestion platform incur energy expenditure as well, apart from producing energy rich biogas. Sources of energy losses include digester heating and pretreatment energy consumption, sludge dewatering after digestion, and energy considerations for sludge hauling. Recent research has established the lowered energy requirement for an anaerobic membrane bioreactor (AnMBR) platform, which aims to achieve direct anaerobic wastewater treatment with valuable resource recovery with methane as the primary energy product, as shown in Figure 5.3, compared to conventional activated sludge with secondary treatment using the anaerobic digestion platform. Further process optimization for the emerging AnMBR platform will focus on

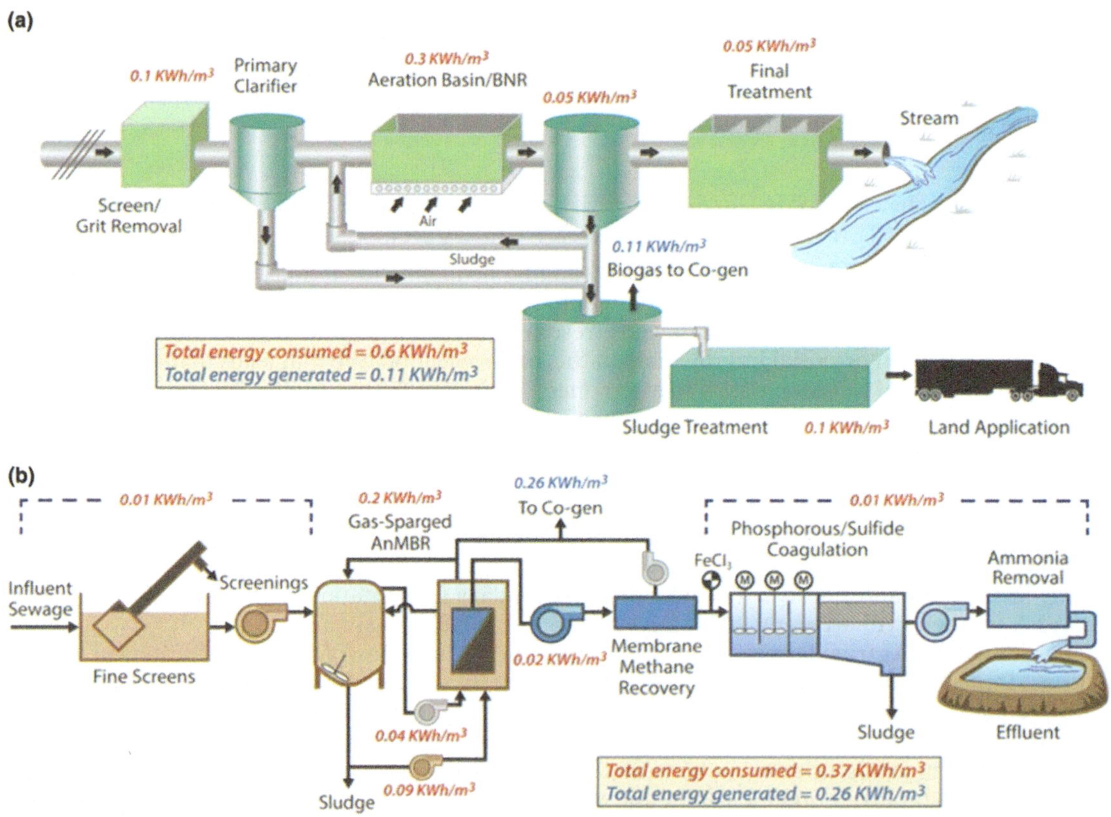

Figure 5.3 Net energy requirement comparison for: (a) conventional activated sludge; (b) AnMBR platform for methane and nutrient capture.

decreasing fouling energy requirements even further by periodic pulse sparging at high flow rates rather than continuous sparging; mixing energy optimization in the primary bioreactor. Anaerobic digester energy optimizations will likely be focused on decreasing pretreatment costs, decreasing mixing energy requirements, enhancing process based sludge dewaterability, and enhancing the overall energy capture efficiency from the produced methane rich biogas.

5.7 TECHNO-ECONOMIC AND LIFE CYCLE ASSESSMENTS FOR SHAPING THE FUTURE OF ANAEROBIC DIGESTION

Traditionally there has been a disconnect between actual experimental data and Life Cycle Assessment (LCA) studies for emerging technologies, with most studies focusing on either aspect discretely, and only a few studies attempt to bridge this gap. Green engineering principles are incorporated late in the design/concept development process, resulting in incremental environmental improvement rather than process pathways that minimize life cycle environmental impacts. Integrated techno-economic and LCA platforms have the ability to proactively guide conceptual designs that are environmentally and economically conscious and aimed at maximizing bioenergy capture and carbon recovery in a higher value form such as carboxylates along with other valuable products like nutrient products and water for indirect/direct potable reuse. The use of the AnMBR process for domestic wastewater treatment presents an opportunity to mitigate environmental, social, and economic impacts currently incurred from energy-intensive conventional aerobic activated sludge processes. A pilot-scale study of the AnMBR and concurrrent Techno Economic Analysis (TEA) and LCA were performed to demonstrate and validate AnMBR technology for more sustainable domestic wastewater treatment compared to aerobic activated sludge.

The feasibility of the proposed AnMBR platform is supported through our preliminary LCA of the AnMBR platform. Figure 5.4(a) shows the environmental impacts of various AnMBR treatment scenarios compared to conventional treatment. The LCA indicated that the most sustainable system was conventional treatment followed by scenario 7, a hybrid AnMBR with vacuum flash tank methane recovery and sulfide removal operating at 15 LMH (liters of treated water per unit membrane surface area per hour) and temperature higher than 25°C. Previous studies have performed detailed evaluations on improving AnMBR process subcomponents to maximize energy recovery and dissolved methane recovery. Few studies have broadly evaluated the role of chemical use, membrane fouling management, and dissolved methane removal technologies. Figure 5.4(b) shows the potential for implementing an AnMBR that achieves an overall environmental impact less than conventional treatment by reducing, if not eliminating, chemical removal of sulfide and phosphorous. The global warming potential (kg CO_2 equivalents) for this AnMBR configuration without chemical coagulation for chemical removal is shown as an offset resulting from methane recovery for bioenergy use. The Sankey diagram of scenario 7 AnMBR (Figure 5.5) shows the nutrient removal component as the major contributor to environmental impacts.

The feasibility of our studied AnMBR system is also corroborated by preliminary TEA. Preliminary analysis show that the levelized operational cost of the hybrid AnMBR with methane and sulfide removal without the volatile fatty acid (VFA) separation is \$0.09/m^3. On a full-scale treatment plant with the capacity to treat 22 730 m^3/day, the annual operation costs are \$ 781 300 respectively. Systems-level optimization has the potential to further reduce the operational costs for the hybrid AnMBR design because of process improvements and associated reductions in chemical and energy use. When compared on a construction cost basis, the AnMBR design has a higher construction cost (\$ 71 582 500) compared to a conventional activated sludge design (\$ 59 991 250). However, it is to be noted that the revenue associated with value-added products that could potentially be recovered in modified AnMBR is currently not accounted for and is expected to bring down the operational costs significantly. Similar scenarios can be developed for the sidestream AD platform as well, although the impacts are projected to be less impactful when compared to direct anaerobic wastewater treatment, although there is promising potential.

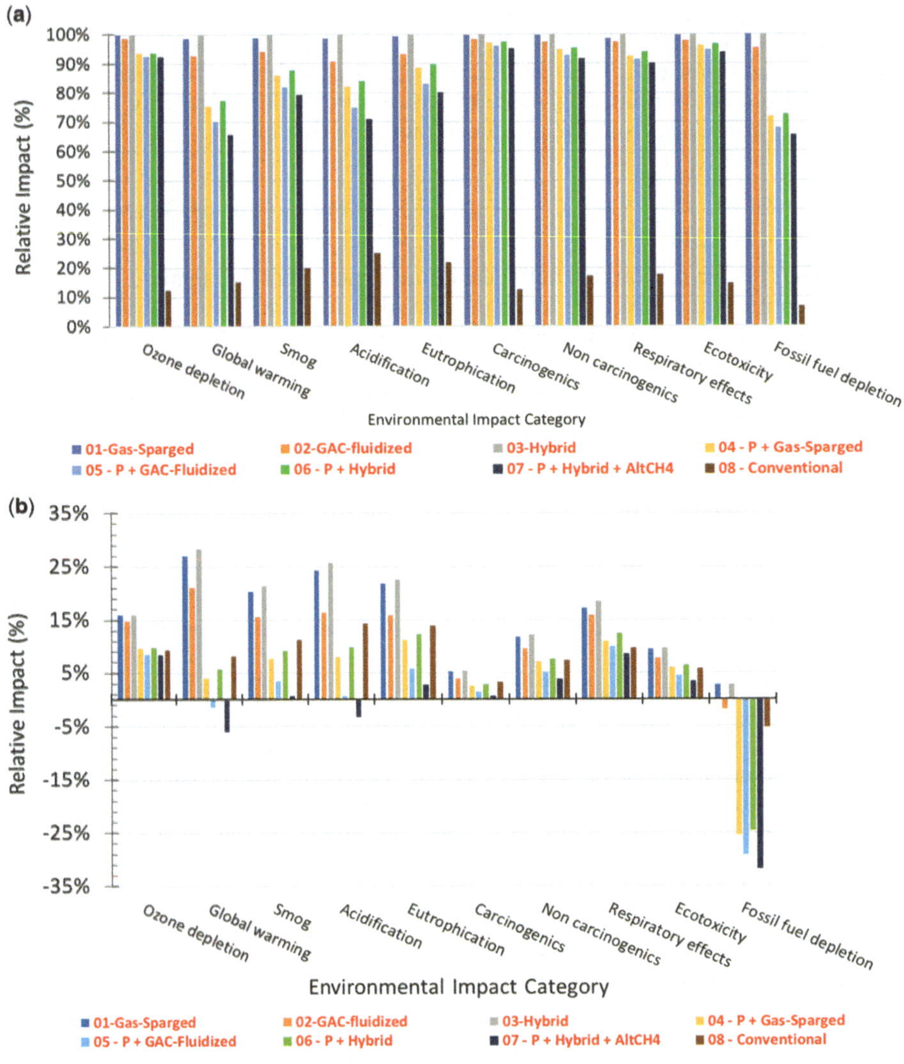

Figure 5.4 Relative impacts of the AnMBR and conventional treatment scenarios at 15 LMH (F2) and >25°C (T1) with (a) and without (b) sulfide and phosphorus removal by chemical coagulation using ferric chloride and ACH. Impacts in *b* are relative to those for Scenarios 1 and 3 in *a*. Source of tables is Harclerode *et al.* (2020).

Conclusion of the TEA and LCA study determined two process subcomponents, sulfide and phosphorus removal and sludge management, that drove chemical use and residuals generation, and in turn the environmental and cost impacts. Furthermore, integrating primary sedimentation and a vacuum degassing tank for dissolved methane removal maximized net energy recovery. Sulfide was generated by anaerobic reduction of naturally occurring sulfate. Previous studies have not considered the cost and environmental impact of sulfide generation and removal via chemical coagulation. The TEA/LCA demonstrated that the AnMBR can have a lower environmental impact and operating cost relative to conventional wastewater treatment if sulfide can be removed biologically rather than with chemical coagulation.

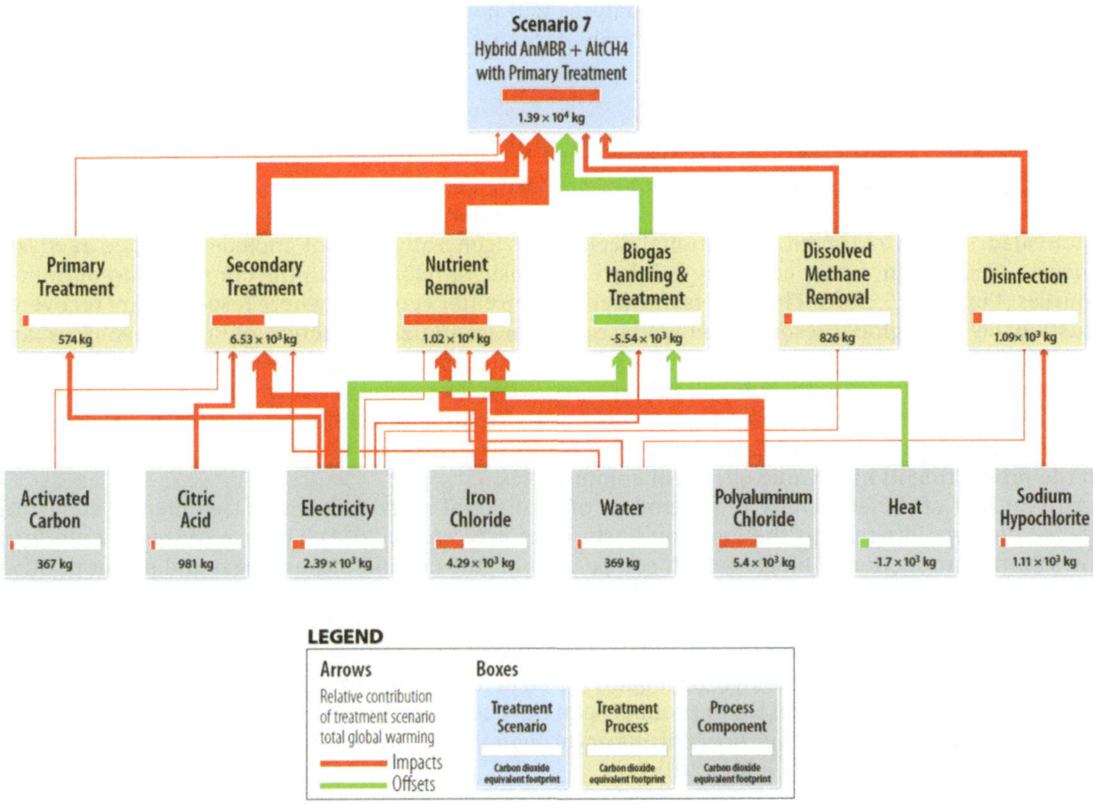

Figure 5.5 Sankey diagrams of the global warming impact (kg CO_2-equivalent/m^3) assessment for scenario 7, a hybrid AnMBR with vacuum flash tank methane recovery and sulfide removal operating at 15 LMH and temperature higher than 25°C. Plots from Harclerode *et al.* (2020).

Even though the environmental impacts of chemicals in our modeled process are significant, we expect these impacts can be reduced significantly based on process optimization studies and when the process is implemented at scale. As such, the final integrated AnMBR platform design has the potential to have much lower environmental impact compared to the conventional process while also recovering value-added products.

5.8 FUTURE STRATEGIES AND ROADMAPS TO DECARBONIZATION

Disruptive pretreatment technologies proven at full-scale have emerged, especially thermal hydrolysis, however there is a greater need to intimately couple decarbonization potential with economic sustainability, while also overcoming technological unintended consequences, such as recalcitrant nitrogen. Favorable technological, economic, and policy breakthroughs are needed to make valorized biomethane generation a widespread option in North American wastewater resource recovery facilities, while practices elsewhere need to consider more sustainable frameworks. Decarbonization potential from the AD platform can be maximized by developing technology platforms that can sequester carbon in a higher and more valuable through the carboxylate platform or other analogous synthesis routes.

Sound techno-economic and life cycle analyses will need to be intimately coupled to current and future AD configurations to maximize not only carbon capture, but overall resource recovery in a holistic fashion.

5.9 ANAEROBIC DIGESTION TECHNIQUES FOR ACHIEVEMENT OF A CIRCULAR ECONOMY

Traditional goals for wastewater treatment are primarily centered around preserving environmental and human health. These goals are met through traditional wastewater treatment techniques, such as activated sludge treatment. Traditional techniques generally do not include resource recovery strategies. The current linear economy is inherently unsustainable due to the consumption of limited resources (Puyol *et al.*, 2016) This begets the need for a circular economy where resources are recovered to sustainably meet global resource demands, reduce or minimize extraction of virgin resources, and reduce environmental impacts. The concept of the circular economy has widened the range of goals for wastewater treatment to include full resource recovery. Through the use of AD in conjunction with additional advanced treatment techniques, wastewater treatment has the potential to be transformed from an energy sink into a profit producing process through electricity generation, nutrient recovery, and the production of high value chemical commodities.

Around half of the electricity requirements of traditional activated sludge treatment systems are dedicated to providing air to the aeration basin (McCarty *et al.*, 2011) The use of the AD platform allows for the production of biogas which can be used for the cogeneration of heat and power to offset operational requirements of the system, and in some cases produce excess power (Batstone & Virdis, 2014). Post-digestion treatment as part of an AD treatment train allows for the recovery of phosphorus and nitrogen. Phosphorus, a non-renewable resource, can be recovered through precipitation or as a salt through crystallization as hydroxyapatite or struvite (Battistoni *et al.*, 2006). Nitrogen, which requires energy intensive processes to produce, can be recovered through the partitioning of nutrients via assimilation from algae or microorganisms or through the adsorption of ammonium to clinoptilolite clay (Batstone *et al.*, 2015; Lim *et al.*, 2019).

ACKNOWLEDGEMENTS

The authors would like to acknowledge help from graduate PhD student in Environmental Engineering, Mr. Arvind Damodara Kannan, for his assistance with formatting the chapter and references. The authors would also like to specifically acknowledge time and effort from graduate and undergraduate students from Dr. Parameswaran's team: Ms. Priyasha Fernando; Ms. Emily Randig; Ms. Marleigh Hutchinson; and Mr. Mason Ericson for their help with compiling Table 5.1 from a very large US dataset. Direct help for Figure 5.1 was obtained from Dr. Cesar Torres, Associate Professor, Chemical Engineering at Arizona State University.

REFERENCES

Agler M. T., Spirito C. M., Usack J. G., Werner J. J. and Angenent L. T. (2012). Chain elongation with reactor microbiomes: upgrading dilute ethanol to medium-chain carboxylates. *Energy & Environmental Science*, **5**(8), 8189–8192. https://doi.org/10.1039/c2ee22101b

Alves H. J., Bley C., Niklevicz R. R., Frigo E. P., Frigo M. S. and Coimbra-Araujo C. H. (2013). Overview of hydrogen production technologies from biogas and the applications in fuel cells. *International Journal of Hydrogen Energy*, **38**(13), 5215–5225. https://doi.org/10.1016/j.ijhydene.2013.02.057

Amha Y. M., Anwar M. Z., Brower A., Jacobsen C. S., Stadler L. B., Webster T. M. and Smith A. L. (2017). Inhibition of anaerobic digestion processes: applications of molecular tools. *Bioresource Technology*, **247**, 999–1014.

Augelletti R., Conti M. and Annesini M. C. (2017). Pressure swing adsorption for biogas upgrading. A new process configuration for the separation of biomethane and carbon dioxide. *Journal of Cleaner Production*, **140**, 1390–1398. https://doi.org/10.1016/j.jclepro.2016.10.013

Batstone D. J. and Virdis B. (2014). The role of anaerobic digestion in the emerging energy economy. *Current Opinion in Biotechnology*, **27**, 142–149. https://doi.org/10.1016/j.copbio.2014.01.013

Batstone D. J., Hulsen T., Mehta C. M. and Keller J. (2015). Platforms for energy and nutrient recovery from domestic wastewater: a review. *Chemosphere*, **140**, 2–11. https://doi.org/10.1016/j.chemosphere.2014.10.021

Battistoni P., Paci B., Fatone F. and Pavan P. (2006). Phosphorus removal from anaerobic supernatants: start-up and steady-state conditions of a fluidized bed reactor full-scale plant. *Industrial & Engineering Chemistry Research*, **45**(2), 663–669. https://doi.org/10.1021/ie050796g

Bhatt A. H., Ren Z. and Tao L. (2020). Value proposition of untapped wet wastes: carboxylic acid production through anaerobic digestion. *Iscience*, **23**(6), 101221. https://doi.org/10.1016/j.isci.2020.101221

Burger G. and Parker W. (2013). Investigation of the impacts of thermal pretreatment on waste activated sludge and development of a pretreatment model. *Water Research*, **47**, 5245–5256. https://doi.org/10.1016/j.watres.2013.06.005

Carlsson M., Lagerkvist A. and Morgan-Sagastume F. (2012). The effects of substrate pretreatment on anaerobic digestion systems: a review. *Waste Management*, **32**, 1634–1650. https://doi.org/10.1016/j.wasman.2012.04.016

Carrère H., Dumas C., Battimelli A., Batstone D. J., Delgenès J. P., Steyer J. P. and Ferrer I. (2010). Pretreatment methods to improve sludge anaerobic degradability: a review. *Journal of Hazardous Materials*, **183**, 1–15. https://doi.org/10.1016/j.jhazmat.2010.06.129

Cecchi F., Traverso P. G., Mata-Alvarez J., Clancy J. and Zaror C. (1988). State of the art of R&D in the anaerobic digestion process of municipal solid waste in Europe. *Biomass*, **16**(4), 257–284. https://doi.org/10.1016/0144-4565(88)90031-5

Chauzy J., Cretenot D., Bausseron A. and Gokelaere X. (2007). Thermal Hydrolysis to Increase Sludge Biodegradability or How to Turn Mesophilic Anaerobic Digestion of Biological Sludge Into an Attractive Process. WEFTEC 2007, 80th Annual Water Environment Federation Technical Exhibition and Conference, 13–17 October, San Diego, CA.

De Vrieze J., Arends J. B. A., Verbeeck K., Gildemyn S. and Rabaey K. (2018). Interfacing anaerobic digestion with (bio)electrochemical systems: potentials and challenges. *Water Research*, **146**, 244–255. https://doi.org/10.1016/j.watres.2018.08.045

Dewil R., Appels L. and Baeyens J. (2006). Energy use of biogas hampered by the presence of siloxanes. *Energy Conversion and Management*, **47**(13–14), 1711–1722. https://doi.org/10.1016/j.enconman.2005.10.016

Eskicioglu C., Kennedy K. J. and Droste R. L. (2006). Characterization of soluble organic matter of waste activated sludge before and after thermal pretreatment. *Water Research*, **40**(20), 3725–3736. https://doi.org/10.1016/j.watres.2006.08.017

Gonzalez A., Hendriks A. T. W. M., van Lier J. B. and de Kreuk M. (2018). Pretreatments to enhance the biodegradability of waste activated sludge: elucidating the rate limiting step. *Biotechnology Advances*, **36**, 1434–1469. https://doi.org/10.1016/j.biotechadv.2018.06.001

Grosser A. and Neczaj E. (2016). Enhancement of biogas production from sewage sludge by addition of grease trap sludge. *Energy Conversion and Management*, **125**, 301–308. https://doi.org/10.1016/j.enconman.2016.05.089

Hagos K., Zong J., Li D., Liu C. and Lu X. (2017). Anaerobic co-digestion process for biogas production: progress, challenges and perspectives. *Renewable and Sustainable Energy Reviews*, **76**(March 2016), 1485–1496. https://doi.org/10.1016/j.rser.2016.11.184

Harclerode M., Doody A., Brower A., Vila P., Ho J. and Evans P. J. (2020). Life cycle assessment and economic analysis of anaerobic membrane bioreactor whole-plant configurations for resource recovery from domestic wastewater. *Journal of Environmental Management*, **269**, 110720. https://doi.org/10.1016/j.jenvman.2020.110720

Haug R. T., Stuckey D. C., Gossett J. M. and Mccarty P. L. (1978). Effect of thermal pretreatment on digestibility and dewaterability of organic sludges. *Journal of Water Pollution Control Federation*, **50**(1), 73–85.

Jiang Y., Lu L., Wang H., Shen R. X., Ge Z., Hou D. X., Chen X., Liang P., Huang X. and Ren Z. J. (2018). Electrochemical control of redox potential arrests methanogenesis and regulates products in mixed culture electro-fermentation. *ACS Sustainable Chemistry & Engineering*, **6**(7), 8650–8658. https://doi.org/10.1021/acssuschemeng.8b00948

Khanal S. K., Grewell D., Sung S. and Van Leeuwen J. (2007). Ultrasound applications in wastewater sludge pretreatment: a review. *Critical Reviews in Environmental Science and Technology*, **37**, 277–313. https://doi.org/10.1080/10643380600860249

Ki D., Parameswaran P., Popat S. C., Rittmann B. E. and Torres C. I. (2015). Effects of pre-fermentation and pulsed-electric-field treatment of primary sludge in microbial electrochemical cells. *Bioresource Technology*, **195**, 83–88. https://doi.org/10.1016/j.biortech.2015.06.128

Kim J., Park C., Kim T. H., Lee M., Kim S., Kim S. W. and Lee J. (2003). Effects of various pretreatments for enhanced anaerobic digestion with waste activated sludge. *Journal of Bioscience and Bioengineering*, **95**(3), 271–275. https://doi.org/10.1016/S1389-1723(03)80028-2

Kim T.-H., Kim T.-H., Yu S., Nam Y. K., Choi D.-K., Lee S. R. and Lee M.-J. (2007). Solubilization of waste activated sludge with alkaline treatment and gamma ray irradiation. *Journal of Industrial and Engineering Chemistry*, **13**(7), 1149–1153.

Kim D., Lee K. and Park K. Y. (2015). Enhancement of biogas production from anaerobic digestion of waste activated sludge by hydrothermal pretreatment. *International Biodeterioration & Biodegradation*, **101**, 42–46. https://doi.org/10.1016/j.ibiod.2015.03.025

Krupp M., Schubert J. and Widmann R. (2005). Feasibility study for co-digestion of sewage sludge with OFMSW on two wastewater treatment plants in Germany. *Waste Management*, **25**(4), 393–399. https://doi.org/10.1016/j.wasman.2005.02.009

Lanzini A. and Leone P. (2010). Experimental investigation of direct internal reforming of biogas in solid oxide fuel cells. *International Journal of Hydrogen Energy*, **35**(6), 2463–2476. https://doi.org/10.1016/j.ijhydene.2009.12.146

Lee I. and Rittmann B. E. (2016). Using focused pulsed technology to remove siloxane from municipal sewage sludge. *Journal of Environmental Engineering*, **142**(1). https://doi.org/10.1061/(asce)ee.1943-7870.0000975

Lee I. S., Parameswaran P., Alder J. M. and Rittmann B. E. (2010). Feasibility of focused-pulsed treated waste activated sludge as a supplemental electron donor for denitrification. *Water Environment Research*, **82**(12), 2316–2324. https://doi.org/10.2175/106143010X12609736967288

Li H., Jin Y., Mahar R., Wang Z. and Nie Y. 2008 Effects and model of alkaline waste activated sludge treatment. *Bioresource Technology*, **99**(11), 5140–5144. https://doi.org/10.1016/j.biortech.2007.09.019

Liang L., Li C., Xu F., He Q., Yan J., Luong T. and Sun N. (2017). Conversion of cellulose rich municipal solid waste blends using ionic liquids: feedstock convertibility and process scale-up. *RSC Advances*, **7**(58), 36585–36593. https://doi.org/10.1039/C7RA06701A

Lim K., Evans P. J. and Parameswaran P. (2019). Long-term performance of a pilot-scale gas-sparged anaerobic membrane bioreactor under ambient temperatures for holistic wastewater treatment. *Environmental Science and Technology*, **53**(13), 7347–7354. https://doi.org/10.1021/acs.est.8b06198

Long J. H., Aziz T. N., Reyes F. L. D. L. and Ducoste J. J. (2012). Anaerobic co-digestion of fat, oil, and grease (FOG): a review of gas production and process limitations. *Process Safety and Environmental Protection*, **90**(3), 231–245. https://doi.org/10.1016/j.psep.2011.10.001

Mata-Alvarez J., Mace S. and Llabres P. (2000). Anaerobic digestion of organic solid wastes. An overview of research achievements and perspectives. *Bioresource Technology*, **74**, 3–16. https://doi.org/10.1016/S0960-8524(00)00023-7

Mata-Alvarez J., Dosta J., Romero-Güiza M. S., Fonoll X., Peces M. and Astals S. (2014). A critical review on anaerobic co-digestion achievements between 2010 and 2013. *Renewable and Sustainable Energy Reviews*, **36**, 412–427. https://doi.org/10.1016/j.rser.2014.04.039

McCarty P. L., Bae J. and Kim J. 2011 Domestic wastewater treatment as a net energy producer – can this be achieved? *Environmental Science and Technology*, **45**(17), 7100–7106. https://doi.org/10.1021/es2014264

Nickel K. and Neis U. (2007). Ultrasonic disintegration of biosolids for improved biodegradation. *Ultrasonics Sonochemistry*, **14**(4), 450–455. https://doi.org/10.1016/j.ultsonch.2006.10.012

Oliveira F. S., Araújo J. M., Ferreira R., Rebelo L. P. N. and Marrucho I. M. (2012). Extraction of L-lactic, L-malic, and succinic acids using phosphonium-based ionic liquids. *Separation and Purification Technology*, **85**, 137–146. https://doi.org/10.1016/j.seppur.2011.10.002

Pavan P., Bolzonella D., Battistoni E. and Cecchi F. (2007). Anaerobic co-digestion of sludge with other organic wastes in small wastewater treatment plants: an economic considerations evaluation. *Water Science and Technology*, **56**(10), 45–53. https://doi.org/10.2166/wst.2007.730

Pavlostathis S. G. and Giraldo-Gomez E. (1991). Kinetics of anaerobic treatment. *Water Science and Technology*, **24**(8), 35–59. https://doi.org/10.2166/wst.1991.0217

Pickworth B., Cranshaw I., Abraham K., Coleman P., Walley P. and Solheim O. E. (2005). Large Scale Reality of Sewage Sludge Pasteurisation and Thermal Hydrolysis. WEFTEC 2005, 78th Annual Water Environment Federation Technical Exhibition and Conference, 29 October–2 November, Washington, DC.

Popat S. C. and Deshusses M. A. (2008). Biological removal of siloxanes from landfill and digester gases: opportunities and challenges. *Environmental Science and Technology*, **42**(22), 8510–8515. https://doi.org/10.1021/es801320w

Puyol D., Batstone D. J., Hulsen T., Astals S., Peces M. and Kromer J. O. (2016). Resource recovery from wastewater by biological technologies: opportunities, challenges, and prospects. *Frontiers in Microbiology*, **7**, 2106.

Righi S., Oliviero L., Pedrini M., Buscaroli A. and Della Casa C. (2013). Life cycle assessment of management systems for sewage sludge and food waste: centralized and decentralized approaches. *Journal of Cleaner Production*, **44**(2013), 8–17. https://doi.org/10.1016/j.jclepro.2012.12.004

Rittmann B. E., Lee H. S., Zhang H. S., Alder J., Banaszak J. E. and Lopez R. (2008a). Full-scale application of focused-pulsed pretreatment for improving biosolids digestion and conversion to methane. *Water Science and Technology*, **58**(10), 1895–1901. https://doi.org/10.2166/wst.2008.547

Salerno M. B., Lee H. S., Parameswaran P. and Rittmann B. E. (2009). Using a pulsed electric field as a pretreatment for improved biosolids digestion and methanogenesis. *Water Environment Research*, **81**(8), 831–839. https://doi.org/10.2175/106143009X407366

Scarlat N., Dallemand J. F. and Fahl F. (2018). Biogas: developments and perspectives in Europe. *Renewable Energy*, **129**, 457–472. https://doi.org/10.1016/j.renene.2018.03.006

Shen Y. W., Linville J. L., Urgun-Demirtas M., Mintz M. M. and Snyder S. W. (2015). An overview of biogas production and utilization at full-scale wastewater treatment plants (WWTPs) in the United States: challenges and opportunities towards energy-neutral WWTPs. *Renewable and Sustainable Energy Reviews*, **50**, 346–362. https://doi.org/10.1016/j.rser.2015.04.129

Shizas I. and Bagley D. M. (2004). Experimental determination of energy content of unknown organics in municipal wastewater streams. *Journal of Environmental Engineering*, **130**(2), 45. https://doi.org/10.1061/(ASCE)0733-9402(2004)130:2(45)

Steinbusch K. J. J., Hamelers H. V. M., Plugge C. M. and Buisman C. J. N. (2011). Biological formation of caproate and caprylate from acetate: fuel and chemical production from low grade biomass. *Energy and Environmental Science*, **4**(1), 216–224. https://doi.org/10.1039/C0EE00282H

Tandukar M. and Pavlostathis S. G. (2015). Co-digestion of municipal sludge and external organic wastes for enhanced biogas production under realistic plant constraints. *Water Research*, **87**, 432–445. https://doi.org/10.1016/j.watres.2015.04.031

Tyagi V. K., Fdez-Güelfo L. A., Zhou Y., Álvarez-Gallego C. J., Garcia L. I. R. and Ng W. J. (2018). Anaerobic co-digestion of organic fraction of municipal solid waste (OFMSW): progress and challenges. *Renewable and Sustainable Energy Reviews*, **93**(April), 380–399. https://doi.org/10.1016/j.rser.2018.05.051

Vlyssides A. G. and Karlis P. K. (2004). Thermal-alkaline solubilization of waste activated sludge as a pretreatment stage for anaerobic digestion. *Bioresource Technology*, **91**(2), 201–206. https://doi.org/10.1016/S0960-8524(03)00176-7

Wolff H. J., Nickel K., Houy A., Lunden A. and Neis U. (2007). Two Years Experience on a Large German STP with Acoustic Disintegration of Waste Activated Sludge for Improved Anaerobic Digestion. 11th IWA World Congress on Anaerobic Digestion, Session PP9C-Biosolids, 23–27 September, Brisbane, Australia.

Xie S., Higgins M. J., Bustamante H., Galway B. and Nghiem L. D. (2018). Current status and perspectives on anaerobic co-digestion and associated downstream processes. *Environmental Science: Water Research and Technology*, **4**(11), 1759–1770. https://doi.org/10.1039/C8EW00356D

Yang Q., Wu B., Yao F., He L., Chen F., Ma Y., Shu X., Hou K., Wang D. and Li X. (2019). Biogas production from anaerobic co-digestion of waste activated sludge: co-substrates and influencing parameters. *Reviews in Environmental Science and Biotechnology*, **18**(4), 771–793. https://doi.org/10.1007/s11157-019-09515-y

Zhang D., Feng Y. M., Huang H. B., Khunjar W. and Wang Z. W. (2020). Recalcitrant dissolved organic nitrogen formation in thermal hydrolysis pretreatment of municipal sludge. *Environment International*, **138**, 105629. https://doi.org/10.1016/j.envint.2020.105629

Zhu X., Leininger A., Jassby D., Tsesmetzis N. and Ren Z. J. (2021). Will membranes break barriers on volatile fatty acid recovery from anaerobic digestion? *Environmental Science and Technology*, **1**(1), 141–153.

doi: 10.2166/9781789061796

Chapter 6

Carbon valorization using the microbial electrochemical technology platform

Jayesh M. Sonawane[1], *Zhiyong Jason Ren*[2] *and Deepak Pant*[2]*

[1]*Department of Chemistry, Alexandre-Vachon Pavilion, Laval University, 1045, avenue de la medicine, local 4064E, Quebec, Canada G1V 0A6*
[2]*Department of Civil and Environmental Engineering & Andlinger Center for Energy and the Environment, Princeton University, Princeton, NJ 08544, USA*
[3]*Separation & Conversion Technology, Flemish Institute for Technological Research (VITO), Boeretang 200, Mol 2400, Belgium*
**Correspondence: deepak.pant@vito.be*

6.1 INTRODUCTION

The world's economic growth is currently dependent upon resources obtained from non-renewable resources such as fossil fuels. These resources generate excessive greenhouse gases resulting in global climate change, leading to floods and droughts, rising sea levels, and more frequent natural disasters (Bhatia *et al.*, 2019; Lu *et al.*, 2018). The global CO_2 emission has increased to approximately 35.5 gigatons (Gt), and the capture and utilization of CO_2 have been expensive (MacDowell *et al.*, 2017). In the meantime, large amounts of organic carbon waste like municipal solid waste (2.01 Gt) and wastewater (around 1000 km^3) have also been a major environmental challenge with the high cost of disposal and treatment (Kaza *et al.*, 2018; Unesco, World Water Assessment Programme, 2012). Alternately, these carbon-abundant waste materials (solid, liquid, and gaseous) can be reused diligently. In that case, value-added energy and products can be generated to increase the value proposition of the process and transform the waste valorization industry. Recently, the US Department of Energy Bioenergy Technologies Office (BETO) reported that the US generates 50 million dry tons of organic waste streams from food waste, manure, oils, fats, greases, and wastewater sludge. Combining the carbon-containing gaseous waste streams, a total 2.6 quadrillion Btu of renewable energy can be recovered (DOE 2017). Different technologies, including biochemical, photochemical, electrochemical, and thermochemical processes, have been developed for carbon valorization, but they all have some advantages and specific challenges (DOE 2017).

Microbial electrochemical technology (MET) is a platform technology in which electroactive microorganisms are used to catalyze bioelectrochemical reactions to generate energy and products from waste carbon materials (Wang & Ren, 2013; Zou & He, 2018). In this process, oxidation and reduction reactions are separated in suitable environmental conditions for the first time. The main asset of the process is that the electrodes can be used as either electron acceptor (anode) or electron donor (cathode) (Jiang & Zeng, 2019; Pandey *et al.*, 2016). Compared to other environmental technologies with only one or two functions, the MET platform is very flexible and has discovered dozens of functions over the years. Almost all MET reactors share one common principle in the anode, in which biodegradable substrates, such as wastewater and food waste, are oxidized by microorganisms and generate electrical current (Wang & Ren, 2013). The current can be captured directly for electricity generation (microbial fuel cells, MFCs) or used to produce H_2 and other value-added chemicals (microbial electrolysis cells, MECs) (Logan, 2008). In addition, such electrons from organic waste carbon can also be used in the cathode chamber to reduce CO_2 and generate organic or inorganic compounds, achieving double benefits of carbon capture and valorization. Microbial electrosynthesis (MES) and microbial electrolytic carbon capture (MECC) are two popular processes in MET that can directly convert cathodic CO_2 and anodic organic waste into products (Lu *et al.*, 2015; Rabaey & Rozendal, 2010), while electrofermentation (EF) is another MET process that uses electro potential to regulate fermentation processes for different products (Nevin *et al.*, 2010).

6.2 THE PRINCIPLES OF MICROBIAL ELECTROCHEMICAL CARBON VALORIZATION

6.2.1 Biocatalytic CO_2 capture and conversion to organic chemicals in MES and EF

The illustration of the working principles of MES and EF is shown in Figure 6.1 (Jiang *et al.* 2019). In MES, electroactive microorganisms use a solid electrode (cathode) as the electron donor and CO_2 as the electron acceptor for electrosynthesis. The electron transfer mechanisms can be direct or mediated by H_2 or other redox agents. Though some electroactive microbes such as *Geobacter* and *Clostridium* have been reported to utilize direct extracellular electron transfer (DEET) via conductive nanowires or c-type cytochromes under environmental conditions (Lovley & Nevin, 2011), electron shuttles such as in situ generated H_2 or other small molecules such as formate and flavin are believed to play important roles as electron transfer mediators (Blanchet *et al.*, 2015).

In contrast, electrodes in EF are not the sole source of electrons; rather, it influences the flux of the self-driven fermentation by regulating the oxidation-reduction potential (ORP) and the NAD^+/NADH ratio (Moscoviz *et al.*, 2016). The extracellular ORP corresponds to the activity of the electrons present in the electrolyte, and the NAD^+/NADH ratio represents intracellular ORP, which controls gene expression and enzyme synthesis for overall metabolic activity. As a result, regulating the redox potential in the reactor can influence fermentation pathways and the product spectrum.

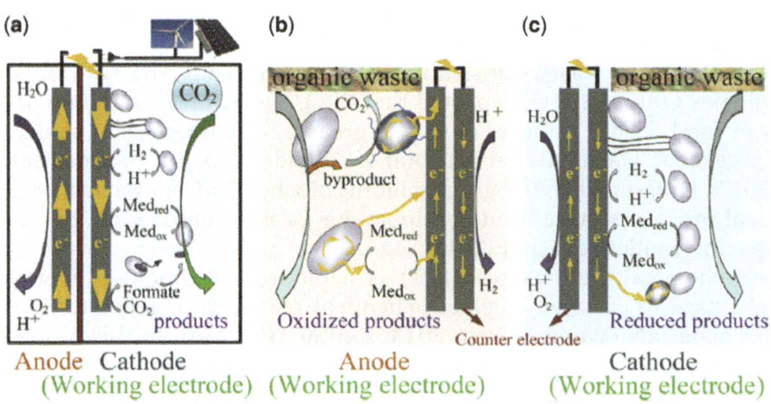

Figure 6.1 Schematic representation of the principle of the microbial ecosystem (MES) and electro-fermentation (EF): (a) The reduction of CO_2 by MES. (b) In anodic EF, the working electrode (anode) accepts electrons from microbes. (c) In cathodic EF, the working electrode (cathode) provides electrons to microbes. Figure reprinted from a previous study with permission from Jiang *et al.* (2019).

EF can be divided into anodic EF and cathodic EF. In an anodic EF, the electrode is an electron acceptor receiving electrons from microbial substrate oxidization. With an EF cathode, the working electrode is an electron donor, which microbes can use to synthesize different products depending on the redox potential. The electron transfer between the electrodes and microbes can be bidirectional. Still, compared with MES, syntrophic interactions between fermentative bacteria and electroactive bacteria were found to dominate and played essential roles in EF reactors, mainly due to the need for degradation of complex organics present in waste materials (Choi & Sang, 2016).

6.2.2 CO_2 capture and mineralization in MECC

MECC reactors share the same anodic reactions as other METs, in which waste organic carbons are oxidized by microorganisms to achieve wastewater treatment and electron extraction (Lu *et al.*, 2015; Zhu & Logan, 2014). Electrons are then accepted by the anode and transferred through an external circuit to the cathode, where they reduce water to produce H_2 and OH^-. Such operation generates a pH gradient between the anode and cathode, as the anode becomes more acidic due to the accumulation of H^+, while the cathode becomes more alkaline due to the accumulation of OH^-. Different reactor designs have been reported to utilize such pH discrepancy for silicate mineral (e.g. wollastonite $CaSiO_3$ or coal fly ash) dissolution that liberates metal ions (Na^+, Ca^{2+}, Mg^{2+}, etc.) in the anode or a separate acid chamber (Lu *et al.*, 2016; Zhu *et al.*, 2014). The metal ions then react with OH^- in the cathode chamber to form the metal hydroxide, whose subsequent reaction with CO_2 leads to spontaneous CO_2 capture and transformation into stable carbonate or bicarbonate products. Figure 6.2 illustrates two reactor designs of MECC; both achieved good CO_2 capture efficiency with concurrent H_2 production and wastewater treatment capability (Huang *et al.*, 2016). The stable metal carbonate and bicarbonate are harvested and used on-site as alkalinity compensation for nitrification and digestion, for improving activated sludge settling properties, or for environmental uses (Sherrard, 1976; Wett *et al.*, 2004).

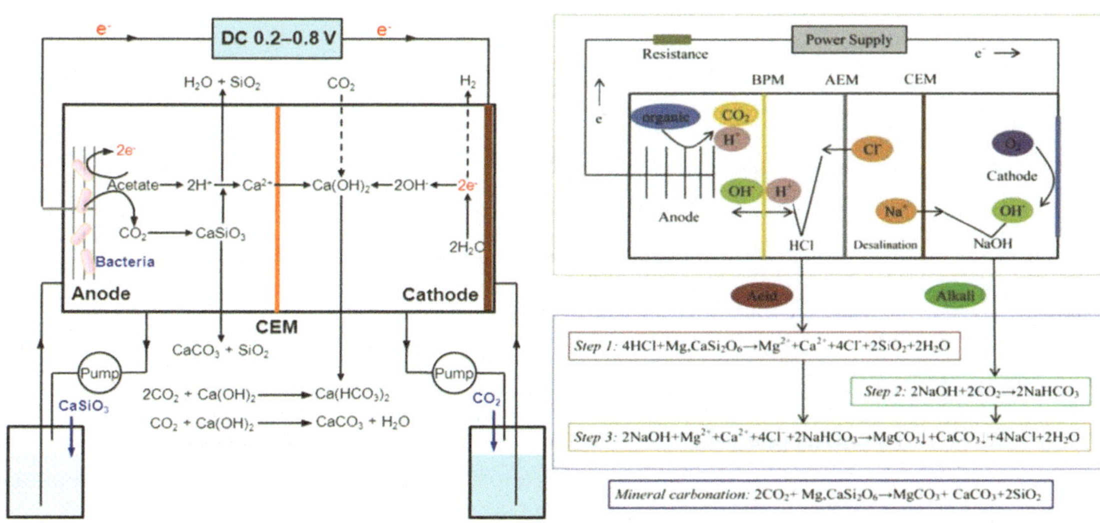

Figure 6.2 Schematic representation of the principle of microbial electrolytic carbon capture (MECC) systems: (a) An integrated system with in situ CO_2 capture and mineralization. (b) A system that separates acid and alkali generations and reactions for carbonization (Lu *et al.* 2015; Zhu & Logan, 2014).

6.3 VALORIZATION OF CARBON COMPOUNDS BY MES

Microbial electrosynthesis utilizes low-cost electron sources from wastewater organics to capture and convert CO_2 into value-added organic products. The MES process generally uses natural bacterial consortia, which has better potential in complex environmental applications than pure culture strain or abiotic electrocatalysis. Additionally, such self-sustaining biocatalysts have high selectivity and yield (>10%) compared with plants and photosynthetic microorganisms (<1–3%) in solar-to-product conversion (Blankenship *et al.*, 2011). It does not require high temperature or pressure, so it demonstrated good potential in CO_2 reduction using renewable electrons, leading to a circular bioeconomy (Bian *et al.*, 2020).

Since the first studies reported the feasibility of producing organic compounds using CO_2 and electrons from electrodes by microbes (Cheng *et al.*, 2009; Nevin *et al.*, 2010), excellent progress had been made in understanding the microbial electron transfer, developing scalable reactors, testing different microbial strains and consortia, and assessing the economic feasibility and environmental impacts for MES carbon valorization. Wastewater has been considered as a major source of feedstock for renewable electrons that couples with CO_2 reduction. Because methanogens and acetogens have been found to be dominant in MES reactors, CH_4 and acetic acid are the primary products of CO_2 reduction. However, many other organic compounds with higher carbon numbers were reported as well. The Wood-Ljungdahl Pathway is known as the primary metabolism in autoelectrophic bacteria, with acetyl-coenzyme A (acetyl-CoA) serving as a key intermediate to produce various organic compounds from CO_2, including formic acid, propionic acid, butyric acid, 2-oxobutyrate, ethanol, isopropanol, butanol, and isobutanol. Figure 6.3 shows the improvement of product titer and production rates of acetic and butyric acids over the years, and it can be found that the highest concentrations of acetic acid and butyric acid were at ~12 and ~3 g/L, respectively, and no significant improvement has been made in the past few years. In contrast, slow improvement was made on the production rate, with the highest rate achieved at around 700 g/m²/d for acetic acid (Prévoteau *et al.*, 2020). Key reasons for lower titer are known to be associated with toxicity exerted by the fermentation products, slow growth of electroactive bacteria, and unoptimized reactor systems (Gildemyn *et al.*, 2015).

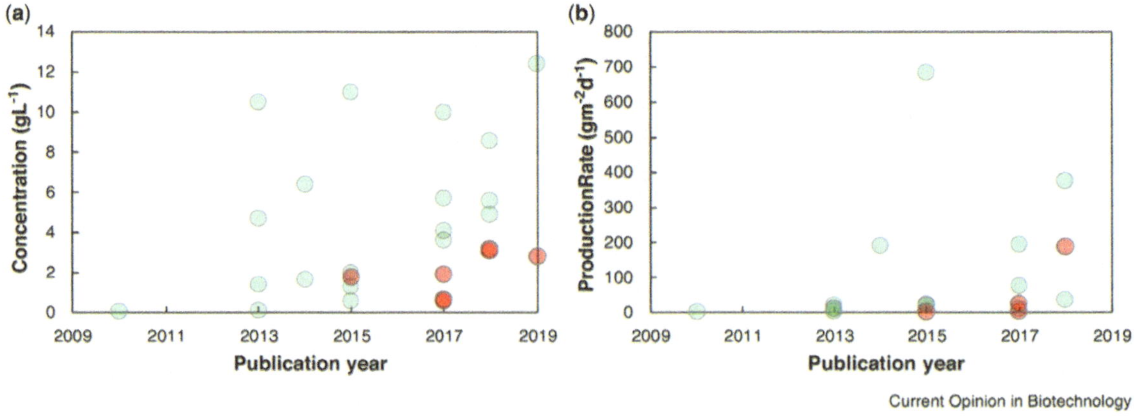

Figure 6.3 Historical evolution of acetic and butyric acid production by MES from CO_2: (a) Maximum concentrations achieved in catholyte. (b) production rates with respect to cathode projected surface area. Green circles (acetic acid), red circles (butyric acid) (Prévoteau *et al.* 2020).

6.3.1 Methane or acetic acid production in MES

Among all conversions, electromethanogenesis was among the earliest CO_2 capture functions discovered in MES, and it has shown good potential in wastewater applications (Cheng *et al.*, 2009). Methane can be produced via CO_2 reduction with either direct electron transfer from the cathode (Equation (6.1)) or indirect electron transfer via an intermediate such as H_2 (Equations (6.2) and (6.3)) (Mateos *et al.*, 2020; Nelabhotla *et al.*, 2021):

Direct electron transfer vis electrode:

$$CO_2 + 8H^+ + 8e^- \rightarrow CH_4 + 2H_2O \tag{6.1}$$

Indirect electron transfer via H_2:

$$2H^+ + 2e^- \rightarrow H_2 \tag{6.2}$$

$$CO_2 + 4H_2 \rightarrow CH_4 + 2H_2O \tag{6.3}$$

A potential application of MES is the combination with anaerobic digestion (AD), because MES can enhance the overall organic removal while in the meantime purifying the biogas generated by AD by converting CO_2 into CH_4 and therefore also increasing the overall CH_4 production yield and efficiency. Studies showed that the gas generated with such a combination had a less than 10% CO_2 content. Figure 6.4 describes an assortment of food waste exhibited by solid lines and a wastewater treatment plant demonstrated by dashed lines along with the employment of a joined AD-MES unit. The rejected water in the wastewater plant contains a COD concentration ranging from 1000 to 8000 mg L^{-1}. The rejected water was recycled to reduce total solids in the inlet feed in the plants for treating food waste. This AD-MES plant could reduce ammonium, sulfide, and COD concentrations from the rejected water. In addition, CO_2 was reduced to CH_4 electrochemically by using optimal cathode potentials and pH. This was done by decreasing the pH of the rejected mass, which enabled the dissolution of the CO_2 present in the biogas. Also, MES had to be designed so that the dissolved CO_2 could efficiently react with the electrons liberated from the surface of the cathode (Nelabhotla & Dinamarca, 2019).

Acetic acid is another popular product from MES del Pilar Anzola Rojas *et al.* (2021). Acetic acid can be produced electrochemically under biologically relevant conditions at −0.28 V (Equation (6.4)), which is slightly lower than the CH_4 generation potential (−0.24 V) (Rabaey & Rozendal, 2010):

$$2HCO_3^- + 9H^+ + 8e^- \rightarrow CH_3COO^- + 4H_2O \tag{6.4}$$

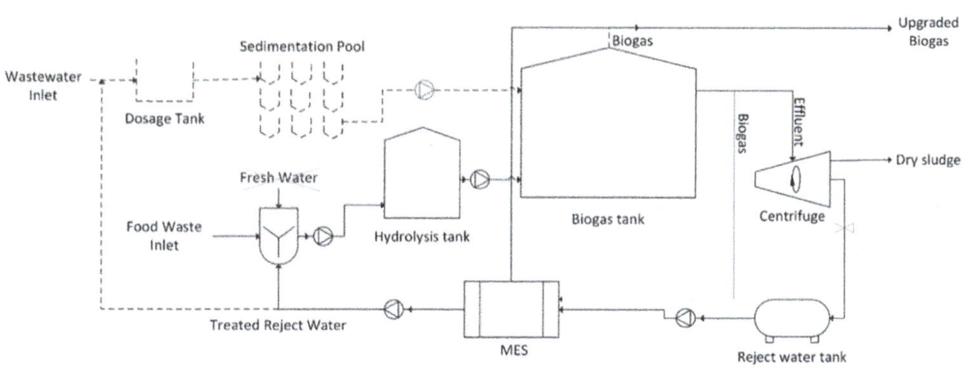

Figure 6.4 AD-MES integrated treatment plant setup (figure reprinted from Nelabhotla & Dinamarca (2019)).

Table 6.1 Conversion of CO_2 to acetic acid using different microbial strains.

Microorganism	Current Density (A m^{-2})	Coulombic Efficiency (%)	Production of Acetate (mM day^{-1})	References
Sporomusa ovata	−0.63	82	1.13	Nie *et al.* (2013)
	−0.48	86	NA	Zhang *et al.* (2013)
	−1.7	89	4.68	Giddings *et al.* (2015)
	−2.1	85	0.17	Nevin *et al.* (2010)
Clostridium ljungdahlii	NA	88	0.013	Nevin *et al.* (2011)
Enriched mixed culture	10	35.46–88	0.12–3.96	Bajracharya *et al.* (2016)
	−44.89	30.25	3.06	Mikkelsen *et al.* (2010)
	10	16.2–49.4	1.3–6.3	Bajracharya *et al.* (2017)
	−200	84	NA	Jourdin *et al.* (2016a), Jourdin *et al.* (2016b)
	−128	29.91	NA	Mohanakrishna *et al.* (2015)
	10	40–50	1.3	Bajracharya *et al.* (2015)
	NA	61	11.67	Gildemyn *et al.* (2015)

The production rate of acetic acid (around 685 gm^{-2} day^{-1}) from CO_2 using MES technology was recently achieved using a newly manufactured electrode and an adaptable microbial culture (Jourdin *et al.*, 2015). A 3D electrophoretic deposition electrode was used as a biocathode, and a multiwalled carbon nanotubes layer was deposited onto reticulated vitreous carbon. A slightly acidic pH (~5.8) increased the rate of the formation of acetate, while the effect of current was carried out separately (Batlle-Vilanova *et al.*, 2015). However, after the biofilm's growth of a particular thickness, a decrease in bacterial growth was observed. The conversion efficiencies of acetate from carbon dioxide and electrons were steady and superior for a mixed culture system, with an average of 98±4 and 100±1%, respectively. A high production rate of the compounds also depends on other factors like the hydraulic retention time. Studies also showed the production rate could be enhanced by increasing the cell voltage or altering different parameters like membrane and electrode resistance, concentrations, pH, and variation in the anode potential (Blanchet *et al.*, 2015). Some found there was a loss of biomass in the continuous mode attributed to the participation of both suspended (planktonic) and biofilm bacteria in the reduction process of CO_2. Moreover, the loss of planktonic bacteria also reduced the production in the reactor operating in continuous mode, which was solved by substituting biofilm-forming microorganisms in place of planktonic bacteria. CO_2 was made available to acetogenic bacteria by continuously passing CO_2 through the culture medium or using bicarbonate as feed. In terms of inoculum, many studies used anaerobic sludge, and some observed a prominent shift toward specific microbial families such as *Clostridiaceae* and *Pseudomonadaceae* on the electrodes' surface (Saratale *et al.*, 2017). Other dominant microbes reported in literature included *Sulfurospirillum, Sporomusa, Clostridium, Tissierella, Arcobacter, Ochrobactrum, Pseudomonas, Sacharolyticum*, and *Desulfovibrio* (Zaybak *et al.*, 2013). Table 6.1 summarizes typical CO_2 to acetic acid MES parameters and identified microbial cultures.

6.3.2 Role of hydrogen in MES

Two mechanisms were reported on MES inward electron transfer to microbial cells. Some electroactive microbes such as *Geobacter* were found produce conductive filaments and/or c-type cytochromes to directly acquire electrons from the electrode at certain redox potentials (Lovley, 2011). However, many studies reported that in situ generated H_2 was the direct electron donor for CO_2 reduction with a higher organic conversion rate, though the exact mechanisms were influenced by the cathode working potential (Blanchet *et al.*, 2015). For example, one study evaluated the function of hydrogen in electron transfer on the methanogenic biofilm formed on the MES cathode. By using a microsensor to detect the in-situ hydrogen generation in conjunction with cyclic voltammetry, Cai *et al.* (2020) analyzed the hydrogen evolution dynamic and confirmed the presence of hydrogen-associated electron transfer near the cathode within a micrometer scale, and they observed colocalized community of archaea and bacteria developed within a 58.10-μm-thick biofilm was correlated with the hydrogen gradient detected by the microsensor.

Hydrogen is increasingly produced via electrolysis powered by low-cost green electricity generated from MET or other renewable sources, and hydrogen serves as a reductant for different chemotropic microbes (Jack *et al.*, 2021). These microbes consist of reversible hydrogenase enzymes that oxidizes molecular hydrogen for CO_2 reduction to organic compounds such as methane, acetic acid, or butyric acid (Ganigué *et al.*, 2015). Besides electrochemically produced hydrogen at the electrodes, hydrogen can also be produced microbially via fermentation, especially in reactors working with mixed microbial communities. Such a source of hydrogen allows the MES to survive periods when no electric power is supplied to the system, thus making intermittent operation possible (del Pilar Anzola Rojas *et al.*, 2018).

Hydrogen-mediated electron transfer has been identified as an important extracellular pathway of sharing reducing equivalents to regulate biofilm activities in MESs and demonstrated higher reactor performance than the direct electron transfer pathway. Direct electron transfer could only provide low current density, but electrocatalytic hydrogen production is tunable. Hydrogen supply from the electrode can be increased with the increase in current densities, but a high rate of hydrogen production may not necessarily lead to high CO_2 conversion by microbes due to their slow metabolism rates compared to the abiotic hydrogen evolution. Therefore, a balance needs to be maintained between hydrogen supply and consumption in MES reactors. Studies reported that biofilm could be eliminated from the electrodes in conditions where vigorous hydrogen evolution occurs at the electrode. Also, though the alkaline condition is desired for the water electrolysis, in MES cathode, the pH needs to be maintained near neutral for biological reactions. As a result, a hybrid MES system can be fabricated by attaching a microbial gas-liquid contactor towards the downstream of the water electrolysis cell.

6.3.3 CO_2 valorization potential from the MES platform

The MES process offers a promising pathway for concurrent CO_2 valorization and wastewater treatment. Though pilot and full-scale systems are still to be tested, preliminary studies evaluated the economic potentials and environmental benefits of producing different chemicals using the MES platform. Figure 6.5(a) and (b) shows that products produced in large volumes, such as methane and acetic acid, have much higher total product value and market size (Jiang *et al.*, 2019). Still, products with higher unit values, such as 2-oxobutyrate, may have higher profit margins due to their higher unit value. In addition, small molecules can be used as a precursor for synthesizing higher valued chemicals via chain elongation or synthesis. In addition to economic benefits, the CO_2 capture potential is shown in Figure 6.5(c) and (d). Based on how many moles of CO_2 will be used to produce unit moles of the different products along with the maximum production rate, it can be seen that products with higher carbon numbers have a higher conversion ratio. The overall CO_2 conversion potential largely depends on the chemical production rate and world market size. Acetate (nearly 10 tonne m^{-3} $year^{-1}$) has a much higher market potential than other chemicals due to a good combination of carbon number and conversion rate. Such analysis provides good insights on the CO_2 capture potential that may lead to additional carbon credit.

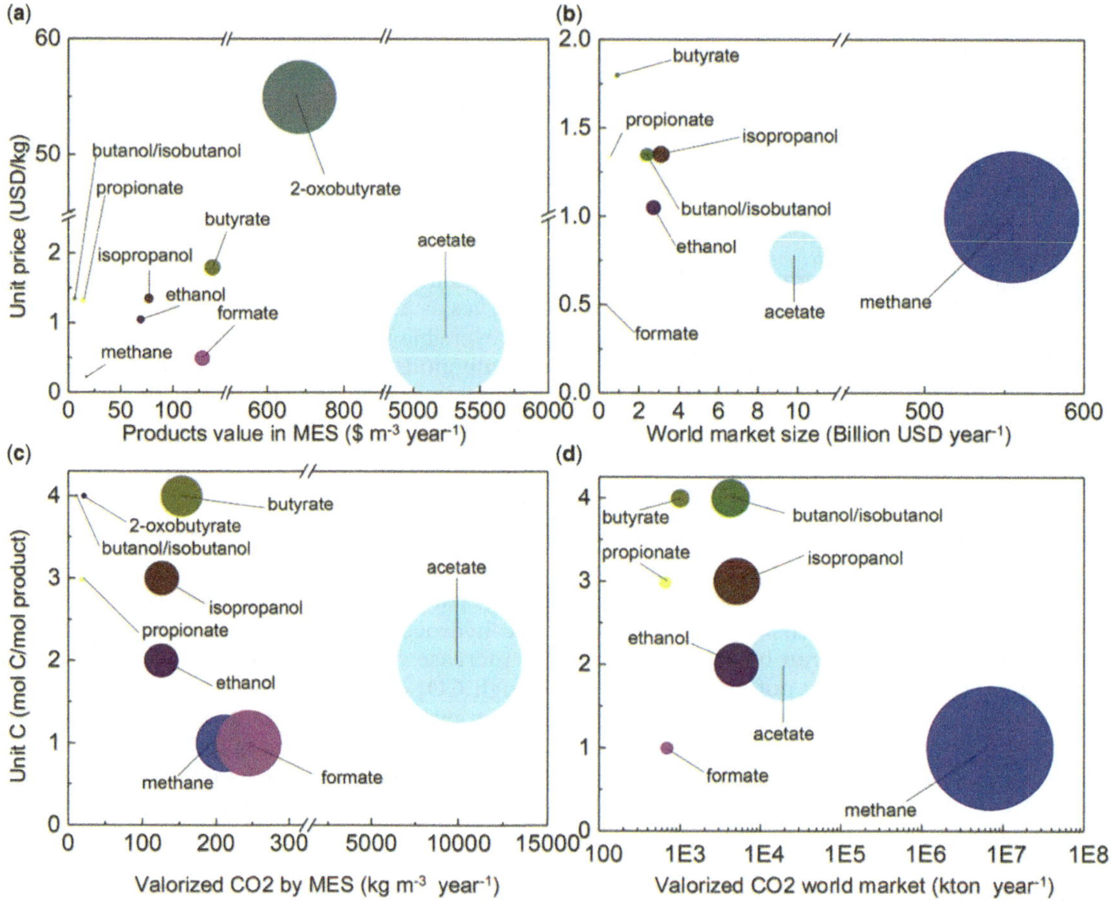

Figure 6.5 Preliminary analysis of the CO_2 valorization potential by MES in terms of economics and carbon utilization. For each circle, the Y-axis value determines the location of the center, while the X-axis value determines the radius. The radius range has no meaning for Y-axis. (a) The product value of different compounds generated in MES reactors, which was calculated by multiplying the unit price with the maximum production rate. (b) The world market size of each produced compound versus their unit price, respectively. (c) CO_2 conversion potential via different products, which was calculated by multiplying the unit conversion ratio with the maximum production rate of each compound. (d) The world market size of CO_2 conversion based on the production of each compound in MES (figure reprinted from Jiang *et al.* (2019).

6.4 VALORIZATION OF CARBON COMPOUNDS BY ELECTROFERMENTATION

Electro-fermentation (EF) is a process that uses electrochemistry to influence microbial metabolism, but different from MES, its main purpose is to regulate fermentation pathways to valorize organic waste carbon to higher-value products. In anodic electro-fermentation, the working electrode (WE) behaves as an anode and accepts electrons generated from organic waste to form oxidized final products. In contrast, the working electrode supplies electrons in cathodic electro-fermentation, behaving as a cathode to furnish a reduced product. The NAD+/NADH ratio represents intracellular ORP because of intracellular redox homeostasis, controlling gene expression and enzyme synthesis for overall metabolic activity. As a result, if the redox potential is artificially tuned, the fermentation pathways can be regulated to generate different ratios of products (Jiang *et al.*, 2018).

6.4.1 Mechanisms of anodic and cathodic EFs

Anodic EF uses the electrode as an electron sink to improve product selectivity and production rate. Theoretically, any biodegradable substrate can be used in the system, and the product profile can vary significantly depending on the feedstock, inocula, and operational condition. Anodic EF can be exploited when there is a need for product upgrade. It has been proposed to use an electrode to replace other electron acceptors such as oxygen for targeted product generation and energy saving. However, since EF is essentially still a fermentation process, the products have been predominantly H_2, alcohol, and volatile fatty acids. With genetically modified strains or mixed cultures, other products such as acetoin, lysine, poly-β-hydroxybutyrate (PHB), and 2-Keto-gluconate have been reported as well (Jiang *et al.*, 2019; Nikhil *et al.*, 2015). For example, H_2 production could be enhanced in anodic EF by regulating the anode potential and the concentration of the autocrine mediators, though the balance between H_2 generation and subsequent consumption by methanogens needs to be carefully balanced (Chandrasekhar *et al.*, 2014). Ethanol could be produced from glycerol using engineered bacterial strains such as *Shewanella oneidensis*, and the titer could be higher than mixed culture fermentation (Russell *et al.*, 2015). Higher value products such as PHBs could also be formed in such systems by the activity of *Ralstonia eutropha* at 0.6 V (Nishio *et al.*, 2013).

In cathodic EF, electrodes are used as direct or indirect electron sources for oxidation reactions to regulate product selectivity and production rate. This leads to an increase in the intracellular NADH content and generate reduced end-product. For example, cathode EF was used to stimulate 1,3-propandiol (1,3-PDO) production from glycerol when a working potential of +0.045 V was applied on *Clostridium pasteurianum* (Choi *et al.*, 2014). A similar product could also be achieved by using a mixed culture, and the applied potential was considered a determinant factor in shaping microbial community and, therefore, metabolite distribution. For instance, by reducing the starting working potential from −0.8 to −1.1 V, the dominant community shifted from *Veillonellaceae* (56–72%) to *Clostridiaceae* (55–57%), which was accompanied by a product shift from propionate to 1,3-PDO (Xafenias *et al.*, 2015; Zhou *et al.*, 2013).

6.4.2 Synergy between EF and anaerobic digestion

Electro-fermentation carries a good potential in improving the value proposition or anaerobic fermentation and digestion. Considering AD is a closed system without external inputs of electron acceptors or energy sources, it tends to reach a thermodynamic equilibrium with the products (CH_4) having the lowest Gibbs energy change per electron than any other organic compound during biological conversion. EF in this case offers a new approach to regulate fermentation pathways by using external electrodes as an alternative source or sink of electrons to control the redox potential of pure or mixed culture systems, alter the electron transfer process, and therefore shape the microbial community and activity to increase the yield and rate of desired products.

If biogas is the targeted product, EF or electrodes can increase the stability of the AD process and accelerate methanogenesis (Figure 6.6). The AD process corresponds to a cascade of oxidation and reduction reactions carried out by a consortia of microorganisms. Interspecies hydrogen transfer (IHT) has been known to play a critical role in connecting organic substrate degradation and methanogenesis. In EF systems for methane production, however, interspecies electron transfer (IET) was observed by many studies. Rather than relying on H_2-mediated electron transfer, electrodes directly facilitate the IET process between syntrophic microorganisms. Fermentation bacteria, methanogens, and electroactive microorganisms constitute the primary syntrophic communities. Recently, studies found that the abundance of hydrogenotrophic methanogens in EF increased 17 times compared to regular AD, and the composition of acetotrophic methanogens remained almost

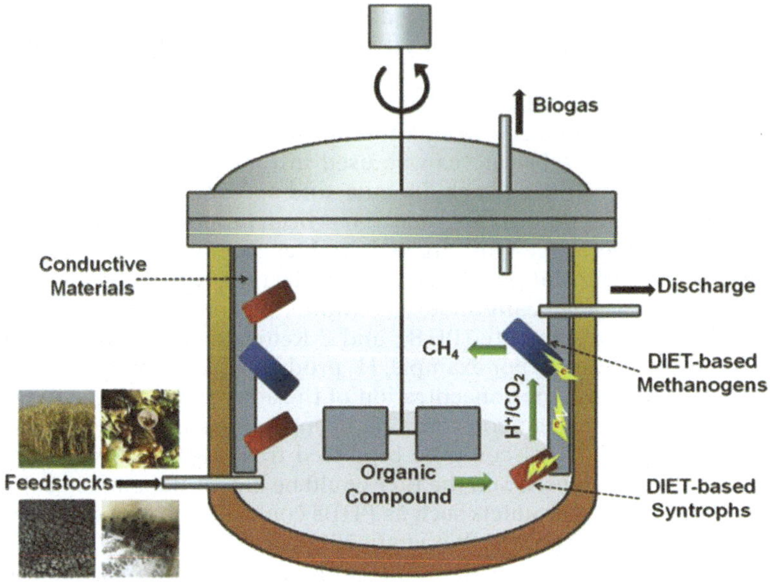

Figure 6.6 Schematic diagram of an anaerobic digestion/fermentation system incorporated with electrodes to enable electro-fermentation (figure reprinted from Zhao *et al.* (2020)).

unchanged (Gajaraj *et al.*, 2017). It was hypothesized that the reduction of CO_2 to CH_4 becomes a major pathway of methanogenesis, in which the electrons are supplied by electroactive bacteria such as *Geobacter* degrading organic acids or collecting electrons from the cathodes. Moreover, IHT can also be enhanced due to the increase in hydrogen generation on the cathodes and subsequent utilization by hydrogenotrophic methanogens (Villano *et al.*, 2017). For example, Liu *et al.* (2019) found that by introducing carbon brush electrodes into anaerobic digestion, VFA concentration drops faster than regular AD control, indicating an accelerated stabilization. Moreover, methane production was increased by 26.3% when a low voltage (0.8 V) was applied, and the content of methane in the headspace also increased by nearly 30%. Community analysis showed the electric current stimulated the growth of hydrogenotrophic methanogens, and *Geobacter* occurred at the cathode with a low abundance. However, acetotrophic *Methanosaeta* still made up a high portion of the archaeal community.

While AD produces renewable biogas, it faces challenges on economic viability and environmental concerns due to the low value of biogas and concerns about its greenhouse gas effects. Recent developments on arrested methanogenesis allow the AD process to be rewired to suppress methanogenesis and promote the production of short chain VFAs and alcohols, because such products not only bring up the values by themselves, they are also chemical precursors for the production of higher-valued chemicals such as PHBs, biofuels, medium chain fatty acids (MCFAs), and single cell protein (SCP) (Zhu *et al.*, 2021). By controlling the redox potential in the EF reactor, the fermentation pathways can be influenced and subsequently regulate the product spectrum. Recent studies confirmed that the electrochemical potential control regulated the product distribution in anaerobic fermentation using natural microbial consortia. Jiang *et al.* (2019) characterized the product spectrum under different working potentials of −1.0, −0.6, and −0.2 V (vs. Ag/AgCl), which spans the electron flow direction from cathodic current to anodic current. It was found when a working potential of −0.2 V was applied; the electrode potential was more positive than the open circuit potential (−0.55 V); therefore, anodic electro-fermentation reactions occurred with electrons flowing toward the electrode. In contrast, when the applied potentials were more negative than the open circuit potential (−0.6 and −1.0 V), cathodic EF conditions were created, where the working electrode became an electron source. Results showed that more negative potential led to higher CH_4 accumulation, while more positive potential showed inhibited methanogenesis activity. For example, increasing the potential from −1.0 to −0.2 V greatly reduced methanogenesis by 68% and acetic acid generation by 58% in neutral pH. Butyric acid production increased by 25%, while propionic acid concentrations remained stable. Laboratory studies showed a range of 61–78% in carbon recovery and 70–87% in electron recovery could be obtained by comparing final fermentation products and the substrate, and the spectrum of each product was tuned based on the difference of working potential. Since EF is regarded as an electrochemically-influenced spontaneous fermentation, the contribution of electron consumption or donation by the working electrode to the electron balances was limited, and energy consumption was low.

6.5 CO_2 MINERALIZATION IN MECC

Different from MES and EF, MECC uses the carbonate chemistry on the cathode to convert CO_2 into carbonate or biocarbonate salts. Most MECC studies have been performed at lab scale to date, but several companies are working on scaling up the systems. The efficiencies and throughputs of MECCs are high compared to biological CO_2 reductions. For example, an in-situ study reported that by using industrial wastewater as the electrolyte, up to 93% of total CO_2 was captured and converted to carbonate salts (Lu *et al.*, 2015). This included the CO_2 derived from organic oxidation in the wastewater and the extergenous CO_2 introduced into the system. This means the MECC can become carbon negative and capture additional CO_2 than it produces. The organic removal was in the range of 56–100%, and a net energy gain of −2 kJ mol^{-1} of CO_2 captured was reported due to high-rate H_2 production on the cathode chamber. Different processes have been tested, including using industrial CO_2 sources (5–15%) and ambient CO_2 from air when combined with an ion exchange resin for pre-concentration. Another study used separate chambers to collect acid and base solutions in two chambers. The acid was used to dissolute silicate minerals for cation sources, and the base solution was used to capture CO_2. The acid and alkali were produced in lab reactors at production efficiencies of 35 and 86%, respectively. Approximately 44% of the absorbed CO_2 was fixed as magnesium or calcium carbonates (Zhu *et al.*, 2014).

The MECC process can be used in municipal wastewater and industrial wastewater, especially those with higher salinity. Unlike traditional microbial electrochemical processes, whose performances are limited by the low conductivity and alkalinity in municipal wastewater, the MECC increases the conductivity and alkalinity during the dissolution of silicate and CO_2, so it does not need alkalinity amendment. Additionally, the process benefits carbonate precipitation as it reduces TDS buildup in the effluent. One ideal example for MECC application is the city of Hong Kong, which uses more than 270 million m^3 of seawater for toilet flushing for six million people, which resulted in a high salinity wastewater discharge. More than 21 cities are considering similar practices to address freshwater shortage (Lee *et al.*, 2015). Other saline wastewaters, such as oil and gas produced water and coal-fired power plant wastewater, can also be ideal entry markets, and they have been tested to have high efficiency due to high conductivity and buffer capacity. High concentrations of metal cations, such as Ca^{2+}, Mg^{2+}, Na^+, Sr^{2+}, and Ba^{2+} present in these wastewaters facilitate carbonate precipitation. Plus, coal fly ash could be used as a cation source to replace silicate and achieve multi-stream waste management for fly ash carbonation, CO_2 capture and mineralization, brine wastewater treatment (Lu *et al.*, 2016).

The MECC process captures the CO_2 generated during the wastewater treatment, and it captures more CO_2 from other sources such as flue gas or even potentially from the air. This brings significant co-benefits for both water resource recovery facilities (WRRFs) and nearby CO_2-emitting industries such as power plants, cement plants, and refineries (Lu *et al.*, 2015, 2018) (Figure 6.7). Interestingly,

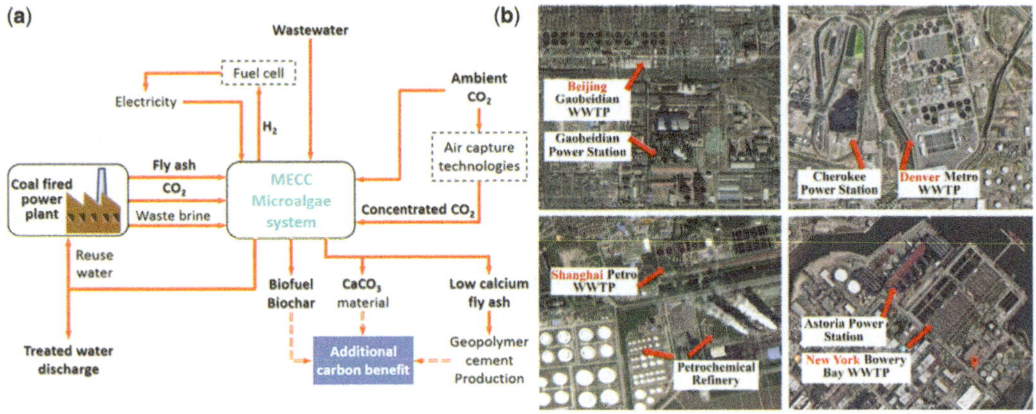

Figure 6.7 (a) schematic of the mutual benefits between an MECC equipped water resource recovery facility (WRRF) and a CO_2 point emission source. (b) example co-locations of CO_2 point sources and WRRFs enable complementary CCU in Beijing, Shanghai, New York city, and Denver (Lu *et al.*, 2015, 2018).

many of these facilities are co-located with or near major WRRFs. The wastewater facility can help capture and sequester the CO_2 emitted from the nearby point source and generate carbon credits, and the emitter can save costs by avoiding the use of expensive and energy-intensive CCS systems yet still meet EPA mandates on carbon pollution reduction. The calcium/magnesium-abundant fly ash generated by the power plant may be used as a silicate supplement to facilitate CO_2 mineralization. For WRRFs that do not have a nearby point source, MECC can help capture the CO_2 from either aerobic or anaerobic treatment processes. Even air capture is feasible when the process is combined with a pre-concentration process by commercially available ion-exchange resins (Huang *et al.*, 2016).

Studies also performed preliminary economic analysis on MECC systems and found that the net cost for mitigating one ton of CO_2 could be \$48/ton (Lu *et al.*, 2015), which was calculated based on a combination of CO_2 capture cost (capital plus operation cost), potential cost offsets (revenue of H_2 and wastewater treatment) and avoided CO_2 emission through reduced fossil fuel consumption for wastewater treatment and commercial H_2 production (nature gas reforming). This net cost is well below the \$70–270/t- CO_2 estimated for coal power plant with geological storage CCS, and also below the cost for direct air CO_2 capture using chemical/thermal methods (on the order of \$1000/t- CO_2) or abiotic electrolytic dissolution of silicate (\$86/t- CO_2) (Cornils, 2020; House *et al.*, 2011; Rau *et al.*, 2013). It should be noted that though MECC has potential for significant energy savings and carbon benefits for the wastewater industry, further work is needed to better understand the technology barriers and to optimize system designs, operational protocols, and applications.

6.6 OUTLOOK

Microbial electrochemical technology provides a versatile platform for simultaneous waste treatment, resource recovery, and CO_2 capture and utilization. While this technology has yet been demonstrated in full scale, this chapter aims to summarizing different processes and opportunities for water and wastewater treatment. We recognize it is challenging to meet and balance multiple objectives and fulfill different treatment needs while in the meantime capturing CO_2 and recovering resources, so this chapter offers several feasible approaches to make such an operation a reality. Rather than building new and separated systems, current reactors such as aeration tanks and anaerobic digestors can be upgraded by installing electrodes to accomplish multi-function and process intensification (ElMekawy *et al.*, 2016). MECC, MES, and EF are based on different mechanisms and generate different products, so they can be retrofitted into different systems with tailored purposes. For example, EF can be used to enhance biogas production, MES can be used to generate VFAs, while MECC will help improve the alkalinity of the wastewater.

It is possible to realize the carbon-negative, revenue-positive wastewater treatment. Still, technological development and implementation as well as more detailed techno-economic, life cycle, and socioeconomic analyses, are required to understand the potential of these technologies. Chapter 1 presented a hypothetical example of MECC plus microalgae to replace traditional anaerobic-anoxic-aerobic activated sludge system, and it shows such conversion can potentially transform wastewater treatment to carbon capture and valorization facilities with a positive revenue flow (Lu *et al.*, 2018).

REFERENCES

Bajracharya S., Ter Heijne A., Dominguez Benetton X., Vanbroekhoven K., Buisman C. J. N., Strik D. P. B. T. B. and Pant D. (2015). Carbon dioxide reduction by mixed and pure cultures in microbial electrosynthesis using an assembly of graphite felt and stainless steel as a cathode. *Bioresource Technology*, **195**, 14–24. https://doi.org/10.1016/j.biortech.2015.05.081

Bajracharya S., Vanbroekhoven K., Buisman C. J. N., Pant D. and Strik D. P. B. T. B. (2016). Application of gas diffusion biocathode in microbial electrosynthesis from carbon dioxide. *Environmental Science and Pollution Research*, **23**, 22292–22308. https://doi.org/10.1007/s11356-016-7196-x

Bajracharya S., Vanbroekhoven K. and Buisman C. J. N. (2017). Bioelectrochemical conversion of CO_2 to chemicals: CO_2 as a next generation feedstock for electricity-driven bioproduction in batch and continuous modes. *Faraday Discussions*, **202**, 433–449. https://doi.org/10.1039/C7FD00050B

Batlle-Vilanova P., Puig S., Gonzalez-Olmos R., Vilajeliu-Pons A., Balaguer M. D. and Colprim J. (2015). Deciphering the electron transfer mechanisms for biogas upgrading to biomethane within a mixed culture biocathode. *RSC Advances*, **5**, 52243–52251. https://doi.org/10.1039/c5ra09039c

Bhatia S. K., Bhatia R. K., Jeon J. M., Kumar G. and Yang Y. H. (2019). Carbon dioxide capture and bioenergy production using biological system – a review. *Renewable and Sustainable Energy Reviews*, **110**, 143–158. https://doi.org/10.1016/j.rser.2019.04.070

Bian B., Bajracharya S., Xu J., Pant D. and Saikaly P. E. (2020). Microbial electrosynthesis from CO_2: challenges, opportunities and perspectives in the context of circular bioeconomy. *Bioresource Technology*, **302**, 122863. https://doi.org/10.1016/j.biortech.2020.122863

Blanchet E. M., Duquenne F., Rafrafi Y., Etcheverry L., Erable B. and Bergel A. (2015). Importance of the hydrogen route in up-scaling electrosynthesis for microbial CO_2 reduction. *Energy and Environmental Science*, **8**, 3731–3744. https://doi.org/10.1039/c5ee03088a

Blankenship R. E., Tiede D. M., Barber J., Brudvig G. W., Fleming G., Ghirardi M., Gunner M. R., Junge W., Kramer D. M., Melis A., Moore T. A., Moser C. C., Nocera D. G., Nozik A. J., Ort D. R., Parson W. W., Prince R. C. and Sayre R. T. (2011). Comparing photosynthetic and photovoltaic efficiencies and recognizing the potential for improvement. *Science*, **332**, 805–809. https://doi.org/10.1126/science.1200165

Chandrasekhar K., Amulya K. and Venkata Mohan S. (2014). Solid phase bio-electrofermentation of food waste to harvest value-added products associated with waste remediation. *Waste Management*, **45**, 57–65. https://doi.org/10.1016/j.wasman.2015.06.001

Cai W., Liu W., Wang B., Yao H., Guadie A. and Wang A. (2020). Semiquantitative detection of hydrogen-associated or hydrogen-free electron transfer within methanogenic biofilm of microbial electrosynthesis. *Applied Environment Microbiology*, **86**, 1–29. https://doi.org/10.1128/aem.01056-20.

Cheng S., Xing D., Call D. F. and Logan B. E. (2009). Direct biological conversion of electrical current into methane by electromethanogenesis. *Environmental Science and Technology*, **43**, 3953–3958, https://doi.org/10.1021/es803531g

Choi O. and Sang B. I. (2016). Extracellular electron transfer from cathode to microbes: application for biofuel production. *Biotechnology for Biofuels*, **9**, 1–14. https://doi.org/10.1186/s13068-016-0426-0

Choi O., Kim T., Woo H. M. and Um Y. (2014). Electricity-driven metabolic shift through direct electron uptake by electroactive heterotroph *Clostridium pasteurianum*. *Scientific Reports* **4**, 6961. https://doi.org/10.1038/srep06961

Cornils B., Herrmann W. A., Xu J. H. and Zanthoff H. W. (2020). Carbon dioxide capture and storage. Catalysis From A to Z:A Concise Encyclopedia. John Wiley & Sons. https://doi.org/10.1002/9783527809080.cataz02826

del Pilar Anzola Rojas M., Zaiat M., Gonzalez E. R., de Wever H. and Pant D. (2018). Effect of the electric supply interruption on a microbial electrosynthesis system converting inorganic carbon into acetate. *Bioresource Technology*, **266**, 203–210. https://doi.org/10.1016/j.biortech.2018.06.074

del Pilar Anzola Rojas M., Zaiat M., González E. R., de Wever H. and Pant D. (2021). Enhancing the gas–liquid mass transfer during microbial electrosynthesis by the variation of CO_2 flow rate. *Process Biochemistry*, **101**, 50–58. https://doi.org/10.1016/j.procbio.2020.11.005

DOE (2017). Biofuels and Bioproducts from Wet and Gaseous Waste Streams: Challenges and Opportunities. Bionergy Technologies Office, Energy Efficiency and Renewable Energy, U.S. Department of Energy, Washington, D.C. https://www.energy.gov/sites/default/files/2017/09/f36/biofuels_and_bioproducts_from_wet_and_gaseous_waste_streams_full_report.pdf.

ElMekawy A., Hegab H. M., Mohanakrishna G., Elbaz A. F., Bulut M. and Pant D. (2016). Technological advances in CO_2 conversion electro-biorefinery: a step toward commercialization. *Bioresource Technology*, **215**, 357–370. https://doi.org/10.1016/j.biortech.2016.03.023

Gajaraj S., Huang Y., Zheng P. and Hu Z. (2017). Methane production improvement and associated methanogenic assemblages in bioelectrochemically assisted anaerobic digestion. *Biochemical Engineering Journal*, **117**, 105–112. https://doi.org/10.1016/j.bej.2016.11.003

Ganigué R., Puig S., Batlle-Vilanova P., Balaguer M. D. and Colprim J. (2015). Microbial electrosynthesis of butyrate from carbon dioxide. *Chemical Communications*, **51**, 3235–3238. https://doi.org/10.1039/c4cc10121a

Giddings C. G. S., Nevin K. P., Woodward T., Lovley D. R. and Butler C. S. (2015). Simplifying microbial electrosynthesis reactor design. *Front Microbiol*, **6**, 1–6. https://doi.org/10.3389/fmicb.2015.00468.

Gildemyn S., Verbeeck K., Slabbinck R., Andersen S. J., Prévoteau A. and Rabaey K. (2015). Integrated production, extraction, and concentration of acetic acid from CO_2 through microbial electrosynthesis. *Environmental Science and Technology Letters*, **2**, 325–328. https://doi.org/10.1021/acs.estlett.5b00212

House K. Z., Baclig A. C., Ranjan M., van Nierop E. A., Wilcox J. and Herzog H. J. (2011). Economic and energetic analysis of capturing CO_2 from ambient air. *Proceedings of the National Academy of Sciences of the United States of America*, **108**, 20428–20433. https://doi.org/10.1073/pnas.1012253108

Huang Z., Jiang D., Lu L. and Ren Z. J. (2016). Ambient CO– capture and storage in bioelectrochemically mediated wastewater treatment. *Bioresource Technology*, **215**, 380–385. https://doi.org/10.1016/j.biortech.2016.03.084

Jack J., Zhu W., Avalos J., Gong J. and Ren Z. J. (2021). Anode co-valorization for scalable and sustainable electrolysis. *Green Chemistry*, **23**, 7917–7936. https://doi.org/10.1039/D1GC02094C

Jiang Y. and Zeng R. J. (2019). Bidirectional extracellular electron transfers of electrode-biofilm: mechanism and application. *Bioresource Technology*, **271**, 439–448. https://doi.org/10.1016/j.biortech.2018.09.133

Jiang Y., Lu L., Wang H., Shen R., Ge Z., Hou D., Chen X., Liang P., Huang X. and Ren Z. J. (2018). Electrochemical control of redox potential arrests methanogenesis and regulates products in mixed culture electro-fermentation. *ACS Sustainable Chemistry & Engineering*, **6**(7), 8650–8658. https://doi.org/10.1021/acssuschemeng.8b00948

Jiang Y., May H. D., Lu L., Liang P., Huang X. and Ren Z. J. (2019). Carbon dioxide and organic waste valorization by microbial electrosynthesis and electro-fermentation. *Water Research*, **149**, 42–55. https://doi.org/10.1016/j.watres.2018.10.092

Jourdin L., Grieger T., Monetti J., Flexer V., Freguia S., Lu Y., Chen J., Romano M., Wallace G. G. and Keller J. (2015). High acetic acid production rate obtained by microbial electrosynthesis from carbon dioxide. *Environmental Science and Technology*, **49**, 13566–13574. https://doi.org/10.1021/acs.est.5b03821

Jourdin L., Freguia S., Flexer V. and Keller J. (2016a). Bringing high-rate, CO_2-based microbial electrosynthesis closer to practical implementation through improved electrode design and operating conditions. *Environmental Science and Technology*, **50**, 1982–1989. https://doi.org/10.1021/acs.est.5b04431

Jourdin L., Lu Y., Flexer V., Keller J. and Freguia S. (2016b). Biologically induced hydrogen production drives high rate/high efficiency microbial electrosynthesis of acetate from carbon dioxide. *ChemElectroChem*, **3**, 581–591. https://doi.org/10.1002/celc.201500530

Kaza S., Yao L., Bhada-Tata P. and Van Woerden F. (2018). What A Waste 2.0: A Global Snapshot of Solid Waste Management to 2050. Urban Development. World Bank, Washington, DC. © World Bank. https://openknowledge.worldbank.org/handle/10986/30317 License: CC BY 3.0 IGO. https://doi.org/10.1596/978-1-4648-1329-0

Lee K. Y., Ng T. W., Li G., An T., Kwan K. K., Chan K. M., Huang G., Yip H. Y. and Wong P. K. (2015). Simultaneous nutrient removal, optimised CO_2 mitigation and biofuel feedstock production by Chlorogonium sp. grown in secondary treated non-sterile saline sewage effluent. *Journal of hazardous materials*, **297**, pp. 241–250.

Liu S., Deng Z., Li H. and Feng K. (2019). Contribution of electrodes and electric current to process stability and methane production during the electro-fermentation of food waste. *Bioresource Technology*, **288**, 121536. https://doi.org/10.1016/j.biortech.2019.121536

Logan B. E. (2008). Microbial Fuel Cells. Wiley Blackwell, Hoboken, New Jersey. https://doi.org/10.1002/9780470258590

Logan B. E., Rossi R., Ragab A. and Saikaly P. E. (2019). Electroactive microorganisms in bioelectrochemical systems. *Nature Reviews Microbiology*, **17**, 307–319. https://doi.org/10.1038/s41579-019-0173-x

Lovley D. R. (2011). Powering microbes with electricity: direct electron transfer from electrodes to microbes. *Environmental Microbiology Reports*, **3**, 27–35, https://doi.org/10.1111/j.1758-2229.2010.00211.x

Lovley D. R. and Nevin K. P. (2011). A shift in the current: new applications and concepts for microbe-electrode electron exchange. *Current Opinion in Biotechnology*, **3**, 441–448. https://doi.org/10.1016/j.copbio.2011.01.009

Lu L., Huang Z., Rau G. H. and Ren Z. J. (2015). Microbial electrolytic carbon capture for carbon negative and energy positive wastewater treatment. *Environmental Science and Technology*, **49**, 8193–8201. https://doi.org/10.1021/acs.est.5b00875

Lu L., Fang Y., Huang Z., Huang Y. and Ren Z. J. (2016). Self-sustaining carbon capture and mineralization via electrolytic carbonation of coal fly ash. *Chemical Engineering Journal*, **306**, 330–335. https://doi.org/10.1016/j.cej.2016.07.060

Lu L., Guest J. S., Peters C. A., Zhu X., Rau G. H. and Ren Z. J. (2018). Wastewater treatment for carbon capture and utilization. *Nature Sustainability*, **1**, 750–758. https://doi.org/10.1038/s41893-018-0187-9

MacDowell N., Fennell P. S., Shah N. and Maitland G. C. (2017). The role of CO_2 capture and utilization in mitigating climate change. *Nature Climate Change*, **7**, 243–249. https://doi.org/10.1038/nclimate3231

Mateos R., Escapa A., San-Martín M. I., de Wever H., Sotres A. and Pant D. (2020). Long-term open circuit microbial electrosynthesis system promotes methanogenesis. *Journal of Energy Chemistry*, **41**, 3–6. https://doi.org/10.1016/J.JECHEM.2019.04.020

Mikkelsen M., Jørgensen M. and Krebs F. C. (2010). The teraton challenge. A review of fixation and transformation of carbon dioxide. *Energy and Environmental Science*, **3**, 43–81. https://doi.org/10.1039/b912904a

Mohanakrishna G., Seelam J. S., Vanbroekhoven K. and Pant D. (2015). An enriched electroactive homoacetogenic biocathode for the microbial electrosynthesis of acetate through carbon dioxide reduction. *Faraday Discuss*, **183**, 445–462. https://doi.org/10.1039/c5fd00041f.

Moscoviz R., Toledo-Alarcón J., Trably E. and Bernet N. (2016). Electro-fermentation: how to drive fermentation using electrochemical systems. *Trends in Biotechnology*, **34**, 11, 856–865. https://doi.org/10.1016/j.tibtech.2016.04.009

Nie H., Zhang T., Cui M., Lu H., Lovley D. R. and Russel T. P. (2013). Improved cathode for high efficient microbial-catalyzed reduction in microbial electrosynthesis cells. *Phys. Chem. Chem. Phys.*, **15**, 14290–14294. https://doi.org/10.1039/c3cp52697f.

Nelabhotla A. B. T. and Dinamarca C. (2019). Bioelectrochemical CO_2 reduction to methane: MES integration in biogas production processes. *Applied Sciences (Switzerland)*, **9**, 16–18. https://doi.org/10.3390/app9061056

Nelabhotla A. B. T., Pant D. and Dinamarca C. (2021). In: Power-to-gas for methanation. Emerging Technologies and Biological Systems for Biogas Upgrading, Aryal N., Ottosen L. D. M., Kofoed M. V. W. and Pant D. (eds), London, UK, pp. 187–221. https://doi.org/10.1016/B978-0-12-822808-1.00008-8

Nevin K. P., Woodard T. L., Franks A. E., Summers Z. M. and Lovley D. R. (2010). Microbial electrosynthesis: feeding microbes electricity to convert carbon dioxide and water to multicarbon extracellular organic compounds. *mBio*, **1**, e00103–10. https://doi.org/10.1128/mBio.00103-10

Nevin K. P., Hensley S. A., Franks A. E., Summers Z. M., Ou J., Woodard T. L., Snoeyenbos-West O. L. and Lovley D. R. (2011). Electrosynthesis of organic compounds from carbon dioxide is catalyzed by a diversity of acetogenic microorganisms. *Applied and Environmental Microbiology*, **77**, 2882–2886. https://doi.org/10.1128/AEM.02642-10

Nikhil G. N., Venkata Subhash G., Yeruva D. K. and Venkata Mohan S. (2015). Synergistic yield of dual energy forms through biocatalyzed electrofermentation of waste: stoichiometric analysis of electron and carbon distribution. *Energy*, **88**, 281–291. https://doi.org/10.1016/j.energy.2015.05.043

Nishio K., Kimoto Y., Song J., Konno T., Ishihara K., Kato S., Hashimoto K. and Nakanishi S. (2013). Extracellular electron transfer enhances polyhydroxybutyrate productivity in *Ralstonia eutropha*. *Environmental Science and Technology Letters*, **1**, 40–43. https://doi.org/10.1021/ez400085b

Pandey P., Shinde V. N., Deopurkar R. L., Kale S. P., Patil S. A. and Pant D. (2016). Recent advances in the use of different substrates in microbial fuel cells toward wastewater treatment and simultaneous energy recovery. *Applied Energy*, **168**, 706–723. https://doi.org/10.1016/j.apenergy.2016.01.056

Prévoteau A., Carvajal-Arroyo J. M., Ganigué R. and Rabaey K. (2020). Microbial electrosynthesis from CO_2: forever a promise? *Current Opinion in Biotechnology*, **62**, 48–57. https://doi.org/10.1016/j.copbio.2019.08.014

Rabaey K. and Rozendal R. A. (2010). Microbial electrosynthesis – revisiting the electrical route for microbial production. *Nature Reviews Microbiology*, **8**, 706–716. https://doi.org/10.1038/nrmicro2422

Rau G. H., Carroll S. A., Bourcier W. L., Singleton M. J., Smith M. M. and Aines R. D. (2013). Direct electrolytic dissolution of silicate minerals for air CO_2 mitigation and carbon-negative H2 production [WWW document]. *Proceedings of the National Academy of Sciences of the United States of America*, **110**(25), 10095–10100. https://doi.org/10.1073/pnas.1222358110

Russell J., Van Ballegooy S., Torvelainen E. and Gulley R. (2015). Consideration of ground variability over an area of geological similarity as part of liquefaction assessment for foundation design. *mBio*, **1**, 1–8. https://doi.org/10.1128/mBio.00190-10.Editor

Saratale G. D., Saratale R. G., Shahid M. K., Zhen G., Kumar G., Shin H. S., Choi Y. G. and Kim S. H. (2017). A comprehensive overview on electro-active biofilms, role of exo-electrogens and their microbial niches in microbial fuel cells (MFCs). *Chemosphere*, **178**, 534–547. https://doi.org/10.1016/j.chemosphere.2017.03.066

Sherrard J. H. (1976). Destruction of alkalinity in aerobic biological wastewater treatment. *Journal of the Water Pollution Control Federation*, **48**, 1834–1839.

Unesco, World Water Assessment Programme (2012). *Taylor & Francis Group*. https://doi.org/10.4324/9781849773355

Villano M., Paiano P., Palma E., Miccheli A. and Majone M. (2017). Electrochemically driven fermentation of organic substrates with undefined mixed microbial cultures. *ChemSusChem*, **10**, 3091–3097. https://doi.org/10.1002/cssc.201700360

Wang H. and Ren Z. J. (2013). A comprehensive review of microbial electrochemical systems as a platform technology. *Biotechnology Advances*, **31**, 1796–1807. https://doi.org/10.1016/j.biotechadv.2013.10.001

Wett B., Eladawy A. and Becker W. (2004). Carbonate addition – an effective remedy against poor activated sludge settling properties and alkalinity conditions in small wastewater treatment plants. *Water Science and Technology*, **48**(11), 411–417. https://doi.org/10.2166/wst.2004.0889

Xafenias N., Anunobi M. O. and Mapelli V. (2015). Electrochemical startup increases 1,3-propanediol titers in mixed-culture glycerol fermentations. *Process Biochemistry*, **50**(10), 1499–1508. https://doi.org/10.1016/j.procbio.2015.06.020

Zaybak Z., Pisciotta J. M., Tokash J. C. and Logan B. E. (2013). Enhanced start-up of anaerobic facultatively autotrophic biocathodes in bioelectrochemical systems. *Journal of Biotechnology*, **168**, 478–485. https://doi.org/10.1016/j.jbiotec.2013.10.001

Zhang T., Nie H., Bain T. S., Lu H., Cui M., Snoeyenbos-West O. L., Franks A. E., Nevin K. P., Russell T. P. and Lovley D. R. (2013). Improved cathode materials for microbial electrosynthesis. *Energy and Environmental Science*, **6**, 217–224. https://doi.org/10.1039/c2ee23350a

Zhao Z., Li Y., Zhang Y. and Lovley D. R. (2020). Sparking anaerobic digestion: promoting direct interspecies electron transfer to enhance methane production. *Iscience*, **23**(12), 101794. https://doi.org/10.1016/j.isci.2020.101794

Zhou M., Chen J., Freguia S., Rabaey K. and Keller J. (2013). Carbon and electron fluxes during the electricity driven 1,3-propanediol biosynthesis from glycerol. *Environmental Science and Technology*, **47**, 11199–11205. https://doi.org/10.1021/es402132r

Zhu X. and Logan B. E. (2014). Microbial electrolysis desalination and chemical-production cell for CO_2 sequestration. *Bioresource Technology*, **159**, 24–29. https://doi.org/10.1016/j.biortech.2014.02.062

Zhu X., Hatzell M. C. and Logan B. E. (2014). Microbial reverse-electrodialysis electrolysis and chemical-production cell for H_2 production and CO_2 sequestration. *Environmental Science and Technology Letters*, **1**, 231–235. https://doi.org/10.1021/ez500073q

Zhu X., Leininger A., Jassby D., Tsesmetzis N. and Ren Z. J. (2021). Will membranes break barriers on volatile fatty acid recovery from anaerobic digestion. *ACS ES&T Engineering*, **1**(1), 141–153. https://doi.org/10.1021/acsestengg.0c00081

Zou S. and He Z. (2018). Efficiently 'pumping out' value-added resources from wastewater by bioelectrochemical systems: A review from energy perspectives. *Water Research*, **131**, 62–73. https://doi.org/10.1016/j.watres.2017.12.026

doi: 10.2166/9781789061796

Chapter 7

Decarbonization potentials in nitrogen management

Kester McCullough[1,2]*, Stephanie Klaus[2] and Charles Bott[2]

[1]Cornell University, Ithaca, NY, USA
[2]Hampton Roads Sanitation District, Virginia Beach, VA, USA
*Correspondence: kkm79@cornell.edu

7.1 INTRODUCTION

Nitrogen management is a critical function of the water sector and is necessary to protect receiving water bodies. This chapter explores the potential for decarbonization of nitrogen removal processes within the water resource recovery facility (WRRF). Nitrogen is present in wastewater primarily in organic forms and as ammonia, and most of the organic nitrogen is hydrolyzed to ammonia in the treatment process. Ammonia is removed from wastewater by conversion to nitrogen gas through oxidation (nitrification) and reduction (denitrification), or through anaerobic ammonia oxidation (anammox). The equations governing these nitrogen transformations can be written as follows:

Autotrophic nitrification:

Nitritation, ammonia oxidizing bacteria (AOB) (yield $=0.15$ gVSS/gNH_4^+-N):

$$\begin{aligned}&1.019NH_4^+ + 1.41O_2 + 0.074CO_2 + 0.019HCO_3^- \\ &\quad\rightarrow 0.019C_5H_7O_2N + NO_2^- + 2H^+ + 0.981H_2O\end{aligned} \tag{7.1}$$

Nitratation, nitrite oxidizing bacteria (NOB) (yield $=0.07$ gVSS/gNO_2^--N):

$$\begin{aligned}&NO_2^- + 0.009NH_4^+ + 0.457O_2 + 0.035CO_2 + 0.009HCO_3^- + 0.009H_2O \\ &\quad\rightarrow 0.009C_5H_7O_2N + NO_3^-\end{aligned} \tag{7.2}$$

Heterotrophic denitrification:

Denitratation (yield $=0.36$ gVSS/gCOD, methanol):

$$\begin{aligned}&NO_3^- + 0.521CH_3OH + 0.056NH_4^+ + 0.056HCO_3^- \\ &\quad\rightarrow 0.056C_5H_7O_2N + NO_2^- + +0.296CO_2 + 0.985H_2O\end{aligned} \tag{7.3}$$

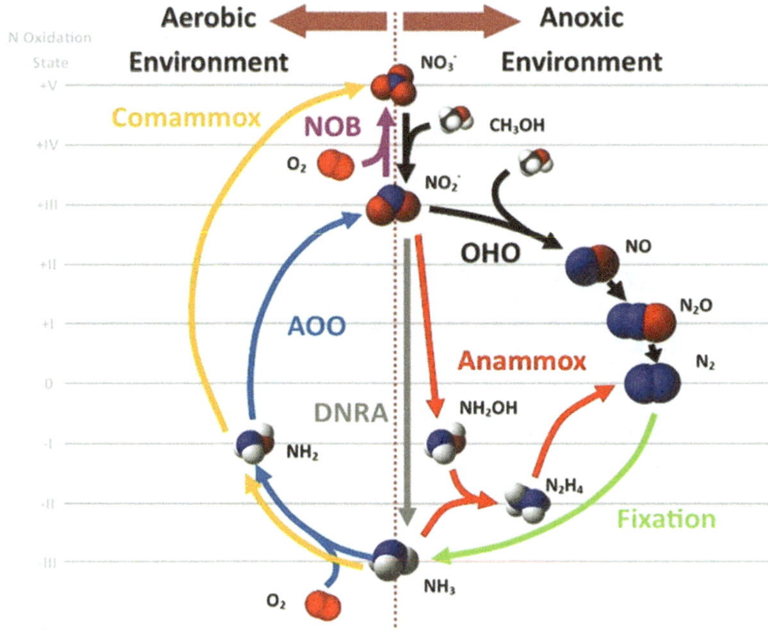

Figure 7.1 The nitrogen cycle. Abbreviations: dissimilatory nitrate reduction to ammonium (DNRA); ammonia oxidizing organisms (AOO) which includes bacteria (AOB) and archaea (AOA); nitrite oxidizing bacteria (NOB); ordinary heterotrophic organisms (OHO).

Denitritation (yield = 0.36 gVSS/gCOD, methanol):

$$\begin{aligned} &NO_2^- + 0.781CH_3OH + 0.084NH_4^+ + 0.084HCO_3^- + H^+ \\ &\quad \rightarrow 0.084C_5H_7O_2N + 0.5N_2 + 0.444CO_2 + 1.978H_2O \end{aligned} \tag{7.4}$$

Anaerobic ammonia oxidation (anammox) (from Strous *et al.* (1998)):

$$\begin{aligned} &NH_4^+ + 1.32NO_2^- + 0.066HCO_3^- + 0.13H^+ = 1.02N_2 + 0.26NO3^- \\ &\quad + 2.03H_2O + 0.066CH_2O_{0.05}N_{0.15} \end{aligned} \tag{7.5}$$

These biological processes are the focus of this chapter (Figure 7.1). The artificial fixation of ammonia from nitrogen gas via the Haber–Bosch process is driven by the need for ammonia fertilizer. Research is currently at lab and pilot-scale to recover ammonia directly from wastewater as a potentially more sustainable alternative (Beckinghausen *et al.*, 2020; Ye *et al.*, 2018). However, no technology currently exists that can compete economically with biological nitrification/denitrification at full-scale facilities (Winkler & Straka, 2019). WRRFs must consider the management of influent (mainstream) nitrogen and nitrogen generated from waste sludges within the facility (sidestream); approaches to decarbonization are specific to each of these contexts.

This introduction provides a broad overview of the carbon costs and decarbonization potentials associated with nitrogen removal in WRRFs, followed by a detailed review and quantitative comparison of technologies and process configurations for both sidestream and mainstream contexts. A review of current research on the topic as well as a case study of a full-scale WRRF implementing these strategies is provided. Finally, paths forward for the industry and the future outlook are considered.

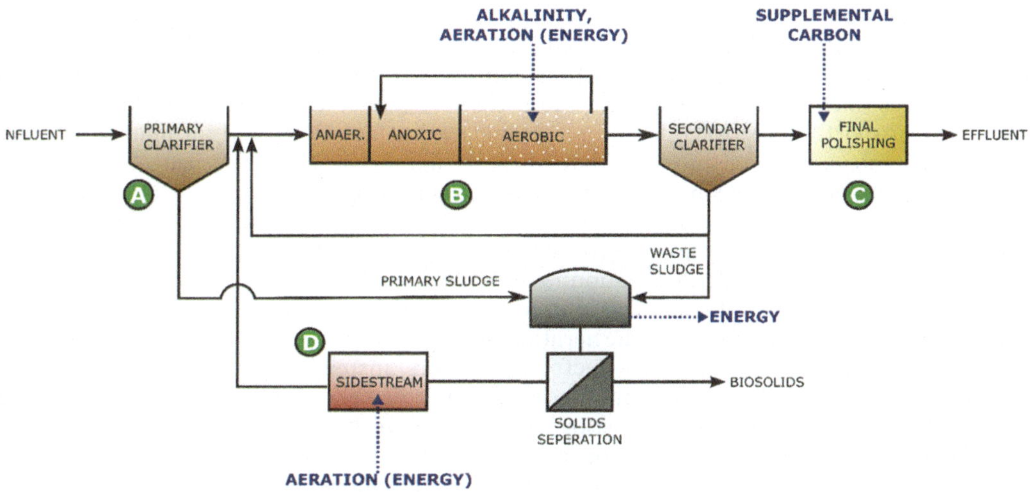

Figure 7.2 Conventional WRRF layout with typical inputs and outputs shown in blue text and locations of potential shortcut nitrogen removal improvements labeled with Green circles.

7.1.1 Carbon footprint costs of nitrogen removal

The carbon footprint for nitrogen removal in a WRRF is driven by energy and chemical inputs for treatment, energy recovery, nitrous oxide emissions, plant capacity, and treatment efficiency. A conventional layout for a WRRF performing biological carbon, nitrogen, and phosphorus removal is shown in Figure 7.2. Locations of typical inputs and outputs of aeration, energy, alkalinity, and supplemental carbon are indicated in blue. These resource requirements can be reduced, and energy outputs increased, via shortcut nitrogen removal. Shortcut nitrogen removal improvements for the facility are indicated by green circles (A) carbon capture/diversion (B) mainstream nitrogen removal (C) mainstream polishing nitrogen removal and (D) sidestream nitrogen removal.

7.1.1.1 Aeration

In both mainstream and sidestream treatment, the energy required for aeration (to provide oxygen for nitrification) generates a significant capital and carbon cost. Reducing the aeration energy required for nitrogen removal involves shortcutting the nitrification/denitrification process and/or utilizing anammox (a collection of strategies broadly referred to as shortcut nitrogen removal). Novel aeration strategies, such as intermittent aeration or low dissolved oxygen, have the potential to decrease both aeration energy demands and improve denitrification through simultaneous nitrification-denitrification or dedicated anoxic periods. Optimized aeration controls, such as ammonia-based aeration control (ABAC) or ammonia vs. NOx (AvN) control, also allow for more efficient aeration and decreased energy usage.

7.1.1.2 Alkalinity

Aerobic nitrification consumes alkalinity, which must be managed to maintain pH and alkalinity concentrations conducive to the growth of the organisms responsible for these processes. Denitrification allows recovery of a fraction of the alkalinity consumed, but alkalinity supplementation is often required, especially with low alkalinity wastewaters. The manufacture, transportation, storage, and use of chemicals for alkalinity adjustment thus contributes to the carbon cost of nitrogen removal. Shortcut nitrogen strategies generally decrease alkalinity demands.

7.1.1.3 Organic carbon

If denitrification is implemented to reduce oxidized nitrogen species, organic carbon (or COD) is required. The COD required for denitrification can come from the influent wastewater or from exogenous sources (methanol, glycerol, etc). The use of exogenous COD incurs not only the cost of manufacture, transport, and storage for those chemicals, but the COD will ultimately be oxidized to CO_2 and generate additional biomass that must be managed. Thus, the use of exogenous (or supplemental) COD incurs a significant carbon cost. Other electron donors have been suggested as more affordable for autotrophic denitrification, including hydrogen and elemental sulfur, but have not been implemented in full-scale wastewater treatment (Di Capua *et al.*, 2019).

The use of influent COD for denitrification is preferable, as the COD needs to be removed in the treatment process anyway. If influent COD is used for denitrification, the excess costs of exogenous COD use are reduced or eliminated, and the aeration demand for influent COD oxidation is decreased, as the oxidized nitrogen species serve as electron acceptors, instead of oxygen. The efficient use of influent COD is thus critical to reduce the overall carbon demands of the process, and a number of different plant configurations and aeration strategies have been implemented to address this. Additionally, the COD demand of the nitrogen removal process is not fixed, and, like aeration and alkalinity demands, can be decreased with shortcut nitrogen removal.

The net COD demand of the nitrogen removal process and the efficiency of influent COD use thus dictate a process-specific minimum carbon-to-nitrogen (C/N) ratio in the influent wastewater. Reducing this required C/N ratio opens the door to upstream carbon capture and redirection. Without first reducing the C/N ratio required for nitrogen removal, upstream carbon diversion is counterproductive, as exogenous COD will be required.

7.1.1.4 WRRF size and capacity

Autotrophs responsible for ammonia oxidation grow much slower than the heterotrophs required for COD and phosphorus removal, and thus dictate the design and operating aerobic solids retention time (SRT) for WRRFs performing nitrogen removal. WRRF size and capacity are directly related to the SRT required for treatment, and lower SRT requirements provide for smaller plant design or increased capacity in existing plants. Lowering SRT requirements thus provides potential for decarbonization by allowing existing plants to increase capacity and avoid additional construction and reduces the tank volumes (and associated material and construction costs) required for new facilities. Furthermore, minimizing operating SRT also minimizes endogenous decay, further decreasing aeration energy requirements and allowing for more transfer of carbon in the form of waste sludge to energy producing processes.

Lower effluent ammonia concentrations require longer SRTs; the relationship between ammonia concentration and growth rate is generally conceptualized by a Monod function. Shortcut nitrogen removal processes that incorporate anammox allow for 20–60% shorter SRTs in the nitrification process, as only a portion (determined by process configuration) of the influent ammonia is oxidized aerobically (McCullough *et al.*, 2021); the remaining fraction of ammonia is oxidized anaerobically along with nitrite via anammox metabolism. While anammox organisms require even longer SRTs than aerobic nitrifiers, they can be retained in the process via attached growth systems, such as filters, moving bed bioreactors (MBBRs), granular sludge, or integrated fixed film activated sludge (IFAS) systems, allowing more compact plant design.

7.1.1.5 Nitrous oxide emissions

Nitrous oxide (N_2O) is a potent greenhouse gas that is emitted during biological nitrogen removal. N_2O can be produced during nitrification via the NH_2OH oxidation pathway or AOB denitrification pathway, and during incomplete (partial) denitrification (Ni & Yuan, 2015). Shortcut nitrogen processes are more likely to produce N_2O than conventional nitrification/denitrification due to low DO concentration, intermittent aeration, high nitrite concentration, and low COD/N ratio (Kampschreur *et al.*, 2009; Wunderlin *et al.*, 2012). It is possible to mitigate N_2O emissions in aerobic processes using strategies such as modifying the aeration strategy and minimizing nitrite concentrations without impacting nitrogen removal performance (Duan *et al.*, 2020; Pijuan *et al.*, 2014). Mitigation strategies for N_2O in anoxic process include minimizing nitrite concentrations which can be controlled via external carbon source dosing (Du *et al.*, 2016; Song *et al.*, 2015). It is important to mitigate N_2O production so that an increase in greenhouse gas production does not offset the carbon savings of shortcut nitrogen removal (Chen *et al.*, 2020).

7.2 CARBON REMOVAL/DIVERSION IN PRIMARY TREATMENT

The following statement is frequently cited in the literature when explaining the benefits of mainstream deammonification: 'partial nitritation and anammox results in 60% less aeration, 90% less sludge production and 100% reduction of organic carbon addition compared to conventional nitrification-denitrification' (Cao *et al.*, 2017; Jetten *et al.*, 1997; Mulder 2003). As Daigger (2014) pointed out, this is completely dependent on carbon removal ahead of the nitrogen removal step. Otherwise, the carbon will get oxidized in aerobic nitrogen removal step, utilizing aeration energy and creating more biomass. In order to realize the benefits of mainstream deammonification, the carbon must be diverted, preferably to an anaerobic digester to recover energy. It should also be noted that in order to remove 100% of the nitrogen, some organic carbon is required to reduce the small amount of nitrate produced by anammox (Daigger 2014). The concept behind the operation of an A/B process for mainstream deammonification is to balance the carbon that is being captured in A-stage with the remaining carbon that is required for nitrogen removal in B-stage. When the mechanism for nitrogen removal is denitrification by OHO (such as in nitrite shunt), it is desirable to have slowly biodegradable COD (sbCOD) in the B-stage influent (Regmi *et al.*, 2014). As more nitrogen is removed through the anammox pathway, less carbon is required, and more carbon can be diverted.

The removal of carbon can be accomplished physically (with or without the addition of chemicals to enhance coagulation/flocculation) or biologically. Primary sedimentation tanks should remove 50–70% of the TSS, 25–40% of the BOD, and 20–35% of the COD (Tchobanoglous *et al.*, 2003). A limitation of primary sedimentation is that soluble and colloidal constituents are not removed. Chemically enhanced primary treatment (CEPT) is the addition of coagulants and/or flocculating agents to the primary settling process to improve physical removal of carbon. Removals of 80–90% TSS including some colloidal particles, 50–80% BOD, and 45–80% COD can be achieved (Tchobanoglous *et al.*, 2003). CEPT can also be used for chemical phosphorus removal. The goal of the high-rate A-stage in an adsorption-bio-oxidation (A/B process) is to provide a controlled carbon loading for B-stage, and by separating the SRTs achieve low-cost COD removal at reduced overall aeration tank volume and aeration energy requirements (Miller *et al.*, 2012). Carbon removal can also be achieved via anaerobic treatment which is an attractive alternative because no aeration is required, methane is produced, and less sludge is produced compared to aerobic processes (Delgado Vela *et al.*, 2015). However, due to slower growth rates of anaerobic microorganisms compared to aerobic, anaerobic treatment requires higher temperatures and therefore it is most typically used in tropical and sub-tropical climates (de Lemos Chernicharo & Von Sperling, 2005).

7.3 APPROACHES FOR CARBON EFFICIENT NITROGEN REMOVAL

7.3.1 Traditional nitrification/denitrification

Pre-anoxic denitrification processes, such as the modified Ludzack–Ettinger (MLE) and A2O process, take advantage of influent carbon for nitrogen removal. Pre-anoxic processes are limited in N removal by the amount and fractionation of influent COD and the maximum internal recycle rate. A way to use influent carbon more efficiently is to utilize a step-feed process in which influent is split among multiple anoxic zones. This process it more complex, but can achieve lower effluent N levels than pre-anoxic denitrification alone. To consistently achieve very low (less than 5 mg/L) effluent N concentrations, depending on influent C:N, post-anoxic denitrification with external carbon addition is normally required. Exceptions can be seen a relatively small plants without primary treatment and with long 'extended aeration' SRTs that maximize endogenous denitrification (e.g., some oxidation ditch processes).

Near complete N removal can be implemented both in the step-feed approach to which supplemental carbon is added to the last anoxic zone, and combined with pre-anoxic denitrification in processes such as a 4- or 5-stage Bardenpho, or UCT, VIP, or Johannesburg processes to which a second anoxic zone is added. Because of the carbon demand for traditional nitrification/denitrification processes, it is desirable to explore shortcut nitrogen removal technologies.

7.3.2 Nitrite shunt and PNA (NOB out-selection)

Two distinctly different processes, nitrite shunt and partial nitritation/anammox (PNA), are combined in the following section because they both rely on NOB out-selection. The produced nitrite is either reduced by OHO, which requires organic carbon, or by anammox which does not. The main challenges of achieving mainstream deammonification are NOB out-selection and anammox retention. NOB repression is easier in sidestream processes due to high free ammonia (FA) concentrations (Anthonisen *et al.*, 1976) and high temperature (Hellinga *et al.*, 1998). Anammox retention becomes more difficult in the mainstream because colder temperatures and lower ammonia concentrations result in slower growth rates (Kartal *et al.*, 2010; Lackner *et al.*, 2015; Ma *et al.*, 2016; Vlaeminck *et al.*, 2012). Strategies developed to give AOB an advantage over NOB in mainstream treatment include: maintaining an ammonia residual in the effluent (Pérez *et al.*, 2014; Poot *et al.*, 2016; Regmi *et al.*, 2014; Welker *et al.*, 2016), transient anoxia (Gilbert *et al.*, 2014a; Kornaros *et al.*, 2010), high DO concentration during intermittent aeration (Al-Omari *et al.*, 2015; Regmi *et al.*, 2014), low DO continuous aeration (for biofilm and granule systems) (Pérez *et al.*, 2014; Poot *et al.*, 2016; Sliekers *et al.*, 2005), seeding of AOB from a sidestream process (Al-Omari *et al.*, 2015), stringent aerobic SRT control (Regmi *et al.*, 2014), and exposure of the mainstream biomass to high levels of nitrous acid (Piculell *et al.*, 2016b; Wang *et al.*, 2014). Transient anoxia can be achieved through intermittent aeration either in time (on/off aeration control) or in space (alternating oxic/anoxic zones). The proposed mechanisms of NOB out-selection from transient anoxia are enzymatic lag (Kornaros *et al.*, 2010), inhibition by intermediates (Courtens *et al.*, 2015; Soler-Jofra *et al.*, 2021), and substrate availability (Gilbert *et al.*, 2014b) (limiting the amount of NO_2^- available aerobically).

Approaches for NOB out-selection depend on the type of system that is being operated and can be broken down into two categories based on the mechanism for anammox retention. Single SRT systems include attached growth biofilm systems (Gilbert *et al.*, 2014b; Gustavsson *et al.*, 2020; Laureni *et al.*, 2016; Liu *et al.*, 2018a) and fully granular systems (Gao *et al.*, 2015; Lotti *et al.*, 2014; Morales *et al.*, 2016; Winkler *et al.*, 2012). Two-SRT systems include: hybrid systems with AOB/NOB/OHO suspended growth and granular anammox (Cao *et al.*, 2013; Han *et al.*, 2016; Wett *et al.*, 2015), and two-phase systems with AOB/NOB/OHO in a separate suspended growth or biofilm reactor followed by a completely anoxic anammox reactor using for example moving bed biofilm (MBBR) or granular sludge (Ma *et al.*, 2011; Regmi *et al.*, 2016). Two-stage shunt processes are possible but once nitrite has been accumulated it should be used by anammox not OHO for denitrification to realize the savings previously mentioned. Aerobic granular sludge and biofilm systems can take advantage of relative diffusion resistance inside and outside of a granule/biofilm to develop and grow different populations simultaneously in a single reactor. However, NOB out-selection can be more challenging because the SRTs of the different populations cannot be separated. Recently, Anoxkaldnes developed plastic media capable of out-selecting NOB spatially by limiting the depth of the biofilm (Piculell *et al.*, 2016a, 2016b). The thickness of the biofilm can be controlled to different depths depending on if it is used in a single stage system, or in the first stage of a two-stage system.

The common mechanism of NOB out-selection to all configurations is maintaining an ammonia residual to maintain high AOB rates by keeping substrate well above limiting conditions. The challenge of suppressing NOB became even more complicated by the discovery in 2015 that certain NOB are capable of oxidizing ammonia directly to nitrate, disrupting the long-accepted dogma of nitrification as a two-step process (van Kessel *et al.*, 2015). The term Comammox (complete ammonia oxidation) was coined to describe the process (van Kessel *et al.*, 2015). The discovery of this pathway may help to explain why NOB out-selection is so difficult (Daims *et al.*, 2016).

7.3.3 Partial denitrification/anammox

Since successful NOB out-selection relies on leaving an ammonia residual, and because there will always be some residual NO_3^-, some sort of polishing is required to meet a stringent effluent nitrogen limit (Le *et al.*, 2019; Regmi *et al.*, 2016). Originally, partial denitrification/anammox (PdNA) was explored as the second stage of a two-stage PNA process with partial nitritation in suspended growth followed by an anoxic MBBR (Regmi *et al.*, 2016). In a two-stage PNA process aerobic ammonia oxidation and anammox processes occur in two separate reactors. Since NOB out-selection was so difficult to achieve, some NO_2^- always will end up oxidized the whole way to NO_3^-. By adding external carbon to this second stage, NO_3^- can be removed in addition to NH_4^+ and NO_2^-. By testing different carbon sources (glycerol, methanol, and acetate) and increasing the nitrogen load it was determined that the PdNA process was robust and could remove substantial amounts of ammonia through the anammox pathway (Campolong *et al.*, 2018). Several research groups simultaneously concluded that PdNA was a viable alternative to sustaining NOB out-selection for partial nitrification for mainstream anammox (Campolong *et al.*, 2018; Du *et al.*, 2017; Le *et al.* 2019).

PdNA requires that part of the ammonia be oxidized upstream, either within a separate zone in a suspended growth system, or within a separate reactor in a polishing process. A benefit of PdNA is that PNA is not entirely excluded. If nitrite accumulation occurs upstream of the PdNA zone, then that is less carbon that needs to be added to the PdNA zone to convert NO_3^- to NO_2^-. In other words, the NOx coming into the PdNA can be any combination of all NO_2^-, all NO_3^-, or anywhere in between. It is important to note that single-stage PNA is different than single stage PdNA. While partial nitritation and anammox can occur in the same aerated reactor if a biofilm can provide a spatial DO gradient, PdNA cannot effectively occur under aerobic conditions so aerobic ammonia oxidation must occur upstream in a separate zone or reactor. There are two types of PdNA configurations: integrated and polishing.

Integrated refers to the PdNA occurring in either a pre- or post-anoxic zone within the BNR process, for example in the second anoxic zone of a Bardenpho process. For the SRT of the anammox bacteria to be sufficiently longer than the rest of the mixed liqor there must be some sort of anammox retention such as a screen or hydrocyclone to retain anammox granules, or anammox can be retained in a biofilm on media (IFAS) (Le *et al.*, 2019). Polishing refers PdNA occurring in a separate reactor downstream of the BNR process (after secondary clarification) such as an MBBR (Campolong *et al.*, 2018) or a deep-bed filter (Cui *et al.*, 2020; Fofana *et al.*, 2021).

7.3.4 Aeration, alkalinity, and COD requirements for mainstream nitrogen removal technologies

The mainstream nitrogen removal processes presented thus far (nitrification/denitrification, nitrite shunt, PNA, and PdNA) present varied opportunities for improved resource efficiency. As mentioned before, it is often assumed that shortcut nitrogen removal (nitrite shunt, PNA, and PdNA) will increase process efficiency, but this is not always the case for mainstream treatment. These potential efficiency improvements are contingent on how influent COD is managed in an upstream process (carbon diversion) and how the COD is utilized in the aerobic/nitrifying process.

Daigger (2014) demonstrated that, although PNA appears to require the least oxygen for nitrogen removal, these benefits disappear when influent COD is used for complete TIN reduction in nitrification/denitrification and nitrite shunt processes, as the additional oxygen required for nitrification is 'recovered'. McCullough *et al.* (2022) expanded this analysis to include alkalinity and supplemental COD requirements and the PdNA process. This analysis, which incorporated yield and assimilation, also included cases where complete TIN removal was not achieved with influent COD, which more accurately reflects most WRRFs. These results are summarized in Table 7.1. Aeration, alkalinity, and supplemental COD requirements for each nitrogen removal process are a function of how efficiently

Table 7.1 Aeration, alkalinity, and supplemental COD requirements for complete nitrogen removal in nitrification/denitrification, nitrite shunt, PdNA, and PNA processes[a].

		Nitrification/ Denitrification	Nitrite Shunt	PdNA	PNA
0% TIN Reduced with Influent COD	Oxygen required (gO_2/gN)	3.46	2.85	2.26	1.81
	Alkalinity required ($gHCO_3/gN$)	5.76	6.24	3.76	3.96
	Supplemental COD (gCOD/gN)	4.96	3.24	1.94	0.67
50% TIN reduced with influent COD	Oxygen required (gO_2/gN)	2.64	2.31	2.03	1.79
	Alkalinity required ($gHCO_3/gN$)	4.66	4.90	3.66	3.75
	Supplemental COD (gCOD/gN)	2.48	1.62	0.90	0.32
100% TIN reduced with influent COD	Oxygen required (gO_2/gN)	1.82	1.78	1.82	1.78
	Alkalinity required ($gHCO_3/gN$)	3.57	3.57	3.57	3.57
	Oxygen required (gO_2/gN)	0.00	0.00	0.00	0.00

[a]Adapted from McCullough *et al.* (2021).

influent COD is used for TIN removal, and process requirements are similar with complete (100%) TIN removal with influent COD. When complete TIN removal cannot be achieved with influent COD, the varied process efficiencies are significant. Thus, decarbonization can be achieved by increasing the efficiency with which influent COD is used for TIN removal and by transitioning to more efficient nitrogen removal processes. PNA is the most efficient of these processes, but also the most difficult to implement due to NOB out-selection. PdNA provides comparable resource efficiency increases but does not require NOB out-selection.

7.3.4.1 Implications of carbon diversion

The potential for mainstream carbon capture is determined by the influent COD and nitrogen concentrations, the COD required per nitrogen removed (determined by the nitrogen removal pathway), and the influent COD anoxic efficiency (the ratio of how much COD is oxidized anoxically, thus used for denitrification, vs aerobically). As more influent COD is oxidized aerobically in the nitrogen removal process (lower influent COD anoxic efficiency), a higher influent C/N ratio is required to maintain nitrogen removal, as less COD is available for denitrification. This can be offset by using a more carbon efficient shortcut nitrogen removal process.

The allowable fraction of COD removal in an upstream carbon capture process (A-Stage or CEPT) is shown in Figure 7.3 as a function of influent COD anoxic efficiency for each nitrogen removal process (assuming denitrification is done with influent COD and a typical wastewater influent C/N ratio = 12.5). For example, in a process with 60% influent COD anoxic efficiency (the other 40% influent COD is oxidized aerobically), conventional nitrification-denitrification, nitrite shunt, PdNA, and PNA would allow for 30, 60, 75, and 90% upstream carbon capture, respectively. As the influent COD anoxic efficiency decreases, the potential for carbon redirection decreases, but is still possible with shortcut nitrogen removal processes, especially PNA. In a conventional nitrification-denitrification process, the influent COD anoxic efficiency must be >40% for carbon capture to be possible, and below this efficiency supplemental COD will be required.

To completely offset the energy costs of conventional treatment, at least 65% of influent COD must be captured upstream (Liu M., 2018; Liu Y.-J., 2018); in the given scenario this is not possible in conventional nitrification-denitrification. Incorporating anammox into the nitrogen removal process makes this possible, and would require only 45% (PdNA) or 15% (PNA) influent COD anoxic efficiency.

Upstream carbon diversion processes also remove nitrogen from influent wastewater, in particulate organic forms (CEPT) and through assimilation of organic nitrogen into biomass (A-Stage). Nitrogen

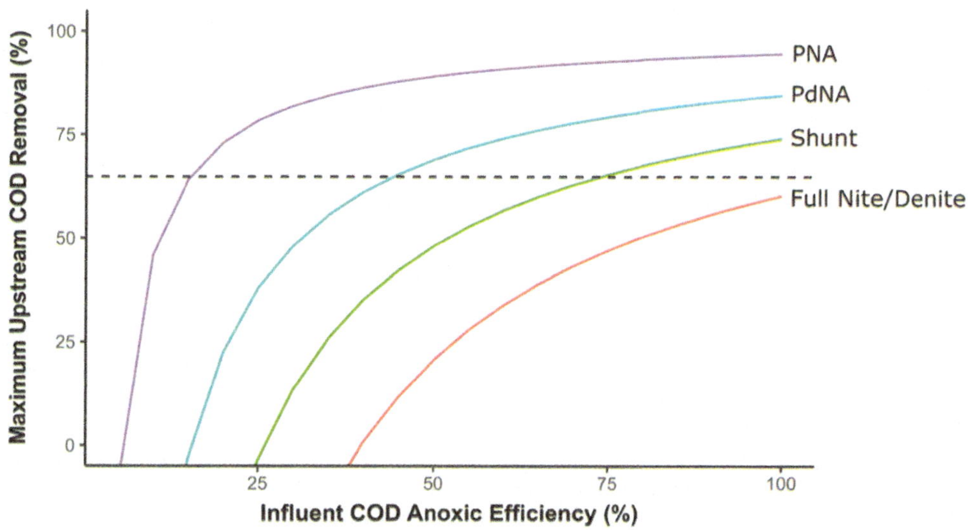

Figure 7.3 Maximum allowable upstream COD removal for complete nitrogen removal in conventional and shortcut nitrogen removal processes, shown as a function of how efficiently influent COD is used for nitrogen reduction in the treatment process. As influent COD is used more efficiently for nitrogen reduction, greater amounts can be removed in the upstream COD diversion process. Shortcut nitrogen removal processes allow for greater COD diversion/capture without compromising nitrogen removal.

removal in the A-Stage process can exceed 10%, and this nitrogen ultimately ends in sludge streams where it can be managed more efficiently in sidestream processes.

Upstream carbon diversion processes also provide the possibility for redirection carbon in the treatment process to where it can be used most beneficially. Fermentation of captured COD can produce an effluent rich in volatile fatty acids (VFA), which can be used for mainstream or sidestream biological phosphorus removal or denitrification.

Shortcut nitrogen removal allows for upstream carbon diversion, providing maximal potential for decarbonization as the aeration, alkalinity, and carbon requirements for nitrogen removal can be drastically reduced, the plant capacity can be increased through SRT reduction, and diverted carbon can be used to generate electricity, heat, or VFA to use as a carbon source elsewhere in the process. Maximizing the efficient use of influent COD for denitrification further decreases the carbon requirements of the process and allows for more carbon redirection. If carbon redirection cannot be implemented, shortcut nitrogen removal processes still provide increased plant capacity and reduce the exogenous carbon costs of nitrogen removal, but the full benefits of these processes will not be realized.

7.3.5 Process control

Ammonia vs. NO_x (AvN) control for nitrite shunt works by controlling the aerobic fraction to meet an NH_4^+ to NO_x ratio in the effluent. In addition to transient anoxia the controller uses high DO, aerobic SRT control, and high residual ammonia concentration to favor AOB over NOB (Regmi *et al.*, 2014). Although AvN control was developed to achieve nitrite shunt through NOB out-selection, AvN and related control approaches have the potential to provide more efficient nitrogen removal than ABAC, even if the goal is not nitrite shunt. By setting ammonia and NO_x equal in the effluent, or by specifying a ratio of NH_4^+/NO_x somewhat less than 1.0 (one) based on the need to comply with an effluent ammonia limit, AvN control oxidizes only the amount of ammonia that can be denitrified utilizing the influent organic carbon that is made available. This maximizes COD utilization efficiency for heterotrophic denitrification using NO_3^- or NO_2^- without the addition of supplemental carbon. This can be achieved with either continuous or intermittent aeration. Another option for AvN is the slope-intercept control concept using the following equation: $NH_4^+ = slope*NO_x + intercept$ where the slope controls the NH_4^+/NO_x ratio and the intercept controls the ammonia effluent limit. When implemented for mainstream deammonification, the slope will be higher as more nitrogen is removed through the anammox pathway. Controlling the ratio of NH_4^+ to NOx is required in the first stage of two-stage PNA and PdNA systems in order to meet the proper stoichiometry for anammox downstream.

7.4 SHORTCUT NITROGEN REMOVAL IMPLEMENTATIONS

7.4.1 Sidestream treatment

Sidestream recycles from dewatered anaerobic digestate have high ammonia, low C/N ratio, and higher temperatures than mainstream. Low C/N and alkalinity/NH_4^+ ratios make conventional nitrification/denitrification extremely inefficient for sidestream nitrogen removal. At the same time, the high ammonia concentration and warmer temperatures make PNA the obvious choice for sidestream treatment due to the ease of NOB out-selection. Any process that utilizes heterotrophic denitrification (traditional nitrification/denitrification, nitrite shunt, PdNA) does not make sense for sidestream treatment because single stage sidestream PNA has proven to be robust and reliable and is the most carbon efficient process. Although sidestream shunt (Hellinga *et al.*, 1998) and PdNA (Sharp *et al.*, 2017) processes do exist, PNA is the preferred treatment option. As of 2014 there were more than 100 full-scale installations of sidestream PNA processes worldwide (Lackner *et al.*, 2014) with many more successful installations to date (Cao *et al.*, 2017).

7.4.2 Mainstream PNA/nitrite shunt

All of this research has been carried out at pilot and bench scale and has led to only two full-scale implementations: a trial of mainstream bioaugmentation of anammox granules from the sidestream with a retention screen on the mainstream WAS (Wett *et al.*, 2015) and the reported discovery of PNA in a step-feed BNR process with low DO in a warm climate (Cao *et al.*, 2018).There are examples of sustained full-scale NOB out-selection but it is typically in warmer climates, and it is not always clear how to repeat the success in other facilities (Cao *et al.*, 2018; Jimenez *et al.*, 2020). It seems clear that the stability, reliability, and overall effectiveness of NOB out-selection are limiting the implementation of full-scale mainstream PNA.

7.4.3 Mainstream PdNA

Due to the difficulty of mainstream NOB out-selection, PdNA is emerging as the most stable approach to shortcut nitrogen removal. During the course of PNA research, an alternative pathway of partial denitrification/anammox, or PdNA, was identified and shown to provide virtually all of the same benefits of reduction in facility size and operational costs, but with considerably greater reliability and ease of implementation (Campolong *et al.*, 2019; Le *et al.* 2019; Ma *et al.* 2016). Although it took a few years from beginning to experiment with PdNA to realizing the full benefits, PdNA is heading to full-scale implementation remarkably fast. The first full-scale installation was at the York River Treatment Plant in 2018 when the deep-bed denitrification filter was transitioned to a PdNA process (Fofana *et al.* in progress).

7.4.4 Partial denitrification/anammox (PdNA) case study

York River Treatment Plant (YRTP) is a 57 000 m^3/d facility with screening, grit removal, primary clarification, and fully aerobic plug flow step feed aeration tanks transitioning in summer 2018 to two-pass step-feed BNR with defined anoxic zones operated in AvN control followed by deep-bed PdNA filters (Figure 7.4). Prior to that the process was full nitrification followed by full denitrification in the filters. A sidestream PNA process was installed in 2012 which reduced the methanol additional

Figure 7.4 The Hampton Roads Sanitation District (HRSD) York River Treatment Plant (YRTP) in Seaford, VA.

to the denitrification filters by approximately 25% (Nifong *et al.*, 2013). AvN control was achieved manually through a combination of aeration and step feed control. After secondary clarification and intermediate pumping are deep-bed filters configured for sensor-driven methanol feed control. The filters were sized for an expansion of the plant and are rated for 114 000 m^3/day, a nitrate load of 0.4 kgN/m^3/day and 5.9 m/hr. At current flows (2021), the filter is underloaded at an average flow rate of 2.4 m/hr. Ammonia removed anoxically in the denitrification filter meant that less ammonia needed to be oxidized upstream in the BNR process. So, when 50% of the aeration volume was turned into anoxic zones with step feed to better utilize influent carbon for denitrification, the PdNA filters were able to make up for this loss of aerobic SRT (Table 7.2). The transition from full nitrification to two-pass step feed resulted in a methanol saving of approximately 60%. The transition from full-denitrification to PdNA, which soon followed the implementation of two-pass step feed, resulted in an additional 50% methanol saving (Fofana *et al.* in progress). This saving in units of methanol added per nitrogen removed is shown in Table 7.2.

7.5 CONCLUSIONS AND FUTURE OUTLOOK

Nitrogen removal, particularly to very stringent discharge limits, is a primary driver for increased energy usage, chemical use (and the embedded energy contained), and capital primarily associated with concrete tanks and associated equipment. The various processes mentioned in this chapter provide discussion on opportunities and challenges for biological N removal with regards to decarbonization. Other processes targeted at the removal or recovery of ammonia such as electrochemical, thermal, or chemical treatment technologies may provide future alternatives, but at this stage are limited to laboratory scale investigations. Further development is needed to understand the potential of these

Table 7.2 YRTP capacity increase and methanol reduction due to PdNA process implementation.

	BNR Process Aerobic SRT (days)	Filter COD Added/N Removed (g/g)
Full nitrification/denitrification	7.6	4.4
AvN control + PdNA	4.8	1.8

processes. Whole plant level optimization is needed to coordinate C, N, P removal and recovery in order to accomplish decarbonization of the sector.

While N recovery through urine separation may be cost-effective and practical, the recovery of N-based fertilizers from wastewater remains challenging, as the cost of biological N removal is declining with shortcut opportunities, and emerging technologies that have the potential to recover N are currently not cost-competitive with Haber–Bosch production of N fertilizers. N recovery is generally limited to the land application of biosolids and biosolids products and the limited amount of N that is precipitated with struvite recovery.

One of the great challenges in preparing a summary of advances in N removal technology is that each facility is very different from a treatment technology standpoint. Other relevant differences include TN and NH_4^+ limits, energy costs, chemical costs, and 'available' capacity. Decarbonization often depends on the starting point or the baseline plant condition, and this cannot be generalized or ignored.

For example, a small plant with stringent TN limits, a five-stage Bardenpho process, and no effective DO control (gross over aeration with DO at say 4–6 mg/L) could gain considerable energy and supplemental carbon chemical benefits by transitioning to ABAC. This seems like an obvious improvement. However, we must be careful not decarbonize wastewater treatment at the expense of plant capacity. Carrying our example further, if this same plant implements ABAC moving to an average DO setpoint of say 0.5 mg/L, then the implication is an inherent loss in nitrification capacity; decreased effective capacity of the plant in terms of aeration tank volume and secondary clarifier area. This is often an unacceptable compromise, even though some optimization of this plant is clearly warranted.

The goal of decarbonizing in the context of N removal must always consider capacity implications. This is critical from the standpoint of gaining utility consensus and buy-in. In fact, as our industry contemplates advanced N removal technologies, the ideal direction involves decarbonizing at the same time as increasing effective capacity. This is collectively known as intensification. It is quite relevant here, because the push towards shortcut N removal can provide the dual benefit of decarbonization with no compromise in capacity, and in fact often depending on the baseline scenario, a dramatic increase in plant capacity, and/or the excess aerobic zone capacity could be dedicated to anaerobic or pre-anoxic zones to better utilize wastewater carbon for N and P removal. The real key to intensification is directing NH_4^+ to anoxic oxidation to N_2 by anammox, allowing for decreased operating/design aerobic SRT while at the same time ensuring that reliably low TN and NH_4^+ limits can be met.

Sidestream PNA processes can be an important and nearly obvious part of that intensification incentive. These processes are now mature and available, but they are not 'plug-and-play,' still requiring considerable operator attention and knowledge for successful operation. Even then, upsets do occur.

After more than 15 years of research, mainstream PNA has been mostly unsuccessful. The few reports of full-scale testing an implementation do not give any indication that significant amounts of influent NH_4^+ are being directed to anammox. Pilot and laboratory testing results have shown promise in some cases, but this has not led to scalable technology. The industry has generally determined that the low growth rate of anammox can be accommodated by selective retention using biofilms and granules. However, consistent and reliable out-selection of NOB has proven very difficult, and this is perhaps made even more complicated by the existence of comammox. With what we know now, there is little promise of legitimate mainstream anammox technology that relies on PNA. That said, additional cost savings could be provided by processes that can periodically take advantage of unsustained and unreliable NO_2^- production from NOB repression by directing residual NH_4^+ and NO_2^- to anammox-based polishing processes.

Mainstream PdNA can provide this opportunity to take advantage of NO_2^- produced either from partial denitrification or from periods of NOB out-selection but, more importantly, it offers performance reliability and the intensification benefit. It is argued here that PdNA does not make sense to be considered for sidestream treatment, but it is immediately valuable for mainstream in the situations where low TN limits are required. PdNA seems best applied in the form of a post polishing process or integrated into a downstream anoxic zone in a step feed BNR or Bardenpho-style process. In these polishing PdNA applications (anoxic zone or post-secondary process), perhaps 10–20% of the influent TKN can be directed to anammox. This is a huge benefit from a capacity standpoint, and it offers considerable operating cost savings benefits. The question remains whether PdNA can be further developed to allow a larger fraction of the influent TKN to be directed to anammox. This of course requires two things – use of wastewater COD for partial denitrification and likely some degree of carbon diversion. Although the benefits may never be quite as good as PNA, the foundation has been laid for mainstream PdNA.

REFERENCES

Al-Omari A., Wett B., Nopens I., De Clippeleir H., Han M., Regmi P., Bott C. and Murthy S. (2015). Model-based evaluation of mechanisms and benefits of mainstream shortcut nitrogen removal processes. *Water Science and Technology*, **71**(6), 840–847. https://doi.org/10.2166/wst.2015.022

Anthonisen A. C., Loehr R. C., Prakasam T. B. S. and Srinath E. G. (1976). Inhibition of nitrification by ammonia and nitrous acid. *Journal (Water Pollution Control Federation)*, **48**(5), 835–852.

Beckinghausen A., Odlare M., Thorin E. and Schwede S. (2020). From removal to recovery: an evaluation of nitrogen recovery techniques from wastewater. *Applied Energy*, **263**, 114616. https://doi.org/10.1016/j.apenergy.2020.114616

Campolong C., Klaus S., Rosenthal A., Sabba F., Baidme M., Wells G., Wett B., Clippeleir H. D., Chandran K. and Bott C. (2019). Comparison of external carbon sources for a polishing partial denitrification/anammox MBBR. Proceedings of the Water Environment Foundation, 56–63.

Cao Y. S., Kwok B. H., Yong W. H., Chua S. C., Wah Y. L. and Yahya A. G. (2013). The main stream autotrophic nitrogen removal in the largest full scale activated sludge process in Singapore: process analysis. Proceedings of WEF/IWA Nutrient Removal and Recovery 2013: Trends in Resource Recovery and Use. Water Environment Federation, Alexandria, Virginia, pp. 28–31.

Cao Y., van Loosdrecht M. C. M. and Daigger G. T. (2017). Mainstream partial nitritation–anammox in municipal wastewater treatment: status, bottlenecks, and further studies. *Applied Microbiology and Biotechnology*, **101**(4), 1365–1383. https://doi.org/10.1007/s00253-016-8058-7

Cao Y., Kwok B. H., van Loosdrecht M. C. M., Daigger G., Png H. Y., Long W. Y. and Eng O. K. (2018). The influence of dissolved oxygen on partial nitritation/anammox performance and microbial community of the 200 000 m^3/d activated sludge process at the Changi water reclamation plant (2011 to 2016). *Water Science and Technology*, **78**(3), 634–643. https://doi.org/10.2166/wst.2018.333

Chen H., Zeng L., Wang D., Zhou Y. and Yang X. (2020). Recent advances in nitrous oxide production and mitigation in wastewater treatment. *Water Research*, **184**, 116168. https://doi.org/10.1016/j.watres.2020.116168

Courtens E. N. P., De Clippeleir H., Vlaeminck S. E., Jordaens R., Park H., Chandran K. and Boon N. (2015). Nitric oxide preferentially inhibits nitrite oxidizing communities with high affinity for nitrite. *Journal of Biotechnology*, **193**, 120–122. https://doi.org/10.1016/j.jbiotec.2014.11.021

Cui B., Yang Q., Liu X., Wu W., Liu Z. and Gu P. (2020). Achieving partial denitrification-anammox in biofilter for advanced wastewater treatment. *Environment International*, **138**, 105612. https://doi.org/10.1016/j.envint.2020.105612

Daigger G. T. (2014). Oxygen and carbon requirements for biological nitrogen removal processes accomplishing nitrification, nitritation, and anammox. *Water Environment Research*, **86**(3), 204–209. https://doi.org/10.2175/106143013X13807328849459

Daims H., Lücker S. and Wagner M. (2016). A new perspective on microbes formerly known as nitrite-oxidizing bacteria. *Trends in Microbiology*, **24**(9), 699–712. https://doi.org/10.1016/j.tim.2016.05.004

Delgado Vela J., Stadler L. B., Martin K. J., Raskin L., Bott C. B. and Love N. G. (2015). Prospects for biological nitrogen removal from anaerobic effluents during mainstream wastewater treatment. *Environmental Science & Technology Letters*, **2**(9), 234–244. https://doi.org/10.1021/acs.estlett.5b00191

Di Capua F., Pirozzi F., Lens P. N. L. and Esposito G. (2019). Electron donors for autotrophic denitrification. *Chemical Engineering Journal*, **362**, 922–937. https://doi.org/10.1016/j.cej.2019.01.069

Du R., Peng Y., Cao S., Wang S. and Niu M. (2016). Characteristic of nitrous oxide production in partial denitrification process with high nitrite accumulation. *Bioresource Technology*, **203**, 341–347.

Du R., Cao S., Li B., Niu M., Wang S. and Peng Y. (2017). Performance and microbial community analysis of a novel DEAMOX based on partial-denitrification and anammox treating ammonia and nitrate wastewaters. *Water Research*, **108**, 46–56.

Duan H., van den Akker B., Thwaites B. J., Peng L., Herman C., Pan Y., Ni B.-J., Watt S., Yuan Z. and Ye L. (2020). Mitigating nitrous oxide emissions at a full-scale wastewater treatment plant. *Water Research*, **185**, 116196. https://doi.org/10.1016/j.watres.2020.116196

Fofana R., Parsons M., Long C., Chandran K., Jones K., Klaus S., Trovato B., Wilson C., De Clippeleir H. and Bott C. (submitted). Full-scale transition from denitrification to Partial denitrification – anammox (PdNA) in Deep-Bed filters: Operational strategies for and benefits of PdNA implementation. *Water Environment Research*.

Gao D.-W., Huang X.-L., Tao Y., Cong Y. and Wang X. (2015). Sewage treatment by an UAFB–EGSB biosystem with energy recovery and autotrophic nitrogen removal under different temperatures. *Bioresource Technology*, **181**, 26–31. https://doi.org/10.1016/j.biortech.2015.01.037

Gilbert E. M., Agrawal S., Brunner F., Schwartz T., Horn H. and Lackner S. (2014a). Response of different nitrospira species to anoxic periods depends on operational DO. *Environmental Science and Technology*, **48**(5), 2934–2941. https://doi.org/10.1021/es404992g

Gilbert E. M., Agrawal S., Karst S. M., Horn H., Nielsen P. H. and Lackner S. (2014b). Low temperature partial nitritation/anammox in a moving bed biofilm reactor treating low strength wastewater. *Environmental Science and Technology*, **48**(15), 8784–8792. https://doi.org/10.1021/es501649m

Gustavsson D. J. I., Suarez C., Wilén B.-M., Hermansson M. and Persson F. (2020). Long-term stability of partial nitritation-anammox for treatment of municipal wastewater in a moving bed biofilm reactor pilot system. *Science of the Total Environment*, **714**, 136342. https://doi.org/10.1016/j.scitotenv.2019.136342

Han M., Vlaeminck S. E., Al-Omari A., Wett B., Bott C., Murthy S. and De Clippeleir H. (2016). Uncoupling the solids retention times of flocs and granules in mainstream deammonification: a screen as effective out-selection tool for nitrite oxidizing bacteria. *Bioresource Technology*, **221**, 195–204. https://doi.org/10.1016/j.biortech.2016.08.115

Hellinga C., Schellen A., Mulder J. W., van Loosdrecht M. C. M. and Heijnen J. J. (1998). The SHARON process: an innovative method for nitrogen removal from ammonium-rich waste water. *Water Science and Technology*, **37**(9), 135–142. https://doi.org/10.2166/wst.1998.0350

Jetten M. S. M., Horn S. J. and van Loosdrecht M. C. M. (1997). Towards a more sustainable municipal wastewater treatment system. *Water Science and Technology*, **35**(9), 171–180. https://doi.org/10.2166/wst.1997.0341

Jimenez J., Wise G., Regmi P., Burger G., Conidi D., Du W. and Dold P. (2020). Nitrite-shunt and biological phosphorus removal at low dissolved oxygen in a full-scale high-rate system at warm temperatures. *Water Environment Research*, **92**(8), 1111–1122. https://doi.org/10.1002/wer.1304

Kampschreur M. J., Temmink H., Kleerebezem R., Jetten M. S. and van Loosdrecht M. C. (2009). Nitrous oxide emission during wastewater treatment. *Water Research*, **43**(17), 4093–4103. https://doi.org/10.1016/j.watres.2009.03.001

Kartal B., Kuenen J. G. and van Loosdrecht M. C. M. (2010). Sewage treatment with anammox. *Science*, **328**(5979), 702–703. https://doi.org/10.1126/science.1185941

Kornaros M., Dokianakis S. N. and Lyberatos G. (2010). Partial nitrification/denitrification can be attributed to the slow response of nitrite oxidizing bacteria to periodic anoxic disturbances. *Environmental Science and Technology*, **44**(19), 7245–7253. https://doi.org/10.1021/es100564j

Lackner S., Gilbert E. M., Vlaeminck S. E., Joss A., Horn H. and van Loosdrecht M. C. M. (2014). Full-scale partial nitritation/anammox experiences – an application survey. *Water Research*, **55**, 292–303. https://doi.org/10.1016/j.watres.2014.02.032

Lackner S., Welker S., Gilbert E. M. and Horn H. (2015). Influence of seasonal temperature fluctuations on two different partial nitritation-anammox reactors treating mainstream municipal wastewater. *Water Science and Technology*, **72**(8), 1358–1363. https://doi.org/10.2166/wst.2015.301

Laureni M., Falås P., Robin O., Wick A., Weissbrodt D. G., Nielsen J. L., Ternes T. A., Morgenroth E. and Joss A. (2016). Mainstream partial nitritation and anammox: long-term process stability and effluent quality at low temperatures. *Water Research*, **101**, 628–639. https://doi.org/10.1016/j.watres.2016.05.005

Le T., Peng B., Su C., Massoudieh A., Torrents A., Al-Omari A., Murthy S., Wett B., Chandran K., DeBarbadillo C., Bott C. and Clippeleir H. D. (2019). Impact of carbon source and COD/N on the concurrent operation of partial denitrification and anammox. *Water Environment Research*, **91**(3), 185–197. https://doi.org/10.1002/wer.1016

Liu M., Krikorian E., Knapp T., Melitas N., Zhao H. and Johnson M. (2018). Mainstream ANITA Mox pilot testing at the joint water pollution control plant. *Proceedings of the Water Environment Federation*, **2018**(5), 169–184. https://doi.org/10.2175/193864718824940385

Liu Y.-J., Gu J. and Liu Y. (2018). Energy self-sufficient biological municipal wastewater reclamation: present status, challenges and solutions forward. *Bioresource Technology*, **269**, 513–519. https://doi.org/10.1016/j.biortech.2018.08.104

Lotti T., Kleerebezem R., Hu Z., Kartal B., Jetten M. S. M. and van Loosdrecht M. C. M. (2014). Simultaneous partial nitritation and anammox at low temperature with granular sludge. *Water Research*, **66**, 111–121. https://doi.org/10.1016/j.watres.2014.07.047

Ma B., Zhang S., Zhang L., Yi P., Wang J., Wang S. and Peng Y. (2011). The feasibility of using a two-stage autotrophic nitrogen removal process to treat sewage. *Bioresource Technology*, **102**(17), 8331–8334. https://doi.org/10.1016/j.biortech.2011.06.017

Ma B., Wang S., Cao S., Miao Y., Jia F., Du R. and Peng Y. (2016). Biological nitrogen removal from sewage via anammox: recent advances. *Bioresource Technology*, **200**, 981–990. https://doi.org/10.1016/j.biortech.2015.10.074

McCullough K., Klaus S., Parsons M., Wilson C. and Bott C. B. (2022). Advancing the Understanding of Mainstream Shortcut Nitrogen Removal: Resource Efficiency, Carbon Redirection, and Plant Capacity. *Water Research* (submitted).

Miller M. W., Bunce R., Regmi P., Hingley D. M., Kinnear D., Murthy S., Wett B. and Bott C. B. (2012). A/B process pilot optimized for nitrite shunt: high rate carbon removal followed by BNR with ammonia-based cyclic aeration control. *Proceedings of the Water Environment Federation*, **2012**(10), 5808–5825. https://doi.org/10.2175/193864712811709607

Morales N., Val del Río Á., Vázquez-Padín J. R., Méndez R., Campos J. L. and Mosquera-Corral A. (2016). The granular biomass properties and the acclimation period affect the partial nitritation/anammox process stability at a low temperature and ammonium concentration. *Process Biochemistry*, **51**(12), 2134–2142. https://doi.org/10.1016/j.procbio.2016.08.029

Mulder A. (2003). The quest for sustainable nitrogen removal technologies. *Water Science and Technology*, **48**(1), 67–75. https://doi.org/10.2166/wst.2003.0018

Ni B.-J. and Yuan Z. (2015). Recent advances in mathematical modeling of nitrous oxides emissions from wastewater treatment processes. *Water Research*, **87**, 336–346. https://doi.org/10.1016/j.watres.2015.09.049

Nifong A., Nelson A., Johnson C. and Bott C. B. (2013). Performance of a full-scale sidestream DEMON® deammonification installation. *Proceedings of the Water Environment Federation*, **2013**(13), 3686–3709. https://doi.org/10.2175/193864713813685700

Fofana R., Bachmann M., Akyon B., Jones K., Delgado J., Klaus S., Parsons M., Bott C., deBarbadillo C. and De Clippeleir H. (2021). Carbon Source Selection for Deep-Bed Partial Denitrification: Anammox (PdNA) Polishing Filters. WEFTEC 2021. Chicago, IL: WEF.

Pérez J., Lotti T., Kleerebezem R., Picioreanu C. and van Loosdrecht M. C. (2014). Outcompeting nitrite-oxidizing bacteria in single-stage nitrogen removal in sewage treatment plants: a model-based study. *Water Research*, **66**, 208–218. https://doi.org/10.1016/j.watres.2014.08.028

Piculell M., Christensson M., Jönsson K. and Welander T. (2016a). Partial nitrification in MBBRs for mainstream deammonification with thin biofilms and alternating feed supply. *Water Science and Technology*, **73**(6), 1253–1260. https://doi.org/10.2166/wst.2015.599

Piculell M., Suarez C., Li C., Christensson M., Persson F., Wagner M., Hermansson M., Jönsson K. and Welander T. (2016b). The inhibitory effects of reject water on nitrifying populations grown at different biofilm thickness. *Water Research*, **104**, 292–302. https://doi.org/10.1016/j.watres.2016.08.027

Pijuan M., Torà J., Rodríguez-Caballero A., César E., Carrera J. and Pérez J. (2014). Effect of process parameters and operational mode on nitrous oxide emissions from a nitritation reactor treating reject wastewater. *Water Research*, **49**, 23–33. https://doi.org/10.1016/j.watres.2013.11.009

Poot V., Hoekstra M., Geleijnse M. A., van Loosdrecht M. C. and Pérez J. (2016). Effects of the residual ammonium concentration on NOB repression during partial nitritation with granular sludge. *Water Research*, **106**, 518–530. https://doi.org/10.1016/j.watres.2016.10.028

Regmi P., Miller M. W., Holgate B., Bunce R., Park H., Chandran K., Wett B., Murthy S. and Bott C. B. (2014). Control of aeration, aerobic SRT and COD input for mainstream nitritation/denitritation. *Water Research*, **57**, 162–171. https://doi.org/10.1016/j.watres.2014.03.035

Regmi P., Holgate B., Miller M. W., Park H., Chandran K., Wett B., Murthy S. and Bott C. B. (2016). Nitrogen polishing in a fully anoxic anammox MBBR treating mainstream nitritation–denitritation effluent. *Biotechnology and Bioengineering*, **113**(3), 635–642. https://doi.org/10.1002/bit.25826

Sharp R., Niemiec A., Khunjar W., Galst S. and Deur A. (2017). Development of a novel deammonification process for cost effective separate centrate and main plant nitrogen removal. *International Journal of Sustainable Development and Planning*, **12**(1), 11–21.

Sliekers A. O., Haaijer S. C. M., Stafsnes M. H., Kuenen J. G. and Jetten M. S. M. (2005). Competition and coexistence of aerobic ammonium- and nitrite-oxidizing bacteria at low oxygen concentrations. *Applied Microbiology and Biotechnology*, **68**(6), 808–817. https://doi.org/10.1007/s00253-005-1974-6

Soler-Jofra A., Pérez J. and van Loosdrecht M. C. M. (2021). Hydroxylamine and the nitrogen cycle: a review. *Water Research*, **190**, 116723. https://doi.org/10.1016/j.watres.2020.116723

Song K., Riya S., Hosomi M. and Terada A. (2015). Effect of carbon sources on nitrous oxide emission in a modified Ludzak Ettinger process. *Water Science and Technology*, **72**(4), 572–578.

Strous M., Heijnen J. J., Kuenen J. G. and Jetten M. S. M. (1998). The sequencing batch reactor as a powerful tool for the study of slowly growing anaerobic ammonium-oxidizing microorganisms. *Applied Microbiology and Biotechnology*, **50**, 589–596. https://doi.org/10.1007/s002530051340

Tchobanoglous G., Burton F.L. and Stensel H.D. (2003). Wastewater Engineering: Treatment and Reuse, McGraw-Hill, Boston.

van Kessel M. A. H. J., Speth D. R., Albertsen M., Nielsen P. H., Op den Camp H. J. M., Kartal B., Jetten M. S. M. and Lücker S. (2015). Complete nitrification by a single microorganism. *Nature*, **528**(7583), 555–559. https://doi.org/10.1038/nature16459

Von Sperling M. and de Lemos Chernicharo C. (2005). Biological Wastewater Treatment in Warm Climate Regions Volume I, IWA Publishing. See https://doi.org/10.2166/9781780402734 (Accessed 25 February 2022).

Vlaeminck S. E., Clippeleir H. D. and Verstraete W. (2012). Microbial resource management of one-stage partial nitritation/anammox. *Microbial Biotechnology*, **5**(3), 433–448. https://doi.org/10.1111/j.1751-7915.2012.00341.x

Wang Q., Ye L., Jiang G., Hu S. and Yuan Z. (2014). Side-stream sludge treatment using free nitrous acid selectively eliminates nitrite oxidizing bacteria and achieves the nitrite pathway. *Water Research*, **55**, 245–255. https://doi.org/10.1016/j.watres.2014.02.029

Welker S., Horn H. and Lackner S. (2016). Substrate contentment: Influence of residual ammonium and dissolved oxygen concentrations on autotrophic nitrogen removal. Proceedings of WEF/IWA Nutrient Removal and Recovery, Denver, USA. Water Environment Federation, Alexandria, Virginia.

Wett B., Podmirseg S. M., Gómez-Brandón M., Hell M., Nyhuis G., Bott C. and Murthy S. (2015). Expanding DEMON sidestream deammonification technology towards mainstream application. *Water Environment Research*, **87**(12), 2084–2089. https://doi.org/10.2175/106143015X14362865227319

Winkler M. K. and Straka L. (2019). New directions in biological nitrogen removal and recovery from wastewater. *Current Opinion in Biotechnology*, **57**, 50–55. https://doi.org/10.1016/j.copbio.2018.12.007

Winkler M.-K. H., Kleerebezem R. and van Loosdrecht M. C. M. (2012). Integration of anammox into the aerobic granular sludge process for main stream wastewater treatment at ambient temperatures. *Water Research*, **46**(1), 136–144. https://doi.org/10.1016/j.watres.2011.10.034

Wunderlin P., Mohn J., Joss A., Emmenegger L. and Siegrist H. (2012). Mechanisms of N_2O production in biological wastewater treatment under nitrifying and denitrifying conditions. *Water Research*, **46**(4), 1027–1037. https://doi.org/10.1016/j.watres.2011.11.080

Ye Y., Ngo H. H., Guo W., Liu Y., Chang S. W., Nguyen D. D., Liang H. and Wang J. (2018). A critical review on ammonium recovery from wastewater for sustainable wastewater management. *Bioresource Technology*, **268**, 749–758. https://doi.org/10.1016/j.biortech.2018.07.111

doi: 10.2166/9781789061796

Chapter 8

Decarbonization potentials in phosphorus management in the water sector

*Annalisa Onnis-Hayden[1], Dongqi Wang[2], Ali Akbari[3], Mi Nguyen[3] and April Z. Gu[3]**

[1]Northeastern University, Boston, MA, USA
[2]Xi'an University of Technology, Xi'an, Shaanxi, China
[3]Cornell University, Ithaca, NY, USA
**Correspondence: aprilgu@cornell.edu*

8.1 OVERVIEW OF GLOBAL PHOSPHORUS CONSUMPTION AND DEMAND IN RELATION TO WATER SUSTAINABILITY

Global food security relies on a sustained phosphorus supply, yet P supplies are limited to identified resources in mines, distributed in a few countries. China, United States, and Morocco are major producers, contributing 53, 10, and 10% of global production, respectively (Jasinski, 2020). There are mixed messages in the literature about the sustainability of currently known global P resources in relation to its increasing demand rate. Some studies have projected the mine reserves of P will be depleted in the next 50–100 years (Dubrovsky *et al.*, 2010). Others have indicated that the currently known reserves, assuming a 4.44% increase in annual production rate per capita (2006–2016), would suffice for several centuries (Vaccari *et al.*, 2019). A recent USGS report suggests that 'there is no imminent shortage of phosphate rock' (Jasinski, 2020). However, the current production pattern will leave Morocco with 71% of the world's known reserves as the only major producer, and that will have socio-economic and political consequences.

P is the vital, irreplaceable element for fertilizer and increasingly expanding biofuel production. Therefore, it is relevant to food and energy security. The limited access and global scarcity of this finite resource are to become amongst the next century's most significant challenges. On the other hand, 80% of mined P is lost due to inefficient utilization and other losses in agricultural runoff and animal wastes. The P loss into the environment further causes widespread eutrophication. Eutrophication is the leading cause for freshwater impairment and is a threat to global water security. For example, in North America, 55% of water resources are eutrophic, imposing substantial direct and indirect costs (Anderson *et al.*, 2002). The direct costs are related to control activities such as wastewater treatment plants and urban and residential runoff management. External, indirect costs have been steadily increased in the last decades (estimated at $2.2 billion just in the United States) and concerns tourism and recreation, commercial fishing, property values, human and animal health, drinking water treatment utilities, mitigation, and restoration activities (Dodds *et al.*, 2009).

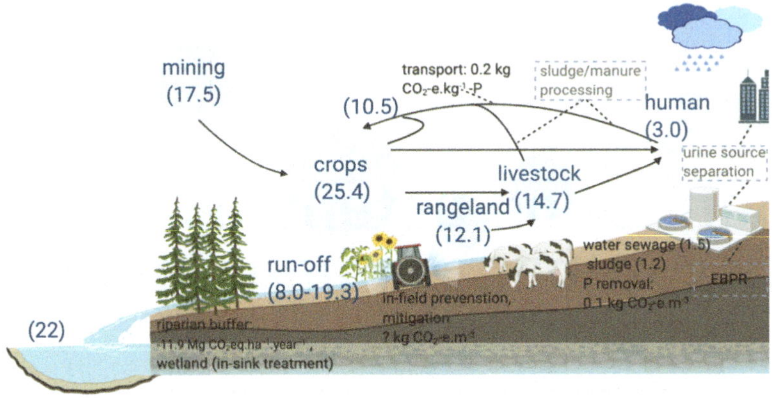

Figure 8.1 Mass flow analysis of major global phosphorus sinks and sources depicting the significant P lost through the cycle. The values in parentheses are annual load in million metric tons and retrieved from Liu *et al.* (2008), Cordell *et al.* (2009) and Rittmann *et al.* (2011). Different places where decarbonization is relevant are included with the associated carbon footprint. The carbon footprint of riparian buffer, and P removal from wastewater are from Coats *et al.* (2011) and Styles *et al.* (2016), respectively. For transport the carbon footprint was calculated assuming 40 km transport of manure by a diesel truck.

The overall global P flow (Figure 8.1) starts with mining, and most of the mined P (80% of mined P, 25.4 Mmmt. yr^{-1} including recycled P) is used for agricultural purposes (Rittmann *et al.*, 2011). The harvested crops are either utilized for livestock or directly used in the food industry. Industrial applications and the production of detergents and P supplements for animal feeding constitute a smaller fraction of P demand (Liu *et al.*, 2008). Due to the extensive losses through various processing stages, only13–17% of mined P (3 $Mmt.yr^{-1}$) is eventually consumed by humans. Losses through non-point sources such as agricultural run-off and erosion constitute at least half of mined P (>8 $Mmt.yr^{-1}$ (Rittmann *et al.*, 2011). The global P load into water resources (22 $Mmt.yr^{-1}$) are mainly originated from non-point sources (about 90%) (Cordell *et al.*, 2009). As high as 30% of P used in croplands are applied to soils with previously high residual P levels which makes soils the largest pool of P in the environments (Alewell *et al.*, 2020; Bouwman *et al.*, 2013). The prolonged but finite accumulation phase of P in soils requires adaptive management strategies to protect water resources. Despite the significant volume of P released from non-point sources, the management of these sources has gained relatively less attention. In comparison, there have been more advancements in removal and recovery technologies from point sources, including municipal wastewater treatment plants or industrial settings such as confined livestock facilities.

Integrated source-watershed management, containment, treatment, and recovery strategies are needed to mitigate the environmental impacts associated with P flow through various processes and environmental compartments. Holistic thinking requires that the associated carbon footprints with various integrated interventions be taken into account.

For example, the phosphorus trading among point and non-point sources is getting more attention and is likely to be implemented increasingly in the near future. In such a scenario, more stringent effluent requirements from point-source treatment units are traded with implementing non-point source management facilities such as constructed wetlands at the sink (area of release to surface waterways). An example, resembling the carbon cap and trade, is a case in Dixie drain facility in Boise, Idaho, where P trading led to a significantly less costly integrated, regional P management (Macintosh *et al.*, 2018). Because of the high carbon footprints of advanced chemical-based P removal technologies needed to achieve the more stringent effluent requirements, integrated trading likely offers a more sustainable resort. Previous studies have quantified significant total direct and indirect carbon footprints of P removal processes in WWTPs (e.g., 0.1 kg CO_2-e.m^{-3} treated wastewater (Coats *et al.*, 2011; Rahman *et al.*, 2016) as a major source of emission in the P management cycle. That underlines the significance of developing technologies with lower carbon footprints. The local, regional, direct, and indirect impacts of P removal strategies should be assessed, and the environmental hotspots identified toward sustainability-oriented improvements.

8.1.1 P management and decarbonization potential pathways

Sustainable P management is only possible through a multi-tier approach, as described in Figure 8.2. At the highest level, management strategies aim to reduce the P demand by improving utilization efficiencies and internal recycling. The avoided burden then contributes to sustainable P management and a reduced direct and indirect carbon footprint. Optimizing nutrient timing, dose, and application techniques, for example, reduces the P demand in agriculture. Manure, crop residues, and sewage are currently partially recycled to croplands, and the total recycled volume is estimated at 60% of annual global mined P. A 100% manure recycling can reduce the global P demand by half (Vaccari *et al.*, 2019). The transport of recycled manure to application sites is one of the carbon-intensive hot spots of the P cycle. Affordable, novel technologies with a low carbon footprint for manure concentration, sludge dewatering, or P extraction would reduce the transport demand and therefore offer an improvement toward the decarbonization of the P management.

At the middle-level are preventive and containment measures. For non-point sources, strategies like field vegetative buffers, diverse cropping systems, water table management, and landscape management practices can reduce the release and transport of P in the environment. In the case of municipal wastewater, source separation from households reduces the volume of wastewater and so the energy demand of processes such as aeration. It further generates concentrated waste streams, enabling more efficient P/resource recovery processes. Urine is only 1% of municipal wastewater volume but contains more than half of its total P content (Maurer *et al.*, 2006). A recent LCA study demonstrated that an integrated urine separation and P recovery imposes lower environmental burdens than conventional wastewater collection, transport, and P recovery units (Hilton *et al.*, 2020). Similarly, for non-point sources, application of alum at source areas which are smaller in size but responsible for the highest P loss (known as critical source area) as opposed to blanket field application significantly reduces the cost (McDowell, 2015; Smith & McDowell, 2016) and likely reduces the carbon footprint of the

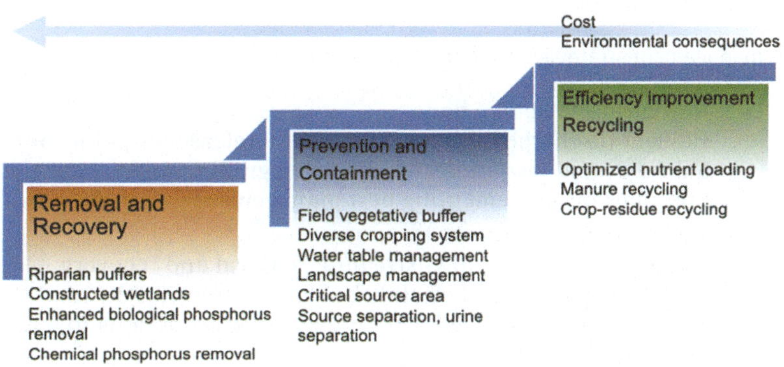

Figure 8.2 Various technologies related to P management from point or nonpoint sources at multiple levels. The cost and environmental consequences are likely to increase from higher-level management practices for utilization efficiency improvement to lower levels of containment treatment and recovery.

mitigation process due to the lower chemical use. Generally, preventive strategies are less expensive and more sustainable alternatives to removal and treatment practices.

At the lowest level is the most expensive and environmentally consequential resort of removal and recovery. These strategies are the last line of defense to protect the water resources and include riparian buffers and wetlands for non-point sources and biological/chemical phosphorus removal and recovery units in wastewater treatment plants. The surface layer in a wetland with high P content provides an opportunity for recycling the lost P to the agriculture field. An external carbon-independent modification of biological enhanced phosphorus removal units has been introduced, offering a lower carbon footprint than conventional technologies (Wang *et al.*, 2019). Novel treatments and recovery technologies are yet to be developed toward more sustainable P management. A systematic comparative sustainability analysis, particularly for non-point source P management practices, is currently missing. Such holistic analyses should introduce and assess multi-tier, integrated management strategies.

8.1.2 Phosphorus management and policy: Current status and practice

As per technologies, we accordingly categorized the existing P management strategies in a multi-tier framework (Table 8.1). These strategies are interdependent and are often being used simultaneously to provide proper P management.

At the highest level, regulatory approaches (also called command-and-control approaches) set standards and limits to control P release into water bodies. For point sources, detergent phosphate bans and effluent phosphorus limits are two main regulatory approaches. The phosphate detergent bans, which are to limit phosphorus load from households to wastewater treatment facilities, have been implemented in several countries and states in the US. In the US, a nationwide voluntary ban of phosphates in laundry detergents began in 1994 (Litke, 1999), followed by a ban of high-phosphate automatic dishwasher detergents in July 2010 in 17 states (Cohen & Keiser, 2017). In Europe, the European Parliament in 2011 ordered a ban of phosphates in laundry detergent by June 2013 and a ban in dishwasher detergent by January 2017 (European Commission, 2011). Other countries that have regulated the concentration of phosphates in detergents include Australia (2014), Brazil (2008), Canada (2010), China (2009), and Japan (1979) (Chong *et al.*, 2019).

Economic incentives in the form of incentive payments, subsidies, or low-interest loans are the voluntary approach to encourage the adoption of practices or actions to reduce the amount of P release into the environment from both point- and non-point sources. In Europe, the EU's Common Agricultural Policy (CAP) provides green payments to farmers who reduce fertilizer use, introduce organic farming measures, and promote biodiversity. A support of 57.9 billion EUR was provided to EU farmers in 2019 (European Commission, 1962). In the US, existing voluntary agricultural

Table 8.1 Existing P management strategies in a multi-tier framework.

Regulatory and incentive approaches	• Detergent phosphate bans • Effluent discharge limit for control of P release for point sources • Subsidies and low-interest loans for upgrading wastewater treatment plants • Subsidies for implementing best management practices to reduce P losses
Preventive and containment practices	• Technologies to prevent P runoff • Strategies to enable downstream P removal and recovery • Partnership between key stakeholders to prevent P losses • Technical assistance, education and outreach actions to address the need of reducing P losses
P removal and recovery strategies	• Technologies to remove and recover P from waste streams • Technical and financial assistance to build stakeholders' capacity in removing and recovering P

programs including the Environmental Quality Incentives Program (EQIP) and the Conservation Technical Assistance Program (CTA) are provided to farmers by the US Department of Agriculture to encourage the implementation of nutrient management practices. The EQIP provides financial and technical assistance to agricultural producers to address natural resource concerns and deliver environmental benefits. The CTA program is to provide technical assistance to land users in planning and implementing conservation systems including nutrient management for improving air, soil and water quality. For municipalities and wastewater treatment plants, the US Clean Water State Revolving Fund (CWSRF) loan program has been offering loans to help fund green projects. Through 2014, the CWSRF program has provided 34 902 project assistance agreements ($105.4 billion) to communities.

At the middle level, preventive and containment measures provide better control of P release into the environment with adopting technologies to prevent P runoff, promoting strategies to enable downstream P removal and recovery, creating key partnerships and/or bridging institutions, and building capacity for stakeholders. Technical and financial assistance to increase the capacity of stakeholders, together with environmental education and outreach activities to raise public awareness can encourage actions to reduce P losses into the environment. For example, the US Department of Agriculture provides technical and financial support to farmers implementing P best management practices to minimize P loss at the source, to mitigate the transport of P and problems associated with excess P in water (Sharpley *et al.*, 2006). For large regional issues, creating partnerships and/or building bridging institutions provide a better platform for coordination among stakeholders. For example, the Chesapeake Bay program as a bridging institution brought together stakeholders from federal, state, local, academic, and nongovernmental organizations to build and adopt policies that support the Bay's restoration (Jones & Tippie, 1983). As part of the same program, states within the Chesapeake Bay drainage area agreed to incorporate Chesapeake Bay issues into school curriculums as part of the Chesapeake 2000 agreement.

The lowest level of P management is to provide P removal and recovery technologies. These management strategies include providing technical and financial assistance for research and adopting improvements in P removal and recovery. Technical and financial assistance is important for building the knowledge and skills required to begin removing and recovering P from the waste streams.

To control P release from wastewater treatment plants into the environment, many countries employed standards to regulate the concentration limit of P in treated municipal wastewater. The main legal act regulating the quality of treated municipal wastewater in EU countries is the Council Directive of May 21, 1991 (or so-called Wastewater Directive) (Council of European 1991). In the US, the National Pollution Discharge Elimination System permit program created in the Clean Water Act is the main regulation for treated wastewater discharged. In general, two major factors determining the limit concentration of phosphorus in treated wastewater effluent are the size of the wastewater treatment plant and the sensitivity to eutrophication of the water bodies receiving treated wastewater. The limit concentration of total P ranges from 0.5 to 2 mg/L for EU members, 0.1–1 mg/L for states in the US, 0.5 mg/L for China (Taihu Lake catchment), and 1 mg/L for Canada (Preisner *et al.*, 2020).

In Europe, efforts to recover P from waste streams based on the circular economy paradigm have been promoted in the European Green Deal and Circular Economy Action Plan (Bianchini & Rossi, 2020). Projects funded by the European Institute of Innovation and Technology Raw Materials and Climate KIC provided recommendations and practical actions for a more sustainable P management in Baltic Sea countries (Bianchini & Rossi, 2020). For example, in case of manure P which is one of the major sources of P release into the environment, removal and recovery technologies can be categorized into: (1) solid–liquid separation technologies; (2) technologies for processing the solid fraction; and (3) technologies for processing the liquid section. A list of selected technologies for P removal and recovery from animal manure is presented in section 8.3. More than 50 P removal and recovery technical approaches are known and have been developed for municipal wastewater (Egle *et al.*, 2016). Based on the access points in wastewater treatment plants, P recovery can be conducted via: (1) direct usage of sludge as soil amendments; (2) recovery from the aqueous phase either before or after sludge dewatering process; (3) recovery from sludge during or after incineration (Egle *et al.*, 2016). A list of selected technologies for P recovery from municipal wastewater is presented in section 8.3.

In general, existing P management approaches were developed and conducted to reduce the negative impacts of P losses into the watershed, which can accelerate freshwater eutrophication and cause water-quality impairment (Litke, 1999; Sharpley & Tunney, 2000). The singular focus on developing technologies for attaining near to complete P removal from waste streams would make it more costly and un-sustainable when overall environmental impacts such as the carbon footprint is overlooked. In the next section, we present the existing and emerging P management practices for water quality improvements in conjunction with possible opportunities for reducing carbon footprint to allow decision makers to select the most sustainable practices.

8.2 DIRECT DECARBONIZATION AND INDIRECT CARBON REDUCTION STRATEGIES FROM POINT SOURCE AND NON-POINT SOURCES OF PHOSPHORUS

There are opportunities for both direct and indirect reduction of carbon emissions at various points and levels in the P source-to-sink flows including agricultural runoff, stormwater, animal manures, food and food-processing wastes, human urine and feces, municipal wastewater and biosolids. Direct decarbonization or carbon input reduction can occur in strategies and technologies that require direct carbon input such as enhanced biological phosphorus removal at wastewater treatment plants, or technologies/strategies that utilize plants to sequester carbon. Indirect carbon reduction refers to those strategies that indirectly lead to carbon footprint reduction throughout the life cycle of a given process, for example, by reducing chemical and energy demand or reducing transportation. The following section will review the available technologies and strategies for removal and recovery of P and identify the decarbonization potential of implementing such technologies at various levels for different P-rich waste streams. Table 8.2 identifies for the different waste stream the current removal/ recovery processes and opportunities for carbon footprint reduction.

8.2.1 Phosphorus in agricultural waste streams

Worldwide, the agriculture sector, including crop and livestock production, forestry and associated land use changes, contribute to up to 30% of the anthropogenic greenhouse gas (GHG) emissions. Indeed, according to the Intergovernmental Panel on Climate Change (IPCC), in 2005 emissions from agriculture alone were 5.1–6.1 Gt CO_2eq yr^{-1}, which represented about 10–12% of the total estimated anthropogenic emissions in the same year.

In addition, agriculture non-point sources are often identified as the greatest source of P to eutrophic water bodies (Dubrovsky *et al.*, 2010). Agriculture produces different P-rich waste streams: agriculture run-off, a nonpoint source pollution from farms caused by surface runoff from fields during rainstorms; also, farms with large livestock and poultry operations are a major source of point source wastewater from activities such as milking, animal washing, as well as flushed spilled feed, urine, and manure. In the United States, these facilities are called concentrated animal feeding operations (CAFO) or confined animal feeding operations. Reducing GHG emission from agriculture, while also controlling P loads to water bodies, is an urgent need; therefore, identifying opportunities of decarbonization within the management of P-rich waste streams from agriculture is of great importance.

8.2.1.1 Agricultural point sources: best management practices for decarbonization

The world's livestock population of 65 billion produces enormous quantities of manure annually, which contain ten times the P annual demand of agriculture (Naidu *et al.*, 2012). Land application of manure has been a recommended management practice, leading to improve soil C sequestration as well as providing an integrated nutrient management strategy (Lal, 2004), as indeed agricultural soil can serve as a potential sink for atmospheric carbon as soil organic carbon (SOC), which also contributes to improve productivity and quality (Kundu *et al.*, 2007; Rudrappa *et al.*, 2006).

Table 8.2 P-rich waste stream, current approaches and opportunities for decarbonization.

	P-rich Waste Stream	Methods for P Removal/Recovery	Opportunities for Carbon Footprint Reduction
Agriculture waste streams	Manure and other agriculture effluents	• Land application of excreta • Thermal gasification • Physical/chemical precipitation of struvite • Anaerobic digestion • Enhanced biological P removal • Oxidation lagoon • Constructed wetland	• Soil carbon capture • Reduced GHG from transportation • Reduced fertilizers application • Recover energy by capturing methane
	Agricultural runoff	• Source control • Ecological ditches • Constructed wetland • Filter Strips and Riparian forest buffers	• Carbon storage in vegetation/tree • Improved water quality
Industrial effluent	Industrial effluents	• Physical/chemical processes (e.g., precipitation of struvite, or calcium phosphate, electrochemical P removal/recovery, etc.) • Anaerobic digestion • Biological processes	• Recover renewable energy and products, such as methane, H2, electricity, and others • Recover nutrients and organic matter to use as fertilizer
Urban waste streams	Municipal wastewater	• Biological processes • Chemical processes • Physical/chemical processes (e.g., precipitation of struvite, or calcium phosphate)	• Reduced external carbon input in EBPR processes • Reduced Energy for aeration and/ or mixing • Reduced chemical addition • Integrated resource recovery (e.g., recover nutrients to use as fertilizer)
	Municipal sewage sludge	• Thermochemical treatment • Anaerobic processes (e.g., anaerobic digestion, anaerobic membrane processes)	• Reduced GHG from transportation and energy use • Recover renewable energy and products, such as methane, H_2, electricity, and others • Recover nutrients and organic matter to use as fertilizer
	Human excrete	• Land application of excreta • Urine diversion	• Reduced GHG from transportation and energy use • Recover nutrients and organic matter to use as fertilizer and chemicals
	Stormwater runoff	• Infiltration beds (grass swales and porous pavements) • Filtration systems (sand filters, vegetated filter strips, etc.) • Retention/Detention Basins (Dry ponds, wet ponds and inline storage) • Constructed Wetlands	• Flood prevention and reduction in municipal pumping demand and energy costs • Reduce heat island effect – heating and cooling energy savings • Carbon storage in vegetation/tree

Table 8.3 Strategies and technologies to recover resources and reduce the emissions of GHG in each step of the manure management chain.

Manure Management Step	Strategy/ Technology	Main Purpose	Main Benefit	Maturity Level
Solid–liquid separation	Screw press Decanter centrifuge Belt filter	Separate slurry manure to liquid and solid portions	• Reduce the formation of ammonia and GHG • Improve nutrient recovery • Facilitate manure handling activities	Commonly applied on farms
Processing of the solid fraction	Pelletizing	Densify manure to form pellets	• Facilitate manure handling activities • Recover nutrients and organic matter to use as fertilizer	Commercially applied to process poultry manure
	Composting	Transform biodegradable organic materials to humic substances	• Recover organic matter to use as soil amendment	Widely applied on farms
Processing of the liquid fraction	Constructed wetland	Utilize natural processes to treat liquid manure	• Improve water quality • Recover nutrients and energy in plant biomass • Carbon storage in plants	Applied on farms where moist conditions can be maintained
	Thermochemical conversion	Use heat to breakdown manure into gases, hydrocarbon fuels and charcoal/ash	• Recover energy and nutrients	Developed and researched at lab-scale level
	Struvite crystallization	Produce struvite	• Recover phosphorus and some nitrogen	Applied at lab- and pilot-scale levels
	Enhanced biological phosphorus removal	Transfer phosphate in the bulk liquid into polyphosphate in a biological process	• Facilitate phosphorus recovery	Applied at lab- and pilot-scale levels

However, soil P surplus and potential pollution of water resources are common consequences of land application of manure in regions with high concentrations of concentrated animal feeding operations (CAFO). Manure transport for distances above ten miles is not economical and usually not practiced. Another common manure management system includes uncovered anaerobic lagoons, which have been identified by the US Environmental Protection Agency as the largest source of methane from farms (Owen & Silver, 2015). These examples clearly indicate the need for management alternatives to resolve agronomic P imbalances for more effective recycling of manure P, and simultaneous carbon footprint reduction. Table 8.3 presents strategies and technologies to recover P and reduce the emission of GHG in the management of manure in addition to land application.

8.2.1.2 Agricultural runoff: best management practices for decarbonization

Agricultural runoff is estimated to be responsible for 50–70% of the total input loading of phosphorus to lakes and streams around the world (Xia *et al.*, 2020). Appropriate management of agricultural runoff, in addition to animal waste is a large concern for the US Environmental Protection Agency (USEPA) and US Department of Agriculture (USDA).

At three levels of prevention, containment, and removal, various technologies have been proposed for agricultural runoff control to reduce both nitrogen and phosphorus loading. The most successful strategies applied to agriculture runoff are conservation tillage and fertilization management (source control – prevention level), ecological ditches (process control – containment level), and finally constructed wetland, buffer strips and riparian forest (end treatment – removal level).

Conservation tillage can sequester carbon, as it is known to increase the soil organic carbon (SOC) content of the surface layer (Lal & Kimble, 1997). At the same time, phosphorus losses in agricultural runoff are significantly reduced with this practice (Liu *et al.*, 2014).

An ecological ditch is an engineered system that has been developed for the removal of agricultural runoff nutrients by sorption, sedimentation, transformation, plant uptake and microbial metabolic activities. Ecological ditches are fundamentally similar to free water surface constructed wetlands (Nsenga Kumwimba *et al.*, 2018) and therefore share the same decarbonization potential of carbon sink within the vegetation in the wetland.

Riparian forest buffers and filter strips and constructed wetland are generally implemented for filtering pollutants from agricultural runoff and preventing their entry into nearby waterbodies. Because forest riparian buffers include trees, which other filter strips often do not, in addition to improving water quality, they are particularly effective for long-term atmospheric carbon sequestration (Rheinhardt *et al.*, 2012).

8.2.2 Phosphorus in industrial effluent: Best management practices for decarbonization

Industries such as paper, fermentation, and wineries, cheese production and other food processing industries, produce large volumes of wastewater which is rich in organic load and nutrients. For example, the US produces 11 103 ML winery effluent annually which contains 116 kt P. They are also responsible for significant GHG emission. As an example, dairy processing, which is among the most energy- and carbon-intensive activities within the global food production industry, has an estimated annual emission of over 128 Mt CO_2 (Xu & Flapper, 2009). The environmental regulation in most countries necessitates the use of appropriate technology to reduce the P content before safe disposal of industrial effluents. Meat and poultry, pulp and paper industry and vegetables, fruits and juices industry wastewater are often treated in lagoons, which are a significant contributor to GHG emission.

EBPR has been demonstrated for phosphorus removal from various types of industrial wastewaters such as food processing wastewater (60–100 mg P L^{-1}) (Mulkerrins *et al.*, 2004) and livestock wastewater (>100 mg P L^{-1}) (Kishida *et al.*, 2009). For wastewaters containing phosphorus and organic carbon at concentrations that are much higher than those in domestic wastewater, an alternative to EBPR is to apply anaerobic treatment upfront to recover bioenergy and then to recover phosphorus through struvite precipitation in the effluent stream of the anaerobic treatment (Yuan *et al.*, 2012). Recovery of P from high-strength industrial effluents is considered a viable strategy for their safe disposal and a source of fertilizer for agricultural use (Altinbas *et al.*, 2002).

8.2.3 Phosphorus in domestic waste streams

Domestic sewage collection and treatment at wastewater treatment plants contribute directly and indirectly to the emission of GHG. Direct emissions are primarily related to the biological processes occurring at the plant (emissions of CO_2 from microbial respiration, N_2O from nitrification and denitrification, and CH_4 from anaerobic digestion), whereas indirect emissions are associated with the energy demand at the plant itself or at associated facilities/operation (e.g., third-party biosolids hauling, production of chemicals and their transportation to the plant, etc.).

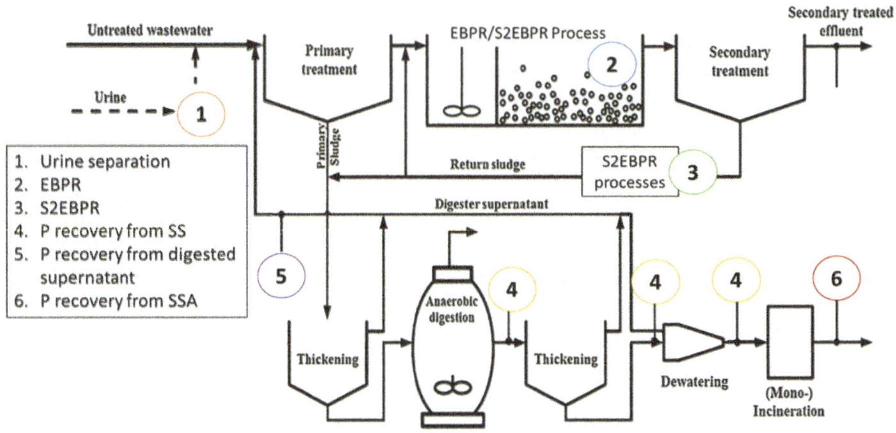

Figure 8.3 Strategies and technologies to recover P resources and reduce the emissions of GHG in each step of the domestic wastewater management chain.

In addition, the discharge of P-rich domestic wastewater significantly contributes to the eutrophication of river and lakes. For example, in the UK up to 70% of P load to rivers has been attributed to sewage discharges (Bunce *et al.*, 2018). This reality has resulted in tightening P discharge standards to reduce P loads entering rivers and lakes, particularly to ecologically sensitive locations, and targeted P-removal has become increasingly common at WWTPs throughout the world. Wastewater treatment plants possess a large but often exploited and inefficiently used potential source of P. Therefore, similar to P-rich agricultural waste streams, strategies for recovery of phosphorus from P-rich streams of domestic wastewater should be evaluated (Cordell *et al.*, 2011; Egle *et al.*, 2016). Opportunities for carbon reduction, to both direct and indirect sources, exist at wastewater treatment plants designed to remove and or recover phosphorus; in addition, supplementary opportunities exist with alternate handling strategies such as urine separation (Figure 8.3). The following sections evaluate the various P treatment/recovery options and associated C reduction opportunities, for domestic P-rich waste streams.

8.2.3.1 Source separated streams: best management practices for decarbonization

According to Cordell, 3–3.3 million metric tons of phosphorus are generated globally in human excreta (i.e., feces and urine) (Cordell *et al.*, 2009); of this amount, about 10–50% are reused annually (Liu *et al.*, 2008). Application of human excreta as organic fertilizer is common both in Asia and in Europe, however, it is far less common elsewhere in the world.

Urine diversion and subsequent nutrient recovery has been increasingly discussed and considered, even though it is inconsistent with well-developed biosolid processing and recycling to farmlands approach. The average human produces 0.8–1.6 L of urine per day which makes up less than 1% of total wastewater flow in developing countries' wastewater collection systems.

Separate urine collection largely improves the potential for nutrient recovery, because the concentrations of both N and P are 100 times higher than in wastewater; in addition, urine diversion, using close-loop sanitation technologies (e.g., urine-separating compost toilets) and P recovery from urine (mainly via Struvite/CaP precipitation) have been shown to be technologically feasible in both the developed and developing world settings (Mihelcic *et al.*, 2011).

A modeling study conducted by Wilsenach and van Loosdrecht (2003) showed that urine separation significantly decreases the energy requirement for wastewater treatment. In their study, the energy requirements (aeration, mixing, sludge dewatering, sludge incineration, pumping) and energy generation (via and methane gas produced by the anaerobic sludge digestion) of an advanced BNR

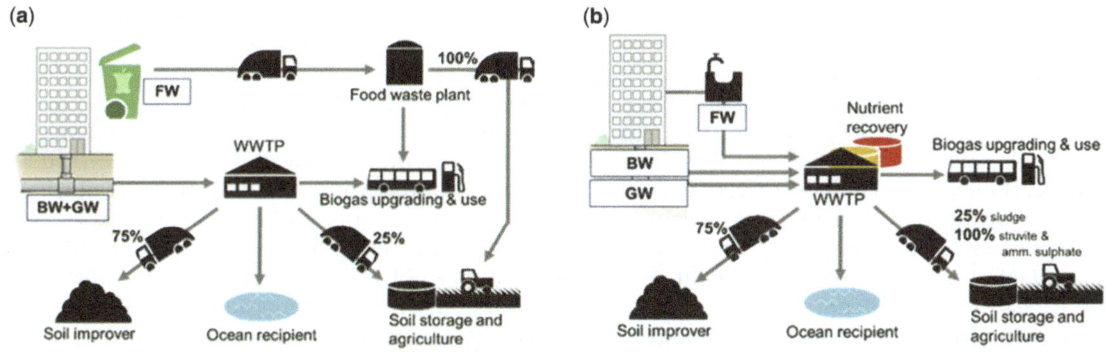

Figure 8.4 Comparison of conventional and source separation systems. BW, GW, and FW refer to blackwater, graywater and food waste, respectively. The figure is from Kjerstadius *et al.* (2017).

process were compared with the integrated processes that treat urine and wastewater separately, and recover P via struvite precipitation. Their findings indicated that, advanced BNR processes significantly increases energy and chemical consumption, requiring around 6 W per person, however, the integrated treatment/recovery process could produce more than 1 W per person.

Numerous LCA studies have been conducted to compare the carbon footprint/environmental impact of different scenarios for removal/recovery/recycling of phosphorus in human excreta or removing them in an advanced WWTP (Bradford-Hartke *et al.*, 2015; Hilton *et al.*, 2020; Kavvada *et al.*, 2017; Spångberg *et al.*, 2014; Xue *et al.*, 2016), and all of the studies have indicated that urine diversion reduced most environmental impacts through a wide range of conditions (reduced flushing, reduce chemical use at BNR plants, better energy recovery from AD, etc.). It can be a particularly effective decarbonization strategy in areas with high levels of nutrient removal, electricity produced primarily from fossil fuels, and relatively little wastewater per capita.

Kjerstadius *et al.* (2017) also conducted an LCA study, but evaluated the decarbonization potential of source separation systems for the management of domestic wastewater and food waste. In the conventional system, blackwater (BW) and graywater (GW) are collected together and treated at a wastewater treatment plant whereas separated food waste is collected by garbage trucks and treated at a dedicated anaerobic digestion plant (see Figure 8.4). In the source separation system considered by the study, BW was collected and then treated together with FW in an anaerobic digestion unit, and nutrient recovery was performed on the digestate effluent by struvite precipitation and ammonia stripping (to produce ammonium sulfate). GW was treated separately in an activated sludge unit. The results for carbon footprint and nutrient recovery (phosphorus and nitrogen) concluded that the source separation system could increase nutrient recovery (0.30–0.38 kg P per capita per year), while decreasing the carbon footprint (-24 to -58 kg CO_2-eq. capita^{-1} year^{-1}), compared to the conventional system. The carbon footprint decreased, mainly due to energy recovery from the increased biogas production, increased replacement of mineral fertilizer in agriculture and less emissions of nitrous oxide from wastewater treatment.

8.2.3.2 Domestic wastewater treatment: best management practices for decarbonization

While studies on source separation of human excreta have clearly indicated high decarbonization opportunities, urine separation is still not widespread and as indicated by Hilton *et al.* (2020), the challenges for the development of large-scale urine collection and processing systems are economy, market, regulatory framework, lack of confidence in performance, and the existence of built conventional systems. Therefore, it is important to investigate existing infrastructure for P removal and recovery and identify decarbonization potential within these existing systems.

Numerous opportunities for direct and indirect C reduction associated with P removal and recovery processes at wastewater treatment plants exists and are evaluated in depth in section 8.3.

8.2.4 Phosphorus in urban runoff and best management practices for decarbonization

Residential lawns and turf areas (e.g., sports fields, golf courses and parks) in urban environments are considered as 'hotspots' of total and dissolved phosphorus input into stormwater (Müller *et al.*, 2020). Moreover, fallen leaves and other detritus are often considered the primary contributors of nutrients to urban stormwater, especially in areas with high overhead tree canopies (George *et al.*, 2012).

A wide variety of stormwater control measures, also known as best management practices (BMPs), are available that can play a significant role in the treatment of urban runoff, such as infiltration beds (grass swales and porous pavements), filtration systems (sand filters, vegetated filter strips, etc.), retention/detention basins (dry ponds, wet ponds and inline storage) and constructed wetlands (Sample *et al.*, 2012).

These strategies are used in conjunction with other measures to reduce the quantity of urban runoff and reduce the impact of urban run-off on water quality (e.g., green roof, stormwater drain, etc). All these strategies are part of large program and initiatives such as the 'Sponge Cities' in China (Zevenbergen *et al.*, 2018), 'Water Sensitive Cities' in Australia (Wong & Brown, 2009), 'Sustainable Urban Drainage Systems' in the UK (Ashley *et al.*, 2015), 'Low Impact Development' in the US (Coffman, 2000) and New Zealand (Shaver, 2000).

Decarbonization opportunities with the implementation of these BMPs and strategies include reduction in municipal pumping demand and energy costs associated with the added flood mitigation; reduction of heat island effect, which results in heating and cooling energy savings; and carbon storage in vegetation or trees in some of the BMPs.

8.3 DECARBONIZATION IN PHOSPHORUS REMOVAL AND RECOVERY PROCESSES

Many WWTPs are facing challenges to achieve lower effluent nutrient levels while reducing chemical/energy requirements and the corresponding carbon footprint with the current technologies and available resources. Innovations in treatment strategies that improve P removal and recovery performance and stability while minimizing the chemical cost, energy consumption, and environmental impacts are therefore urgently required. Table 8.4 summarizes decarbonization strategies in P removal and recovery processes, which are described in detail in the following sections.

8.3.1 Carbon requirements in enhanced biological phosphorus removal processes

An enhanced biological phosphorus removal (EBPR) process has been successfully applied to achieve low phosphorus (P) concentrations in effluents from the WWTPs. EBPR mainly relies on polyphosphate accumulating organisms (PAOs) which are capable of anaerobically assimilating carbon substrate (e.g., volatile fatty acids (VFAs)) as intracellular metabolites (e.g., polyhydroxyalkanoates (PHAs)), and aerobically utilizing the stored polymers for luxury P uptake and polyphosphate (PolyP) synthesis (Oehmen *et al.*, 2007). P is removed from the EBPR process through removing the excess activated sludge with high PolyP content. The available and suitable carbon substrate is essential for PAOs' metabolisms and the corresponding EBPR performance. The influent carbon to phosphorus (C/P) ratio is positively correlated with the efficiency and stability of EBPR systems, and the minimum readily biodegradable COD (rbCOD) to P ratio was generally recommended as 15:1–25:1. However, in practice, many conventional EBPR facilities suffer from the insufficient and fluctuating carbon source in influent, which is mainly related to local sewage characteristics, or high level of sewage in-line biodegradation and leakage (such as the cases in China described by Cao *et al.* 2019), eventually leading to an undesirable nutrient removal performance. Therefore, external carbon addition via commercial sources or on-site primary sludge fermentation has often been practiced in many WWTPs for enhancing EBPR.

Table 8.4 Summary of advantages and challenges of different decarbonization methods in P removal and recovery processes.

	Method		Advantages	Challenges
Decarbonization in EBPR processes	Decarbonization via operational strategies for EBPR	Advanced aeration control	• 40% less of energy requirement • May favor PAOs at low DO	• Require better understanding of PAO-GAO competition
		Optimizing carbon source and chemical addition	• Reduce carbon footprint without compromise in performance	• Not eliminate all the carbon footprint inputs
	Decarbonization via new pathway/ process for EBPR	S2EBPR	• Improved performance without external carbon supplement • Improved carbon utilization efficiency for EBPR • Less susceptibility to fluctuating influent impacts • Improved denitrification • Less energy for O&M and sludge treatment	• Additional fermentation unit is sometimes required • Larger anaerobic zone is sometimes required
		DPAO-based process	• Both N and P removal with a minimized COD utilization and oxygen requirement • 20–30% less of sludge production	• DPAO activity varied largely • Potential accumulation of N_2O
	Combined EBPR with innovations in N removal process with decarbonization potential	Combined EBPR with nitritation/ denitritation (nitrite shunt)	• 25% less of oxygen requirement • 40% less of carbon requirement	• No well-established process in full-scale
		Combined EBPR with partial nitritation/ anammox	• 60% less of oxygen requirement • 90% less of carbon requirement • 75% less of sludge production	• No well-established process in full-scale • Potential accumulation of N_2O
		Coupled aerobic-anoxic nitrous decomposition operation (CANDO) + P	• Similar reductions in carbon, aeration demands and sludge production compared to nitrite shunt process • N_2O conversion to N_2 with energy recovery	• No well-established process in full-scale

(Continued)

Table 8.4 Summary of advantages and challenges of different decarbonization methods in P removal and recovery processes (*Continued*).

Method			Advantages	Challenges
		Recovery from the liquid phase	• Lower emissions • Lower energy demands • Reduction of chemicals • Recovery of nitrogen	• Low rates of P recovery
Additional Technologies for P removal and recovery from wastewater streams	Methods for P recovery (via precipitation and/ or crystallization with or without membrane, ion exchange technologies, electrochemical adsorption or thermal treatment)[a]	Recovery from sludge (solid phase)	• Recovery of nitrogen	• Large amount of chemicals used
		Recovery from sludge ash	• Reduction of gaseous emissions • Reduction of energy demand • Heavy metal decontamination • Destruction of organic micropollutants	• higher recycling rate
	Methods for carbon sequestration	Constructed wetland	• Capturing CO_2 to biomass • Joint nutrient removal • Less or no energy, carbon and chemical demands • Flood control	• Limited P removal capacity due to media saturation • Potential greenhouse gas emission
		Microalgae cultivation	• Capturing CO_2 to biomass • Joint nutrient removal • Less or no energy, carbon and chemical demands • Biofuel production • Soil amendment • Bioplastic production	• Limited performance at low temperature and light supply
	Genetically modified PAOs for enhanced bio-P adsorption		• Improved P removal and recovery performance • Less chemical and energy inputs	• No well-established process in full-scale

A variety of carbon sources have been applied in EBPR systems for achieving high P removal efficiency and stability (Table 8.5) which increases the overall treatment cost. The main drawbacks of adding external carbon source in EBPR systems include:

1. The increase in the overall carbon dioxide emissions for the production and transport of added commercial carbon sources leading to uneconomical and unsustainable operational practices.
2. Safety issues associated with the transport, handling, and storage of the external carbon sources.

Table 8.5 Summary of advantages and limitations of different carbon sources for EBPR.

Carbon Type			**Example**	**Advantages**	**Drawbacks**
External carbon sources	Single carbon source	VFA	Acetate, propionate	Promote EBPR performance in most cases	High cost; high carbon footprint; storage problems; overdosage often induces GAO-like metabolism
		Alcohol	Ethanol, glycerol	More economical compared to VFAs	High carbon footprint; an adaptation period is sometimes required; longer anaerobic phase is sometimes required
		Sugar	Glucose	May favor PAOs with fermentation capacity (e.g. *Tetrasphaera*)	High cost; high carbon footprint; an adaptation period is sometimes required; glucose has been reported to favor GAOs over PAOs
		Amino acid	Aspartate, glutamate, glycine	Favor different groups of PAOs selectively	High cost; high carbon footprint; deterioration in EBPR performance was sometimes observed for unknown reason
	Complex carbon source		Casein hydrolysate, yeast extract, peptone	Favor *Tetrasphaera* PAOs; provide diverse organic compounds for PAOs	High cost; high carbon footprint
	Industrial wastes		Crude glycerol, agro-food industrial wastewaters	The utilization of waste materials reduces cost and carbon footprint	Pre-fermentation may be required
Internal carbon sources/ on-site sludge fermentation	Primary sludge		Primary sludge fermentate; A-stage sludge	No external carbon addition; sludge reduction	Additional treatment unit is required; odor
	Activated sludge		Return activated sludge; mixed liquor	Improved P removal performance without external carbon addition; fewer odors; sludge reduction	Additional treatment unit or larger anaerobic zone is sometimes required

3. Long adaptation periods required in the startup process to acclimate the bacterial/PAO community for preferential utilization specific carbon source.
4. The increased sludge production rate and the operational cost of water treatment and sludge processing as a result of addition of external organic carbon.

Alternatively, the practice of on-site primary sludge fermentation is widely established in many WWTPs, which reduces the overall carbon input in EBPR process. However, the carbon supply from primary sludge fermentation is usually not adequate to ensure efficient P removal, as the VFA production is often affected by several environmental and operational factors, such as influent properties, process configuration, SRT, HRT, pH, temperature, and so on. Other drawbacks of implementing on-site primary sludge fermentation in EBPR systems include:

- infeasible particularly for facilities with no primary treatment unit;
- increased footprint from additional construction and operation costs;
- potential odors from fermenter;
- reduced energy recovery via anaerobic digestion;
- potential effects of recalcitrant organic compounds and nutrients derived from fermentation step on EBPR.

8.3.2 Carbon footprint reducing via operational strategies for EBPR

To meet the increasingly stringent P limits, a great number of facilities without consistent and effective EBPR performance are forced to increasingly rely on chemical flocculants (e.g., Al, Fe salts) or more advanced tertiary treatments (e.g., coagulation combined with membrane or media filtration) as backup or routine method for P removal. These technologies are independent of the influent carbon shortage, but inevitably have higher operation cost and carbon footprint in WWTPs. Dosage of sporadic metal salt also reduces the P recovery efficiency and increases sludge production, further increasing carbon footprints in WWTPs. Therefore, the modifications and optimizations of EBPR process that simultaneously and efficiently removes nutrients and reduces carbon footprint are highly desired. This section provides the possible operational strategies for decarbonization in tne EBPR process.

8.3.2.1 Advanced aeration control

In an EBPR system, aeration for aerobic P uptake and microbial growth is the major energy intensive step. A proper (dissolved oxygen) DO and aeration control can save up to 40% of the energy demand. Previous studies have shown that Accumulibacter PAOs have an advantage over Competibacter GAOs at low DO levels, as PAOs had a higher oxygen affinity and thus largely maintained their aerobic activity (Carvalheira *et al.*, 2014). Therefore, an appropriate aeration control can potentially improve EBPR performance, decrease the energy costs and offer decarbonization potential in WWTPs. Currently, advanced aeration control strategies (e.g., ammonia-based aeration control (ABAC), ammonia vs. nitrate (AVN) control) along with the model-based predictive and optimization (e.g., bioprocess intelligent optimization system (BIOS)) have been successfully implemented in some biological nitrogen removal processes (Maktabifard *et al.*, 2018). Similar innervations can also be applied for EBPR system control, under the premise of better understanding of the PAO-GAO competition and their metabolism.

8.3.2.2 Optimizing carbon source and chemical addition

The amount of external carbon required in WWTPs could be efficiently and economically reduced with online monitoring of influent wastewater quality and from a developed numerical model. This monitoring practice thus reduces the potential carbon footprint without compromise in EBPR performance. Similarly, the dosing of chemical salts in the chemical P removal (CPR) process could also be optimized. However, the optimized carbon or chemical dosing strategies could not eliminate the significant energy consumption and greenhouse gas emissions in chemical treatment processes.

8.3.3 Carbon footprint reducing via new pathway/process for EBPR

8.3.3.1 Innovations in P removal process – S2EBPR

Recently, a modified EBPR technology, side-stream EBPR (S2EBPR), has been developed and applied in over 80 full-scale WWTPs. The S2EBPR demonstrated significantly improved process stability (Barnard *et al.*, 2017; Gu *et al.*, 2019). In S2EBPR configuration, a portion or all of return activated sludge (RAS) or mixed liquor is diverted through a sidestream anaerobic reactor, to the mainstream anoxic or aerobic zone receiving influent. The continuous supply of internal carbon (e.g., VFAs) via sidestream hydrolysis/fermentation of RAS or mixed liquor can be utilized for mainstream nutrient removal, therefore reducing or eliminating the needs for external carbon addition and minimizing chemical usage (Onnis-Hayden *et al.*, 2020; Srinivasan *et al.*, 2021; Wang *et al.*, 2019). The advantages of S2EBPR process include:

- Improved P removal and recovery performance without external carbon supplement and improved carbon utilization efficiency for EBPR.
- Less and no-direct dependence on influent carbon loads therefore less susceptibility to fluctuating influent impacts.
- Improved denitrification resulting from influent carbon diversion to anoxic denitrification zones.
- Easy implementation for a variety of existing WWTPs configurations.

A full-scale comparative study conducted by the recent WERF project (U1R13) has demonstrated that S2EBPR configuration improved P removal performance and stability more than the conventional A2O configuration (Gu *et al.*, 2019; Srinivasan *et al.*, 2021; Wang *et al.*, 2019). Under the special selection conditions in the sidestream anaerobic reactor, a more diverse EBPR microbiome can be obtained. The diversified microbiome offer functional redundancy and complementation and therefore better resilience to perturbations. The extended anaerobic retention time in S2EBPR processes can provide a competitive advantage to PAOs over GAOs and allow more efficient carbon utilization by PAOs. Minimizing the growth of GAOs improves the efficiency of organic carbon and oxygen usage for P removal, thus resulting in lower P effluent concentration and carbon footprint. The longer anaerobic/anoxic zones and/or intermittent mixing implemented in the S2EBPR processes require less energy for operation and maintenance, exhibiting a decarbonization potential compared to conventional EBPR configurations. In addition, the sidestream anaerobic sludge hydrolysis/fermentation reduces the daily sludge production and reduces the corresponding energy consumption in the sludge treatment process.

8.3.3.2 Combined EBPR with innovations in N removal process

More efficient and cost-effective methods than well-established conventional nutrient removal methods (i.e., denitrification with exogenous carbon addition to remove N as well as chemical precipitation to remove P) with less chemical/energy input are required in WWTPs. Innovative N removal bioprocesses that 'short-circuit' the conventional nitrification–denitrification paradigm offer the opportunity to dramatically decrease aeration and carbon requirements for N removal, thereby conserving energy and offering the opportunity to route additional carbon to energy production.

8.3.3.3 Combined EBPR with nitritation/denitritation

Combined EBPR with innovations in the shortcut N removal process, including nitritation coupled to denitritation (nitrite shunt), offer a route to low-energy, low-carbon BNR. Nitrite shunt implies the reduction of oxygen consumption by 25% and consequently reduces the total energy required by 60%. Additionally, it eliminates the use of electron donor (organic carbon) by 40% compared to denitrification, which makes it suitable for wastewater with low carbon to nitrogen ratio. Nitrite shunt also results in a significant decrease in the sludge production. Recently, Roots *et al.* (2019) demonstrated the efficient and reliable combined shortcut N, P, and organic matter removal in a lab-scale SBR treating real mainstream wastewater without exogenous chemicals.

8.3.3.4 *Combined EBPR with partial nitritation/anammox*

The most promising short-circuit N removal process leverages the combined microbial processes of partial nitritation and anammox (PN/A, or so called 'deammonification'). In the PN/A process, the need for organic carbon decreases by 90%, aeration requirements decrease by 60% and sludge production decreases by 75%, thus, profoundly decreasing the associated carbon footprint. The carbon in influent can be harvested for biogas production. Therefore, the implementation of a mainstream PN/A process would bring WWTPs close to energy autarky. Achieving EBPR along with the PN/A process is challenging because EBPR relies on a favorable C/P ratio and alternating anaerobic–aerobic/anoxic cycling which cannot easily be provided by the typical PN/A process configurations. However, recent research breakthroughs and successful implementation of S2EBPR opens the possibility to achieve shortcut N removal and influent carbon-independent EBPR simultaneously. A recent study by Campolong *et al.* (2018) has demonstrated the successful operation of a PN/A + S2EBPR process treating real wastewater, providing a promising pathway for reliable and environmental friendly nutrient removal in both sidestream (e.g., high-strength anaerobic digester liquor) and mainstream (e.g., municipal wastewater). However, some lab-scale studies with shortcut N removal systems including PN/A process indicated increased N_2O emission as a result of incomplete nitrification/denitrification and low dissolved oxygen compared to conventional N removal biotechnologies. These observations warrant further scrutiny of the PN/A systems and design and operation modifications for N_2O mitigation given its significant global warming potential (~300 times of CO_2).

8.3.3.5 *DPAO-based processes*

Denitrifying polyphosphate-accumulating organisms (DPAOs) are a subgroup of PAOs capable of removing P by using nitrate and/or nitrite as electron acceptor. They are preferred over other denitrifiers for simultaneous removal of N and P with a minimized COD utilization and oxygen requirement. In addition, DPAOs produce 20–30% less sludge than PAOs (Kuba *et al.*, 1996). However, the ecology of DPAO are not well understood and no DPAO-based process has been established in full-scale. Although DPAO activities and populations are observed in many full-scale EBPR facilities, the traditional aerobic zone is still required since the DPAO contribution to P removal is shown to vary largely from 15 to 100%. Therefore, DPAOs activities alone cannot ensure reliable nutrient removal. Accumulation of N_2O has also been observed in the DPAO-based studies, which requires further study.

8.3.3.6 *Coupled aerobic–anoxic nitrous decomposition operation (CANDO)*

Although N_2O is treated as an undesired by-product in the nitrogen removal process, it could be utilized as a renewable energy source in propulsion and automotive applications if properly captured and combusted. This innovative process, known as Coupled Aerobic–anoxic Nitrous Decomposition Operation (CANDO), involves three steps: (1) nitritation of NH_4^+ to NO_2^-; (2) partial anoxic reduction of NO_2^- to N_2O; and (3) N_2O conversion to N_2 with energy recovery. CANDO can offer a great improvement with reductions in carbon, aeration demand, and sludge production, which is comparable with nitrite shunt. It may also be combined with EBPR, as it has an alternating anaerobic/anoxic cycling condition. A lab-scale CANDO + P process has been successfully demonstrated by Gao *et al.* (2017), indicating its potential applications for bioenergy production and nutrient removal.

8.3.4 Additional technologies for phosphorus removal and recovery from wastewater streams with carbon footprint reduction potential

8.3.4.1 *Phosphorus recovery technologies at WWTP*

As shown in Figure 8.3, there are various opportunities for P recovery from different waste streams at wastewater treatment plants, such as: secondary treated effluent, digester supernatant, sewage sludge (SS) and sewage sludge ash (SSA) (Montag & Pinnekamp, 2008). These streams differ widely in terms of volume, P concentration, the form of P (dissolved as orthophosphate or biologically/chemically bound), the characteristic of the source (liquid, liquid/solid, solid), pollutant content, requiring different recovery technologies and therefore different decarbonization resorts.

More than 30 processes to recover phosphorus from waste streams have been identified. The available solutions mainly focus on: struvite (crystals of magnesium ammonium phosphate) precipitation from liquid fraction from different steps of wastewater treatment and they include: (1) sludge after digestion; (2) wet chemical P recovery through an acid attack of ash leaching phosphates; (3) thermal solubilization of phosphates in ash with simultaneous reduction of heavy metals; and (4) use of ash for fertilizer manufacturing (Smol *et al.*, 2020). Several life cycle assessments case studies have identified opportunities and burdens associated with the advanced P recovery from the different streams in wastewaters treatment plants. Amann *et al.* (2018) compared the environmental impacts and GHG emissions of 18 P recovery technologies, described in Egle *et al.* (2015). Amann *et al.* (2018) concluded that the recovery from the liquid phase generates less emissions and has lower energy demands, but offers low rates of recovery, while recovery from sludge (solid phase) has relatively higher emissions and higher energy demands. The recovery of phosphorus from sludge ash, on the other hand, is the most promising option. It presents a higher recycling rate, the possibility of heavy metal decontamination, and reduction of gaseous emissions and energy demand (see Figure 8.5 extracted from Amann *et al.* 2018). According to Bradford-Hartke *et al.* (2015), phosphorus recovery in an advanced BNR centralized water reclamation facility led to a net reduction of 5 $kgCO_2eq/kgP$. The net carbon footprint reduction is due to avoided N_2O emissions, lower power consumption, and reduced chemical usage for pH control (due to reduced nitrification), which offset power and chemicals demands of dewatering liquid required for struvite production.

Pradel and Aissani (2019) compared the environmental impacts of sludge-based phosphate fertilizer production to producing mineral fertilizers from phosphate rocks. Their results indicated

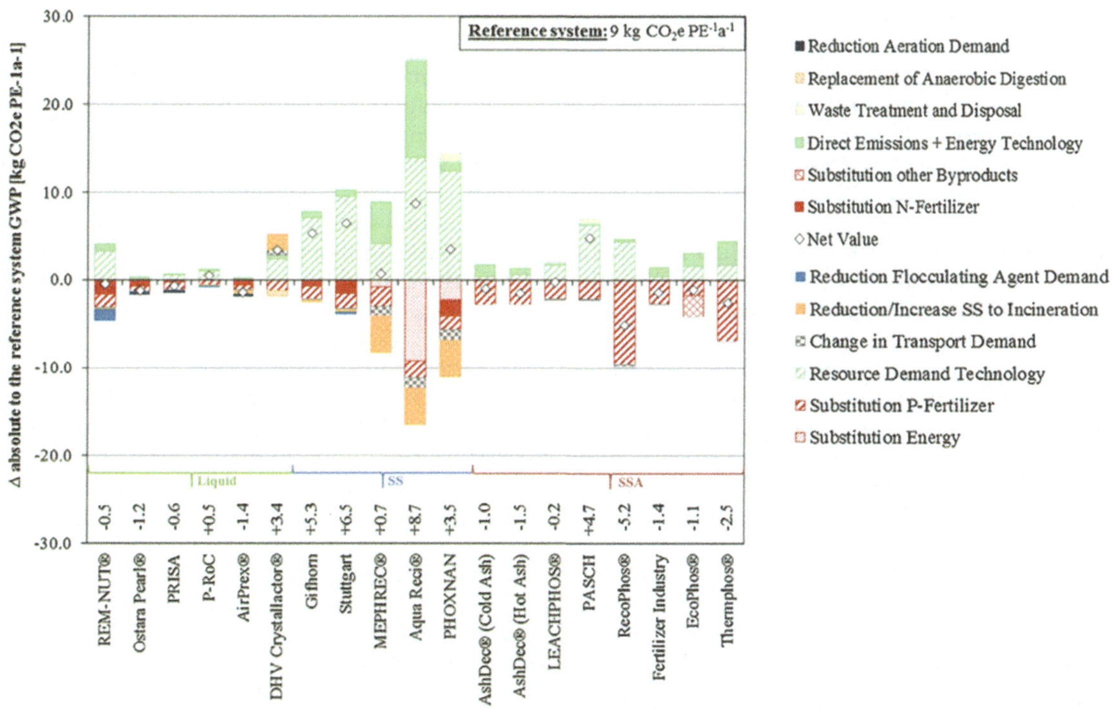

Figure 8.5 Changes in carbon footprints of various P removal technologies. The figure is from previously published study (Amann *et al.*, 2018).

that sludge-based phosphate fertilizers appeared to have higher environmental impacts than mineral phosphate fertilizers production, mainly due to their consumption of large amounts of electricity and reactants needed to recover phosphorus, and their low phosphorus content in comparison with phosphate rocks. Qualitatively similar conclusions were reported by Golroudbary *et al.* (2019). On the contrary, Tonini *et al.* (2019) suggested that the environmental impacts of recovering sewage derived P may be up to 81% less than mining P containing rock. They attribute the inconsistency of their results with previous studies to the different assumptions, and that they have included the external costs associated with all relevant emissions (including dissipated phosphate).

Other studies have indicated benefits from P recovery which are generally overlooked, for example, the reduction in eutrophication potential (reduced phosphate rock mining and therefore lower P water release from mining (Remy & Jossa, 2015), mitigation of Cd and U input into agricultural soils (Bigalke *et al.*, 2017), reduction of heavy metal input compared to conventional agricultural sewage sludge application (Lederer & Rechberger, 2010), and decreased nitrogen emissions for technologies which also recover nitrogen (Johansson *et al.*, 2008). Overall, in spite of all inconsistences, these results suggest that not all P recovery technologies offer decarbonization potential. Thorough, holistic assessments of phosphorus recovery technologies are required.

8.3.4.2 Constructed wetland

Constructed wetland (CW) is a sustainable wastewater treatment technology. CW systems integrate vegetation, soils and microbial ecosystems to treat a variety of waste streams (e.g., municipal or industrial wastewater, greywater or stormwater runoff, etc.), while capturing CO_2 to plant biomass. The greenhouse gas emissions are related to the construction and operation, wastewater and sludge transportation. The energy, carbon, and chemical demands in conventional WWTPs would be eliminated in a well-functioned CW, yielding lower carbon footprints. In addition, CW often provide multifunctions such as flood control, biomass production, biodiversity, and recreational and educational services. However, some forms of CWs also emit large quantities of CH_4 and N_2O, especially in the CWs with a denitrification zone. The P removal performance in CWs is often limited by the capacity of the media to adsorb, bind or precipitate the incoming P. Once the media is saturated or blocked, the P removal performance in CWs will vastly drop. So, an important research goal of CWs is to seek specialized substrates with conducive physico-chemical properties to improve nutrient removal (especially P). One effective and cost-effective media is the dewatered alum sludge, a residual by-product of drinking water treatment facilities. The alum sludge-based constructed wetland systems have been successfully applied in full-scale and achieved enhanced P removal performance. Additionally, the planted vegetation as well as the PAOs in the rhizosphere may also play important roles in both P and C sequestration.

8.3.4.3 Microalgae cultivation for joint nutrient removal and energy production

The incorporation of microalgae cultivation is a cost-effective and sustainable measure in WWTPs, as microalgae can fix exogenous CO_2 during autotrophic growth while assimilating N, P and metal in wastewater. Harvested lipid-rich microalgae could be used for generation of biofuels (e.g., biodiesel) with the potential to reduce greenhouse gas emissions through replacement of fossil fuels. The other uses of algae biomass include carbon- and nutrient-rich soil amendment, animal feed, and bioplastic production. Microalgae have been widely studied for CO_2 capture and utilization, and extensive research has been carried out on their use in large-scale (>5000 acres) cultivation systems.

8.3.4.4 Genetically modified PAOs for enhanced bio-P adsorption

The selection and enrichment of active PAOs is one of the prerequisites for successful EBPR, however, it could be influenced in practice by many operating and environmental factors, leading to less effective and stable performance. Removal of P from water using high-affinity phosphate-specific bacterial proteins has recently attracted research interest. In one study, *Escherichia coli* was genetically modified to overexpress phosphate-binding proteins (PBPs, also known as PstS or PhoS), resulting in a highly improved P removal and recovery performance. Implementation of recombinant-plasmid bacteria systems for selective P adsorption in actual wastewater treatment applications is a challenging but attractive approach with relatively less chemical/energy input for configuration, modification and maintenance compared to other processes/technologies, therefore deserving further exploration.

8.4 QUANTIFICATION OF DECARBONIZATION POTENTIAL FROM PHOSPHORUS REMOVAL AND RECOVERY PROCESSES

A range of P management and removal and recovery technologies for point and non-point sources as briefly described above are available to decision-makers. An environmentally sound process design and selection requires holistic comparative analyses that provide tangible metrics for environmental impacts associated with the entire cradle to grave life cycle of each alternative. Material and energy flow, generated wastes, and emissions at the process level are considered in life cycle analysis (LCA). The cumulative impacts of each process per functional unit in mid-point impact categories such as eutrophication and global warming are calculated with proper weighting schemes. Therefore, LCA is well suited for educated decision making about decarbonization alternatives, identifying the environmental hot posts, trade-offs, and opportunities for improvements based on quantitative measures of board environmental consequences.

Such a holistic perspective is often missing in developing new interventions for P management, removal, and recovery. For example, there have been mixed messages in the literature concerning the sustainable treatment level (Foley *et al.*, 2010; Lundie *et al.*, 2004; Renzoni & Germain, 2007). A lower P effluent lowers the risks of eutrophication. On the other hand, an energy-intensive P removal intervention may offset the net environmental benefits of an incremental decrease in nutrient loads to water resources.

Similarly, more studies are needed to address the environmental sustainability of P recovery from various sources. For example, a recent study has suggested that with increasing P demands and consumption overtime and with current mining, processing, and recovery technologies, the carbon footprint of P recycling will increase exponentially and will exceed that of processing and mining (Golroudbary *et al.*, 2019).

8.4.1 LCA studies for the quantification of decarbonization potential for non-point sources

A broad understanding of the environmental impacts of the life cycle of non-point source P management strategies is missing. There have been P-oriented LCA studies addressing eutrophication potential from agricultural sources (Ortiz-Reyes & Anex, 2018). Those studies have focused on estimating the P transport and discharge from agricultural sources and overlooked the indirect impacts of control and mitigation strategies and their associated carbon footprint.

Few studies have provided critical reviews on non-point source management practices in relation to practicality, cost-effectiveness, and regulatory requirements (Dinnes, 2004; Macintosh *et al.*, 2018). It has been suggested that source-oriented practices offer a cheaper and more effective solution than endpoint alternatives such as wetlands. No study has compared the sustainability of containment practices at the source with in-sink treatment practices or with combinational approaches. Wetlands are considered low-tech with low energy demands solution. More than 80% of the environmental impacts of the life cycle of wetlands are from the construction phase (Resende *et al.*, 2019). Riparian buffers can retain as high as 97% of P from run-off and eroded soil (Fox & Penn, 2013) and at the same time offer a net saving (11.9 Mg CO_2eq ha^{-1} $year^{-1}$) in global warming potential (Styles *et al.*, 2016). Due to the diverse range and diffused nature of non-point sources, in-field management practices may not be universally effective, cost-effective, or the optimum solution in relation to environmental consequences. Also, there are often continuous P releases into the environment from accumulated legacy P content of the soil (Powers *et al.*, 2016), even when effective source containment strategies are in place, necessitating combinational strategies at the source and in the sink for protecting watersheds. The selection of a proper integrated management strategy requires a holistic picture of the environmental net-benefit of various alternatives. Therefore, LCA is a key tool for a sustainable P management of non-point sources in the water sector. Environmental release and transports of P varies from site to site, and in the same site varies with time scales. Thus, future LCA studies should, beyond common practices, consider such site-specific temporal variations.

8.4.2 LCA studies for P removal and recovery processes in WWTPs and quantification of decarbonization potential

To maximize the potential of wastewater resources, a robust and integrative approach is needed to quantitatively compare the environmental attributes of diverse technology options for P removal and recovery technologies. In recent years few publications have discussed the environmental impacts of nutrient removal technologies (Coats *et al.*, 2011; Foley *et al.*, 2010; Rahman *et al.*, 2016), or P recovery technologies (Amann *et al.*, 2018; Bradford-Hartke *et al.*, 2015), but little research has been published on such a comparative sustainability assessment of both recovery and removal of P from wastewater.

An assessment of 27 nutrient removal technologies was carried out by Rahman *et al.* (2016); life cycle impact assessment (LCIA) of the representative treatment process configurations with different levels of treatment for both nitrogen and phosphorus was performed. Results showed that advanced technologies that achieve high-level nutrient removal significantly decreased local eutrophication potential, while chemicals and electricity use for these advanced treatments, particularly multistage enhanced tertiary processes and reverse osmosis, increased indirect eutrophication potential and contributed to other impacts including human and ecotoxicity, global warming potential, ozone depletion, and acidification.

Regardless of the required effluent limit, when biological phosphorus removal processes are compared to chemical processes in terms of environmental impacts, it would appear that best practices would center wastewater treatment first on the biological process (Coats *et al.*, 2011). The EBPR process also produces significantly fewer biosolids and no chemical sludge, which allows for further reduction of carbon footprint due to avoided transport/handling of these byproducts.

The LCA studies published that focused on P recovery schemes and technologies from urine and various wastewater streams have highlighted the potential decarbonization associated with these technologies, as already discussed in previous sections, but also identified some inconsistencies. The typical approach in the published studies is the comparison of environmental impacts (often not consistent among studies) in different scenarios involving different technologies and case-specific waste streams. Consequently, quantitative results about environmental impact are strictly related to the specific application and should not be used as a basis for deriving general conclusions about the implications of P recovery. Further in-depth, comprehensive analyses about P recovery systems are necessary to know their broad environmental impacts.

8.5 FUTURE OUTLOOK AND RESEARCH NEEDS

It is vital to promote P management approaches for improving water quality in conjunction with possible opportunities for reducing carbon footprint and contributing to net-zero circular economy. The results from current studies on P recovery technologies seem to indicate that not all P recovery technologies, when evaluated on a system level, offer opportunities for decarbonization; however, some inconsistences still exist among studies, due to local impact, as well as differences in approaches and assumption. Therefore, evaluating the decarbonization potential associated with phosphorus recovery technologies, should be an important aspect for further studies.

Regional, integrated P management practices are likely to be considered more increasingly, which, for example, include urban point sources with non-point sources such as farmlands. In the same context, P trading policies should be developed. Implementing such policies should be contingent on holistic life cycle analyses of net environmental impacts associated with each intervention. This consideration would allow decision makers to select the most appropriate practices that meet both water and air quality goals.

REFERENCES

Alewell C., Ringeval B., Ballabio C., Robinson D. A., Panagos P. and Borrelli P. (2020). Global phosphorus shortage will be aggravated by soil erosion. *Nature Communications*, **11**(1), 4546, https://doi.org/10.1038/s41467-020-18326-7

Altinbas M., Ozturk I. and Aydin A. F. (2002). Ammonia recovery from high strength agro industry effluents. *Water Science and Technology*, **45**(12), 189–195, https://doi.org/10.2166/wst.2002.0426

Amann A., Zoboli O., Krampe J., Rechberger H., Zessner M. and Egle L. (2018). Environmental impacts of phosphorus recovery from municipal wastewater. *Resources, Conservation and Recycling*, **130**, 127–139, https://doi.org/10.1016/j.resconrec.2017.11.002

Anderson D. M., Glibert P. M. and Burkholder J. M. (2002). Harmful algal blooms and eutrophication: nutrient sources, composition, and consequences. *Estuaries*, **25**(4), 704–726, https://doi.org/10.1007/BF02804901

Ashley R., Walker L., D'Arcy B., Wilson S., Illman S., Shaffer P., Woods-Ballard B. and Chatfield P. (2015). UK Sustainable drainage systems: past, present and future. *Proceedings of the Institution of Civil Engineers – Civil Engineering*, **168**(3), 125–130, https://doi.org/10.1680/cien.15.00011

Barnard J. L., Dunlap P. and Steichen M. (2017). Rethinking the mechanisms of biological phosphorus removal. *Water Environment Research*, **89**(11), 2043–2054, https://doi.org/10.2175/106143017X15051465919010

Bianchini A. and Rossi J. (2020). An integrated industry-based methodology to unlock full-scale implementation of phosphorus recovery technology. *Sustainability*, **12**(24), 10632, https://doi.org/10.3390/su122410632

Bigalke M., Ulrich A., Rehmus A. and Keller A. (2017). Accumulation of cadmium and uranium in arable soils in Switzerland. *Environmental Pollution*, **221**, 85–93, https://doi.org/10.1016/j.envpol.2016.11.035

Bouwman L., Goldewijk K. K., Hoek K. W. V. D., Beusen A. H. W., Vuuren D. P. V., Willems J., Rufino M. C. and Stehfest E. (2013). Exploring global changes in nitrogen and phosphorus cycles in agriculture induced by livestock production over the 1900–2050 period. *Proceedings of the National Academy of Sciences*, **110**(52), 20882–20887, https://doi.org/10.1073/pnas.1012878108

Bradford-Hartke Z., Lane J., Lant P. and Leslie G. (2015). Environmental benefits and burdens of phosphorus recovery from municipal wastewater. *Environmental Science and Technology*, **49**(14), 8611–8622, https://doi.org/10.1021/es505102v

Bunce J. T., Ndam E., Ofiteru I. D., Moore A. and Graham D. W. (2018). A review of phosphorus removal technologies and their applicability to small-scale domestic wastewater treatment systems. *Frontiers in Environmental Science*, **6**, 8, https://doi.org/10.3389/fenvs.2018.00008

Campolong C., Ferguson L., Klaus S., Wilson C., Wett B., Murthy S. and Bott C. B. (2018). Achieving mainstream deammonification with biological phosphorus removal via sidestream RAS fermentation in an A/B process. *Proceedings of the Water Environment Federation*, **2018**(13), 2634–2639, https://doi.org/10.2175/193864718825137142

Cao Y., Tang J., Henze M., Yang X., Gan Y., Li J., Kroiss H., Van Loosdrecht M., Zhang Y. and Daigger G. (2019). The leakage of sewer systems and the impact on the 'black and odorous water bodies' and WWTPs in China. *Water Science and Technology*, **79**(2), 334–341, https://doi.org/10.2166/wst.2019.051

Carvalheira M., Oehmen A., Carvalho G., Eusébio M. and Reis M. A. M. (2014). The impact of aeration on the competition between polyphosphate accumulating organisms and glycogen accumulating organisms. *Water Research*, **66**, 296–307, https://doi.org/10.1016/j.watres.2014.08.033

Chong J., Dominish E., Tjandraarmadja G., Prentice E., Retamal M. and Mitrovic S. (2019). Ecosystem Impacts of Phosphorus and Surfactants in Consumer Products. Institute for Sustainable Futures. Report prepared by the Institute for Sustainable Futures, University of Technology Sydney for Stewart Investors.

Coats E. R., Watkins D. L. and Kranenburg D. (2011). A comparative environmental life-cycle analysis for removing phosphorus from wastewater: biological versus physical/chemical processes. *Water Environment Research*, **83**(8), 750–760, https://doi.org/10.2175/106143011X12928814444619

Coffman L. S. (2000). Low-impact Development Design: A New Paradigm for Stormwater Management Mimicking and Restoring the Natural Hydrologic Regime, an Alternative Stormwater Management Technology. Prince

George's County Department of Environmental Resources and US EPA (Report EPA 841-B-00-003), Maryland County, USA. Retrieved from: http://www.epa.gov/owow/NPS/lidnatl.pdf (accessed 23 January 2021)

Cohen A. and Keiser D. A. (2017). The effectiveness of incomplete and overlapping pollution regulation: evidence from bans on phosphate in automatic dishwasher detergent. *Journal of Public Economics*, **150**, 53–74, https://doi.org/10.1016/j.jpubeco.2017.03.005

Cordell D., Drangert J.-O. and White S. (2009). The story of phosphorus: global food security and food for thought. *Global Environmental Change*, **19**(2), 292–305, https://doi.org/10.1016/j.gloenvcha.2008.10.009

Cordell D., Rosemarin A., Schröder J. J. and Smit A. (2011). Towards global phosphorus security: a systems framework for phosphorus recovery and reuse options. *Chemosphere*, **84**(6), 747–758, https://doi.org/10.1016/j.chemosphere.2011.02.032

Dinnes D. L. (2004). Assessments of Practices to Reduce Nitrogen and Phosphorus Nonpoint Source Pollution of Iowa's Surface Waters. USDA-ARS, National Soil Tilth Laboratory, Ames, IA, p. 366.

Dodds W. K., Bouska W. W., Eitzmann J. L., Pilger T. J., Pitts K. L., Riley A. J., Schloesser J. T. and Thornbrugh D. J. (2009). Eutrophication of U.S. Freshwaters: analysis of potential economic damages. *Environmental Science and Technology*, **43**(1), 12–19, https://doi.org/10.1021/es801217q

Dubrovsky N., Burow K., Clark G. M., Gronberg J. M., Hamilton P. A., Hitt K. J., Mueller D. K., Munn M., Nolan B. T., Puckett L. J., Rupert M. G., Short T. M., Spahr N. E., Sprague L. A. and Wilber W. G. (2010). The quality of our nation's waters – nutrients in the nation's streams and groundwater, 1992–2004. *US Geological Survey Circular*, **1350**(2), 174.

Egle L., Rechberger H. and Zessner M. (2015). Overview and description of technologies for recovering phosphorus from municipal wastewater. *Resources, Conservation and Recycling*, **105**, 325–346, https://doi.org/10.1016/j.resconrec.2015.09.016

Egle L., Rechberger H., Krampe J. and Zessner M. (2016). Phosphorus recovery from municipal wastewater: an integrated comparative technological, environmental and economic assessment of P recovery technologies. *Science of the Total Environment*, **571**, 522–542, https://doi.org/10.1016/j.scitotenv.2016.07.019

European Commission. (1962). The Common Agricultural Policy at A Glance. European Commission, Brussels. Available at: https://ec.europa.eu/info/food-farming-fisheries/key-policies/common-agricultural-policy/cap-glance_en (accessed 23 January 2021)

European Commission. (2011). EP Supports Ban of Phosphates in Consumer Detergents. European Commission, Brussels. Available at: https://ec.europa.eu/commission/presscorner/detail/en/IP_11_1542

Foley J., Haas D. D., Hartley K. and Lant P. (2010). Comprehensive life-cycle inventories of alternative wastewater treatment systems. *Water Research*, **44**(5), 1654–1666.

Fox G. A. and Penn C. J. (2013). Empirical model for quantifying total phosphorus reduction by vegetative filter strips. *Transactions of the ASABE*, **56**(4), 1461–1469.

Gao H., Liu M., Griffin J. S., Xu L., Xiang D., Scherson Y. D., Liu W.-T. and Wells G. F. (2017). Complete nutrient removal coupled to nitrous oxide production as a bioenergy source by denitrifying polyphosphate-accumulating organisms. *Environmental Science and Technology*, **51**(8), 4531–4540, https://doi.org/10.1021/acs.est.6b04896

George H., Terril N., Unruh J. B., Laurie T. and Jerry S. (2012). Potential unintended consequences associated with urban fertilizer bans in Florida – a scientific review. *HortTechnology Hortte*, **22**(5), 600–616, https://doi.org/10.21273/HORTTECH.22.5.600

Golroudbary S. R., Wali M. E. and Kraslawski A. (2019). Environmental sustainability of phosphorus recycling from wastewater, manure and solid wastes. *Science of the Total Environment*, **672**, 515–524, https://doi.org/10.1016/j.scitotenv.2019.03.439

Gu A., Tooker N., Onnis-Hayden A., Wang D., Srinivasan V., Li G., Takács I. and Vargas E. (2019). Optimization and Design of A Side-Stream EBPR Process as A Sustainable Approach for Achieving Stable and Efficient Phosphorus Removal. The Water Research Foundation, DC, USA.

Hilton S. P., Keoleian G. A., Daigger G. T., Zhou B. and Love N. G. (2020). Life-cycle assessment of urine diversion and conversion to fertilizer products at the city scale. *Environmental Science and Technology*, **55**(1), 593–603, https://doi.org/10.1021/acs.est.0c04195

Jasinski S. M. (2020). Phosphate Rock Data Sheet – Mineral Commodity Summaries 2020. US Geological Survey, Reston, Virginia.

Johansson K., Perzon M., Fröling M., Mossakowska A. and Svanström M. (2008). Sewage sludge handling with phosphorus utilization–life-cycle assessment of four alternatives. *Journal of Cleaner Production*, **16**(1), 135–151, https://doi.org/10.1016/j.jclepro.2006.12.004

Jones G. A. and Tippie V. K. (1983). Chesapeake Bay Program: Findings and Recommendations. US Environmental Protection Agency, Philadelphia.

Kavvada O., Tarpeh W. A., Horvath A. and Nelson K. L. (2017). Life-cycle cost and environmental assessment of decentralized nitrogen recovery using ion exchange from source-separated urine through spatial modeling. *Environmental Science and Technology*, **51**(21), 12061–12071, https://doi.org/10.1021/acs.est.7b02244

Kishida N., Tsuneda S., Kim J. H. and Sudo R. (2009). Simultaneous nitrogen and phosphorus removal from high-strength industrial wastewater using aerobic granular sludge. *Journal of Environmental Engineering*, **135**(3), 153–158, https://doi.org/10.1061/(ASCE)0733-9372(2009)135:3(153)

Kjerstadius H., Bernstad Saraiva A., Spångberg J. and Davidsson Å. (2017). Carbon footprint of urban source separation for nutrient recovery. *Journal of Environmental Management*, **197**, 250–257, https://doi.org/10.1016/j.jenvman.2017.03.094

Kuba T., Van Loosdrecht M. and Heijnen J. (1996). Phosphorus and nitrogen removal with minimal COD requirement by integration of denitrifying dephosphatation and nitrification in a two-sludge system. *Water Research*, **30**(7), 1702–1710, https://doi.org/10.1016/0043-1354(96)00050-4

Kundu S., Bhattacharyya R., Prakash V., Ghosh B. N. and Gupta H. S. (2007). Carbon sequestration and relationship between carbon addition and storage under rainfed soybean–wheat rotation in a sandy loam soil of the Indian Himalayas. *Soil and Tillage Research*, **92**(1), 87–95, https://doi.org/10.1016/j.still.2006.01.009

Lal R. (2004). Soil carbon sequestration impacts on global climate change and food security. *Science*, **304**(5677), 1623–1627, https://doi.org/10.1126/science.1097396

Lal R. and Kimble J. M. (1997). Conservation tillage for carbon sequestration. *Nutrient Cycling in Agroecosystems*, **49**(1), 243–253, https://doi.org/10.1023/A:1009794514742

Lederer J. and Rechberger H. (2010). Comparative goal-oriented assessment of conventional and alternative sewage sludge treatment options. *Waste Management*, **30**(6), 1043–1056, https://doi.org/10.1016/j.wasman.2010.02.025

Litke D. (1999). Review of Phosphorus Control Measures in the United States and Their Effects on Water Quality. US Geological Survey, Reston, Virginia.

Liu Y., Villalba G., Ayres R. U. and Schroder H. (2008). Global phosphorus flows and environmental impacts from a consumption perspective. *Journal of Industrial Ecology*, **12**(2), 229–247, https://doi.org/10.1111/j.1530-9290.2008.00025.x

Liu K., Elliott J. A., Lobb D. A., Flaten D. N. and Yarotski J. (2014). Conversion of conservation tillage to rotational tillage to reduce phosphorus losses during snowmelt runoff in the Canadian prairies. *Journal of Environmental Quality*, **43**(5), 1679–1689, https://doi.org/10.2134/jeq2013.09.0365

Lundie S., Peters G. M. and Beavis P. C. (2004). Life-cycle assessment for sustainable metropolitan water systems planning. *Environmental Science and Technology*, **38**(13), 3465–3473, https://doi.org/10.1021/es034206m

Macintosh K. A., Mayer B. K., McDowell R. W., Powers S. M., Baker L. A., Boyer T. H. and Rittmann B. E. (2018). Managing diffuse phosphorus at the source versus at the sink. *Environmental Science and Technology*, **52**(21), 11995–12009, https://doi.org/10.1021/acs.est.8b01143

Maktabifard M., Zaborowska E. and Makinia J. (2018). Achieving energy neutrality in wastewater treatment plants through energy savings and enhancing renewable energy production. *Reviews in Environmental Science and Bio/Technology*, **17**(4), 655–689, https://doi.org/10.1007/s11157-018-9478-x

Maurer M., Pronk W. and Larsen T. A. (2006). Treatment processes for source-separated urine. *Water Research*, **40**(17), 3151–3166, https://doi.org/10.1016/j.watres.2006.07.012

McDowell R. W. (2015). Treatment of pasture topsoil with alum to decrease phosphorus losses in subsurface drainage. *Agriculture, Ecosystems and Environment*, **207**, 178–182, https://doi.org/10.1016/j.agee.2015.04.017

Mihelcic J. R., Fry L. M. and Shaw R. (2011). Global potential of phosphorus recovery from human urine and feces. *Chemosphere*, **84**(6), 832–839, https://doi.org/10.1016/j.chemosphere.2011.02.046

Montag D. M. and Pinnekamp J. (2008). Phosphorus Recovery During Wastewater Treatment: Development of a Process for Integration in Municipal Wastewater Treatment Plants. PhD thesis, RWTH Aachen University, Aachen, Germany.

Mulkerrins D., O'Connor E., Lawlee B., Barton P. and Dobson A. (2004). Assessing the feasibility of achieving biological nutrient removal from wastewater at an Irish food processing factory. *Bioresource Technology*, **91**(2), 207–214, https://doi.org/10.1016/S0960-8524(03)00173-1

Müller A., Österlund H., Marsalek J. and Viklander M. (2020). The pollution conveyed by urban runoff: a review of sources. *Science of the Total Environment*, **709**, 136125, https://doi.org/10.1016/j.scitotenv.2019.136125

Nsenga Kumwimba M., Meng F., Iseyemi O., Moore M. T., Zhu B., Tao W., Liang T. J. and Ilunga L. (2018). Removal of non-point source pollutants from domestic sewage and agricultural runoff by vegetated drainage ditches (VDDs): design, mechanism, management strategies, and future directions. *Science of the Total Environment*, **639**, 742–759, https://doi.org/10.1016/j.scitotenv.2018.05.184

Naidu R., Lamb D. T., Bolan N. S. and Gawandar J.. (2012). Recovery and reuse of phosphorus from wastewater sources. Advanced nutrient management: gains from the past–goals for the future. Occasional Report **25**. See http://flrc.massey.ac.nz/publications.html.

Oehmen A., Lemos P. C., Carvalho G., Yuan Z., Keller J., Blackall L. L. and Reis M. A. (2007). Advances in enhanced biological phosphorus removal: from micro to macro scale. *Water Research*, **41**(11), 2271–2300, https://doi.org/10.1016/j.watres.2007.02.030

Onnis-Hayden A., Srinivasan V., Tooker N. B., Li G., Wang D., Barnard J. L., Bott C., Dombrowski P., Schauer P. and Menniti A. (2020). Survey of full-scale sidestream enhanced biological phosphorus removal (S2EBPR) systems and comparison with conventional EBPRs in North America: process stability, kinetics, and microbial populations. *Water Environment Research*, **92**(3), 403–417, https://doi.org/10.1002/wer.1198

Ortiz-Reyes E. and Anex R. P. (2018). A life-cycle impact assessment method for freshwater eutrophication due to the transport of phosphorus from agricultural production. *Journal of Cleaner Production*, **177**, 474–482, https://doi.org/10.1016/j.jclepro.2017.12.255

Owen J. J. and Silver W. L. (2015). Greenhouse gas emissions from dairy manure management: a review of field-based studies. *Global Change Biology*, **21**(2), 550–565, https://doi.org/10.1111/gcb.12687

Powers S. M., Bruulsema T. W., Burt T. P., Chan N. I., Elser J. J., Haygarth P. M., Howden N. J. K., Jarvie H. P., Lyu Y., Peterson H. M., Sharpley A. N., Shen J., Worrall F. and Zhang F. (2016). Long-term accumulation and transport of anthropogenic phosphorus in three river basins. *Nature Geoscience*, **9**(5), 353–356, https://doi.org/10.1038/ngeo2693

Pradel M. and Aissani L. (2019). Environmental impacts of phosphorus recovery from a 'product' life-cycle assessment perspective: allocating burdens of wastewater treatment in the production of sludge-based phosphate fertilizers. *Science of the Total Environment*, **656**, 55–69, https://doi.org/10.1016/j.scitotenv.2018.11.356

Preisner M., Neverova-Dziopak E. and Kowalewski Z. (2020). An analytical review of different approaches to wastewater discharge standards with particular emphasis on nutrients. *Environmental Management*, **66**(4), 694–708, https://doi.org/10.1007/s00267-020-01344-y

Rahman S. M., Eckelman M. J., Onnis-Hayden A. and Gu A. Z. (2016). Life-cycle assessment of advanced nutrient removal technologies for wastewater treatment. *Environmental Science and Technology*, **50**(6), 3020–3030, https://doi.org/10.1021/acs.est.5b05070

Remy C. and Jossa P. (2015). Life-cycle Assessment of Selected Processes for P Recovery From Sewage Sludge, Sludge Liquor, or Ash. P-Rex Deliverable 9 Report. Kompetenzzentrum Wasser, Berlin.

Renzoni R. and Germain A. (2007). Life-cycle assessment of water: from the pumping station to the wastewater treatment plant. *The International Journal of Life-Cycle Assessment*, **12**(2), 118–126, https://doi.org/10.1065/lca2005.12.243

Resende J. D., Nolasco M. A. and Pacca S. A. (2019). Life-cycle assessment and costing of wastewater treatment systems coupled to constructed wetlands. *Resources, Conservation and Recycling*, **148**, 170–177, https://doi.org/10.1016/j.resconrec.2019.04.034

Rheinhardt R. D., Brinson M. M., Meyer G. F. and Miller K. H. (2012). Carbon storage of headwater riparian zones in an agricultural landscape. *Carbon Balance and Management*, **7**(1), 41–54, https://doi.org/10.1186/1750-0680-7-4

Rittmann B. E., Mayer B., Westerhoff P. and Edwards M. (2011). Capturing the lost phosphorus. *Chemosphere*, **84**(6), 846–853, https://doi.org/10.1016/j.chemosphere.2011.02.001

Roots P., Sabba F., Rosenthal A. F., Wang Y., Yuan Q., Rieger L., Yang F., Kozak J. A., Zhang H. and Wells G. F. (2019). Integrated low-energy and low carbon shortcut nitrogen removal with biological phosphorus removal for sustainable mainstream wastewater treatment. *bioRxiv*, 772004.

Rudrappa L., Purakayastha T. J., Singh D. and Bhadraray S. (2006). Long-term manuring and fertilization effects on soil organic carbon pools in a typic haplustept of semi-arid sub-tropical India. *Soil and Tillage Research*, **88**(1), 180–192, https://doi.org/10.1016/j.still.2005.05.008

Sample D. J., Grizzard T. J., Sansalone J., Davis A. P., Roseen R. M. and Walker J. (2012). Assessing performance of manufactured treatment devices for the removal of phosphorus from urban stormwater. *Journal of Environmental Management*, **113**, 279–291, https://doi.org/10.1016/j.jenvman.2012.08.039

Sharpley A. and Tunney H. (2000). Phosphorus research strategies to meet agricultural and environmental challenges of the 21st century. *Journal of Environmental Quality*, **29**(1), 176–181, https://doi.org/10.2134/jeq2000.00472425002900010022x

Sharpley A. N., Daniel T., Gibson G., Bundy L., Cabrera M., Sims T., Stevens R., Lemunyon J., Kleinman P. and Parry R. (2006). Best Management Practices to Minimize Agricultural Phosphorus Impacts on Water Quality. USDA Agricultural Research Service Publication, Washington, DC.

Shaver E. (2000). Low Impact Design Manual for the Auckland Region. Regional Council Technical Report TR2008/045, Auckland, NZ.

Srinivasan V. N., Li G., Wang D., Tooker N. B., Dai Z., Onnis-Hayden A., Bott C., Dombrowski P., Schauer P., Pinto A. and Gu A. Z. (2021). Oligotyping and metagenomics reveal distinct Candidatus accumulibacter communities in side-stream versus conventional full-scale enhanced biological phosphorus removal (EBPR) systems. *Water Research*, **206**, 117725, https://doi.org/10.1016/j.watres.2021.117725

Smith L. C. and McDowell R. W. (2016). The use of alum to decrease phosphorus loss from dairy farm laneways in southern New Zealand. *Soil Use and Management*, **32**(1), 69–71, https://doi.org/10.1111/sum.12252

Smol M., Preisner M., Bianchini A., Rossi J., Hermann L., Schaaf T., Kruopienė J., Pamakštys K., Klavins M. and Ozola-Davidane R. (2020). Strategies for sustainable and circular management of phosphorus in the Baltic Sea Region: the holistic approach of the InPhos Project. *Sustainability*, **12**(6), 2567, https://doi.org/10.3390/su12062567

Spångberg J., Tidåker P. and Jönsson H. (2014). Environmental impact of recycling nutrients in human excreta to agriculture compared with enhanced wastewater treatment. *Science of the Total Environment*, **493**, 209–219, https://doi.org/10.1016/j.scitotenv.2014.05.123

Styles D., Börjesson P., D'Hertefeldt T., Birkhofer K., Dauber J., Adams P., Patil S., Pagella T., Pettersson L. B., Peck P., Vaneeckhaute C. and Rosenqvist H. (2016). Climate regulation, energy provisioning and water purification: quantifying ecosystem service delivery of bioenergy willow grown on riparian buffer zones using life-cycle assessment. *AMBIO*, **45**(8), 872–884, https://doi.org/10.1007/s13280-016-0790-9

Tonini D., Saveyn H. G. and Huygens D. (2019). Environmental and health co-benefits for advanced phosphorus recovery. *Nature Sustainability*, **2**(11), 1051–1061, https://doi.org/10.1038/s41893-019-0416-x

Urban Waste Water Treatment Directive (UWWTD). (1991). Council of the European communities. Council directive of 21 May 1991 concerning urban waste water treatment (91/271/EEC). *Official Journal*, **L**, 135–140.

Vaccari D. A., Powers S. M. and Liu X. (2019). Demand-driven model for global phosphate rock suggests paths for phosphorus sustainability. *Environmental Science and Technology*, **53**(17), 10417–10425, https://doi.org/10.1021/acs.est.9b02464

Wang D., Tooker N. B., Srinivasan V., Li G., Fernandez L. A., Schauer P., Menniti A., Maher C., Bott C. B. and Dombrowski P. (2019). Side-stream enhanced biological phosphorus removal (S2EBPR) process improves system performance-A full-scale comparative study. *Water Research*, **167**, 115109, https://doi.org/10.1016/j.watres.2019.115109

Wilsenach J. and van Loosdrecht M. (2003). Impact of separate urine collection on wastewater treatment systems. *Water Science and Technology*, **48**(1), 103–110, https://doi.org/10.2166/wst.2003.0027

Wong T. H. and Brown R. R. (2009). The water sensitive city: principles for practice. *Water Science and Technology*, **60**(3), 673–682, https://doi.org/10.2166/wst.2009.436

Xia Y., Zhang M., Tsang D. C. W., Geng N., Lu D., Zhu L., Igalavithana A. D., Dissanayake P. D., Rinklebe J., Yang X. and Ok Y. S. (2020). Recent advances in control technologies for non-point source pollution with nitrogen and phosphorous from agricultural runoff: current practices and future prospects. *Applied Biological Chemistry*, **63**(1), 8, https://doi.org/10.1186/s13765-020-0493-6

Xu T. and Flapper J. (2009). Energy use and implications for efficiency strategies in global fluid-milk processing industry. *Energy Policy*, **37**(12), 5334–5341, https://doi.org/10.1016/j.enpol.2009.07.056

Xue X., Hawkins T. R., Schoen M. E., Garland J. and Ashbolt N. J. (2016). Comparing the life-cycle energy consumption, global warming and eutrophication potentials of several water and waste service options. *Water*, **8**(4), 154.

Yuan Z., Pratt S. and Batstone D. J. (2012). Phosphorus recovery from wastewater through microbial processes. *Current Opinion in Biotechnology*, **23**(6), 878–883, https://doi.org/10.1016/j.copbio.2012.08.001

Zevenbergen C., Fu D. and Pathirana A. (2018). Sponge Cities: Emerging Approaches, Challenges and Opportunities. MDPI, Basel, Switzerland.

doi: 10.2166/9781789061796

Chapter 9

Decarbonization potentials using photobiological systems

*Lara Méndez[1,2], Cristian A. Sepúlveda-Muñoz[1,2], María del Rosario Rodero[1,2], Ignacio de Godos[1,2] and Raúl Muñoz[1,2]**

[1]*Institute of Sustainable Processes, University of Valladolid, Dr. Mergelina s/n., Valladolid 47011, Spain*
[2]*Department of Chemical Engineering and Environmental Technology, University of Valladolid, Dr. Mergelina s/n., Valladolid 47011, Spain*
**Correspondence: mutora@iq.uva.es*

9.1 INTRODUCTION

With a current focus on the need to limit the effects of climate change, greenhouse gas emissions (GHG) associated with the wastewater treatment (WWT) sector need to be reduced. In this context, significant reductions can be attained by using phototrophic microorganisms as a platform for WWT combined with anaerobic digestion and biogas upgrading for bioenergy production. Microorganisms like microalgae or purple photosynthetic bacteria (PPB) are capable of supporting a cost-effective WWT with a lower energy consumption (with the associated reduction in indirect CO_2 emissions), an enhanced nutrient recovery and an *in-situ* assimilation of the CO_2 produced during pollutant oxidation compared to conventional WWT technologies. Thus, PPB constitute a promising biological platform for the treatment of high strength wastewaters based on their high growth rates, ability to use infrared radiation and tolerance to high salinity and low temperatures. Similarly, algal-bacterial symbiosis supports complex interactions that contribute to a sustainable assimilation of carbon and nutrients from multiple types of wastewaters. The photosynthetic biomass generated during WWT can be further valorized into added-value products, including biogas. In this regard, biogas upgrading into biomethane can be integrated in photosynthetic wastewater treatment schemes based on the ability of PPB and microalgae to capture and utilize CO_2 and use H_2S as an electron donor, which can further decarbonize WWT. Finally, the recent advances in photobioreactor design that have boosted the biodegradation potential of photobiological-based systems, while lowering their energy demand, will be critically discussed in this book chapter.

Greenhouse gas (GHG) emissions, mainly carbon dioxide (CO_2) from the combustion of fossil fuels, have increased since the beginning of the industrial revolution with a higher incidence since the 1950s. The accumulation of GHG in the atmosphere has caused harmful effects on the environment and climate change. The European Commission has recently targeted a GHG emissions reduction of at least 55% by 2030 in comparison to 1990 emissions. Therefore, there is an urgent need to limit all GHG emissions and further decarbonize both the energy system and all industrial sectors by boosting and implementing innovative and sustainable production and waste management technologies (Qiao *et al.*, 2020).

In the context of the wastewater treatment (WWT) sector, the implementation of new technologies based on phototrophic microorganisms in combination with anaerobic digestion (AD) systems have proven to be cost-efficient for WW bioremediation and energy production (Fouilland *et al.*, 2014; Maity *et al.*, 2014; Qiao *et al.*, 2020). These technologies do not require external aeration, which significantly reduces the energy demand and CO_2 footprint of WWT. Microalgae are capable of assimilating multiple pollutants like nitrogen compounds and phosphates present in aqueous effluents or CO_2 from off-gases with the consequent generation of oxygen and biomass during the photosynthetic process. The implementation of microalgal-based systems allows reducing direct and indirect CO_2 emissions, supports a cost-effective WWT with lower energy consumption, and enhances nutrient recovery compared to conventional WWT technologies. Moreover, the recovered microalgae biomass can be further valorized into other added value compounds such as biofertilizers or employed as biofuels substrates, that is biodiesel, integrating this process into the biorefinery concept and circular economy. Likewise, purple phototrophic bacteria (PPB) have emerged as a novel and promising platform for the removal of pollutants, which can be operated as a standalone technology or in combination with other technologies such as algal-bacterial systems. PPB-based systems can use organic matter and infrared spectra from solar radiation as energy source. These microorganisms have been reported to tolerate and grow in high strength waters with a high salinity or toxicity and support high rates of organic matter and nutrient assimilation (Batstone *et al.*, 2015; Hülsen *et al.*, 2014).

These photosynthetic platforms can be integrated with AD. AD is a biological process driven by symbiotic microbial communities in the absence of oxygen, in which biodegradable organic matter is bioconverted into biogas and nutrients are mineralized in the digestate. Indeed, AD can cope with multiple organic residues, which are used as feedstocks for biogas generation (Čater *et al.*, 2015; González-Fernández *et al.*, 2008; Mendez *et al.*, 2014a; Nkemka & Murto, 2013; Rani *et al.*, 2012). Biogas is a mixture of gases typically composed of CH_4 and CO_2, and other components such as H_2S, O_2 or N_2 at lower concentrations. This biogas can be employed for the generation of renewable heat and electricity in CHP engines or being upgraded to biomethane by removing biogas impurities. Thus, biomethane is a purified form of raw biogas that can be injected into the natural gas grid or use as a vehicle fuel substitute. The removal of CO_2, H_2O, H_2S and other impurities is required during biogas upgrading according to most biomethane standards. In this context, biogas upgrading can be integrated to WWT in microalgal or PPB ponds based on the ability of these photosynthetic organisms to fix gaseous CO_2 using the residual nutrients present in digestates (López *et al.*, 2013). For instance, CO_2 is fixed as organic carbon by the photosynthetic apparatus of microalgae, with the concomitant production of oxygen. This O_2 can be used by sulfur oxidizing bacteria to remove H_2S from biogas (Muñoz *et al.*, 2015).

The so called digestate, which is the liquid effluent of AD, is rich in phosphorous and nitrogen. This digestate, prior solid–liquid separation, can be used as culture media in the microalgal pond for the generation of an algal biomass that can be further valorized as a feedstock for biostimulant or biofertilizer production. This will ultimately reduce the energy demand (and CO_2 footprint) of digestate management (Guilayn *et al.*, 2020).

Overall, the lower energy demand of photosynthetic systems during WWT, along with the potential integration of biogas upgrading in photobioreactors fed with digestates and the inherent capture of the CO_2 generated during organic matter mineralization, bring new opportunities to decarbonize wastewater treatment and enhance nutrient recovery in areas with high irradiations and moderate temperatures.

9.2 PHOTOSYNTHETIC WASTEWATER TREATMENT

9.2.1 Microalgae

Anthropogenic activities result in the generation of a broad variety of pollutants in domestic, livestock, agro-industrial and industrial wastewaters. These effluents have been traditionally treated using physical, chemical and biological processes such as filtration, sedimentation, chemicals addition, aerobic activated sludge-based treatments, anaerobic digestion, and so on. (Englande *et al.*, 2015). However, conventional treatment configurations entail a high-energy consumption, high CO_2 footprint, low nutrient recovery and environmental impacts, despite providing satisfactory reductions of carbon, nitrogen and phosphorous concentrations in the treated water (Posadas *et al.*, 2017a).

Microalgal technologies are eco-friendly processes that exhibit low environmental impacts, while reducing the operating costs of conventional WWT technologies. Microalgal-based WWT is carried out by photoautotrophic microorganisms that support a photosynthetic carbon dioxide fixation, where sunlight energy is converted into chemical energy (biomass) by a light-driven redox reaction using H_2O as electron donor. This process results in the assimilation of CO_2 and nutrients in the form of algal-biomass concomitantly with the release of oxygen as side-product, which can be used by bacteria to oxidize organic matter to CO_2 and ammonium to nitrate/nitrite (Masojídek *et al.*, 2013; Rochaix 2016). In this context, algal-bacterial symbiotic bioprocesses entail lower operational costs than conventional activated sludge systems (Figure 9.1) and a more sustainable WWT (Barreiro-Vescovo *et al.*, 2020; Posadas *et al.*, 2017a). Since the photosynthetic oxygen released can replace mechanical aeration, nutrient recovery is enhanced as a result of microalgae growth, and the overall (direct and indirect) CO_2 emissions are reduced. The so called 'microalgae' are photoautotrophic microorganisms that comprise both eukaryotic microalgae and prokaryotic blue green algae or cyanobacteria (Singh & Dhar, 2019), which possess heterotrophic and autotrophic metabolisms.

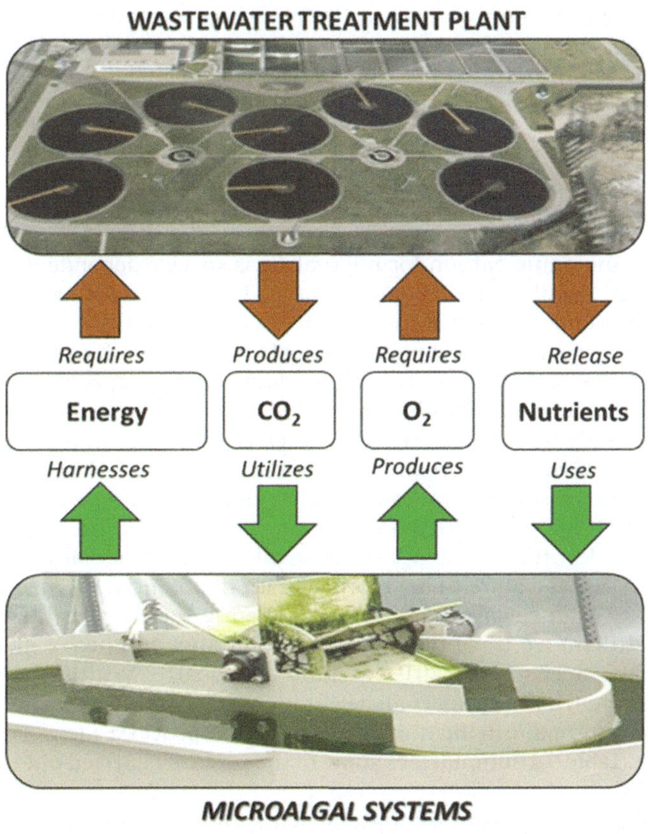

Figure 9.1 Schematic diagram showing the synergy between microbial aerobic treatment on a WWTP and a microalgae system.

Biological CO_2 fixation is typically carried out by terrestrial plants, which are able to remove only 3–6% of the CO_2 supplied. In this context, the uncomplicated cellular structures and rapid growth of microalgae endow them with higher photosynthetic and CO_2 fixation efficiencies that enable 10–50 times faster CO_2 fixation rates (Cuellar-Bermudez *et al.*, 2015; Iasimone *et al.*, 2017). A typical carbon content of microalgal biomass averages 50% of their dry weight, which entails 1.8 kg of CO_2 demand per kg of microalgae produced (Curtis 2010; Molazadeh *et al.*, 2019; Posadas *et al.*, 2017a; Schediwy *et al.*, 2019). Optimal growth of algae requires several elements in the culture broth, mainly C/N/P, in stoichiometric proportion to that found in the composition of the algal biomass, which is determined by the Redfield ratio (106:16:1 C/N/P). Most WWs contain a lower C/N/P ratio than that needed for microalgal growth, leading to carbon limitation and therefore hindering biomass growth and nutrient recovery (Toledo-Cervantes *et al.*, 2018). Hence, the supplementation of external C sources is a common strategy to sustain an active microalgae growth. This additional CO_2 can be obtained from alternative emission sources such as power plants flue gas, industrial off-gases or biogas. A direct injection of the CO_2 laden gas stream into the microalgae culture via fine bubble diffusers improves the mass transfer of CO_2, increasing the concentration of inorganic carbon, and thus enhancing microalgal biomass productivity (Rezvani *et al.*, 2016; Toledo-Cervantes *et al.*, 2018).

Microalgae growth is governed by environmental conditions. The most important factors determining microalgae growth in WW are nutrients concentration, pH and alkalinity, light and temperature. The pH of the cultivation broth is highly dependent on photosynthetic activity, alkalinity, and microbial respiration (Posadas *et al.*, 2017a). Photosynthetic activity increases pH as a result of microalgal CO_2 uptake from the cultivation broth. This pH can reach values of up to 11, which can inhibit microalgal growth, although a pH range of 7–8 is considered optimum for microalgae growth. pH can also modify the equilibrium of the nutrient species available in the cultivation broth and impact on the gas–liquid CO_2 mass transfer. Equation (9.1) shows carbon distribution in aqueous medium as a function of the pH:

$$CO_2 + H_2O \overset{Pk_1=3.6}{\Longleftrightarrow} H_2CO_3 \overset{Pk_2=6.3}{\Longleftrightarrow} HCO_3^- + H^+ \overset{Pk_3=10.3}{\Longleftrightarrow} CO_3^{2-} + 2H^+ \tag{9.1}$$

The preferred form of inorganic carbon for microalgae is species dependent. Many species are able to use both CO_2 and HCO_3^-, while some others are limited to only one of them (Markou *et al.*, 2014). In this context, since the pH of microalgae cultivation broths can range from 6.5 to 10, bicarbonate is the dominant inorganic carbon species in most photobioreactors (Canon-Rubio *et al.*, 2016; Chi *et al.*, 2011).

In addition, alkalinity has a significant role in inorganic carbon speciation and governs the CO_2 gas-liquid mass transfer rate to the cultivation broth. The CO_2 present in gas streams such as biogas is absorbed into the aqueous algal broth and reacts with OH^- and water to form carbonate-bicarbonate ions, while increasing total dissolved inorganic carbon (DIC) (Canon-Rubio *et al.*, 2016; Markou *et al.*, 2014). Therefore, the external supplementation of CO_2 does not only support pH control but also increases DIC availability (Choi *et al.*, 2019; Posadas *et al.*, 2015a, 2017a).

Most biotechnologies devoted to WWT also involve the remediation of the nitrogen and phosphorous present in wastewaters. The bioremediation potential of microalgae for inorganic nitrogen and phosphorus from sewage has been consistently reported in literature, along with the capacity of microalgae to remove trace organic micropollutants and heavy metals (González *et al.*, 2008; Ji *et al.*, 2013; López-Serna *et al.*, 2019; Mendez *et al.*, 2016; Whitton *et al.*, 2015; Yang *et al.*, 2015). The assimilation of nutrients by microalgae entails significantly lower energy consumptions and CO_2 emissions than conventional nitrification–denitrification processes (e.g. 1.5 kwh kg $N_{removed}^{-1}$) or phosphate precipitation. Table 9.1 summarizes some of the research carried out in different wastewater phytoremediation studies.

Nitrogen and phosphorous are also essential nutrients for the development of microalgal cultures and both (phosphorous to a lesser extent) are limiting factors in the growth of algae (Curtis, 2010). Nitrogen approximately represents around 5–10% of algal composition (Markou *et al.*, 2014) and

Table 9.1 Summary of microalgal bioremediation on WW effluents.

Wastewater	Microalgae	Reactor	Removal Efficiencies	References
Swine manure	*Oocystis* sp., *Microspora* sp., *Nitzschia* sp., *Chlorella* sp., *Chlamydomonas* sp., *Protoderma* sp.	HRAP	COD: 76%; TKN: 83%; P: 10%	de Godos *et al.* (2009)
Swine wastewater + fish wastewater	*Tribonema*sp.; *Chlorella zofingiensis*	Photobioreactor (1L)	TN: 86.4%; TP: 84.7%	Cheng *et al.* (2020)
Urban wastewater	*Nitzschia* spp. *Gomphonema parvulum, Cyclotella meninghiana, Melosira varians, Oscillatoria* sp., *Phormidium sp.*	AFW (Algal floway)	N removal rate:2.52 g m^{-2} d^{-1}; P removal rate: 1.25 g m^{-2} d^{-1}	Marella *et al.* (2019)
Anaerobic cattle wastewater	*Scenedesmus obliquus*	VAPs (vertical alveolar panel photobioreactors)	COD: 61%; NH_4^+: 96%; PO_4^{3-}: 70%	de Mendonça *et al.* (2018)
Domestic wastewater	Activated sludge + *Chlorella vulgaris*	Photobioreactor (1L)	TN:97.58%; (45% of N-NH_4^+ lost from air stripping)	Leong *et al.* (2018)
Municipal wastewater	*Chlorella* sp. IM−01	–	NH_4: 98.4%; NO_3^-: 97.8%; TP: 89.39%	Kiran *et al.* (2014)
Swine slurry	*Chlorella vulgaris, Scenedesmus obliquus, Chlamydomonas reindhardtii*	Photobioreactor (1L)	NH_4: 99% (N uptake:64.3%); PO_4^{3-}:82% (summer)	Molinuevo-Salces *et al.* (2016)
Domestic wastewater	Activated sludge + *Scenedesmus* sp.	HRAP	COD: 84%; TN: 79%; TP: 57%	Posadas *et al.* (2015a)
Municipal wastewater	*Chlorella vulgaris*	Photobioreactor (5L)	COD: 76.5%; N-NH_3: 91.5%	Arun *et al.* (2017)

phosphorous is around 1% (Solovchenko *et al.*, 2016). Nitrogen concentration markedly affects the composition of the microalgae. A limitation of the nitrogen source available in the culture media implies the use of intracellular nitrogen to carry out metabolic functions (Pancha *et al.*, 2014). However, high concentration of nitrogen can lead to inhibitory effects (He *et al.*, 2013). Likewise, pH regulates the balance NH_4^+/NH_3 concentrations as shown in Equation (9.2):

$$NH_4^+ + OH^- \xLeftrightarrow{pk=9.25} NH_3 + H_2O \tag{9.2}$$

It has been reported that NH_4^+ inhibits photosynthetic activity in some microalgae species at concentrations higher than 100 mg N-$NH_4^+ \cdot L^{-1}$ and pH >8 as a result of NH_3 toxicity since NH_3 is highly inhibitory to microalgae growth (Abeliovich & Azov, 1976; Posadas *et al.*, 2014). Besides, NH_3 stripping occurs in open photobioreactors operated at high pH. Despite being soluble in water, NH_3 is a highly volatile compound that is dominant at high pH, which can lead to the loss of N-NH_3 by volatilization. This fact is particularly relevant during WWT with mechanical aeration (Cai *et al.*, 2013; Jamieson *et al.*, 2003; Mendez *et al.*, 2016), which increases the overall CO_2 footprint of the WWT plant since NH_3 is a precursor of N_2O. Additionally, nitrification–denitrification processes can be implemented in microalgal-bacterial systems involving the removal of nitrogen by the oxidation of N-NH_4^+ into N-NO_2^- and N-NO_3^-, and their further conversion to N_2 (García *et al.*, 2017). These processes are particularly relevant when treating WW effluents with high NH_4^+ concentration such as centrate or livestock effluents that contain up to 800 and 3000 mg N-$NH_4^+ \cdot L^{-1}$, respectively (Molinuevo-Salces *et al.*, 2012; Morales-Amaral *et al.*, 2015; Posadas *et al.*, 2013). These effluents can be diluted to provide NH_4^+ concentrations well-suited for algal growth or supplied at low loading rates in order to avoid microalgae inhibition during WWT.

Biological phosphorous assimilation rate is governed by environmental factors such as temperature, pH or the availability of ions (K^+, Na^+, Mg_2^+), which can influence the transport of phosphate into the cells (Cembella *et al.*, 1982; Correll 1998). Microalgae can accumulate an excess of phosphorus via 'luxury uptake' mechanisms via polyphosphate synthesis (Eixler *et al.*, 2006), the chemical species of phosphorous that can be used in situations of nutrient limitation. Additionally, phosphorous from WW can be removed by precipitation and sinking in the form of struvite or hydroxyapatite when PO_4^{3-} ions are combined with Ca_2^+ and Mg_2^+ present in the WW (De-Bashan & Bashan, 2004; Mendez *et al.*, 2016; Posadas *et al.*, 2017a).

Finally, it should be stressed that microalgae cultivation during WWT supports a cost-effective bioremediation of pollutants from WW with a concomitant production of a valuable biomass. Microalgae are a potential source of a broad range of high-value products suitable for exploitation in different biotechnological fields such as cosmetics, nutraceuticals, and pharmaceuticals (Chu, 2012). However, when using WW as a water and nutrient source, the potential applications of the alga-bacterial biomass generated are limited by the inherent risk of contamination by various pollutants or pathogens present in the WW. In this context, the algal-bacterial biomass produced from WW can be used for the production of low-added value products for aquaculture feed, biofertilizers, bioactive substances or feedstock for renewable biofuels such as biogas (Mohd Udaiyappan *et al.*, 2017; Singh & Dhar, 2019; Whitton *et al.*, 2015).

9.2.2 Purple phototrophic bacteria

Purple phototrophic bacteria (PPB) are a diverse bacterial group composed of purple sulfur bacteria (PSB) and purple non-sulfur bacteria (PNSB) (Capson-Tojo *et al.*, 2020), which differ mainly in their ability to tolerate high and low hydrogen sulfide (H_2S) concentrations, respectively. PPB are phototrophic organisms capable of obtaining energy from solar radiation via anoxygenic photosynthesis and with the ability of fixing CO_2 transforming it into cell material, which confer them the ability to decrease the CO_2 footprint of WWT (Capson-Tojo *et al.*, 2020). This process is carried out due to the presence of pigments in these types of microorganisms, most of them belonging to the group of bacteriochlorophylls and carotenoids. PPB mainly synthesize carotenoids such as spirilloxanthin, rhodopsin, spheroidene and lycopene (Hunter *et al.*, 2009). Additionally, PPB exhibit a very versatile metabolism and are capable of growing under autotrophic, heterotrophic and mixotrophic mode (Sepúlveda-Muñoz *et al.*, 2020a), which confer them the ability to adapt to different environments. Ecological niches where PPB can be found are mainly soils, natural water bodies and wastewater. Additionally, PPB have been described to live in extreme environmental conditions such as high salinity, pH and low temperature (Hülsen *et al.*, 2016a, 2019). Likewise, it has been consistently reported that PPB are efficient for WWT and exhibit a great potential for the recovery of carbon and nutrients while synthesizing high added value compounds (Lu *et al.*, 2019b). Overall, PPB represent a promising but unexplored group of microorganisms capable of supporting WWT based energy from solar radiation with a low CO_2 footprint.

PPB exhibit the greatest metabolic diversity among microorganisms (Larimer *et al.*, 2004). Thus, PPB can grow under anaerobic conditions photoheterotrophically and photoautotrophically, and aerobically under chemoheterotrophic and chemoautotrophic conditions (Figure 9.2). Under phototrophic mode the energy for PPB growth is obtained mainly from light energy (solar radiation), while under chemotrophic mode energy derives from the degradation of organic compounds. The prevailing environmental conditions or cultivation media determine the metabolism in PPB, which

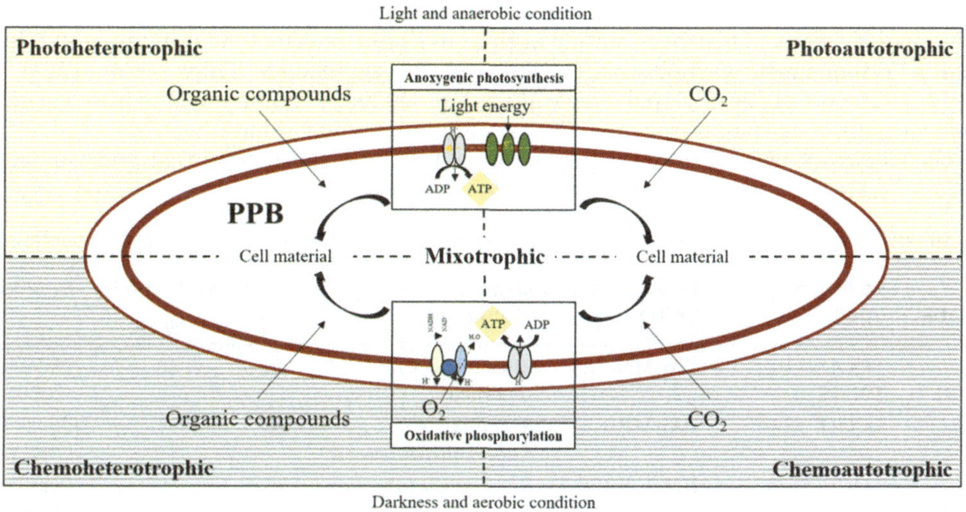

Figure 9.2 Simplified metabolic diagram of purple phototrophic bacteria based on the metabolism of *Rhodopseudomonas palustris*.

can even grow simultaneously utilizing light energy and organic substrates under mixotrophic growth. Under anaerobic conditions in the presence of light energy, phototrophic growth is favored and results in the production of energy by anoxygenic photosynthesis and biosynthesis of adenosine triphosphate (ATP) by photophosphorylation. At this point, it should be stressed that ATP is the main molecule used by PPB to preserve energy to be used in different metabolic routes. On the other hand, under aerobic conditions and absence of light, the predominant metabolism will be chemotrophic since the presence of oxygen inhibits the synthesis of bacteriochlorophyll and affects the photosynthetic capacity of PPB (Izu *et al.*, 2001). Therefore, oxidative phosphorylation for ATP synthesis is favored under aerobic conditions (Lu *et al.*, 2011). Under aerobic conditions, PPB use oxygen as an electron acceptor and obtain energy from the proton motive force generated by consumption of NADH (molecules synthesized from the degradation of organic compounds).

PPB can use CO_2 under autotrophic mode or organic compounds under heterotrophic mode as a carbon source. Indeed, PPB are capable of metabolizing different carbon sources due to their great metabolic plasticity. For instance, PPB can use as a carbon source small molecules of some carbohydrates, fatty acids and alcohols via photoheterotrophic metabolism (Lu *et al.*, 2019b) using multiple metabolic pathways such as tricarboxylic acid cycle, Embden–Meyerhof pathway, pentose phosphate route or fatty acid metabolism (Larimer *et al.*, 2004). Likewise, PPB can fix CO_2 via the Calvin–Benson–Bassham cycle, which can partially support the decarbonization of WWT (Lo *et al.*, 2018). For instance, an efficient metabolism in the reuse of carbon has been described in *R. palustris*, which is able to use the CO_2 produced by catabolic routes and fix it to synthesize biomass (Navid *et al.*, 2019). On the other hand, PPB can use CO_2 as an electron acceptor under photoheterotrophic mode. PPB can metabolize all forms of inorganic nitrogen like NH_4^+, NO_2^-, NO_3^- and fix atmospheric N_2 (Sepúlveda-Muñoz *et al.*, 2020b) and organic compounds containing nitrogen (i.e amino acids or proteins). PPB can use NH_3 as an electron donor under chemoautotrophic mode and NO_3^- as an electron acceptor under chemoheterotrophic mode.

This inherent ability of PPB to assimilate most types of carbon and nitrogen at high biomass yields has attracted great interest in PPB-based WWT (Capson-Tojo *et al.*, 2020) as an alternative to conventional biological treatments such as anaerobic digestion or activated sludge system where most carbon and nitrogen present in wastewater is released to the atmosphere. In this context, the presence of PNSB like *R. palustris*, *R. sphaeroides* and *R. capsulatus* has been reported in WWT systems due to the low concentrations of hydrogen sulfide in this environment. The WWT ability of PPB has been confirmed in both domestic wastewaters (Hülsen *et al.*, 2014) and also in high strength WW such as piggery WW (García *et al.*, 2019; Sepúlveda-Muñoz *et al.*, 2020a) or poultry WW (Hülsen *et al.*, 2018). High removal rates of carbon, nitrogen and other pollutants removal have been achieved, exceeding removal efficiencies of 90% (Table 9.2). In addition, PPB have been proposed and validated as a

Table 9.2 Summary of wastewater treatment in batch photobioreactors with purple phototrophic bacteria.

Type of Wastewater	Pollutants (mg L^{-1})	Reactor type Volume (L)	Strain of PPB (Dominant)	Pollutant Removal		References
				C (%)	N (%)	
Domestic wastewater	TCOD: 430 TN: 43	PAnMBR 2 L	PPB mixed (not specified)	95	86	Dalaei *et al.* (2020)
Synthetic wastewater	TCOD: 349	PAnMBR	*R. palustris Rhodospirillaceae*	88	97	de las Heras *et al.* (2020)
Piggery wastewater	TOC: 15775–1131 TN: 5028–366	Batch PBR 0.5 L	*Rhodopseudomonas*	75	39	Sepúlveda-Muñoz *et al.* (2020b)
Brewery wastewater	COD: 2200–3200 NH_4^+: 50–70	MBR system 200 L	PPB mixed	99	–	Lu *et al.* (2019a)
VFA-rich food industry wastewater	COD: 5122 TN: 298	Batch PBR 0.8 L	*R. palustris*	89	91	Liu *et al.* (2019)
Saline domestic wastewater	TCOD: 418 TN: 12	PAnMBR 2 L	PPB mixed	86	62	Hülsen *et al.* (2019)
Piggery wastewater	TOC: 574 TN: 166	Open PBR 3 L	*Rhodoplanes*	87	83	García *et al.* (2019)
Domestic wastewater	TCOD:370–540 TN: 48–56	PAnMBR 2 L	PPB mixed (not specified)	93	91	Dalaei *et al.* (2019)
Piggery wastewater	TOC: 1989–10318 TN: 563–2209	Batch PBR 0.4 L	PPB mixed (not specified)	78	13	Marín *et al.* (2019b)
Poultry processing wastewater	TCOD: 4000 TKN: 200	PAnMBR 2 L	PPB mixed (not specified)	92	64	Hülsen *et al.* (2018)
Brewery wastewater	COD: 3300	MBR system 10 mL	PPB mixed (not specified)	96	–	Yang *et al.* (2018)
Digested piggery wastewater	COD: 4792 NH_4^+-N: 913	Batch PBR 0.2 L	*Rhodobacter blasticus* and *R. capsulatus*	83	–	Wen *et al.* (2016)
Acidic food industry wastewater	COD: 3350 TN: 200	PBR 0.8 L	*R. palustris*	90	92	Liu *et al.* (2016)
Synthetic sugar wastewater	COD: 6000 NH_4^+-N: 400	PBR	*Rhodopseudomonas*	95	–	Zhou *et al.* (2015)
Brewery wastewater	COD: 8000–10000 TN: 1	Batch PBR 0.6 L	*R. Sphaeroides*	94	–	Lu *et al.* (2013)

platform for biogas upgrading coupled with WWT, reporting high efficiencies of CO_2, H_2S and organic matter (Marín *et al.*, 2019b).

Finally, the PPB biomass generated from WWT is rich in compounds with added value such as pigments (bacteriochlorophylls and carotenoids) and other molecules like coenzyme Q10, single-cell protein, nutrients, polyhydroxyalkanoates, pantothenic acid and 5-aminolevulinic acid (5-ALA) (Capson-Tojo *et al.*, 2020; Lu *et al.*, 2019b). These molecules can be extracted from the PPB biomass in the context of a circular economy and WWT biorefinery.

9.2.3 Photobioreactors for wastewater treatment and resource recovery

9.2.3.1 Microalgae reactors

Photobioreactors used for microalgae cultivation are typically classified according to the contact with atmosphere as open and closed reactors. Open photobioreactor configurations present low operational and investment costs and are easily scaled-up. The main disadvantages reported in open photobioreactors are the risk of microbial contamination and water losses (Acién *et al.*, 2013), which are not critical during WWT. Therefore, open systems are considered the most close-to-market approach to be applied at industrial scale. On the other hand, enclosed systems have been tested for WWT only at lab and pilot scale. These intensive algal cultures systems are designed to support high areal irradiations and offer a better control of environmental conditions (less risk of contamination, temperature control, etc.) but the energy consumption and material costs are significantly higher than those of their open counterparts (Acién *et al.*, 2017; Ibrahim *et al.*, 2020).

Wastewater treatment in microalgae reactors is determined: by (1) microalgae photosynthetic activity, which directly depends on the impinging irradiation in the photobioreactor and promotes oxygenation and nutrient up-take; and (2) abiotic mechanisms such as NH_3 stripping and phosphate precipitation that result in a nitrogen and phosphorous concentration decrease in the final effluent. Both factors are affected by photobioreactor configuration and operation mode.

9.2.3.1.1 Open raceways

Open algae systems for wastewater treatment were originally developed in the 1950s and 1960s by Oswald and co-workers (Oswald, 1978). Shallow mixed ponds, with a depth varying between 0.20 and 0.40 m, constructed in meandering configuration of simple or multiple channels, provide a large volume of irradiated algal broth (Figure 9.3). This photobioreactor configuration emerged

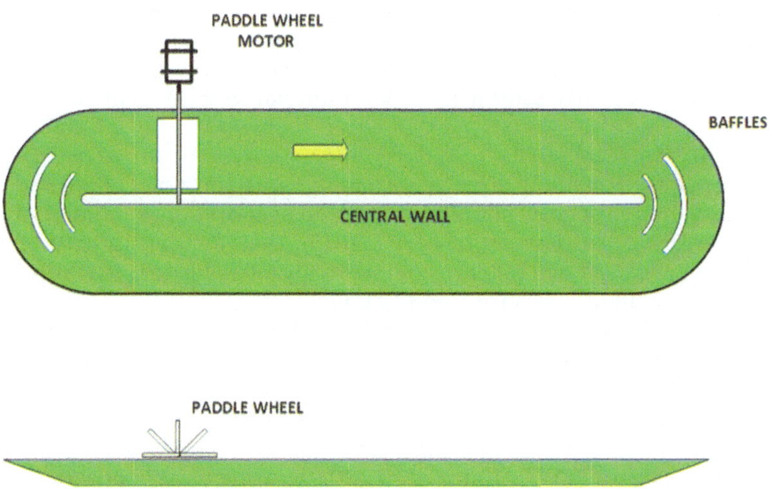

Figure 9.3 Schematic diagram of a raceway photobioreactor.

as a modification of the widespread stabilization ponds where algae only developed on the surface (Spellman & Drinan, 2014). Therefore, shorter hydraulic retention time (HRT) between 3 and 10 days can be applied in open raceways. Continuous mixing by paddle wheels allows for higher algae biomass production and fully aerobic conditions in the cultivation broth. During the central hours of the day, sunlight strikes perpendicularly to the cultivation surface, resulting in the higher treatment capacities of the day. Indeed, oxygen oversaturation has been commonly reported under favorable environmental conditions and moderate loading rates (Arbib *et al.*, 2017; Hamouri, 2009). Biomass harvesting in open ponds devoted to WWT has been carried out in settlers, settling ponds, lamella settlers or DAF units (dissolved oxygen flotations) (Craggs *et al.*, 2012; de Godos *et al.*, 2016).

Open raceways are simple to build in horizontal surfaces delimited with walls or earth slopes (Craggs *et al.*, 2012). The original system designed by Oswald and co-workers was implemented in the 1970s and 1980s in real scale facilities treating urban wastewater. Hence, 2- and 5-hectare open raceways were operated for years in St. Helena and Hollister wastewater treatment plants in California (Park *et al.*, 2013). These demonstration units were constructed in combination with facultative ponds and maturation ponds to optimize the process. Subsequent studies have used different integration of the raceway reactors (also called high-rate algae ponds) into domestic WWT plants. The open algal ponds are normally preceded by pretreatment units (mainly primary settlers) to remove suspended materials (Craggs *et al.*, 2012; Hamouri *et al.*, 2003). Other demonstration units are based on the use of anaerobically treated wastewater (Hamouri, 2009; Hamouri *et al.*, 2003). Recently, a European demonstration project (ALLGAS) based on algae production using urban wastewater as cultivation media reached a positive energy production by combining anaerobic pretreatment of WW and biogas production with algae biomass (de Godos *et al.*, 2017). A total surface of 3.6 hectares was implemented and bioenergy was produced as biomethane, which is used as a vehicle fuel resulting in a ratio of energy return on investment of 2 (www.all-gas.eu). A recent techno-economic evaluation performed under the framework of this project showed a reduction in domestic wastewater treatment costs from 0.22 to 0.15 \$ m^{-3} and a reduction in the energy demand by a factor of 4, which results in a significant reduction in the indirect CO_2 emissions (Acién *et al.*, 2017)

The potential of open systems has been also evaluated for the treatment of livestock, industrial, agro-industrial effluents (Mulbry *et al.*, 2010, 2005; Olguín *et al.*, 2007). In the case or piggery and cattle wastewaters, dilution is applied in order to reduce turbidity of culture media and inhibition mediated by elevated ammonia levels present in these livestock effluents (Godos *et al.*, 2010). Dilution is typically preceded by solid separation units such as sieves, settlers, and coagulation-flocculation units (Barlow *et al.*, 1975; González-Fernández *et al.*, 2010). However, recent studies showed the feasibility of algae growth in undiluted anaerobically digested swine manure using selected and acclimatized microalgae species (Ayre *et al.*, 2017). Algae grown in diluted cattle effluents has been recently studied as a protein and HUFA (highly unsaturated fatty acid) feedstock for animal feed (Murry *et al.*, 2019).

9.2.3.1.2 Enclosed photobioreactors

Enclosed photobioreactors do not allow direct gas exchange with the atmosphere and entails a number of advantages, such as a limited species contamination and evaporation, smaller footprint, and higher gas to liquid mass transfer rates (Karemore *et al.*, 2016), thus increasing carbon dioxide capture. Three main categories of enclosed reactors are commercially available for algae production and WWT treatment: bubble columns, tubular reactors, and flat panels.

Bubble column photobioreactors consist of transparent cylinders made of methacrylate or glass mixed by continuous aeration. As a result of their vertical configuration, sunlight capture by cells at the central hours of the day is minimum. These systems are size limited since the height can only reach a few meters. Therefore, no large scale WWT experiences have been reported using bubble columns. However, the integration of microalgae cultures in new processes such as simultaneous production of biomolecules and WWT (Kalra *et al.*, 2020), or biomass recovery through membrane filtration (Syahirah *et al.*, 2020), has been investigated in bubble columns.

Tubular photobioreactors are made of transparent tubular pipes provided with a continuous liquid recirculation (via centrifugal pumps or airlift units) to maintain cells suspension and light distribution to the bulk algal broth. Different configurations have been developed for biomass production applications: two plane tubular, near-horizontal tubular, the helical bubble reactor and α-tubular photobioreactor (Molina-Grima *et al.*, 2010). The arrangement of the solar collector tubes determines the irradiated surface and light received by algae cells. The high construction and operational costs compared to raceways have hampered the use of this complex technology for WWT purposes (Ibrahim *et al.*, 2020). In this context, de Godos *et al.* (2017) compared the performance of tubular reactors with raceway systems and observed that although high nutrient removals were reached in the enclosed unit (98% for N and P), collapse of algae cultures was reported after 30 days of operation due to biofouling.

Flat panels are vertical translucent plates containing the culture broth and receiving illumination on both sides. Mixing is typically provided by continuous aeration. The first designs used glass sheets connected with rubber, resulting in an expensive system difficult to scale-up (Samson & Leduy, 1985). Researchers have also proposed the use of plastic bags installed inside metal frames in order to reduce the installation costs (Tredici & Materassi, 1992). Similarly to bubble columns and tubular photobioreactors, no large-scale experiences using residual effluents as culture medium have been reported with flat panels. The performance of a lab-scale flat panel photobioreactor for nutrient removal from secondary effluents was studied by Ruiz *et al.* (2012), who achieved removal rates of 89 and 84% for nitrogen and phosphorous, respectively.

9.2.3.2 Purple photosynthetic bacteria

PPB photobioreactors for treatment of polluted effluents is an emerging technology. Most of the reported experiences are batch cultures performed in enclosed photobioreactors under artificial infrared illumination (Budiman *et al.*, 2014; Choorit *et al.*, 2002; Madukasi & Zhang, 2010; Zhou *et al.*, 2016). Domestic, agro-industrial and industrial wastewaters have been treated in PPB photobioreactors under laboratory-controlled conditions (Puyol *et al.*, 2020). Two main photobioreactor configurations have been documented for the continuous treatment of wastewater: vertical and horizontal systems.

Vertical reactors are similar to algae flat panels and consist of two panels placed opposite each other that contain the PPB broth. A narrow light path (distance between panels) guarantees an effective light exposure of the PPB cells. Most of the reported experiences dealing with vertical systems include membranes for biomass separation: MBR (membrane reactor) or PAnMBR (photoanaerobic membrane bioreactor) (Hülsen *et al.*, 2018) (Figure 9.4). Biomass separation allows the concentration of biomass and reduces the operational HRTs down to 8–24 h, while the solid retention time can be maintained between 2 and 20 days. MBR support high removal rates of organic matter and nutrients, ranging from 88 to 99% for total COD, 77–92% for nitrogen and 77–98% for phosphorous (Hülsen *et al.*, 2016b, 2018). A similar performance has been achieved by Nagadomi *et al.* (2000) using flat panels with immobilized cells in ceramic supports. However, the different illumination conditions applied do not allow general conclusions to be drawn. In this context, while some authors applied light cycles of infrared radiation with intensities ranging between 45 and 133 W/m^2, other studies have been performed using the complete light spectrum (visible and infrared) (Puyol *et al.*, 2020). Multiple light sources have been applied: IR-LEDs, incandescent bulbs and fluorescent lamps with and without visible filters (Hiraishi *et al.*, 1989; Hülsen *et al.*, 2019).

Horizontal systems are similar to raceway photobioreactors used for algae cultivation. Biomass separation and recirculation can be included to control PPB concentration in the bioreactor. In this regard, Sepúlveda-Muñoz *et al.* (2020a) compared open and enclosed configurations in lab-scale horizontal systems treating piggery wastewater with consistent nitrogen and carbon removals under long-term operation (more than one year). The open photobioreactor configuration achieved higher organic carbon and nitrogen removals than the enclosed photobioreactor due to the larger contribution of abiotic mechanisms of inorganic carbon and ammonia volatilization and the additional contribution of aerobic biodegradation mechanisms. Compared to vertical systems, horizontal photobioreactors

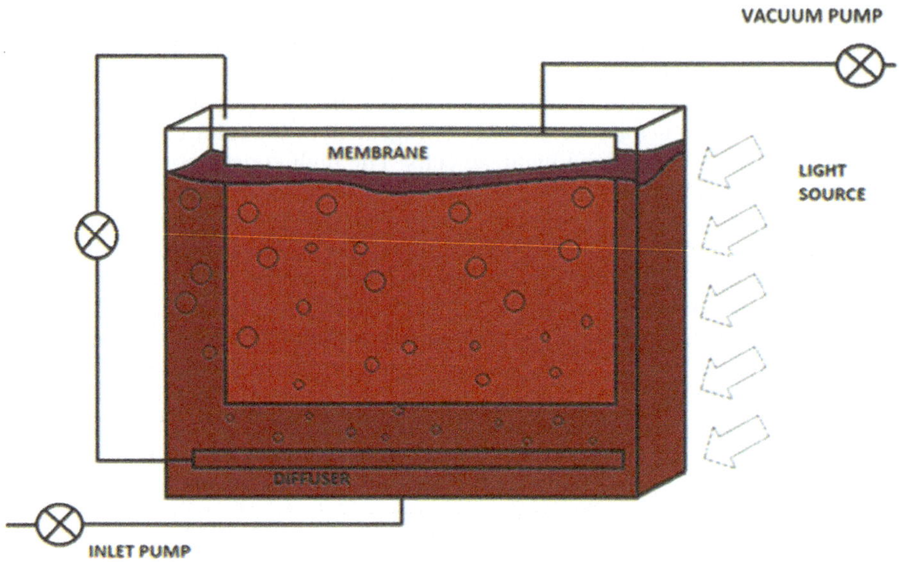

Figure 9.4 Schematic diagram of photoanaerobic membrane bioreactor for purple bacteria culture in wastewater.

are easily scaled-up and some experiences have been reported in outdoor conditions using larger volumes of culture. For instance, 19 m^3 pilot raceways were operated for more than one year using domestic wastewater and molasses by University Nova de Lisboa and the company FCC Aqualia. Visible light filters were placed in the complete surface to avoid microalgae growth and the process was operated to promote the synthesis de polyhydroxyalkanoates.

9.3 ENHANCED BIOGAS PRODUCTION FROM MICROALGAE AND PPB FOR WWT DECARBONIZATION

Circular bioeconomy has emerged in recent years as an essential component of sustainable and green industrial activity. This approach focuses on utilizing all the potential of natural resources through cascading biomass use and recycling, while ensuring that natural capital is preserved (Rajesh Banu *et al.*, 2020) In this context, photosynthetic biomass biorefining is presented as a promising approach to convert algal or PBB biomass to value-added products, biofuels, and chemicals. Thus, microalgae or PPB from WWT or residues of these biomass sources, for instance those from microalgal lipid extraction for biodiesel production (Uggetti *et al.*, 2017), can be used as a substrate of anaerobic digestion processes, which in turn can provide digestate for microalgal growth based on the high nutrient and inorganic carbon content of these AD effluents (Figure 9.5).

Anaerobic digestion involves a series of biological reactions where the breakdown of complex organic matter is carried out in the absence of oxygen, nitrate, nitrite or sulfate (typically used as electron acceptors). Anaerobic digestion occurs as a result of four sequential steps: hydrolysis, acidogenesis, acetogenesis and methanogenesis. A complex network of anaerobic bacteria, archaea and fungi are responsible for this process, which results in the production of biogas and the release of nutrients to the anaerobic broth (Choi *et al.*, 2019; Sanz *et al.*, 2017; Uggetti *et al.*, 2017). Biogas is a mixture of gases, mainly CH_4 and CO_2 at concentrations in the range of 50–80 and 15–50%, respectively, with other compounds such as water, H_2S, NH_3, N_2 and O_2 present at lower concentrations. The methane produced in AD can be used on-site as fuel gas in WWT plants to generate heat in boilers or heat and

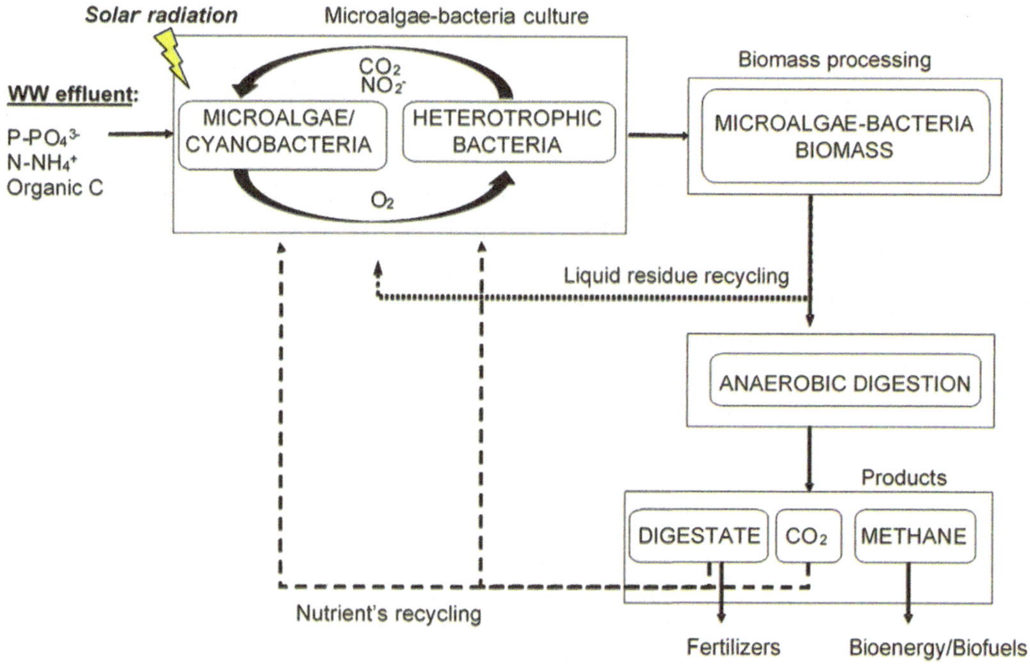

Figure 9.5 Integration of algal-bacterial WWT system coupled with AD.

power generation in gas engines or turbines, which allows for the decrease in the overall CO_2 footprint of the WWT. Alternative uses such as substitute of natural gas or biofuel for vehicles engines require biogas upgrading for the removal of impurities, to increase the methane content up to 90% CH_4 (Muñoz *et al.*, 2015; Uggetti *et al.*, 2017). Biogas upgrading contributes to an increase in the calorific value of methane while reducing transportation costs and minimize corrosion in pipelines, engines and biogas storage structures produced by some contaminants (Marín *et al.*, 2019a).

Anaerobic digestion is a mature technology already established in activated sludge-based WWT plants to obtain biogas from the treatment of the sludge generated during primary and secondary treatment. However, the versatility of AD allows biogas production from a large range of substrates including organic waste such as agricultural residues, animal manure, energy crops, micro and microalgae and even PPB. Thus, biogas generation entails less technical complexity and environmental impacts than the production of other biofuels since the extraction of specific components of the biomass is not required, which increases the efficiency of the overall process.

Eukaryotic microalgae are composed of a semi-rigid structure or cell wall that confers cell protection against physical, chemical and biological agents. The composition of microalgae cell wall is species-specific and based on complex structures as cellulose, hemicellulose, pectin and glycoproteins, which makes cell walls highly resistant structures (Gonzalez-Fernandez *et al.*, 2017). The composition and structure of microalgae cell wall influences AD performance and consequently the potential methane yield. Indeed, the cell wall hinders the anaerobic biodegradability of most microalgae species due to its high recalcitrance and resistance to microbial attack (Uggetti *et al.*, 2017). In this context, a broad variety of pretreatments can be found in literature for the disruption of microalgal cells in order to increase the final yields of biogas production (Gonzalez-Fernandez *et al.*, 2017; Mahdy *et al.*, 2014a, 2014b; Mendez *et al.*, 2014b; Passos *et al.*, 2014, 2015a). However, the differences among composition and structural characteristics of different microalgal species requires tailoring the selection of the optimal method for each microalgal biomass. Pretreatments can be classified as chemical, thermal, mechanical or biological. Thermal pretreatments entail the application of heat for organic matter disruption and solubilization. Several studies have consistently proven the efficiency of thermal pretreatments, however, the application of high temperature to microalgae biomass may lead to the formation of recalcitrant compounds that could eventually inhibit anaerobic digestion (Atelge *et al.*, 2020; Carrère *et al.*, 2016). Therefore, the application of thermal methods requires the optimization of the operational conditions to each particular substrate. For instance, Mendez *et al.* (2015) evaluated the efficiency of thermal pretreatments using *C. vulgaris* as a feedstock for methane production in batch and semicontinuous mode in a CSTR reactor. Hence, the anaerobic digestion of raw and thermally pretreated algal biomass at 120°C for 40 min was compared, and a significant increase from 138 and 85 mL CH_4 g COD_{in}^{-1} with raw microalga up to 266 and 126 mL CH_4 g COD_{in}^{-1} with thermally pretreated algae was recorded under batch and semi-continuous mode, respectively. Similar results were obtained by Schwede *et al.* (2013) in terms of increase in methane yield using *Nannochloropsis salina* pretreated at 100 and 120°C for 2 and 8 h. The methane yields of the pretreated microalgae compared with raw microalga increased by 2 to 2.85-fold in CSTR and batch mode, respectively. Chemical pretreatments are commonly used to solubilize polymers. These pretreatments are also associated with the potential formation of by-products that may induce inhibition of anaerobic microbial communities. On the other hand, mechanical pretreatments support cell wall disruption by structural fragmentation of the recalcitrant organic matter using mechanical forces. Mechanical methods imply the application of shear forces, pressure or energy using bead milling, homogenization, ultrasonication and microwaves. Mechanical pretreatments have been widely used in microalgae, showing efficient results independently on the algal species employed (Barragán-Trinidad & Buitrón, 2020; Passos *et al.*, 2015b). The energy requirements of these pretreatments are sometimes substantially higher compared to the energy recovered as methane, which represents their main limitation. For instance, Passos *et al.* (2013) reported a maximum biogas yield of 307 mL biogas·g VS^{-1}, which corresponded to a 78% increase compared with raw microalga when using microwaves at different specific energies on microalgae cultivated with WW effluent from a hydrolytic up-flow sludge blanket (HUSB) reactor fed with urban WW.

Biological pretreatments involve the use of enzymes responsible for the solubilization of recalcitrant biomass. Enzymes are molecules that will bond to a specific target of the cell to perform the lysis and solubilization of the organic compounds. Enzymatic methods represent a promising alternative to high energy-consuming pretreatments and have recently generated great interest since hydrolytic enzymes perform cell lysis under mild reaction conditions, without the generation of inhibitory side-products, and with a low energy demand. However, despite the use of enzymes is an effective alternative to thermal, chemical and physical methods, enzymes are costly molecules that need to be continuously supplied to the feedstock (Barragán-Trinidad & Buitrón, 2020). The selection of adapted anaerobic inocula to the substrate, in this particular case microalgal biomass, has been recently proposed as a low-cost alternative to the use of pretreatments methods. For instance, Gonzalez-Fernandez *et al.* (2018) reported differences in methane yields with inocula adapted to degrade sewage sludge and to digest microalgae biomass. In this study, microalgal adapted sludge showed 79.2 ± 3.1 and 108.2 ± 1.9 mL CH_4 g COD in^{-1} under mesophilic and thermophilic conditions, respectively, when digesting *Scenedesmus* sp. biomass, while non-adapted anaerobic sludge only supported to 63.1 ± 3.1 mL CH_4 g COD in^{-1} in mesophilic conditions. However, there was no remarkable difference on the final yield (which ranged from 105 to 114 mL CH_4 g COD in^{-1} for the tested sludges) when digesting *Chlorella sorokiniana*. This study confirmed the relevance of the previous adaptation of the sludge inocula to the biomass to be digested as well as the key role of the microalgae biomass digested.

Despite PPB having been consistently proven to efficiently remove WW pollutants and operate as a promising emerging tool for bioremediation technologies, the AD of PPB biomass has not been largely assessed and literature related to digestibility of this biomass is scarce. Some authors have suggested that since PPB are also anaerobic biomass, these organisms will be tolerant to the reducing conditions of the digestor, which may limit the AD performance. Additionally, PPB are rich in proteins which can eventually limit biomass degradability due to hydrolytic limitations. Hülsen *et al.* (2020) reported anaerobic VS degradations of approximately 55% and methane yield of 330 ± 4.3 and 315 ± 2.1 $mLCH_4$ gVS^{-1} under mesophilic and thermophilic conditions, respectively, in a continuous digester treating PPB grown in domestic WW. However, the relatively low economic profits obtained from biogas and the low digestibility exhibited by PPB grown in WW recommend anaerobic digestion only when alternative PPB valorization strategies are not feasible (Capson-Tojo *et al.*, 2020).

9.4 CO_2 CAPTURE AND BIOGAS UPGRADING USING PHOTOSYNTHETIC SYSTEMS DURING WWT

Biogas, the most valuable by-product from the anaerobic digestion, is a bioenergy vector able to reduce the current dependence on these unsustainable sources and to increase the overall sustainability of WWT (Ryckebosch *et al.*, 2011; Sarkodie *et al.*, 2020). However, the presence of some contaminants such as CO_2 decreases biogas calorific value, while other compounds such as H_2S are toxic and generate corrosion on biogas pipelines and combustion engines, thus limiting the widespread use of biogas (Awe *et al.*, 2017). The removal of these biogas pollutants (upgrading) is a requirement for its injection into natural gas grids or its use as a vehicle fuel in order to fulfil most international biomethane standards: $CH_4 \geq 90\%$, $CO_2 \leq 2$–4%, $O_2 \leq 1\%$ and $H_2S + COS < 5$ mg/Nm^3 (Muñoz *et al.*, 2015; UNE-EN 16723, 2017). In addition, biogas upgrading brings about new opportunities to capture additional CO_2 from WWT. Although physical/chemical technologies for CO_2 removal, such as scrubbing, membrane separation or adsorption, have been widely applied due to their high efficiency and commercially availability, only their biological counterparts exhibit a low environmental impact due to their lower energy demand and CO_2 fixation mechanisms (Muñoz *et al.*, 2015).

Photosynthetic biogas upgrading consists of CO_2 biofixation by eukaryotic microalgae and/or prokaryotic cyanobacteria via photosynthesis prior CO_2 transfer from raw biogas to the cultivation broth (Ángeles *et al.*, 2020a, 2020b). As a result, the CO_2 is not only removed from the biogas (thus increasing its energy content), but it is recovered as a microalgal biomass that can be used as feedstock for the generation of added value bioproducts (thus reducing the associated operational costs) (Ángeles *et al.*, 2020a).

Biogas upgrading in algal-bacterial photobioreactors is typically coupled to domestic wastewater or anaerobic effluents treatment, which provides the nutrients and water required for microalgal growth. This recovery of nutrients ultimately increases the environmental sustainability of photosynthetic biogas upgrading (Rodero *et al.*, 2019). In fact, the supply of CO_2 from biogas in algal-bacterial broth overcomes the carbon limitation that hinders most microalgae-based wastewater treatment in photobioreactors due to the low C/N and C/P ratios of most domestic wastewaters and digestates compared to the optimal ratios to support microalgae growth (C/N 6:1; C/P 48:1), thus increasing nutrients recovery and biomass productivity (Posadas *et al.*, 2015a; Woertz *et al.*, 2009). In addition, this green technology can support the simultaneous removal of H_2S from biogas via oxidation to SO_4^{2-}/S^0 by aerobic sulfide oxidizing bacteria (SOB) using the oxygen photosynthetically produced by microalgae (Toledo-Cervantes *et al.*, 2016). Although H_2S can provoke inhibition on microalgae activity at H_2S concentrations of ≥ 100 ppm_v, the rapid H_2S oxidation mediated chemically by the high dissolved oxygen concentrations prevailing in microalgal broth or biologically by aerobic alkaliphilic SOB involves a low exposure to H_2S (Meier *et al.*, 2018).

This rapid H_2S oxidation and its three times higher aqueous solubility in comparison to that of CO_2 according to their dimensionless Henry's law constants (C_L/C_G of ≈ 2.44 and ≈ 0.83 at 25°C for H_2S and CO_2, respectively), results in higher H_2S removal efficiencies compared to those of CO_2. Therefore, the CO_2 gas–liquid mass transfer always represents the critical step during photosynthetic biogas upgrading. High pH values (9–10) in the cultivation broth favor CO_2 and H_2S gas–liquid concentration gradients due to their acidic nature and consequently increase their mass transfer from biogas to the algal-bacterial broth. For instance, the increase of the pH from 7 to 10 in an indoors high-rate algal pond (HRAP) interconnected to an absorption column resulted in an increase of CO_2 removal from less than 20% to almost 100% (Bahr *et al.*, 2014). Although most microalgae show a maximum activity at pH 7–8, some microalgae/cyanobacteria species such as *Anabaena*, *Spirulina*, *Chlorella*, *Chlorococcum* and *Scenedesmus* are suitable for photosynthetic biogas upgrading due to their ability to grow at high pH and CO_2 concentrations (Bose *et al.*, 2019). A high buffer capacity mediated by a high alkalinity or inorganic carbon (IC) concentration in the cultivation broth is also necessary to prevent an elevated pH decrease in the absorption column as a result of the overload of these acidic gases or biological processes such as nitrification, which tend to acidify the algal-bacterial broth of the photobioreactor. In this context, the use of the liquid fraction of digestates instead of domestic wastewater as a low-cost nutrient source during photosynthetic biogas upgrading is preferable due to its higher pH and alkalinity (Rodero *et al.*, 2019). Nevertheless, IC concentrations in the cultivation broth of >2400 mg $C{\cdot}L^{-1}$ involve a high salt content (carbonates), which exerts a negative impact on photosynthetic activity along with an increase in CO_2 stripping to the atmosphere from the photobioreactor surface (Rodero *et al.*, 2020b). In this regard, process operation with an optimum alkalinity in the cultivation broth is a must to avoid acidification without compromising the environmental benefits of this biotechnology.

Environmental factors such as temperature and dissolved oxygen (DO) in the cultivation broth also have influence on both biomethane quality and the subsequent CO_2 uptake. Although the optimal temperature for microalgae growth often ranges between 28 and 35°C, low temperatures support higher CO_2 and H_2S removals due to a higher solubility of the gases in the cultivation broth. However, this effect is minimum at medium-high alkaline cultivation conditions (Park *et al.*, 2011; Rodero *et al.*, 2018). Otherwise, large amounts of DO in the cultivation broth result in a high O_2 desorption from the liquid to the biogas as well as the inhibition of photosynthetic activity (Pawlowski *et al.*, 2015; Posadas *et al.*, 2015b).

On the other hand, an optimum design and operation of the process is also necessary to enhance upgraded biogas quality. In this context, the mass transfer of CO_2 and H_2S from the biogas to the cultivation broth can take place in the photobioreactor or in an external absorption column interconnected to the photobioreactor (i.e. HRAP or tubular), the latter configuration being preferred since it prevents a high oxygen stripping from the cultivation broth to the biogas besides entailing a more effective gas–liquid mass transfer due to the larger biogas bubble residence times (Figure 9.6) (Meier *et al.*, 2015). The liquid to biogas (L/G) ratio is a key operational parameter determining

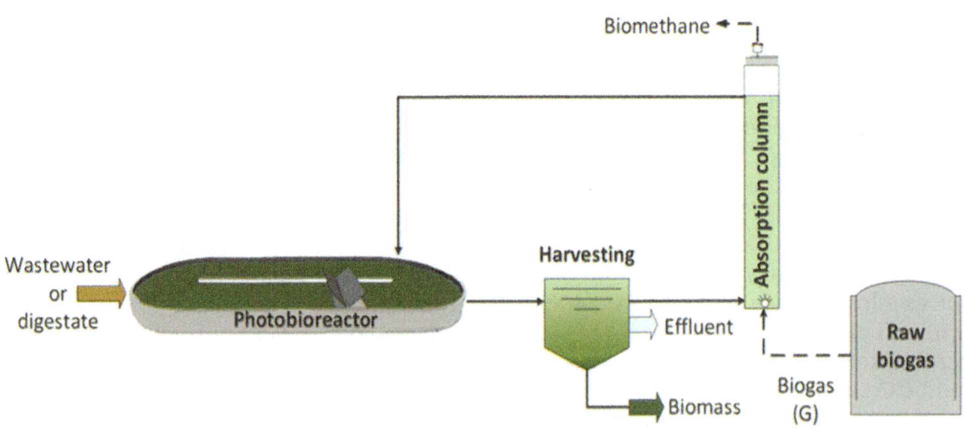

Figure 9.6 Schematic diagram of the algal–bacterial process for the simultaneous biogas upgrading and wastewater/digestate treatment.

Table 9.3 Experimental studies on photosynthetic biogas upgrading under different configurations.

Experimental Set-up Design	CO_2-RE (%)	Upgraded Biogas Composition (%)	Operational Parameters	Microalgae Population	References
180 L HRAP interconnected to a 0.8 L bubble column	40–95	O_2<1 H_2S:0	Indoors pH: 7–10 L/G ratio: 0.4–1.6 Nutrient source: synthetic medium and diluted centrate	*Spirulina platensis*, *Phormidium*, *Oocystis*, *Microspora* sp.	Bahr *et al.* (2014)
180 L HRAP interconnected to a 2.5 L bubble column	80	O_2:0.3–3 N_2:6–10 H_2S:0	Indoors pH: 8 L/G ratio: 0.5–67 Nutrient source: diluted anaerobically digested vinasse	*Chlorella* sp., *Pseudanabaena* sp., *Chloromonas* sp., *Geitlerinema* sp., *Microspora* sp., *Stigeoclonium* sp., *Planktolyngbya* sp.	Serejo *et al.* (2015)
180 L HRAP interconnected to a 2.5 L bubble column	72–79	CH_4:81 CO_2:6.8–9.2 O_2:0.7–1.2 N_2:5.9–7.2 H_2S:0	Indoors pH: 8 L/G ratio: 10.7 Nutrient source: diluted anaerobically digested vinasse and diluted raw vinasse	*Geitlerinema* sp., *Limnothrix planktonica*, *Pseudanabaena minima*, *Stigeoclonium tenue*, *Leptolyngbya benthonica*, *Planktolyngvya brevicellularis*, *Staurosira* sp.	Posadas *et al.* (2015b)
75 L HRAP interconnected to a 0.7 L bubble column	93	CO_2:1.9 O_2:1.2	Indoors pH: 7.5–8 Nutrient source: synthetic medium	*Nannochloropsis gaditana*	Meier *et al.* (2015)
180 L HRAP interconnected to a 2.5 L bubble column	97–99	CH_4:95–96 CO_2:0.1–2 O_2:0.1–1 N_2:1–4 H_2S:0	Indoors pH: 10.2 L/G ratio: 0.3–1.0 Nutrient source: digestate	*Chlorella minutissima*	Toledo-Cervantes *et al.* (2017)
25 L HRAP interconnected to a 0.35 L bubble column	89–94	O_2:2.6 H_2S:0	Indoors pH: 9.3–9.7 L/G ratio: 5 Nutrient source: synthetic medium	*Picochlorum* sp., *Halospirulina* sp.	Franco-Morgado *et al.* (2017)
Closed photobioreactor of 1 L	100	CO_2:0 O_2:10–24	Indoors pH: 9.5 Nutrient source: synthetic medium	*Spirulina platensis*	Converti *et al.* (2009)
60 L closed photobioreactor interconnected to a 3.5 L bubble column	97–98	CH_4:82.6–83.6 CO_2:0.4–1.8 O_2:8.3–9.6 N_2:6.0–7.7 H_2S:0–0.01	Indoors pH: 10–10.7 L/G ratio: 5–10 Nutrient source: synthetic medium	*Acutodesmus obliquus*	Toledo-Cervantes *et al.* (2018)

(*Continued*)

Table 9.3 Experimental studies on photosynthetic biogas upgrading under different configurations (*Continued*).

Experimental Set-up Design	CO_2-RE (%)	Upgraded Biogas Composition (%)	Operational Parameters	Microalgae Population	References
45.6 L tubular photobioreactor interconnected to an 84 L mixing chamber and a 2.6 L bubble column	85–96	CH_4:90.8–97.2 CO_2:1.1–6.2 O_2:0.4–0.8 N_2: 1.6–2.0 H_2S:0	Indoors pH: 8.9–9.3 L/G ratio: 0.5 Nutrient source: synthetic medium	*Aphanothece* sp., *Chlorella* sp., *Chlorella vulgaris*, *Mayamaea* sp., *Chlorella homosphaera*, *Pseudanabaena* sp.	Ángeles *et al.* (2020a)
Closed photobioreactor of 1 L	98	CH_4:50–53 CO_2:1.2–2.5 O_2:18.3–23.4 H_2S:0	Indoors pH: 5.5–7 Nutrient source: synthetic medium	*Chlorella vulgaris*	Mann *et al.* (2009)
180 L HRAP interconnected to a 2.5 L bubble column	50–95	CH_4:72–93 CO_2:4–12 O_2:0.1–2.0 N_2: 0.6–5.0 H_2S:0	Outdoors pH: 9–10 L/G ratio: 0.5 Nutrient source: centrate	*Chlorella* sp., *Chloroidium saccharophilum*, *Pseudanabaena* sp.	Posadas *et al.* (2017b)
180 L HRAP interconnected to a 2.5 L bubble column	64–96	CH_4:85–98; CO_2:0.8–11.9 O_2:0–3.8 N_2:0.6–5.8 H_2S:0	Outdoors pH: 9.2–9.8 L/G ratio: 1.0 Nutrient source: centrate	*Chlorella vulgaris*, *Pseudanabaena* sp., *Chlorella kessieri*, *Leptolyngbya lagerheimii*	Marín *et al.* (2018)
11.7 m³ semi-closed photobioreactor interconnected to a 45 L bubble column	>91	CH_4:94–99 CO_2:0.1–1.4 N_2+O_2:0.9–5.9 H_2S:0	Outdoors pH: 8–9 L/G ratio: 0.5	*Chlorella vulgaris*, *Stigeoclonium tenue*, *Nitzschia closterium*, *Navicula amphora*	Marín *et al.* (2019a)
9.6 m³ HRAP interconnected to a 150 L bubble column	68–96	CH_4:88–97 CO_2:1.5–12 O_2<1 H_2S:0	Outdoors pH: 9.05–9.50 L/G ratio: 0.8–2.4 Nutrient source: centrate	Microalgal consortium	Rodero *et al.* (2020a)

the CO_2 and H_2S gas-liquid mass transfer and O_2/N_2 stripping in the absorption column. In fact, a control strategy to guarantee a biomethane quality over time under the typical daily and seasonal variations in environmental conditions based on the optimization of the L/G ratio was successfully validated under semi-industrial scale (Rodero *et al.*, 2020a). High L/G ratios enhance CO_2 and H_2S removal efficiencies due to lower acidification of the scrubbing liquid along the absorption column at the expense of increasing O_2 and N_2 stripping from the liquid to the upgraded biogas (Rodero *et al.*, 2019; Serejo *et al.*, 2015). The gas–liquid flow configuration in the absorption column also determines the upgraded biogas composition. Co-current flow operation is preferred due to lower O_2 and N_2 desorption and no S accumulation in the diffuser, although counter-current gas–liquid configuration favors the gas–liquid mass transfer rates (Toledo-Cervantes *et al.*, 2017). On the other hand, innovative operational strategies to enhance biomethane quality have been recently evaluated. In this context, the installation of a hollow fibre membrane prior to the biogas absorption column and the increase in the operational pressure in the biogas absorption column to minimize N_2 and O_2 content in the upgraded biogas have been validated at pilot scale as promising approaches (Ángeles *et al.*, 2020b, 2020c). Moreover, the location of an HRAP inside a greenhouse and mechanical CO_2 stripping during winter conditions have been demonstrated to increase CO_2 removal under unfavorable climatic conditions (Marín *et al.*, 2021).

Despite CO_2 capture biotechnology still being under demo-scale validation, it constitutes an attractive alternative to the conventional technologies for biogas upgrading based on its cost-effectiveness and environmentally friendliness. In this context, CO_2 and H_2S removal efficiencies as high as 99 and 100%, respectively, with a final CH_4 content in the upgraded biogas as high as 98% being reported (Table 9.3). Furthermore, this biotechnology exhibits low CH_4 losses ($\leq$5%) as a result of the poor aqueous solubility of this potent greenhouse gas (Posadas *et al.*, 2017b).

ACKNOWLEDGEMENT

The Spanish Ministry of Science, Innovation and Universities (FJC 2018-038402-I), the Regional Government of Castilla y León and the EU-FEDER programme (CLU 2017-09, VA281P18 and UIC 71) and CONICYT (PFCHA/DOCTORADO BECAS CHILE/2017-72180211) are gratefully acknowledged.

REFERENCES

Abeliovich A. and Azov Y. (1976). Toxicity of ammonia to algae in sewage oxidation ponds. *Applied and Environmental Microbiology*, **31**, 801–806, https://doi.org/10.1128/aem.31.6.801-806.1976

Acién F. G., Fernández-Sevilla J. M. and Molina Grima E. (2013). Photobioreactors for the production of microalgae. *Reviews in Environmental Science and Bio/Technology*, **12**, 131–151, https://doi.org/10.1007/s11157-012-9307-6

Acién F. G., Molina E., Fernández-Sevilla J. M., Barbosa M., Gouveia L., Sepúlveda C., Bazaes J., and Arbib Z. (2017). Economics of microalgae production. In: C. Gonzalez-Fernandez and R. Muñoz (eds). *Microalgae-Based Biofuels and Bioproducts: From Feedstock Cultivation to End-Products*. Elsevier Inc. Amsterdam, pp. 485–503.

Ángeles R., Arnaiz E., Gutiérrez J., Sepúlveda-Muñoz C. A., Fernández-Ramos O., Muñoz R. and Lebrero R. (2020a). Optimization of photosynthetic biogas upgrading in closed photobioreactors combined with algal biomass production. *Journal of Water Process Engineering*, **38**, 101554, https://doi.org/10.1016/j.jwpe.2020.101554

Ángeles R., Marín D., Rodero M. del R., Pascual C., González-Sanchez A., de Godos I., Lebrero R. and Muñoz R. (2020b). Biogas treatment for H_2S, CO_2, and other contaminants removal. In: Biofiltration to Promising Options in Gaseous Fluxes Biotreatment, G. Soreanu and É. Dumont (eds.), Elsevier, Amsterdam, pp. 153–176.

Ángeles R., Rodríguez Á, Domínguez C., García J., Prádanos P., Muñoz R. and Lebrero R. (2020c). Strategies for N_2 and O_2 removal during biogas upgrading in a pilot algal-bacterial photobioreactor. *Algal Research*, **48**, 101920, https://doi.org/10.1016/j.algal.2020.101920

Arbib Z., de Godos Crespo I., Corona E. L. and Rogalla F. (2017). Understanding the biological activity of high rate algae ponds through the calculation of oxygen balances. *Applied Microbiology and Biotechnology*, **101**, 5189–5198, https://doi.org/10.1007/s00253-017-8235-3

Arun J., Shreekanth S. J., Sahana R., Raghavi M. S., Gopinath K. P. and Gnanaprakash D. (2017). Studies on influence of process parameters on hydrothermal catalytic liquefaction of microalgae (chlorella vulgaris) biomass grown in wastewater. *Bioresource Technology*, **244**, 963–968, https://doi.org/10.1016/j.biortech.2017.08.048

Atelge M. R., Atabani A. E., Rajesh Banu J., Krisa D., Kaya M., Eskicioglu C., Kumar G., Lee C., Yildiz Y. Ş., Unalan S., Mohanasundaram R. and Duman F. (2020). A critical review of pretreatment technologies to enhance anaerobic digestion and energy recovery. *Fuel*, **270**, 117494, https://doi.org/10.1016/j.fuel.2020.117494

Awe O. W., Yaqian Z., Ange N., Doam Pham M. and Nathalie L. (2017). A review of biogas utilisation, purification and upgrading technologies. *Waste and Biomass Valorization*, **8**, 267–283, https://doi.org/10.1007/s12649-016-9826-4

Ayre J. M., Moheimani N. R. and Borowitzka M. A. (2017). Growth of microalgae on undiluted anaerobic digestate of piggery effluent with high ammonium concentrations. *Algal Research*, **24**, 218–226, https://doi.org/10.1016/j.algal.2017.03.023

Bahr M., Díaz I., Dominguez A., González Sánchez A. and Muñoz R. (2014). Microalgal-biotechnology as a platform for an integral biogas upgrading and nutrient removal from anaerobic effluents. *Environmental Science and Technology*, **48**, 573–581, https://doi.org/10.1021/es403596m

Barlow E. W. R., Boersma L., Phinney H. K. and Miner J. R. (1975). Algal growth in diluted pig waste. *Agriculture and Environment*, **2**(4), 339–355, https://doi.org/10.1016/0304-1131(75)90040-5

Barragán-Trinidad M. and Buitrón G. (2020). Hydrogen and methane production from microalgal biomass hydrolyzed in a discontinuous reactor inoculated with ruminal microorganisms. *Biomass and Bioenergy*, **143**, 105825, https://doi.org/10.1016/j.biombioe.2020.105825

Barreiro-Vescovo S., González-Fernández C., Ballesteros M. and de Godos I. (2020). Activity determination of an algal-bacterial consortium developed during wastewater treatment based on oxygen evolution. *Journal of Water Process Engineering*, **36**, 101278, https://doi.org/10.1016/j.jwpe.2020.101278

Batstone D. J., Hülsen T., Mehta C. M. and Keller J. (2015). Platforms for energy and nutrient recovery from domestic wastewater: A review. *Chemosphere*, **140**, 2–11, https://doi.org/10.1016/j.chemosphere.2014.10.021

Bose A., Lin R., Rajendran K., O'Shea R., Xia A. and Murphy J. D. (2019). How to optimise photosynthetic biogas upgrading: a perspective on system design and microalgae selection. *Biotechnology Advances*, **37**, 107444, https://doi.org/10.1016/j.biotechadv.2019.107444

Budiman P. M., Wu T. Y., Ramanan R. N. and Hay J. X. W. (2014). Treatment and reuse of effluents from palm oil, pulp, and paper mills as a combined substrate by using purple nonsulfur bacteria. *Industrial & Engineering Chemistry Research*, **53**(39), 14921–14931, https://doi.org/10.1021/ie501798f

Cai T., Park S. Y. and Li Y. (2013). Nutrient recovery from wastewater streams by microalgae: Status and prospects. *Renewable & Sustainable Energy Reviews*, **19**, 360–369, https://doi.org/10.1016/j.rser.2012.11.030

Canon-Rubio K. A., Sharp C. E., Bergerson J., Strous M. and De la Hoz Siegler H. (2016). Use of highly alkaline conditions to improve cost-effectiveness of algal biotechnology. *Applied Microbiology and Biotechnology*, **100**, 1611–1622, https://doi.org/10.1007/s00253-015-7208-7

Capson-Tojo G., Batstone D. J., Grassino M., Vlaeminck S. E., Puyol D., Verstraete W., Kleerebezem R., Oehmen A., Ghimire A., Pikaar I., Lema J. M. and Hülsen T. (2020). Purple phototrophic bacteria for resource recovery: challenges and opportunities. *Biotechnology Advances*, **43**, 107567, https://doi.org/10.1016/j.biotechadv.2020.107567

Carrère H., Antonopoulou G., Affes R., Passos F., Battimelli A., Lyberatos G. and Ferrer I. (2016). Review of feedstock pretreatment strategies for improved anaerobic digestion: from lab-scale research to full-scale application. *Bioresource Technology*, **199**, 386–397, https://doi.org/10.1016/j.biortech.2015.09.007

Čater M., Fanedl L., Malovrh Š. and Logar R. M. (2015). Biogas production from brewery spent grain enhanced by bioaugmentation with hydrolytic anaerobic bacteria. *Bioresource Technology*, **186**, 261–269, https://doi.org/10.1016/j.biortech.2015.03.029

Cembella A. D., Antia N. J. and Harrison P. J. (1982). The utilization of inorganic and organic phosphorous compounds as nutrients by eukaryotic microalgae: a multidisciplinary perspective: part 1. *Critical Reviews in Microbiology*, **10**, 317–391, https://doi.org/10.3109/10408418209113567

Cheng P., Cheng J. J., Cobb K., Zhou C., Zhou N., Addy M., Chen P., Yan X. and Ruan R. (2020). Tribonema sp. and Chlorella zofingiensis co-culture to treat swine wastewater diluted with fishery wastewater to facilitate harvest. *Bioresource Technology*, **297**, 122516, https://doi.org/10.1016/j.biortech.2019.122516

Chi Z., O'Fallon J. V. and Chen S. (2011). Bicarbonate produced from carbon capture for algae culture. *Trends in Biotechnology*, **29**(11), 537–541, https://doi.org/10.1016/j.tibtech.2011.06.006

Choi Y. Y., Patel A. K., Hong M. E., Chang W. S. and Sim S. J. (2019). Microalgae bioenergy with carbon capture and storage (BECCS): An emerging sustainable bioprocess for reduced CO_2 emission and biofuel production. *Bioresource Technology Reports*, **7**, 100270, https://doi.org/10.1016/j.biteb.2019.100270

Choorit W., Thanakoset P., Thongpradistha J., Sasaki K. and Noparatnaraporn N. (2002). Identification and cultivation of photosynthetic bacteria in wastewater from a concentrated latex processing factory. *Biotechnology Letters*, **24**, 1055–1058, https://doi.org/10.1023/A:1016026412361

Chu W.-L. (2012). Biotechnological applications of microalgae. *International Journal of Medical Science and Education*, **6**, 24–37.

Converti A., Oliveira R. P. S., Torres B. R., Lodi A. and Zilli M. (2009). Biogas production and valorization by means of a two-step biological process. *Bioresource Technology*, **100**, 5771–5776, https://doi.org/10.1016/j.biortech.2009.05.072

Correll D. L. (1998). The role of phosphorus in the eutrophication of receiving waters: A review. *Journal of Environmental Quality*, **27**(2), 261–266, https://doi.org/10.2134/jeq1998.00472425002700020004x

Craggs R., Sutherland D. and Campbell H. (2012). Hectare-scale demonstration of high rate algal ponds for enhanced wastewater treatment and biofuel production. *Journal of Applied Phycology*, **24**, 329–337, https://doi.org/10.1007/s10811-012-9810-8

Cuellar-Bermudez S. P., Garcia-Perez J. S., Rittmann B. E. and Parra-Saldivar R. (2015). Photosynthetic bioenergy utilizing CO_2: an approach on flue gases utilization for third generation biofuels. *Journal of Cleaner Production*, **98**, 53–65, https://doi.org/10.1016/j.jclepro.2014.03.034

Curtis T. P. (2010). Low-energy wastewater treatment: strategies and technologies. In: Environmental Microbiology, R. Mitchell and J.-D. Gu (eds.), John Wiley & Sons, Inc., Hoboken, NJ, USA, pp. 301–318.

Dalaei P., Ho D., Nakhla G. and Santoro D. (2019). Low temperature nutrient removal from municipal wastewater by purple phototrophic bacteria (PPB). *Bioresource Technology*, **288**, 121566, https://doi.org/10.1016/j.biortech.2019.121566

Dalaei P., Bahreini G., Nakhla G., Santoro D., Batstone D. and Hülsen T. (2020). Municipal wastewater treatment by purple phototropic bacteria at low infrared irradiances using a photo-anaerobic membrane bioreactor. *Water Research*, **173**(4), 115535, https://doi.org/10.1016/j.watres.2020.115535

De-Bashan L. E. and Bashan Y. (2004). Recent advances in removing phosphorus from wastewater and its future use as fertilizer (1997–2003). *Water Research*, **38**, 4222–4246, https://doi.org/10.1016/j.watres.2004.07.014

de Godos I., Blanco S., García-Encina P. A., Becares E. and Muñoz R. (2009). Long-term operation of high rate algal ponds for the bioremediation of piggery wastewaters at high loading rates. *Bioresource Technology*, **100**, 4332–4339, https://doi.org/10.1016/j.biortech.2009.04.016

de Godos I., Arbib Z., Lara E. and Rogalla F. (2016). Evaluation of high rate algae ponds for treatment of anaerobically digested wastewater: effect of CO_2 addition and modification of dilution rate. *Bioresource Technology*, **220**, 253–261, https://doi.org/10.1016/j.biortech.2016.08.056

de Godos I., Arbib Z., Lara E., Cano R., Muñoz R. and Rogalla F. (2017). Wastewater treatment in algal systems. In: Innovative Wastewater Treatment & Resource Recovery Technologies: Impacts on Energy, Economy and Environment, J. Lema and S. Suarez (eds.), IWA Publishing, London, pp. 76–95.

de las Heras I., Molina R., Segura Y., Hülsen T., Molina M. C., Gonzalez-Benítez N., Melero J. A., Mohedano A. F., Martínez F. and Puyol D. (2020). Contamination of N-poor wastewater with emerging pollutants does not affect the performance of purple phototrophic bacteria and the subsequent resource recovery potential. *Journal of Hazardous Materials*, **385**, 121617, https://doi.org/10.1016/j.jhazmat.2019.121617

de Mendonça H. V., Ometto J. P. H. B., Otenio M. H., Marques I. P. R. and dos Reis A. J. D. (2018). Microalgae-mediated bioremediation and valorization of cattle wastewater previously digested in a hybrid anaerobic reactor using a photobioreactor: comparison between batch and continuous operation. *Science of the Total Environment*, **633**, 1–11, https://doi.org/10.1016/j.scitotenv.2018.03.157

Eixler S., Karsten U. and Selig U. (2006). Phosphorus storage in Chlorella vulgaris (Trebouxiophyceae, Chlorophyta) cells and its dependence on phosphate supply. *Phycologia*, **45**, 53–60, https://doi.org/10.2216/04-79.1

Englande A. J., Krenkel P. and Shamas J. (2015). Wastewater treatment and water reclamation. Reference Module in Earth Systems and Environmental Sciences, 639–670, https://doi.org/10.1016/B978-0-12-409548-9.09508-7

Fouilland E., Vasseur C., Leboulanger C., Le Floc'h E., Carré C., Marty B., Steyer J.-P. P. and Sialve B. (2014). Coupling algal biomass production and anaerobic digestion: production assessment of some native temperate and tropical microalgae. *Biomass and Bioenergy*, **70**, 564–569, https://doi.org/10.1016/j.biombioe.2014.08.027

Franco-Morgado M., Alcántara C., Noyola A., Muñoz R. and González-Sánchez A. (2017). A study of photosynthetic biogas upgrading based on a high rate algal pond under alkaline conditions: influence of the illumination regime. *Science of the Total Environment*, **592**, 419–425, https://doi.org/10.1016/j.scitotenv.2017.03.077

García D., Alcántara C., Blanco S., Pérez R., Bolado S. and Muñoz R. (2017). Enhanced carbon, nitrogen and phosphorus removal from domestic wastewater in a novel anoxic-aerobic photobioreactor coupled with biogas upgrading. *Chemical Engineering Journal*, **313**, 424–434, https://doi.org/10.1016/j.cej.2016.12.054

García D., de Godos I., Domínguez C., Turiel S., Bolado S. and Muñoz R. (2019). A systematic comparison of the potential of microalgae-bacteria and purple phototrophic bacteria consortia for the treatment of piggery wastewater. *Bioresource Technology*, **276**, 18–27, https://doi.org/10.1016/j.biortech.2018.12.095

Godos I. D., Vargas V. A., Blanco S., González M. C. G., Soto R., García-Encina P. A., Becares E. and Muñoz R. (2010). A comparative evaluation of microalgae for the degradation of piggery wastewater under photosynthetic oxygenation. *Bioresource Technology*, **101**(14), 5150–5158, https://doi.org/10.1016/j.biortech.2010.02.010

González C., Marciniak J., Villaverde S., García-Encina P. A. and Muñoz R. (2008). Microalgae-based processes for the biodegradation of pretreated piggery wastewaters. *Applied Microbiology and Biotechnology*, **80**, 891–898, https://doi.org/10.1007/s00253-008-1571-6

González-Fernández C., León-Cofreces C. and García-Encina P. A. (2008). Different pretreatments for increasing the anaerobic biodegradability in swine manure. *Bioresource Technology*, **99**, 8710–8714, https://doi.org/10.1016/j.biortech.2008.04.020

González-Fernández C., Molinuevo-Salces B. and García-González M. C. (2010). Open and enclosed photobioreactors comparison in terms of organic matter utilization, biomass chemical profile and photosynthetic efficiency. *Ecological Engineering*, **36**, 1497–1501, https://doi.org/10.1016/j.ecoleng.2010.07.007

Gonzalez-Fernandez C., Mendez L., Ballesteros M. and Tomas-Pejó E. (2017). Hydrothermal processing of microalgae. In: Hydrothermal Processing in Biorefineries: Production of Bioethanol and High Added-Value Compounds of Second and Third Generation Biomass, A. Ruiz Leza, M. Thomsen and H. L. Trajano (eds.), Springer International Publishing, New York, pp. 483–500.

Gonzalez-Fernandez C., Barreiro-Vescovo S., De Godos I., Fernandez M., Zouhayr A. and Ballesteros M. (2018). Biochemical methane potential of microalgae biomass using different microbial inocula. *Biotechnology for Biofuels*, **11**, 1–11, https://doi.org/10.1186/s13068-018-1188-7

Guilayn F., Rouez M., Crest M., Patureau D. and Jimenez J. (2020). Valorization of digestates from urban or centralized biogas plants: a critical review. *Reviews in Environmental Science and Bio/Technology*, **19**, 419–462, https://doi.org/10.1007/s11157-020-09531-3

Hamouri B. E. (2009). Rethinking natural, extensive systems for tertiary treatment purposes: The high-rate algae pond as an example. *Desalination and Water Treatment*, **4**, 128–134, https://doi.org/10.5004/dwt.2009.367

Hamouri B. E., Rami A. and Vasel J. L. (2003). The reasons behind the performance superiority of a high rate algal pond over three facultative ponds in series. *Water Science and Technology*, **48**(2), 269–276, https://doi.org/10.2166/wst.2003.0130

He P. J., Mao B., Shen C. M., Shao L. M., Lee D. J. and Chang J. S. (2013). Cultivation of Chlorella vulgaris on wastewater containing high levels of ammonia for biodiesel production. *Bioresource Technology*, **129**, 177–181, https://doi.org/10.1016/j.biortech.2012.10.162

Hiraishi A., Shi J. L. and Kitamura H. (1989). Effects of organic nutrient strength on the purple nonsulfur bacteria content and metabolic activity of photosynthetic sludge for wastewater treatment. *Journal of Fermentation and Bioengineering*, **68**(4), 269–276, https://doi.org/10.1016/0922-338X(89)90028-7

Hülsen T., Batstone D. J. and Keller J. (2014). Phototrophic bacteria for nutrient recovery from domestic wastewater. *Water Research*, **50**, 18–26, https://doi.org/10.1016/j.watres.2013.10.051

Hülsen T., Barry E. M., Lu Y., Puyol D. and Batstone D. J. (2016a). Low temperature treatment of domestic wastewater by purple phototrophic bacteria: performance, activity, and community. *Water Research*, **100**, 537–545, https://doi.org/10.1016/j.watres.2016.05.054

Hülsen T., Barry E. M., Lu Y., Puyol D., Keller J. and Batstone D. J. (2016b). Domestic wastewater treatment with purple phototrophic bacteria using a novel continuous photo anaerobic membrane bioreactor. *Water Research*, **100**, 486–495, https://doi.org/10.1016/j.watres.2016.04.061

Hülsen T., Hsieh K., Tait S., Barry E. M., Puyol D. and Batstone D. J. (2018). White and infrared light continuous photobioreactors for resource recovery from poultry processing wastewater – A comparison. *Water Research*, **144**, 665–676, https://doi.org/10.1016/j.watres.2018.07.040

Hülsen T., Hsieh K. and Batstone D. J. (2019). Saline wastewater treatment with purple phototrophic bacteria. *Water Research*, **160**, 259–267, https://doi.org/10.1016/j.watres.2019.05.060

Hülsen T., Lu Y., Rodríguez I., Segura Y., Martínez F., Puyol D. and Batstone D. J. (2020). Anaerobic digestion of purple phototrophic bacteria – The release step of the partition-release-recover concept. *Bioresource Technology*, **306**, 123125, https://doi.org/10.1016/j.biortech.2020.123125

Hunter C. N., Daldal F., Thurnauer M. C. and Beatty T. J. (2009). The Purple Phototrophic Bacteria, 1st edn. Springer, The Netherlands.

Iasimone F., De Felice V., Panico A. and Pirozzi F. (2017). Experimental study for the reduction of CO_2 emissions in wastewater treatment plant using microalgal cultivation. *Journal of CO_2 Utilization*, **22**, 1–8, https://doi.org/10.1016/j.jcou.2017.09.004

Ibrahim G., Muñoz R., Llamas B. and de Godos I. (2020). Carbon dioxide capture from carbon dioxide-rich gases by microalgae. In: From Biofiltration to Promising Options in Gaseous Fluxes Biotreatment, G. Soreanu and E. Dumont (eds.), Elsevier, Amsterdam, pp. 373–396.

Izu K., Nakajima F., Yamamoto K. and Kurisu F. (2001). Aeration conditions affecting growth of purple nonsulfur bacteria in an organic wastewater treatment process. *Systematic and Applied Microbiology*, **24**, 294–302, https://doi.org/10.1078/0723-2020-00027

Jamieson T. S., Stratton G. W., Gordon R. and Madani A. (2003). The use of aeration to enhance ammonia nitrogen removal in constructed wetlands. *Canadian Biosystems Engineering*, **45**, 9–14.

Ji M.-K., Abou-Shanab R. A. I., Kim S.-H., Salama E.-S., Lee S.-H., Kabra A. N., Lee Y.-S., Hong S. and Jeon B.-H. (2013). Cultivation of microalgae species in tertiary municipal wastewater supplemented with CO_2 for nutrient removal and biomass production. *Ecological Engineering*, **58**, 142–148, https://doi.org/10.1016/j.ecoleng.2013.06.020

Kalra R., Gaur S. and Goel M. (2020). Microalgae bioremediation: A perspective towards wastewater treatment along with industrial carotenoids production. *Journal of Water Process Engineering*, **40**, 101794, https://doi.org/10.1016/j.jwpe.2020.101794

Karemore A., Ramalingam D., Yadav G., Subramanian G. and Sen R. (2016). Photobioreactors for improved algal biomass production: analysis and design considerations. In: Algal Biorefinery: An Integrated Approach, D. Das (ed.), Springer International Publishing, New York, pp. 103–124.

Kiran B., Pathak K., Kumar R. and Deshmukh D. (2014). Cultivation of chlorella sp. IM-01 in municipal wastewater for simultaneous nutrient removal and energy feedstock production. *Ecological Engineering*, **73**, 326–330, https://doi.org/10.1016/j.ecoleng.2014.09.094

Larimer F. W., Chain P., Hauser L., Lamerdin J., Malfatti S., Do L., Land M. L., Pelletier D. A., Beatty J. T., Lang A. S., Tabita F. R., Gibson J. L., Hanson T. E., Bobst C., Torres Y Torres J. L., Peres C., Harrison F. H., Gibson J. and Harwood C. S. (2004). Complete genome sequence of the metabolically versatile photosynthetic bacterium Rhodopseudomonas palustris. *Nature Biotechnology*, **22**, 55–61, https://doi.org/10.1038/nbt923

Leong W. H., Lim J. W., Lam M. K., Uemura Y., Ho C. D. and Ho Y. C. (2018). Co-cultivation of activated sludge and microalgae for the simultaneous enhancements of nitrogen-rich wastewater bioremediation and lipid production. *Journal of the Taiwan Institute of Chemical Engineers*, **87**, 216–224, https://doi.org/10.1016/j.jtice.2018.03.038

Liu S., Zhang G., Zhang J., Li X. and Li J. (2016). Performance, carotenoids yield and microbial population dynamics in a photobioreactor system treating acidic wastewater: effect of hydraulic retention time (HRT) and organic loading rate (OLR). *Bioresource Technology*, **200**, 245–252, https://doi.org/10.1016/j.biortech.2015.10.044

Liu S., Daigger G. T., Kang J. and Zhang G. (2019). Effects of light intensity and photoperiod on pigments production and corresponding key gene expression of Rhodopseudomonas palustris in a photobioreactor system. *Bioresource Technology*, **294**, 122172, https://doi.org/10.1016/j.biortech.2019.122172

Lo K. J., Lin S. S., Lu C. W., Kuo C. H. and Liu C. T. (2018). Whole-genome sequencing and comparative analysis of two plant-associated strains of Rhodopseudomonas palustris (PS3 and YSC3). *Scientific Reports*, **8**, 12769, https://doi.org/10.1038/s41598-018-31128-8

López J. C., Quijano G., Souza T. S. O., Estrada J. M., Lebrero R. and Muñoz R. (2013). Biotechnologies for greenhouse gases (CH_4, N_2O, and CO_2) abatement: state of the art and challenges. *Applied Microbiology and Biotechnology*, **97**, 2277–2303, https://doi.org/10.1007/s00253-013-4734-z

López-Serna R., Posadas E., García-Encina P. A. and Muñoz R. (2019). Removal of contaminants of emerging concern from urban wastewater in novel algal-bacterial photobioreactors. *Science of the Total Environment*, **662**, 32–40, https://doi.org/10.1016/j.scitotenv.2019.01.206

Lu H., Zhang G. and Dong S. (2011). Quantitative study of PNSB energy metabolism in degrading pollutants under weak light-micro oxygen condition. *Bioresource Technology*, **102**, 4968–4973, https://doi.org/10.1016/j.biortech.2011.01.027

Lu H., Zhang G., Dai X., Schideman L., Zhang Y., Li B. and Wang H. (2013). A novel wastewater treatment and biomass cultivation system combining photosynthetic bacteria and membrane bioreactor technology. *Desalination*, **322**, 176–181, https://doi.org/10.1016/j.desal.2013.05.007

Lu H., Peng M., Zhang G., Li B. and Li Y. (2019a). Brewery wastewater treatment and resource recovery through long term continuous-mode operation in pilot photosynthetic bacteria-membrane bioreactor. *Science of the Total Environment*, **646**, 196–205, https://doi.org/10.1016/j.scitotenv.2018.07.268

Lu H., Zhang G., Zheng Z., Meng F., Du T. and He S. (2019b). Bio-conversion of photosynthetic bacteria from non-toxic wastewater to realize wastewater treatment and bioresource recovery: A review. *Bioresource Technology*, **278**, 383–399, https://doi.org/10.1016/j.biortech.2019.01.070

Madukasi E. I. and Zhang G. (2010). Microaerobic biodegradation of high organic load wastewater by phototrophic bacteria. *African Journal of Biotechnology*, **9**(25), 3852–3860.

Mahdy A., Mendez L., Ballesteros M. and González-Fernández C. (2014a). Autohydrolysis and alkaline pretreatment effect on chlorella vulgaris and scenedesmus sp. Methane production. *Energy*, **78**, 48–52, https://doi.org/10.1016/j.energy.2014.05.052

Mahdy A., Mendez L., Blanco S., Ballesteros M. and González-Fernández C. (2014b). Protease cell wall degradation of chlorella vulgaris: effect on methane production. *Bioresource Technology*, **171**, 421–427, https://doi.org/10.1016/j.biortech.2014.08.091

Maity J. P., Bundschuh J., Chen C. Y. and Bhattacharya P. (2014). Microalgae for third generation biofuel production, mitigation of greenhouse gas emissions and wastewater treatment: present and future perspectives – A mini review. *Energy*, **78**, 104–113, https://doi.org/10.1016/j.energy.2014.04.003

Mann G., Schlegel M., Schumann R. and Sakalauskas A. (2009). Biogas-conditioning with microalgae. *Agronomy Research*, **7**(1), 33–38.

Marella T. K., Datta A., Patil M. D., Dixit S. and Tiwari A. (2019). Biodiesel production through algal cultivation in urban wastewater using algal floway. *Bioresource Technology*, **280**, 222–228, https://doi.org/10.1016/j.biortech.2019.02.031

Marín D., Posadas E., Cano P., Pérez V., Blanco S., Lebrero R. and Muñoz R. (2018). Seasonal variation of biogas upgrading coupled with digestate treatment in an outdoors pilot scale algal-bacterial photobioreactor. *Bioresource Technology*, **263**, 58–66, https://doi.org/10.1016/j.biortech.2018.04.117

Marín D., Ortíz A., Díez-Montero R., Uggetti E., García J., Lebrero R. and Muñoz R. (2019a). Influence of liquid-to-biogas ratio and alkalinity on the biogas upgrading performance in a demo scale algal-bacterial photobioreactor. *Bioresource Technology*, **280**, 112–117, https://doi.org/10.1016/j.biortech.2019.02.029

Marín D., Posadas E., García D., Puyol D., Lebrero R. and Muñoz R. (2019b). Assessing the potential of purple phototrophic bacteria for the simultaneous treatment of piggery wastewater and upgrading of biogas. *Bioresource Technology*, **281**, 10–17, https://doi.org/10.1016/j.biortech.2019.02.073

Marín D., Carmona-Martínez A. A., Blanco S., Lebrero R. and Muñoz R. (2021). Innovative operational strategies in photosynthetic biogas upgrading in an outdoors pilot scale algal-bacterial photobioreactor. *Chemosphere*, **264**, 128470, https://doi.org/10.1016/j.chemosphere.2020.128470

Markou G., Vandamme D. and Muylaert K. (2014). Microalgal and cyanobacterial cultivation: the supply of nutrients. *Water Research*, **65**, 186–202, https://doi.org/10.1016/j.watres.2014.07.025

Masojídek J., Torzillo G. and Koblízek M. (2013). Photosynthesis in microalgae. In: Handbook of Microalgal Culture: Applied Phycology and Biotechnology, A. Richmond and Q. Hu (eds.), John Wiley & Sons, Ltd., New Jersey, pp. 21–36.

Meier L., Pérez R., Azócar L., Rivas M. and Jeison D. (2015). Photosynthetic CO_2 uptake by microalgae: An attractive tool for biogas upgrading. *Biomass and Bioenergy*, **73**, 102–109, https://doi.org/10.1016/j.biombioe.2014.10.032

Meier L., Stará D., Bartacek J. and Jeison D. (2018). Removal of H_2S by a continuous microalgae-based photosynthetic biogas upgrading process. *Process Safety and Environmental Protection*, **119**, 65–68, https://doi.org/10.1016/j.psep.2018.07.014

Mendez L., Mahdy A., Ballesteros M. and González-Fernández C. (2015). Biomethane production using fresh and thermally pretreated chlorella vulgaris biomass: A comparison of batch and semi-continuous feeding mode. *Ecological Engineering*, **84**, 273–277, https://doi.org/10.1016/j.ecoleng.2015.09.056

Mendez L., Mahdy A., Ballesteros M. and González-Fernández C. (2014a). Methane production of thermally pretreated chlorella vulgaris and scenedesmus sp. Biomass at increasing biomass loads. *Applied Energy*, **129**, 238–242, https://doi.org/10.1016/j.apenergy.2014.04.110

Mendez L., Mahdy A., Demuez M., Ballesteros M. and González-Fernández C. (2014b). Effect of high pressure thermal pretreatment on Chlorella vulgaris biomass: organic matter solubilisation and biochemical methane potential. *Fuel*, **117**, 674–679, https://doi.org/10.1016/j.fuel.2013.09.032

Mendez L., Sialve B., Tomás-Pejó E., Ballesteros M., Steyer J. P. and González-Fernández C. (2016). Comparison of chlorella vulgaris and cyanobacterial biomass: cultivation in urban wastewater and methane production. *Bioprocess and Biosystems Engineering*, **39**, 703–712, https://doi.org/10.1007/s00449-016-1551-7

Mohd Udaiyappan A. F., Abu Hasan H., Takriff M. S. and Sheikh Abdullah S. R. (2017). A review of the potentials, challenges and current status of microalgae biomass applications in industrial wastewater treatment. *Journal of Water Process Engineering*, **20**, 8–21, https://doi.org/10.1016/j.jwpe.2017.09.006

Molazadeh M., Ahmadzadeh H., Pourianfar H. R., Lyon S. and Rampelotto P. H. (2019). The use of microalgae for coupling wastewater treatment with CO_2 biofixation. *Frontiers in Bioengineering and Biotechnology*, **7**(48), 1–12.

Molina-Grima E. M., Fernández Sevilla J. M. and Acién Fernández F. G. (2010). Microalgae, mass culture methods. In: Encyclopedia of Industrial Biotechnology, M. C. Flickinger (ed.), Wiley, New Jersey, pp. 1–24.

Molinuevo-Salces B., González-Fernández C., Gómez X., García-González M. C. and Morán A. (2012). Vegetable processing wastes addition to improve swine manure anaerobic digestion: evaluation in terms of methane yield and SEM characterization. *Applied Energy*, **91**, 36–42, https://doi.org/10.1016/j.apenergy.2011.09.010

Molinuevo-Salces B., Mahdy A., Ballesteros M. and González-Fernández C. (2016). From piggery wastewater nutrients to biogas: microalgae biomass revalorization through anaerobic digestion. *Renewable Energy*, **96**, 1103–1110, https://doi.org/10.1016/j.renene.2016.01.090

Morales-Amaral M. del M., Gómez-Serrano C., Acién F. G., Fernández-Sevilla J. M. and Molina-Grima E. (2015). Production of microalgae using centrate from anaerobic digestion as the nutrient source. *Algal Research*, **9**, 297–305, https://doi.org/10.1016/j.algal.2015.03.018

Mulbry W., Westhead E. K., Pizarro C. and Sikora L. (2005). Recycling of manure nutrients: use of algal biomass from dairy manure treatment as a slow release fertilizer. *Bioresource Technology*, **96**(4), 451–458, https://doi.org/10.1016/j.biortech.2004.05.026

Mulbry W., Kangas P. and Kondrad S. (2010). Toward scrubbing the bay: nutrient removal using small algal turf scrubbers on chesapeake Bay tributaries. *Ecological Engineering*, **36**, 536–541, https://doi.org/10.1016/j.ecoleng.2009.11.026

Muñoz R., Meier L., Díaz I., Jeison D., Diaz I. and Jeison D. (2015). A review on the state-of-the-art of physical/chemical and biological technologies for biogas upgrading. *Reviews in Environmental Science and Biotechnology*, **14**, 727–759, https://doi.org/10.1007/s11157-015-9379-1

Murry M. A., Murinda S. E., Huang S. and Mark A. (2019). Bioconversion of agricultural wastes from the livestock industry for biofuel and feed production. In: Advanced Bioprocessing for Alternative Fuels, Biobased Chemicals, and Bioproducts, M. Hosseini (ed.), Elsevier Inc., Amsterdam, pp. 225–247.

Nagadomi H., Kitamura T., Watanabe M. and Sasaki K. (2000). Simultaneous removal of chemical oxygen demand (COD), phosphate, nitrate and H_2S in the synthetic sewage wastewater using porous ceramic immobilized photosynthetic bacteria. *Biotechnology Letters*, **22**(17), 1369–1374, https://doi.org/10.1023/A:1005688229783

Navid A., Jiao Y., Wong S. E. and Pett-Ridge J. (2019). System-level analysis of metabolic trade-offs during anaerobic photoheterotrophic growth in Rhodopseudomonas palustris. *BMC Bioinformatics*, **20**, 233, https://doi.org/10.1186/s12859-019-2844-z

Nkemka V. N. and Murto M. (2013). Biogas production from wheat straw in batch and UASB reactors: the roles of pretreatment and seaweed hydrolysate as a co-substrate. *Bioresource Technology*, **128**, 164–172, https://doi.org/10.1016/j.biortech.2012.10.117

Olguín E. J., Sánchez-Galván G. and Pérez-Pérez T. (2007). Assessment of the phytoremediation potential of Salvinia minima Baker compared to Spirodela polyrrhiza in high-strength organic wastewater. *Water Air and Soil Pollution*, **181**(1), 135–147, https://doi.org/10.1007/s11270-006-9285-9

Oswald W. J. (1978). The engineering aspects of microalgae. In: CRC Handbook of Microbiology, 2nd edn. (II) Fungi, Algae, Protozoa, and Viruses, A. I. Laskin and H. A. Lechevalier (eds.), CRC Press, West Palm Beach, Florida, pp. 519–552.

Pancha I., Chokshi K., George B., Ghosh T., Paliwal C., Maurya R. and Mishra S. (2014). Nitrogen stress triggered biochemical and morphological changes in the microalgae Scenedesmus sp. CCNM 1077. *Bioresource Technology*, **156**, 146–154, https://doi.org/10.1016/j.biortech.2014.01.025

Park J. B. K., Craggs R. J. and Shilton A. N. (2011). Wastewater treatment high rate algal ponds for biofuel production. *Bioresource Technology*, **102**, 35–42, https://doi.org/10.1016/j.biortech.2010.06.158

Park J. B. K., Craggs R. J. and Shilton A. N. (2013). Enhancing biomass energy yield from pilot-scale high rate algal ponds with recycling. *Water Research*, **47**, 4422–4432, https://doi.org/10.1016/j.watres.2013.04.001

Passos F., Solé M., García J. and Ferrer I. (2013). Biogas production from microalgae grown in wastewater: effect of microwave pretreatment. *Applied Energy*, **108**, 168–175, https://doi.org/10.1016/j.apenergy.2013.02.042

Passos F., Uggetti E., Carrère H. and Ferrer I. (2014). Pretreatment of microalgae to improve biogas production: a review. *Bioresource Technology*, **172**, 403–412, https://doi.org/10.1016/j.biortech.2014.08.114

Passos F., Carretero J. and Ferrer I. (2015a). Comparing pretreatment methods for improving microalgae anaerobic digestion: thermal, hydrothermal, microwave and ultrasound. *Chemical Engineering Journal*, **279**, 667–672, https://doi.org/10.1016/j.cej.2015.05.065

Passos F., Uggetti E., Carrère H. and Ferrer I. (2015b). Algal biomass: physical pretreatments. In: Pretreatment of Biomass, A. Pandey, S. Negi, P. Binod and C. Larroche (eds.), Elsevier, Amsterdam, pp. 195–226.

Pawlowski A., Mendoza J. L., Guzmán J. L., Berenguel M., Acién F. G. and Dormido S. (2015). Selective pH and dissolved oxygen control strategy for a raceway reactor within an event-based approach. *Control Engineering Practice*, **44**, 209–218, https://doi.org/10.1016/j.conengprac.2015.08.004

Posadas E., García-Encina P. A., Soltau A., Domínguez A., Díaz I. and Muñoz R. (2013). Carbon and nutrient removal from centrates and domestic wastewater using algal-bacterial biofilm bioreactors. *Bioresource Technology*, **139**, 50–58, https://doi.org/10.1016/j.biortech.2013.04.008

Posadas E., Bochon S., Coca M., García-González M. C., García-Encina P. A. and Muñoz R. (2014). Microalgae-based agro-industrial wastewater treatment: a preliminary screening of biodegradability. *Journal of Applied Phycology*, **26**, 2335–2345, https://doi.org/10.1007/s10811-014-0263-0

Posadas E., Morales M. del M., Gomez C., Acién F. G. G. and Muñoz R. (2015a). Influence of pH and CO_2 source on the performance of microalgae-based secondary domestic wastewater treatment in outdoors pilot raceways. *Chemical Engineering Journal*, **265**, 239–248, https://doi.org/10.1016/j.cej.2014.12.059

Posadas E., Serejo M. L., Blanco S., Pérez R., García-Encina P. A. and Muñoz R. (2015b). Minimization of biomethane oxygen concentration during biogas upgrading in algal-bacterial photobioreactors. *Algal Research*, **12**, 221–229, https://doi.org/10.1016/j.algal.2015.09.002

Posadas E., Alcántara C., García-Encina P. A., Gouveia L., Guieysse B., Norvill Z., Acién F. G., Markou G., Congestri R., Koreiviene J. and Muñoz R. (2017a). Microalgae cultivation in wastewater. In: Microalgae-Based Biofuels Bioproducts From Feedstock Cultivation to End-Products, R. Raul Muñoz and C. Gonzalez-Fernandez (eds.), Elsevier, Amsterdam, pp. 67–91.

Posadas E., Marín D., Blanco S., Lebrero R. and Muñoz R. (2017b). Simultaneous biogas upgrading and centrate treatment in an outdoors pilot scale high rate algal pond. *Bioresource Technology*, **232**, 133–141, https://doi.org/10.1016/j.biortech.2017.01.071

Puyol D., Monsalvo V. M., Marin E., Rogalla F., Melero J. A., Martínez F., Hülsen T. and Batstone D. J. (2020). Purple phototrophic bacteria as a platform to create the next generation of wastewater treatment plants: energy and resource recovery. In: Wastewater Treatment Residues as Resources for Biorefinery Products and Biofuels, J. A. Olivares, D. Puyol, J. A. Melero and J. Dufour (eds.), Elsevier, Amsterdam, Netherlands, pp. 255–280.

Qiao S., Hou C., Wang X. and Zhou J. (2020). Minimizing greenhouse gas emission from wastewater treatment process by integrating activated sludge and microalgae processes. *Science of the Total Environment*, **732**, 139032, https://doi.org/10.1016/j.scitotenv.2020.139032

Rajesh Banu J., Preethi Kavitha S., Gunasekaran M. and Kumar G. (2020). Microalgae based biorefinery promoting circular bioeconomy-techno economic and life-cycle analysis. *Bioresource Technology*, **302**, 122822, https://doi.org/10.1016/j.biortech.2020.122822

Rani R. U., Kumar S. A., Kaliappan S., Yeom I.-T., Banu J. R., Uma Rani R., Adish Kumar S., Kaliappan S., Yeom I.-T. and Rajesh Banu J. (2012). Low temperature thermo-chemical pretreatment of dairy waste activated sludge for anaerobic digestion process. *Bioresource Technology*, **103**, 415–424, https://doi.org/10.1016/j.biortech.2011.09.124

Rezvani S., Moheimani N. R. and Bahri P. A. (2016). Techno-economic assessment of CO_2 bio-fixation using microalgae in connection with three different state-of-the-art power plants. *Computers & Chemical Engineering*, **84**, 290–301, https://doi.org/10.1016/j.compchemeng.2015.09.001

Rochaix J. D. (2016). The dynamics of the photosynthetic apparatus in algae. In: Applied Photosynthesis – New Progress, M. Najafpour (ed.), IntechOpen, London, pp. 23–52.

Rodero M. del R., Posadas E., Toledo-Cervantes A., Lebrero R. and Muñoz R. (2018). Influence of alkalinity and temperature on photosynthetic biogas upgrading efficiency in high rate algal ponds. *Algal Research*, **33**, 284–290, https://doi.org/10.1016/j.algal.2018.06.001

Rodero M. del R., Lebrero R., Serrano E., Lara E., Arbib Z., García-Encina P. A. A. and Muñoz R. (2019). Technology validation of photosynthetic biogas upgrading in a semi-industrial scale algal-bacterial photobioreactor. *Bioresource Technology*, **279**, 43–49, https://doi.org/10.1016/j.biortech.2019.01.110

Rodero M. del R., Carvajal A., Arbib Z., Lara E., de Prada C., Lebrero R., Muñoz R., Rodero R., Carvajal A., Arbib Z., Lara E. and Prada C. D. (2020a). Performance evaluation of a control strategy for photosynthetic biogas upgrading in a semi-industrial scale photobioreactor. *Bioresource Technology*, **307**, 123207, https://doi.org/10.1016/j.biortech.2020.123207

Rodero M. del R., Severi C. A., Rocher-Rivas R., Quijano G. and Muñoz R. (2020b). Long-term influence of high alkalinity on the performance of photosynthetic biogas upgrading. *Fuel*, **281**, 118804, https://doi.org/10.1016/j.fuel.2020.118804

Ruiz J., Arbib Z., Barragán J. and Perales J. A. (2012). Performance of a flat panel reactor in the continuous culture of microalgae in urban wastewater: prediction from a batch experiment. *Bioresource Technology*, **127**, 456–463, https://doi.org/10.1016/j.biortech.2012.09.103

Ryckebosch E., Drouillon M. and Vervaeren H. (2011). Techniques for transformation of biogas to biomethane. *Biomass and Bioenergy*, **35**, 1633–1645, https://doi.org/10.1016/j.biombioe.2011.02.033

Samson R. and Leduy A. (1985). Multistage continuous cultivation of blue-green alga spirulina maxima in the flat tank photobioreactors with recycle. *The Canadian Journal of Chemical Engineering*, **63**, 105–112, https://doi.org/10.1002/cjce.5450630117

Sanz J. L., Rojas P., Morato A., Mendez L., Ballesteros M. and González-Fernández C. (2017). Microbial communities of biomethanization digesters fed with raw and heat pre-treated microalgae biomasses. *Chemosphere*, **168**, 1013–1021, https://doi.org/10.1016/j.chemosphere.2016.10.109

Sarkodie S. A., Adams S. and Leirvik T. (2020). Foreign direct investment and renewable energy in climate change mitigation : does governance matter? *Journal of Cleaner Production*, **263**, 121262, https://doi.org/10.1016/j.jclepro.2020.121262

Schediwy K., Trautmann A., Steinweg C. and Posten C. (2019). Microalgal kinetics – a guideline for photobioreactor design and process development. *Engineering in Life Sciences*, **19**, 830–843, https://doi.org/10.1002/elsc.201900107

Schwede S., Rehman Z.-U., Gerber M., Theiss C. and Span R. (2013). Effects of thermal pretreatment on anaerobic digestion of nannochloropsis salina biomass. *Bioresource Technology*, **143**, 505–511, https://doi.org/10.1016/j.biortech.2013.06.043

Sepúlveda-Muñoz C. A., Ángeles R., de Godos I. and Muñoz R. (2020a). Comparative evaluation of continuous piggery wastewater treatment in open and closed purple phototrophic bacteria-based photobioreactors. *Journal of Water Process Engineering*, **38**, 101608, https://doi.org/10.1016/j.jwpe.2020.101608

Sepúlveda-Muñoz C. A., de Godos I., Puyol D. and Muñoz R. (2020b). A systematic optimization of piggery wastewater treatment with purple phototrophic bacteria. *Chemosphere*, **253**, 126621, https://doi.org/10.1016/j.chemosphere.2020.126621

Serejo M. L., Posadas E., Boncz M. A., Blanco S., Garcia-Encina P. and Muñoz R. (2015). Influence of biogas flow rate on biomass composition during the optimization of biogas upgrading in microalgal-bacterial processes. *Environmental Science & Technology*, **49**, 3228–3236, https://doi.org/10.1021/es5056116

Singh J. and Dhar D. W. (2019). Overview of carbon capture technology: microalgal biorefinery concept and state-of-the-art. *Frontiers in Marine Science*, **6**(29), 1–10. https://doi.org/10.3389/fmars.2019.00029

Solovchenko A., Verschoor A. M., Jablonowski N. D. and Nedbal L. (2016). Phosphorus from wastewater to crops: An alternative path involving microalgae. *Biotechnology Advances*, **34**, 550–564, https://doi.org/10.1016/j.biotechadv.2016.01.002

Spellman F. R. and Drinan J. E. (2014). Wastewater Stabilization Ponds. CRC Press, Boca Raton.

Syahirah N., Aron M., Shiong K., Wayne K., Veeramuthu A., Chang J. and Loke P. (2020). Microalgae cultivation in wastewater and potential processing strategies using solvent and membrane separation technologies. *Journal of Water Process Engineering*, **39**, 101701.

Toledo-Cervantes A., Serejo M. L., Blanco S., Pérez R., Lebrero R. and Muñoz R. (2016). Photosynthetic biogas upgrading to bio-methane: boosting nutrient recovery via biomass productivity control. *Algal Research*, **17**, 46–52, https://doi.org/10.1016/j.algal.2016.04.017

Toledo-Cervantes A., Madrid-Chirinos C., Cantera S., Lebrero R. and Muñoz R. (2017). Influence of the gas-liquid flow configuration in the absorption column on photosynthetic biogas upgrading in algal-bacterial photobioreactors. *Bioresource Technology*, **225**, 336–342, https://doi.org/10.1016/j.biortech.2016.11.087

Toledo-Cervantes A., Morales T., González Á, Muñoz R. and Lebrero R. (2018). Long-term photosynthetic CO_2 removal from biogas and flue-gas: exploring the potential of closed photobioreactors for high-value biomass production. *Science of the Total Environment*, **640–641**, 1272–1278, https://doi.org/10.1016/j.scitotenv.2018.05.270

Tredici M. R. and Materassi R. (1992). From open ponds to vertical alveolar panels: the Italian experience in the development of reactors for the mass cultivation of phototrophic microorganisms. *Journal of Applied Phycology*, **4**, 221–231, https://doi.org/10.1007/BF02161208

Uggetti E., Passos F., Solé M., Garfí M. and Ferrer I. (2017). Recent achievements in the production of biogas from microalgae. *Waste and Biomass Valorization*, **8**, 129–139, https://doi.org/10.1007/s12649-016-9604-3

UNE-EN 16723 (2017). Natural Gas and Biomethane for Use in Transport and Biomethane for Injection in the Natural Gas Network. Asociación Española de Normalización. https://www.une.org/encuentra-tu-norma/busca-tu-norma/norma/?c=N0058455 (Accessed 16 February 2022).

Wen S., Liu H., He H., Luo L., Li X., Zeng G., Zhou Z., Lou W. and Yang C. (2016). Treatment of anaerobically digested swine wastewater by Rhodobacter blasticus and Rhodobacter capsulatus. *Bioresource Technology*, **222**, 33–38, https://doi.org/10.1016/j.biortech.2016.09.102

Whitton R., Ometto F., Pidou M., Jarvis P., Villa R. and Jefferson B. (2015). Microalgae for municipal wastewater nutrient remediation: mechanisms, reactors and outlook for tertiary treatment. *Environmental Technology Reviews*, **4**(1), 133–148, https://doi.org/10.1080/21622515.2015.1105308

Woertz I., Feffer A., Lundquist T. and Nelson Y. (2009). Algae grown on dairy and municipal wastewater for simultaneous nutrient removal and lipid production for biofuel feedstock. *Journal of Environmental Engineering*, **135**, 1115–1122, https://doi.org/10.1061/(ASCE)EE.1943-7870.0000129

Yang J., Cao J., Xing G. and Yuan H. (2015). Lipid production combined with biosorption and bioaccumulation of cadmium, copper, manganese and zinc by oleaginous microalgae Chlorella minutissima UTEX2341. *Bioresource Technology*, **175**, 537–544, https://doi.org/10.1016/j.biortech.2014.10.124

Yang A., Zhao W., Peng M., Zhang G., Zhi R. and Meng F. (2018). A special light-aerobic condition for photosynthetic bacteria-membrane bioreactor technology. *Bioresource Technology*, **268**, 820–823, https://doi.org/10.1016/j.biortech.2018.08.008

Zhou Q., Zhang P. and Zhang G. (2015). Biomass and pigments production in photosynthetic bacteria wastewater treatment: effects of light sources. *Bioresource Technology*, **179**, 505–509, https://doi.org/10.1016/j.biortech.2014.12.077

Zhou Q., Zhang G., Lu Y. and Wu P. (2016). Feasibility study and process optimization of citric acid wastewater treatment and biomass production by photosynthetic bacteria. *Desalination and Water Treatment*, **57**(14), 6261–6267, https://doi.org/10.1080/19443994.2015.1005687

doi: 10.2166/9781789061796

Chapter 10

Sludge management and utilization for decarbonization

Meltem Urgun-Demirtas[1], Rachel Dalke[1] and Krishna R. Pagilla[2]*

[1]Applied Materials Division, Argonne National Laboratory, 9700 S Cass Ave, Lemont, IL, 60439, USA
[2]Civil and Environmental Engineering, University of Nevada, Reno, 1664 N Virginia Street, Reno, NV, 89557, USA
**Correspondence: demirtasmu@anl.gov*

10.1 OVERVIEW OF CURRENT SLUDGE MANAGEMENT PRACTICES

In addition to the constantly growing global population, stringent water treatment regulations also increase global annual production of sewage sludge (biosolids) (Table 10.1). As a result, biosolids treatment has become a primary issue alongside wastewater treatment (Figure 10.1). Biosolids disposal has been noted as a significant factor affecting treatment plant feasibility, with management accounting for a large share (25–65%) of operational costs (Arias *et al.*, 2021; Mu'azu *et al.*, 2019). Recycling, reuse, conversion, and nutrient or energy recovery have been identified as components of biosolids management (Shaddel *et al.*, 2019). Environmental regulations on the management of waste combined with a potential for nutrient recovery are push factors for wastewater resource recovery facilities (WRRFs) to utilize advanced treatment techniques to manage biosolids wastes. Significant research has been devoted to reducing environmental impacts and cost and enhancing the energy efficiency of biosolids management at WRRFs. Strategic selection of technologies and development of new ones are needed to manage biosolids sustainably in the future and also enable decarbonization of the water sector.

Sludge (biosolids) management strategies vary around the world (Table 10.2). In the United States (US), land application, composting, incineration, anaerobic digestion (AD), and landfilling are the most commonly used biosolids management methods (Lee *et al.*, 2020). Figure 10.2 shows the distribution of biosolids management methods from publicly-owned treatment works (POTWs) in the US as reported by the US Environmental Protection Agency (EPA). In China, biosolids management methods include composting, co-combustion, thermal drying incineration, and cement manufacturing (Arias *et al.*, 2021). Wastewater treatment generated approximately eight million tons of biosolids in the European Union (EU) in 2016 (European Commission, 2020). The main methods of biosolids management in the EU are agricultural use and incineration (Kacprzak *et al.*, 2017); however, regulations for land application and level of application vary widely in different countries. In Egypt, it is estimated that 2.1 million tons of dry solids is produced annually, with approximately 85% improperly disposed due to a lack of facilities capable of stabilizing the waste (Abdel Wahaab *et al.*, 2020). Poland generated

Table 10.1 Global sludge production data for various countries.

Country	Sludge Production (thousand dry metric tons)	Year
EU–28[a]	8000	2016
US[b]	12 555	2017
China[b]	6455	2017
Japan[b]	2405.82	2017
Germany[c]	1800	2016
Norway[d]	147.6	2018
Czechia[d]	228.22	2018
Sweden[d]	210.9	2018
Hungary[d]	217.842	2018
Austria[d]	234.481	2018
Romania[d]	247.76	2018
Turkey[d]	318.50	2018
Netherlands[d]	341.03	2018
Poland[d]	583.07	2018
Croatia[d]	19.23	2018
Cyprus[d]	8.406	2018
Latvia[d]	24.591	2018
Lithuania[c]	44.192	2018
Albania[d]	94.5	2018
Slovakia[d]	55.93	2018
Slovenia[d]	38.1	2018
Serbia[d]	9.6	2018
Bosnia and Herzegovina[d]	9.5	2018
Malta[d]	8.28	2018
Switzerland[d]	177	2017
Bulgaria[d]	68.6	2017
Ireland[d]	58.773	2017

[a]*European Commission (2020).*
[b]*Wei et al. (2020).*
[c]*Roskosch and Heidecke (2018).*
[d]*Eurostat (2020).*

a substantial amount of sludge each year (>500 000 tons/yr from 2009 to 2018) that was primarily managed through landfilling (Eurostat, 2020; Rosiek, 2020). In 2016 a law was passed prohibiting landfilling of sludge; thereby creating a need for the development of adequate alternative management strategies. In 2016, up to 34 and 30% of sludge generated in Poland was managed via agricultural use (soil formation, fertilizer) and thermal transformation (incineration, co-incineration, gasification, pyrolysis, wet oxidation), respectively (Przydatek & Wota, 2020).

The objective of this chapter is to review and summarize current knowledge about biosolids management and provide a basis for future practices. The chapter focuses on how current management practices help decarbonization, the role of biosolids management strategies in achieving the decarbonization targets of utilities, and how challenges (e.g., emerging contaminants, odors, public scrutiny and upset) in management can be addressed in meeting such targets. Since Chapter XX

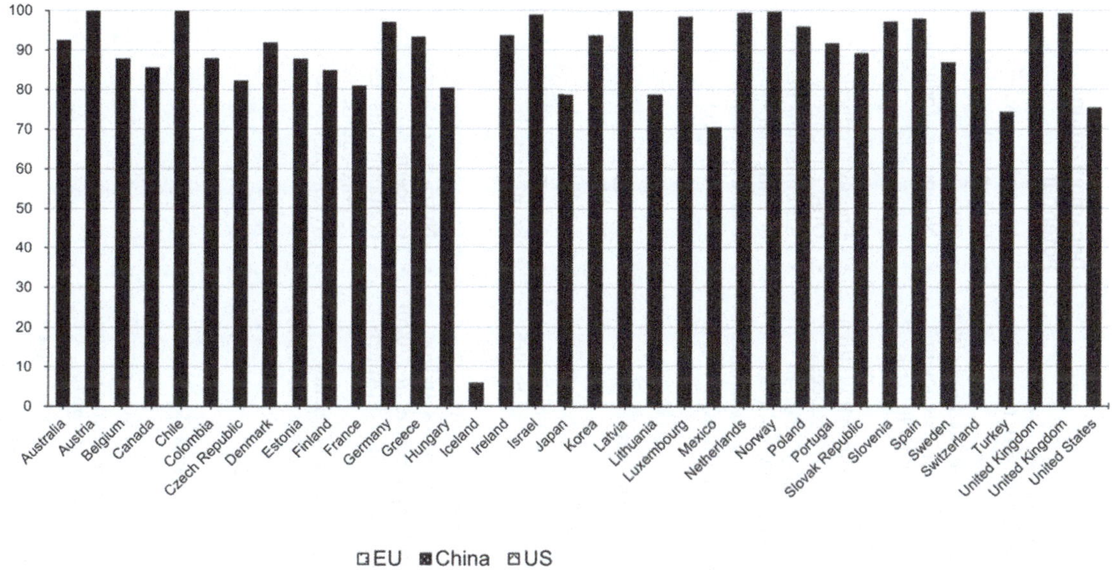

Figure 10.1 Percent of resident population connected to urban wastewater collecting system. (OECD.Stat 2021).

discusses AD application in detail, this chapter focuses on sludge treatment alternatives beyond AD. It should be noted that residual digestate after AD of sludge also needs additional treatment and disposal. New emerging concepts, namely Water-Energy Nexus, Circular Economy, and Nutrient Trading, are important vehicles for decarbonization in shaping the future sludge management practices. These concepts significantly help to reduce the financial burdens of sludge management on societies and overcome ecological issues and resource scarcity. New technologies and approaches need to be developed to extract energy and nutrients from sludge and improve process and energy efficiency. Recovered energy and nutrients help utilities become a source of revenue generation, overcoming their reputation as pollution mitigation entities. In turn, they will become entities contributing to reduction in carbon emissions and achieve decarbonization of the water sector. Renewable energy production and resource recovery are presented as areas of sludge management to close the linearity of waste production and implement a circular economy of waste management. The chapter closes with a section on implementation of decarbonization at the utilities and future strategies and pathways.

10.2 SLUDGE TO ENERGY/PRODUCTS

Treatment methods for sludge can be categorized into biological or thermal types. AD is a biological process with high resource recovery potential for biosolids management due to low energy inputs (Xu *et al.*, 2020) and high extraction efficiency of organic energy (Nakkasunchi *et al.*, 2021). High solids-AD of biosolids combined with other components of organic waste (e.g. food waste, yard waste) has been identified as a suitable management method for biosolids with low impacts in areas of global warming potential, acidification, eutrophication, ecotoxicity, and overall cost (Lee *et al.*, 2020). Furthermore, AD does not utilize aeration, allowing for significantly reduced plant operating costs (Chen *et al.*, 2020; Seiple *et al.*, 2020). AD is largely more effective than incineration for energy recovery, but is subject to poor conversion efficiencies and generates a large volume of residual wet digestate that requires processing (Chen *et al.*, 2020). Anaerobic sludge treatment has potential benefits both economically and ecologically due to the ability to produce and utilize biogas (Nakkasunchi

Table 10.2 Advantages and disadvantages of common sludge management practices ((Wiechmann *et al.*, 2013) unless otherwise stated).

Management Practice	Advantages	Disadvantages	Decarbonization Potential
Farming, landscaping and other types of material recovery and reuse	• Allows for the use of nutrients and phosphorus • Most cost efficient of the available methods	• Sewage sludge acts as a sink for pollutants and pathogens • Pollutants accumulate in the food chain • Lack of confidence in consistency in nutrient availability and health risks (Shaddel *et al.*, 2019)	• Nutrient recovery • Uncontrolled GHG emissions • No energy production
Mono-incineration	• Allows for long-term wastewater management planning • Eliminates organic pollutants in sewage sludge • Allows for energy recovery • Allows for P recovery from ash • Incineration with P recovery reduces resource use and opens up new markets	• Phosphorus recovery from ash is still a complex and cost-intensive process • Transport can potentially result in additional environmental pollution • The most cost-intensive of the available wastewater disposal methods	• Energy production • Only P recovery • Air emissions need to be mitigated • P bioavailability needs to be considered
Co-incineration	• Eliminates all pathogens and organic pollutants from sewage sludge • Allows for energy recovery • Less cost-intensive than mono-incineration • Reduces resource use in that it requires less fuel and is an alternative to aggregates	• Sewage sludge nutrients are unrecoverable • Does not allow for P recovery from ash • Long haulage routes can entail deleterious environmental and health effects	• Energy production • Building materials production • Air emissions need to be mitigated
Anaerobic digestion (AD)	• Most energy-efficient means to capture energy from biosolids (Wei *et al.*, 2020) • Land application of digested sludge lowers fertilizer consumption (Wei *et al.*, 2020) • Improved sludge dewaterability (Shaddel *et al.*, 2019) • Additional income from product sales (Shaddel *et al.*, 2019) • Co-digestion can promote hydrolysis and improve methane yield (Zhang *et al.*, 2020)	• Subject to digester upsets from inappropriate substrate ratios and/or operating conditions (Chow *et al.*, 2020) • Long retention times required (Appels *et al.*, 2011) • Low overall degradation efficiency (hydrolysis-limited) (Appels *et al.*, 2011) • Low level of biogas utilization in many places (Seiple *et al.*, 2020)	• Energy production • Nutrient recovery • Concern contaminants removal require additional treatment methods
Composting	• Increases soil fertility and biodiversity (Bruni *et al.*, 2020) • Decreases the need for chemical fertilizers (Bruni *et al.*, 2020) • Contributes to awareness and promotion of community-level waste management practices (Bruni *et al.*, 2020)	• Higher space requirements (Jędrczak, 2018) • Compressed mixtures have lower porosity and moisture content and cause unpleasant odors (Jędrczak, 2018) • Produces higher amount of greenhouse gas emissions (Jędrczak, 2018)	• Nutrient recovery • Air emissions need to be mitigated

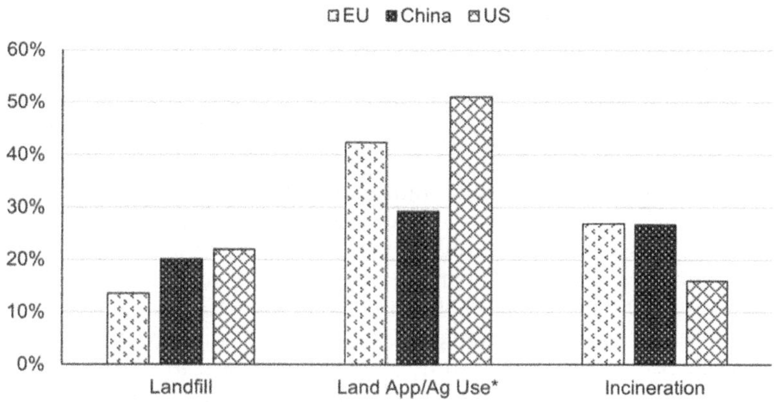

Figure 10.2 Sludge management practice utilization in the United States, European Union, and China (Đurđević *et al.*, 2019; US EPA, 2019; Wei *et al.*, 2020).

et al., 2021). However, it has been reported in the US that biogas from sludge processing has a low level of utilization with less than half of WRRFs with influent flow rates above one million gallons per day (MGD) or 3785 m^3/day using AD, and few using the biogas for heat or electricity generation (Seiple *et al.*, 2020; Shen *et al.*, 2015). Additionally, growing legal use or disposal requirements for the solid products of common biological methods such as AD or composting complicate their application; thus, efforts are being focused on alternative methods (Świerczek *et al.*, 2021).

Sludge incineration in conjunction with phosphorus (P) recovery has been gaining a lot of attention. In Germany, large wastewater treatment plants have been using sludge incineration to recover P in ashes to meet the country's goal of obtaining at least 20% of raw phosphate from sewage sludge. Total authorized sludge incineration capacity in Germany is approximately 1.5 M dry metric tons/yr and 80% of this capacity in use (Wiechmann *et al.*, 2013). Co-incineration of sludge at coal/lignite fired power plants and cement plants has many advantages including reduced fossil fuel and carbon emissions and lowered cost; however, P is not recoverable during the process. Sludge incineration has been also used for the production of building materials such as cement. This method is one of the best options if sewage contains a high concentration of heavy metals (HM) since it transforms bioavailable HMs to more stable forms, thereby alleviating their leaching toxicities into the receiving environments (Cao *et al.*, 2020). Although sludge incineration maximizes solids reduction (thereby lowering land requirements) and yields energy recovery with a stable ash production, it has high capital and operation costs. Toxicity of the ash also limits the overall feasibility of the management strategy (Arias *et al.*, 2021). In most cases, air emissions are an important issue and require additional treatment to meet air quality specifications.

An alternative process for treating raw biosolids is pyrolysis, an oxygen-deficient thermal degradation process. Pyrolysis is an inexpensive and robust procedure that has shown promise as a treatment tool for raw biosolids by generating a liquid and solid stream that can be used for liquid fuels and soil amendment, respectively (Callegari & Capodaglio, 2018). Thermal hydrolysis prior to (an)aerobic digestion or composting has been reported to increase feedstock biodegradability, solids loading capability (9 vs. 6%), and biosolid dewaterability, all while maintaining a better energy balance (Flores-Alsina *et al.*, 2021). The high temperatures and pressures of pyrolysis improve the soil quality through the inactivation of pathogens, thereby expanding the applicability of the finished product (landfill vs. land application) (Flores-Alsina *et al.*, 2021). Biochar is one product generated through pyrolysis that has garnered attention from researchers as a means to improve soil quality and enhance nutrient levels. Applications of biochar are seen in adsorption of antibiotics, heavy metals, dyes, and phenolic compounds in wastewater effluent. It also aids in N retainment in soils after agricultural application of the product (Singh *et al.*, 2020). Pyrolysis has also been utilized to convert biosolids into granular activated carbon (GAC). Producing GAC via pyrolysis is reported to provide a financial benefit by reducing facility operating costs by 20–40% (Mu'azu *et al.*, 2019). However, GAC has reportedly less effective adsorption of heavy metals (Singh *et al.*, 2020).

Recently the efficacy of implementing thermal treatment methods has been studied; however, their current utilization is limited due to economic limitations. It has been shown that thermal treatment methods have benefits within the framework of a circular economy, allowing for energy and natural resource recovery, as well as valuable byproduct generation (Tsybina & Wuensch, 2018). A model study conducted in South Africa found using a centralized sludge management technique with thermal pretreatment and AD resulted in 36.5% conversion of COD in the influent into methane and 41 and 65% N and P in the biosolids, respectively. This is a substantial improvement from the disposal practice utilized in the study, which was landfilling (Flores-Alsina *et al.*, 2021).

Hydrothermal liquefaction (HTL) is another process that has the potential to treat and valorize biosolids via conversion into biocrude oil. A significant advantage to utilizing HTL over pyrolysis is the capability to process wet feedstocks, thereby eliminating the need for a cost-intensive drying step. A modeling study in the US determined that facilities with capacity greater than 17 400 m^3 per day could supply approximately 10 million metric tons of feedstock and produce 3.7 million m^3 of biocrude/year (Seiple *et al.*, 2020). It has been shown that the type of feedstock processed and operation conditions selected influence the elemental composition of biocrude. WRRF biosolids have been shown to produce high biocrude yield (45%), as well as 55–80% carbon recovery in the oil product. Additionally, it has been shown that the majority of inorganics (including P) become concentrated in the solids product of post-HTL (weight percentages >70%), where they may be separated out and used as fertilizer (Conti *et al.*, 2020). HTL has shown promise as a technology suitable for the destruction of micropollutants in sludge; however, the complex composition of the resultant process water limits its scale-up applications (Silva Thomsen *et al.*, 2020). While raw process water has been identified as a health hazard due to a high concentration of toxic compounds, treated effluent has the potential for valorization through energy and nutrient recovery (Watson *et al.*, 2020). While HTL is a highly efficient strategy for biomass conversion into fuels, process applications are hindered by expensive upgrading techniques required to meet biofuel standards. Catalysts have been proposed as a means to overcome this challenge by increasing biocrude quality, but many are non-recoverable and thus present additional economic challenges. Heterogeneous catalysts have shown potential as a viable alternative that reduce costs by not only increasing biocrude yield and quality but allowing for catalysis recovery and reuse (Scarsella *et al.*, 2020). Biocrude from HTL has a lower oxygen and moisture content and higher heating value relative to pyrolysis, resulting in lower fixed and operational costs and a subsequent competitive advantage as a bioconversion technology (Dimitriadis & Bezergianni, 2017).

10.3 LAND APPLICATION AND DEDICATED LANDFILLING

Landfilling has traditionally been a commonly used sludge disposal method due to its low cost and overall operational simplicity; however, it has several disadvantages, including the large land requirement, environmental pollution (soil, air, surface and groundwater), and lost potential to recover energy and nutrients from sludge. Limited land availability and stringent air quality requirements have led to many regions disallowing sludge acceptance for disposal. For example, landfills in California will ban 75% of organics waste (including sludge) in 2025.

Sludge has been land applied for a variety of purposes, such as soil conditioner, partial fertilizer, surface cover, filler, and ingredient in materials formulation. Land application of sludge has been regulated in many countries. In the US, sewage sludge is regulated under EPA Rule 40 CFR Part 503, which includes restrictions for pollutant concentrations and application rates to reduce pathogens, vector attraction and heavy metals in sludge applied to the land or placed on a surface disposal site (US EPA, 1992, 1995;

Walker *et al.*, 1994). Most land applications in the US are on agricultural land (pastures and cropland), disturbed areas (e.g., brownfields), plant nurseries, forests, recreational areas (e.g., parks, golf courses), lawns and gardens, cemeteries, highways, and airport runway medians (US EPA, 2003a, 2003b). The use of sludge in agriculture within the EU is currently regulated only by the limits of heavy metals (Cd, Cu, Hg, Ni, Pb and Zn) listed in Council Directive 86/278/EEC. Low doses of sludge applications have shown beneficial effects on microbial biomass, organic carbon, and soil microbial activity. In some cases, excessive application of sewage sludge with high heavy metals concentrations into soil has been found to increase the bioavailability of heavy metals (Hudcová *et al.*, 2019).

10.4 RECLAMATION OF BROWNFIELDS

Brownfields are abandoned, idled, or underused lands due to contamination and deteriorated soil conditions. Many brownfields have little or no value to the surrounding area. Biosolids have been used to reclaim brownfields to improve soil qualities, allowing beneficial use of these underutilized properties. The application rates of biosolids onto brownfield sites are often higher than agricultural applications. In such applications, balancing the carbon to nitrogen ratio (C:N) is essential to maximize the organic content in the land and minimize the potential nitrate loss into the water table. In some applications, there is a need to add to residual material such as hardwood leaves, straw, compost, and paper mill fines with a high carbon to nitrogen ratio (40:1) to provide excess carbon and immobilize the nitrogen from nitrogen-rich biosolids (Brown, 2001). Biosolids applications have also been an effective treatment procedure for acid mine drainage. In a typical mine reclamation site, biosolids and carbon sources have been applied at a rate of 10–25 and 100–150 dry tons, respectively. However, these application rates are subject to change depending on site characteristics and soil depths needed for reclamation (Cogger, 2000). Biosolids application not only improves soil carbon content, but also chemically binds metals within a matrix that limits bioavailability and leaching of toxic metals in acid mine drainage. This is an inexpensive method for incorporating biosolids into highly contaminated topsoil in acid mine drainage.

There are many great examples of sludge applications for brownfield reclamation in the US. For example, the Metropolitan Water Reclamation District (MWRD) of Chicago, USA has been reclaiming 6070 hectares of land left from coal mining. With MWRD's Prairie Plan, sludge application leveled the land and enriched the carbon, nitrogen (N), phosphorus (P), and micro-nutrient content of soil. The brownfield has been converted to fertile land for agricultural purposes (MWRD, 2021). Sludge application (40 tons/hectare biosolids) and lime amendment revegetated the barren tailings areas of Upper Arkansas River Site, Leadville in Colorado. This application reduced the availability of concerned metal contaminants (ITRC, 2010). It also improved soil quality by increasing pH, total organic carbon, water-holding capacity, total nutrient concentration, and plant and soil microbial activity. Results showed the plant community established, while the soil microbial community started to recover, one year after treatment

10.5 COMPOSTING

Composting is a biochemical process involving the degradation of organic matter by microorganisms under natural or controlled conditions (Bruni *et al.*, 2020; Onwosi *et al.*, 2017; Sánchez *et al.*, 2017). Composting has been identified as a valuable waste stabilization technique due to its wide environmental compatibility (Onwosi *et al.*, 2017). The success of implementing composting is dependent on the concentrations of heavy metals in the waste material. When properly managed, composting reduces greenhouse gas emissions, increases soil fertility and biodiversity, and reduces the need for chemical fertilizers (Bruni *et al.*, 2020). However, challenges exist with leachate generation, gas emissions, and lack of uniformity in compost modeling, which is used to determine how control measures affect the overall composting process. Several methods have been applied to reduce the negative effects of composting, including the addition of bulking agents such as sawdust, rice straw, wood chips, and cotton gin waste (Onwosi *et al.*, 2017). Biochar has also been used as a bulking agent during composting, wherein the addition lowered concentrations of metals and arsenic in soil as well as stronger adsorption and microbial community activity (Ye *et al.*, 2019). A study in Poland showed the maturation of composting of biosolids impacts the concentrations of organic carbon, nutrients, and heavy metals (but not the percentage of their mobile or bioavailable forms) and is an effective stabilization method (Bożym & Siemiątkowski, 2018). Alongside Poland, Italy is making large strides to implement improved biosolids management practices (Mininni *et al.*, 2019). The country utilizes numerous composting facilities within their waste management framework. Extensive efforts to source-separate organic waste materials have helped to maximize recovery and meet the country's sustainability goals. Sewage sludge is the second main fraction in composting plants in Italy. However, its percentage decreased from 17% in 2004 to 10.6% in 2017. Most of the compost products are used in agriculture (approximately 70%) (Bruni *et al.*, 2020).

10.6 RESOURCE RECOVERY

Phosphorus (P) is an essential element in formulation of most fertilizers and P reserves in most countries are estimated to be depleted within 100 years (Falk *et al.*, 2020). As a non-renewable resource with no nutritional (agricultural) substitute (Shaddel *et al.*, 2019), P recovery has become a topic of great interest to the scientific community. WRRF biosolids are rich in P, which can be found in both the biosolids itself as well as leachate and biosolids ashes (Cieślik & Konieczka, 2017). P in sewage sludge is present in both inorganic and organic forms. The chemical composition of inorganic P depends on the treatment processes (e.g. Fe coagulation) present in the plant layout (Falk *et al.*, 2020). It has been reported that 90–95% of incoming phosphorus into a WRRF is incorporated into the biosolids, and P recovery from the aqueous portion is limited to 20–40% (Shaddel *et al.*, 2019). Main methods for P recovery are struvite precipitation, calcium phosphate (Ca-P) precipitation, and phosphoric acid reduction (Shaddel *et al.*, 2019). Struvite precipitation is expensive and requires external chemicals to be added, making it cost-prohibitive for smaller plants. Ca-P precipitation is a more effective method for removing P, as the chemicals used in Ca-P recovery are more accessible than those for struvite precipitation. Furthermore, Ca-P precipitation forms hydroxyapatite, which lends itself to a wider range of uses than struvite (Law & Pagilla, 2019). The implementation of any of these various management techniques is further complicated by the fact that utilizing phosphate rock and existing phosphate-based fertilizers has become increasingly cheaper relative to recovering P from WRRFs (Law & Pagilla, 2018).

The plant availability of P in ashes from thermal processes depends on the applied technology. The average P solubility of 24 German sewage sludge ashes in a neutral ammonium citrate solution showed relatively poor P-plant availability (25.6%) (Kruger & Adam, 2015). An inventory of sewage sludge ashes generated in the mono sludge incineration plants in Poland showed that 26 756 Mg of ashes in 2018 were produced from 11 sludge mono-incineration plants operated with a total capacity of 160 300 Mg dry weight of sludge annually. Total recovered phosphorus was about 1614 Mg (13% of the total ashes) with an average of 33.9% bioavailability (Smol *et al.*, 2020). These results indicate that there is a need for increasing plant availability of P by pre- or post-treatment methods to modify the ash composition and structure, so ashes can be used as a fertilizer. The fate of phosphorus from the combustion of sewage sludge was studied in fluidized and fixed bed reactors to determine distribution, elemental composition, and crystallinity of P, hence enabling more efficient P recovery from combustion ashes (Falk *et al.*, 2020).

N-recovery from the sludge has received less attention than P-recovery due to less economic return and incentives. However, newly emerged concepts such as nutrient trading and circular economy can help to develop cost effective N-recovery technologies and maximize resource recovery from sludge.

10.7 SLUDGE STABILZIATION FOR REMOVAL OF EMERGING CONTAMINANTS

Most conventional wastewater treatment operations have limited capability to entirely remove emerging contaminants (ECs) (hormones, antibiotics, personal care products, etc.), and the partially removed ECs from wastewater treatment processes end up in large sludge volumes. The presence of high ECs concentrations in sludge may raise additional concerns if they are released into the environment without proper treatment. The type of sludge treatment determines the fate of ECs in the receiving environment (Dubey *et al.*, 2021). Biodegradation and sorption are the primary mechanisms for the removal of ECs. AD and composting are usually successful in the removal of many ECs. A recent study conducted by the Water Environment Research Foundation (WERF) highlights the importance of studying the transformation of ECs by different sludge stabilization methods. The results showed that only 18% of estrogenicity was removed by aerobic digestion while the estrogenicity increased in AD due to the transformation of tested ECs to a more estrogenic form. There is also a need to integrate sludge pretreatment (e.g., sonication, ozonation, thermal hydrolysis) with sludge stabilization to increase the removal of recalcitrant ECs. More research is needed on the fate, transformation, and removal mechanisms of ECs during conventional and advanced sludge treatment methods.

10.8 CENTRALIZED VS. DISTRIBUTED SLUDGE MANAGEMENT PRACTICES

There is a vast difference between urban and rural areas regarding the adoption of sludge management strategies in developed countries. This difference is exacerbated between developed and developing countries because of differences in political structures, national priorities, socio-economic conditions, cultural traits, access to public health and safety services, and financial resources. Centralized waste management strategies have been more viable options for highly populated urban areas in developed countries because of their convenience and efficiencies. With new emerging concepts, decentralization started to be a more sustainable alternative for developing countries, rural areas and even small communities in the urban areas in developed countries (Righi *et al.*, 2013). As urban areas in developed countries face increased challenges in sludge use for land application or landfilling due to scarce land, the transportation costs and the environmental impact may also drive those areas to adopt decentralized strategies for sludge management. The overall carbon footprint of sludge management is likely to be lower even though the onsite sludge processing to convert into useful products may be more expensive.

10.9 ENVIRONMENTAL AND ECONOMIC LIFE CYCLE ASSESSMENT OF SLUDGE MANAGEMENT TECHNOLOGIES

Currently, many WRRFs successfully demonstrate simultaneous energy efficiency and nutrient recovery. A combination of two or more sludge management technologies (e.g. AD and incineration) is a very common practice at energy efficient plants such as the Mainz WRRF (Gretzschel *et al.*, 2020) and Hamburg's Köhlbrandhöft WRRF (Mills *et al.*, 2014; Laurich, 2011). The implementation of both AD with biogas utilization and biosolids incineration with electricity generation at the WRRFs in Texas led to an estimated 83% reduction in electricity consumption (Stillwell *et al.*, 2010). Surplus heat and electricity at WRRFs have been fed to the district grid system. A new piloting application is the storage of heat surplus at the Hamburg WRRF during the summer in an aquifer to compensate for seasonal fluctuations and save heating energy in the winter (Schafer *et al.*, 2020). Aquifer thermal energy storage (ATES) systems are open systems in which groundwater is used as the heat transfer medium between the external energy source and the aquifer. Groundwater is heated up and cooled down depending on the season.

There are few case studies capable of tracking carbon flow in utilities with multipurpose sludge utilization for quantifying the impact of increasing application of sludge as biogenic carbon sources.

An assessment of the environmental and economic sustainability of each sludge management option including landfilling, composting, incineration, digestion, and land application is needed. Life Cycle Assessment (LCA) and Life Cycle Cost Analysis (LCCA) have been used to provide a comparative assessment of environmental and economic impacts of sludge management alternatives (Arias *et al.*, 2021; Lee *et al.*, 2020; Yoshida *et al.*, 2018).

10.10 IMPLEMENTATION: CHALLENGES AND OPPORTUNITIES, REGULATORY AND SOCIAL ISSUES

Legal stipulations and environmental regulations greatly influence the management techniques for biosolids or sludge. WRRFs should undertake decarbonization more holistically. The whole carbon lifecycle assessment should ensure the potential carbon emissions of each sludge treatment method involved in the plant layout. The mitigation steps required to offset the carbon impact should also be factored into the overall framework. In future sludge management applications, stringent regulations (e.g., air quality, presence of ECs, sludge application rate) drive the need for transforming current sludge treatment strategies to further reduce carbon emissions. The development of performance-enhancing and cost-reducing sludge treatment strategies is essential for carbon reductions. These require a detailed assessment of carbon implications of each process and their potential impact on sludge management, as well as opportunities for cost-effective, less carbon-intensive renewable energy generation and resource recovery.

While the switch towards on-site heat and power production has gradually reduced the carbon intensity of wastewater treatment operations, this 'decarbonization' is far from adequate to achieve the carbon reductions necessary to decarbonize the wastewater industry. Further understanding is needed on how future wastewater treatment processes and discharge criteria will affect sludge management strategies to determine the least-carbon solutions. This approach can both reduce the volume and concentrations of sludge and contaminants and reduce carbon intensity and transform conventional sludge treatment processes into low carbon intensity sustainable processes.

There is no straightforward recipe for the decarbonization of sludge treatment operations. The involvement of key stakeholders across the supply chain, including dischargers, utilities, regulators, farmers, and customers (public acceptance), is critical to determine the best sustainable sludge management practices. There is a need to develop decision-making tools to evaluate a large number of parameters and their interrelations with criteria, sub-criteria, and alternatives to select a sustainable sludge management strategy and fulfill all requirements for proper waste management. The multicriteria decision-making (MCDM) method has been mostly used in the presence of multiple, and in most cases, conflicting criteria (Đurđević *et al.*, 2020). SWOT (strengths, weaknesses, opportunities, threats) analysis has also been considered to determine an optimal route for sustainable sludge management. Recently, there has been an emerging effort using a combination of various models, such as analytic hierarchy process model with LCA and LCCA analysis (Đurđević *et al.*, 2020; Turunen *et al.*, 2018). Close interlinks between different energy production and nutrient recovery methods should be established during the selection of sludge treatment and disposal process. In this effort, pre- and post-treatment routes should also be built into the selection process. There is a need for a forecast of organic matter and nutrient material flows and expected sludge management costs based on a transparent methodology composed of different scenarios in line with circular economy and industrial ecology for agronomic applications of the final products. A detailed analysis of all the essential components, including regulatory, local, public, economic, operational, and technical aspects in municipal sludge disposal needs to be considered in this framework.

10.11 FUTURE STRATEGIES AND ROADMAPS

The sludge management law/regulations are more than 30 years old and outdated. There is a need for revisitation of applicable laws and regulations to achieve decarbonization at the utilities. Future sludge

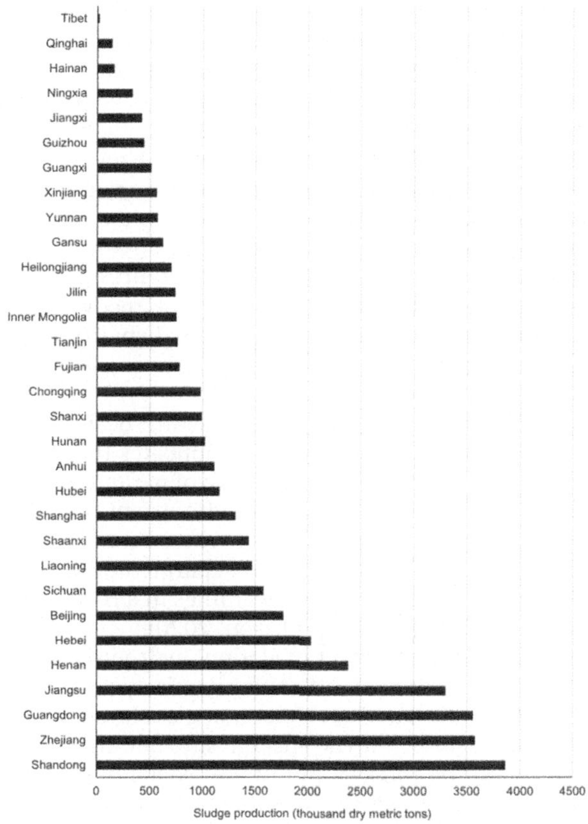

Figure 10.3 Sludge production in different provinces of China in 2019 (Wei *et al.*, 2020).

management strategies should also include the proper sludge treatment/disposal methods to minimize greenhouse emissions in addition to pathogen inactivation and concerned contaminants removal. China's sludge management strategy dramatically changed recently. As a result of 18 standards and 12 regulations promulgated in China, land application of sludge reduced from 60.9% in 2009 to 21.9% in 2017 (Wei *et al.*, 2020) (Figures 10.2 and 10.3).

Circular agreements and deals are emerging in the European Union as an alternative form of governance to meet stakeholders' needs across the value chain, hence increasing the circulation level of materials. Through such agreements, individuals are provided an opportunity to redistribute risks and responsibilities in ways more appropriate for achieving a circular economy (Johansson, 2021). Sweden recently developed a certification program for wastewater treatment plants called REVAQ to alleviate concerns about the land application of sludge for agricultural purposes. There is a continuous effort to reduce concern contaminants in wastewater streams before intake to meet REVAQ certification specifications. Approximately 45% of sludge produced in Sweden meets REVAQ quality standards set for farm applications (Dagerskog & Olsson, 2020). Another example is the implementation of the Green Deal in the Netherlands, which has promoted incinerator bottom ash as a construction aggregate (Government of the Netherlands, 2016).

To meet increasingly strict quality requirements and alleviate P shortage, sludge incineration for energy production with P recovery is becoming the most commonly applied process. In current wastewater treatment applications, most P in wastewater has been captured in sludge as a result of the chemical precipitation techniques used to meet discharge criterion for P removal. For decarbonization of sludge management practices, it is crucial to close the nutrient recovery loop (Table 10.2). Nitrogen recovery has significant potential in the decarbonization of wastewater treatment operations. Recovered nitrogen can be used as a feedstock to produce fertilizers which require very energy-intensive processes from fossil-based feedstocks. N_2O emissions resulted from wastewater treatment operations can be minimized or eliminated with this approach. However, nitrogen capture in sludge streams is relatively small (~25%) since a large portion of nitrogen ends up in treated water and the atmosphere during the wastewater treatment operations. To maximize nitrogen recovery, there is a need to identify nitrogen-rich streams. One nitrogen-rich stream is recycled water from sludge dewatering processes. Separate collection and treatment of nitrogen-rich streams also provide complete nitrogen recovery at decentralized plants. Nutrient recovery and materials production (e.g. building/construction materials, adsorbents) with minimal environmental impacts compared to their fossil-driven production routes are critical to lower GHG emissions, and hence are helpful in achieving decarbonization targets.

Centralized wastewater treatment plants have been usually designed to treat gray water with stormwater and industrial wastewater. This operation complicates the sludge management practices because sludge quality and quantity depend on a wide range of parameters. Decentralized wastewater treatment plants play a critical role in developing a closed nutrient recovery loop since they can be easily designed to allow upstream source-separation of wastewater prior to intake. A new source-separation approach has been developed in Sweden to separate and dry out the urine at the point source, thereby diverting the bulk of nitrogen from sewer lines in a relatively easy way with minimum retrofitting of the existing pipelines (Dagerskog & Olsson, 2020).

Various new alternative processes such as wet oxidation, hydrolysis, hydrothermal carbonization, and supercritical water oxidation, currently at either embryonic stage or pilot-scale, will be scaled up to large-scale applications. Implementation of such technologies should not only be looked at in terms of economic and environmental feasibility at the specific site, but also in terms of carbon emissions over the life cycle of sludge. In addition to LCA and LCCA, social impact analysis (e.g., employment, income) will be widely used tools to guide the facilities in selecting the best sludge management methods while focusing on decarbonization and sustainability goals. Useful products that recover carbon from sludge in a sustainable manner to decarbonize the water sector is the ultimate goal.

REFERENCES

Abdel Wahaab R., Mahmoud M. and van Lier J. B. (2020). Toward achieving sustainable management of municipal wastewater sludge in Egypt: the current status and future prospective. *Renewable and Sustainable Energy Reviews*, **127**, 109880, 1–11. https://doi.org/10.1016/j.rser.2020.109880

Appels L., Lauwers J., Degrève J., Helsen L., Lievens B., Willems K., Van Impe J. and Dewil R. (2011). Anaerobic digestion in global bio-energy production: potential and research challenges. *Renewable and Sustainable Energy Reviews*, **15**(9), 4295–4301. https://doi.org/10.1016/j.rser.2011.07.121

Arias A., Feijoo G. and Moreira M. T. (2021). Benchmarking environmental and economic indicators of sludge management alternatives aimed at enhanced energy efficiency and nutrient recovery. *Journal of Environmental Management*, **279**(1), 111594, 1–10. https://doi.org/10.1016/j.jenvman.2020.111594

Bożym M. and Siemiątkowski G. (2018). Characterization of composted sewage sludge during the maturation process: a pilot scale study. *Environmental Science and Pollution Research*, **25**(34), 34332–34342. https://doi.org/10.1007/s11356-018-3335-x

Brown S. and Henry C. (2001). Using Biosolids for Reclamation/Remediation of Disturbed Soils. University of Washington, pp. 1–26. Available at https://www.epa.gov/sites/default/files/2015-05/documents/biosolidswhitepaper-uwash.pdf (Accessed 23 February 2022).

Bruni C., Akyol Ç., Cipolletta G., Eusebi A. L., Caniani D., Masi S., Colón J. and Fatone F. (2020). Decentralized community composting: past, present and future aspects of Italy. *Sustainability*, **12**(8), 3319, 1–20. https://doi.org/10.3390/su12083319

Callegari A. and Capodaglio A. (2018). Properties and beneficial uses of (bio)chars, with special attention to products from sewage sludge pyrolysis. *Resources*, **7**(1), 20, 1–22. https://doi.org/10.3390/resources7010020

Cao X., Ma R., Zhang Q., Wang W., Liao Q., Sun S., Zhang P. and Liu X. (2020). The factors influencing sludge incineration residue (SIR)-based magnesium potassium phosphate cement and the solidification/stabilization characteristics and mechanisms of heavy metals. *Chemosphere*, **261**, 127789, 1–9. https://doi.org/10.1016/j.chemosphere.2020.127789

Chen W.-T., Haque Md. A., Lu T., Aierzhati A. and Reimonn G. (2020). A perspective on hydrothermal processing of sewage sludge. *Current Opinion in Environmental Science & Health*, **14**, 63–73. https://doi.org/10.1016/j.coesh.2020.02.008

Chow W. L., Chong S., Lim J. W., Chan Y. J., Chong M. F., Tiong T. J., Chin J. K. and Pan G.-T. (2020). Anaerobic co-digestion of wastewater sludge: a review of potential co-substrates and operating factors for improved methane yield. *Processes*, **8**(1), 39, 1–21. https://doi.org/10.3390/pr8010039

Cieślik B. and Konieczka P. (2017). A review of phosphorus recovery methods at various steps of wastewater treatment and sewage sludge management. The concept of 'no solid waste generation' and analytical methods. *Journal of Cleaner Production*, **142**, 1728–1740. https://doi.org/10.1016/j.jclepro.2016.11.116

Conti F., Toor S. S., Pedersen T. H., Seehar T. H., Nielsen A. H. and Rosendahl L. A. (2020). Valorization of animal and human wastes through hydrothermal liquefaction for biocrude production and simultaneous recovery of nutrients. *Energy Conversion and Management*, **216**, 112925, 1–11. https://doi.org/10.1016/j.enconman.2020.112925

Dagerskog L. and Olsson O. (2020). Swedish Sludge Management at the Crossroads, Report, Stockholm Environment Institute, Stockholm, Sweden.

Dimitriadis A. and Bezergianni S. (2017). Hydrothermal liquefaction of various biomass and waste feedstocks for biocrude production: a state of the art review. *Renewable and Sustainable Energy Reviews*, **68**, 113–125. https://doi.org/10.1016/j.rser.2016.09.120

Dubey M., Mohapatra S., Tyagi V. K., Suthar S. and Kazmi A. A. (2021). Occurrence, fate, and persistence of emerging micropollutants in sewage sludge treatment. *Environmental Pollution*, **273**, 116515, 1–18. https://doi.org/10.1016/j.envpol.2021.116515

Đurđević D., Blecich P. and Jurić Ž. (2019). Energy recovery from sewage sludge: the case study of Croatia. *Energies*, **12**(10), 1927, 1–19. https://doi.org/10.3390/en12101927

Đurđević D., Trstenjak M. and Hulenić I. (2020). Sewage sludge thermal treatment technology selection by utilizing the analytical hierarchy process. *Water*, **12**(5), 1–16. https://doi.org/10.3390/w12051255

European Commission (2020). Waste Streams: Sewage Sludge. Available at https://ec.europa.eu/environment/topics/waste-and-recycling/sewage-sludge_en (accessed 29 September 2021)

Eurostat (2020). Sewage Sludge Production and Disposal. Available at https://ec.europa.eu/eurostat/databrowser/view/env_ww_spd/default/table?lang=en (accessed 29 September 2021)

Falk J., Skoglund N., Grimm A. and Öhman M. (2020). Fate of phosphorus in fixed bed combustion of biomass and sewage sludge. *Energy & Fuels*, **34**(4), 4587–4594. https://doi.org/10.1021/acs.energyfuels.9b03976

Flores-Alsina X., Ramin E., Ikumi D., Harding T., Batstone D., Brouckaert C., Sotemann S. and Gernaey K. V. (2021). Assessment of sludge management strategies in wastewater treatment systems using a plant-wide approach. *Water Research*, **190**, 116714, 1–15. https://doi.org/10.1016/j.watres.2020.116714

Government of the Netherlands (2016). Green deals overview, Report, Government of the Netherlands, Netherlands. Available at https://www.greendeals.nl/sites/default/files/uploads/2016/03/Progress_report_Green_Deals_ENG.pdf (accessed 29 September 2021)

Gretzschel O., Schäfer M., Steinmetz H., Pick E., Kanitz K. and Krieger S. (2020). Advanced wastewater treatment to eliminate organic micropollutants in wastewater treatment plants in combination with energy-efficient electrolysis at WWTP Mainz. *Energies*, **13**(14), 3599, 1–27. https://doi.org/10.3390/en13143599

Hudcová H., Vymazal J. and Rozkošný M. (2019). Present restrictions of sewage sludge application in agriculture within the European Union. *Soil and Water Research*, **14**(2), 104–120. https://doi.org/10.17221/36/2018-SWR

ITRC (2010). Chemical Stabilization: Phosphate and Biosolids Treatment. Mining Waste Treatment Technology Selection. Available at https://frtr.gov/matrix/documents/Solidification-and-Stabilization/2010-Chemical-Stabilization-Phosphate-and-Biosolids-Treatment.pdf (accessed 29 September 2021)

Jędrczak A. (2018). Composting and fermentation of biowaste – advantages and disadvantages of processes. *Civil and Environmental Engineering Reports*, **28**(4), 71–87. https://doi.org/10.2478/ceer-2018-0052

Johansson N. (2021). Circular agreements – exploring the role of agreements and deals as a political tool for a circular economy. *Circular Economy and Sustainability*, **1**, 499–505. https://doi.org/10.1007/s43615-021-00004-5

Kacprzak M., Neczaj E., Fijałkowski K., Grobelak A., Grosser A., Worwag M., Rorat A., Brattebo H., Almås Å. and Singh B. R. (2017). Sewage sludge disposal strategies for sustainable development. *Environmental Research*, **156**, 39–46. https://doi.org/10.1016/j.envres.2017.03.010

Krüger O. and Adam C. (2015). Recovery potential of German sewage sludge ash. *Waste Management*, **45**, 400–406. https://doi.org/10.1016/j.wasman.2015.01.025

Laurich, F. (2011). Hamburg Wasser's efforts for a self-sufficient energy supply of Kohlbrandhoft/Dradenau WWTP. International Water Week, Amsterdam.

Law K. P. and Pagilla K. R. (2018). Phosphorus recovery by methods beyond struvite precipitation. *Water Environment Research*, **90**(9), 840–850. https://doi.org/10.2175/106143017X15131012188006

Law K. P. and Pagilla K. R. (2019). Reclaimed phosphorus commodity reserve from water resource recovery facilities – a strategic regional concept towards phosphorus recovery. *Resources, Conservation and Recycling*, **150**, 104429, 1–12. https://doi.org/10.1016/j.resconrec.2019.104429

Lee E., Oliveira D. S. B. L., Oliveira L. S. B. L., Jimenez E., Kim Y., Wang M., Ergas S. J. and Zhang Q. (2020). Comparative environmental and economic life cycle assessment of high solids anaerobic co-digestion for biosolids and organic waste management. *Water Research*, **171**, 115443, 1–12. https://doi.org/10.1016/j.watres.2019.115443

Mills N., Pearce P., Farrow J., Thorpe R. B. and Kirkby N. F. (2014). Environmental and economic life cycle assessment of current and future sewage sludge to energy technologies. *Waste Management*, **34**(1), 185–195. https://doi.org/10.1016/j.wasman.2013.08.024

Mininni G., Mauro E., Piccioli B., Colarullo G., Brandolini F. and Giacomelli P. (2019). Production and characteristics of sewage sludge in Italy. *Water Science and Technology*, **79**(4), 619–626. https://doi.org/10.2166/wst.2019.064

Mu'azu N. D., Essa M. H., Aga O. and Jarrah N. (2019). Life cycle assessment approach to sustainable sewage sludge management for water pollution control. *Journal of Physics: Conference Series*, **1349**, 012145, 1–8. https://doi.org/10.1088/1742-6596/1349/1/012145

MWRD (2021). Biosolids Utilization. Available at https://legacy.mwrd.org/irj/go/km/docs/documents/MWRD/internet/protecting_the_environment/Water_Reclamation_Plants/htm/Biosolids_Disposal.htm (accessed 29 September 2021)

Nakkasunchi S., Hewitt N. J., Zoppi C. and Brandoni C. (2021). A review of energy optimization modelling tools for the decarbonisation of wastewater treatment plants. *Journal of Cleaner Production*, **279**, 123811, 1–20. https://doi.org/10.1016/j.jclepro.2020.123811

OECD.Stat (2021). Wastewater Treatment (% Population Connected). Available at https://stats.oecd.org/Index.aspx?DataSetCode=WATER_TREAT (accessed 29 September 2021)

Onwosi C. O., Igbokwe V. C., Odimba J. N., Eke I. E., Nwankwoala M. O., Iroh I. N. and Ezeogu L. I. (2017). Composting technology in waste stabilization: on the methods, challenges and future prospects. *Journal of Environmental Management*, **190**, 140–157. https://doi.org/10.1016/j.jenvman.2016.12.051

Przydatek G. and Wota A. K. (2020). Analysis of the comprehensive management of sewage sludge in Poland. *Journal of Material Cycles and Waste Management*, **22**(1), 80–88. https://doi.org/10.1007/s10163-019-00937-y

Righi S., Oliviero L., Pedrini M., Buscaroli A. and Della Casa C. (2013). Life cycle assessment of management systems for sewage sludge and food waste: centralized and decentralized approaches. *Journal of Cleaner Production*, **44**, 8–17. https://doi.org/10.1016/j.jclepro.2012.12.004

Rosiek K. (2020). Directions and challenges in the management of municipal sewage sludge in Poland in the context of the circular economy. *Sustainability*, **12**(9), 3686, 1–28. https://doi.org/10.3390/su12093686

Roskosch A. and Heidecke P. (2018). Sewage Sludge Disposal in the Federal Republic of Germany, Report, German Environment Agency, Dessau-Roßlau, Germany. Available at https://www.umweltbundesamt.de/sites/default/files/medien/1410/publikationen/190116_uba_fb_klaerschlamm_engl_bf.pdf (accessed 29 September 2021)

Sánchez Ó. J., Ospina D. A. and Montoya S. (2017). Compost supplementation with nutrients and microorganisms in composting process. *Waste Management*, **69**, 136–153. https://doi.org/10.1016/j.wasman.2017.08.012

Scarsella M., de Caprariis B., Damizia M. and De Filippis P. (2020). Heterogeneous catalysts for hydrothermal liquefaction of lignocellulosic biomass: a review. *Biomass and Bioenergy*, **140**, 105662, 1–15. https://doi.org/10.1016/j.biombioe.2020.105662

Schäfer M., Gretzschel O. and Steinmetz H. (2020). The Possible Roles of Wastewater Treatment Plants in Sector Coupling. *Energies*, **13**(8), 2088, 1–20. https://doi.org/10.3390/en13082088

Seiple T. E., Skaggs R. L., Fillmore L. and Coleman A. M. (2020). Municipal wastewater sludge as a renewable, cost-effective feedstock for transportation biofuels using hydrothermal liquefaction. *Journal of Environmental Management*, **270**, 110852, 1–9. https://doi.org/10.1016/j.jenvman.2020.110852

Shaddel S., Bakhtiary-Davijany H., Kabbe C., Dadgar F. and Østerhus S. (2019). Sustainable sewage sludge management: from current practices to emerging nutrient recovery technologies. *Sustainability*, **11**(12), 3435, 1–13. https://doi.org/10.3390/su11123435

Shen Y., Linville J. L., Urgun-Demirtas M., Mintz M. M. and Snyder S. W. (2015). An overview of biogas production and utilization at full-scale wastewater treatment plants (WWTPs) in the United States: challenges and opportunities towards energy-neutral WWTPs. *Renewable and Sustainable Energy Reviews*, **50**, 346–362. https://doi.org/10.1016/j.rser.2015.04.129

Silva Thomsen L. B., Carvalho P. N., dos Passos J. S., Anastasakis K., Bester K. and Biller P. (2020). Hydrothermal liquefaction of sewage sludge; energy considerations and fate of micropollutants during pilot scale processing. *Water Research*, **183**, 116101, 1–8. https://doi.org/10.1016/j.watres.2020.116101

Singh S., Kumar V., Dhanjal D. S., Datta S., Bhatia D., Dhiman J., Samuel J., Prasad R. and Singh J. (2020). A sustainable paradigm of sewage sludge biochar: valorization, opportunities, challenges and future prospects. *Journal of Cleaner Production*, **269**, 122259, 1–16. https://doi.org/10.1016/j.jclepro.2020.122259

Smol M., Adam C. and Anton Kugler S. (2020). Inventory of Polish municipal sewage sludge ash (SSA) – mass flows, chemical composition, and phosphorus recovery potential. *Waste Management*, **116**, 31–39. https://doi.org/10.1016/j.wasman.2020.07.042

Stillwell A., Hoppock D. and Webber M. (2010). Energy recovery from wastewater treatment plants in the United States: a case study of the energy-water nexus. *Sustainability*, **2**(4), 945–962. https://doi.org/10.3390/su2040945

Świerczek L., Cieślik B. M. and Konieczka P. (2021). Challenges and opportunities related to the use of sewage sludge ash in cement-based building materials – a review. *Journal of Cleaner Production*, **287**, 125054, 1–12. https://doi.org/10.1016/j.jclepro.2020.125054

Tsybina A. and Wuensch C. (2018). Analysis of sewage sludge thermal treatment in the contect of circular economy. *Detritus*, **2**(1), 1–15. https://doi.org/10.31025/2611-4135/2018.13668

Turunen V., Sorvari J. and Mikola A. (2018). A decision support tool for selecting the optimal sewage sludge treatment. *Chemosphere*, **193**, 521–529. https://doi.org/10.1016/j.chemosphere.2017.11.052

US EPA (1992). Technical Support Document for Reduction of Pathogens and Vector Sludge Attraction in Sewage Sludge, Report EPA 822/R-93–004, Office of Water, US EPA, Washington, DC, USA. Available at https://www.epa.gov/sites/production/files/2019-12/documents/tsd-reduction-pathogens-vector-1992.pdf (accessed 29 September 2021)

US EPA (1995). Part 503 Implementation Guidance, Report EPA 833-R-95–001, Office of Water, US EPA, Washington, DC, USA. Available at https://www3.epa.gov/npdes/pubs/owm0237.pdf (accessed 29 September 2021)

US EPA. (2003a). Biosolids Technology Fact Sheet, Fact Sheet No. 832-F-03–012, pp. 1–8). Office of Water, US EPA, Washington, DC, USA. Available at https://www.epa.gov/sites/default/files/2018-11/documents/biosolids-technology-factsheet.pdf (Accessed 23 February 2022).

US EPA. (2003b). Environmental Regulations and Technology: Control of Pathogens and Vector Attraction in Sewage Sludge (625/R-92/013; pp. 1–186). US EPA Office of Research and Development, US EPA, Washington, DC, USA. Available at https://www.epa.gov/sites/production/files/2015-07/documents/epa-625-r-92-013.pdf (Accessed 23 February 2022).

US EPA (2019). Basic Information about Biosolids [Other Policies and Guidance]. Available at https://www.epa.gov/biosolids/basic-information-about-biosolids (accessed 29 September 2020)

Walker J., Knight L. and Stein L. (1994). A Plain English Guide to the EPA Part 503 Biosolids Rule, Report EPA/832/R-93/003, Office of Wastewater Management, US EPA, Washington, DC, USA. Available at https://www.epa.gov/sites/production/files/2018-12/documents/plain-english-guide-part503-biosolids-rule.pdf (accessed 29 September 2021)

Watson J., Wang T., Si B., Chen W.-T., Aierzhati A. and Zhang Y. (2020). Valorization of hydrothermal liquefaction aqueous phase: pathways towards commercial viability. *Progress in Energy and Combustion Science*, **77**, 100819, 1–45. https://doi.org/10.1016/j.pecs.2019.100819

Wei L., Zhu F., Li Q., Xue C., Xia X., Yu H., Zhao Q., Jiang J. and Bai S. (2020). Development, current state and future trends of sludge management in China: based on exploratory data and CO_2-equivalent emissions analysis. *Environment International*, **144**, 106093, 1–12. https://doi.org/10.1016/j.envint.2020.106093

Wiechmann B., Dienemann C., Kabbe C., Simone B., Ines V. and Andrea R. (2013). Sewage Sludge Management in Germany, Report. Umweltbundesamt, Germany.

Xu Y., Lu Y., Zheng L., Wang Z. and Dai X. (2020). Perspective on enhancing the anaerobic digestion of waste activated sludge. *Journal of Hazardous Materials*, **389**, 121847, 1–16. https://doi.org/10.1016/j.jhazmat.2019.121847

Ye S., Zeng G., Wu H., Liang J., Zhang C., Dai J., Xiong W., Song B., Wu S. and Yu J. (2019). The effects of activated biochar addition on remediation efficiency of co-composting with contaminated wetland soil. *Resources, Conservation and Recycling*, **140**, 278–285. https://doi.org/10.1016/j.resconrec.2018.10.004

Yoshida H., ten Hoeve M., Christensen T. H., Bruun S., Jensen L. S. and Scheutz C. (2018). Life cycle assessment of sewage sludge management options including long-term impacts after land application. *Journal of Cleaner Production*, **174**, 538–547. https://doi.org/10.1016/j.jclepro.2017.10.175

Zhang Q., Zhang L., Guo B. and Liu Y. (2020). Mesophiles outperform thermophiles in the anaerobic digestion of blackwater with kitchen residuals: insights into process limitations. *Waste Management*, **105**, 279–288. https://doi.org/10.1016/j.wasman.2020.02.018

doi: 10.2166/9781789061796

Chapter 11

Decarbonization potentials in intensified water and wastewater systems using membrane-related technologies

*Boyan Xu, Shujuan Huang, Chuansheng Wang, Tze Chiang Albert Ng and How Yong Ng**

National University of Singapore Environmental Research Institute, 5A Engineering Drive 1, 117411, Singapore
**Correspondence: howyongng@nus.edu.sg*

11.1 INTRODUCTION

The foremost role of the wastewater treatment plants (WWTPs) is to remove pollutants in wastewater to produce a high-quality effluent in order to protect the environment, safeguard human health and/or achieve water reuse. However, it may be insufficient for some conventional WWTPs to provide qualified or reusable water for industrial or domestic recycling. For example, conventional activated sludge (CAS) could poorly remove emerging micro-pollutants in wastewater such as pharmaceutically active compounds (PhACs) owing to the rapid developing pharmaceutical industries (Radjenović *et al*., 2009). To meet the more stringent effluent discharge standards and the increasing need of wastewater reclamation (Melin *et al*., 2006), additional treatment processes are called for to polish the produced wastewater by removing undesirable compounds. Membrane technology, which is at least half a century old, can selectively remove undesirable components over a wide range of molecular weights. In addition to the high-quality effluent, the small footprint was another advantage of membrane technologies, especially considering the growing tension in urban land use. For example, membrane bioreactors (MBRs) technology (Ng & Ng, 2010), in which membranes were immerged into activated sludge, could achieve high biochemical efficacy and high quality of treated wastewater.

The increasing demand and consumption of freshwater due to development and further societal growth leads to an insufficient availability of freshwater for many countries worldwide. Therefore, desalination processes which extract portable water from non-conventional sources such as seawater and brackish water become necessary, especially for some water-stressed countries such as Singapore. Historically, it has been performed through evaporation using thermal energy, and in the late 1970s the reverse osmosis (RO) membranes started to be implemented for seawater desalination (Figure 11.1a) (Judd, 2017). The RO technology is not only light-weight, highly compact and productive, but also less energy intensive than thermal desalination such as multi-stage flash (MSF), multiple-effect distillation (MED) and thermal vapor compression (TVC), thus leading to the replacement of thermal desalination technologies in many parts of the world (Ali *et al*., 2018). In general, it could thus be said that the

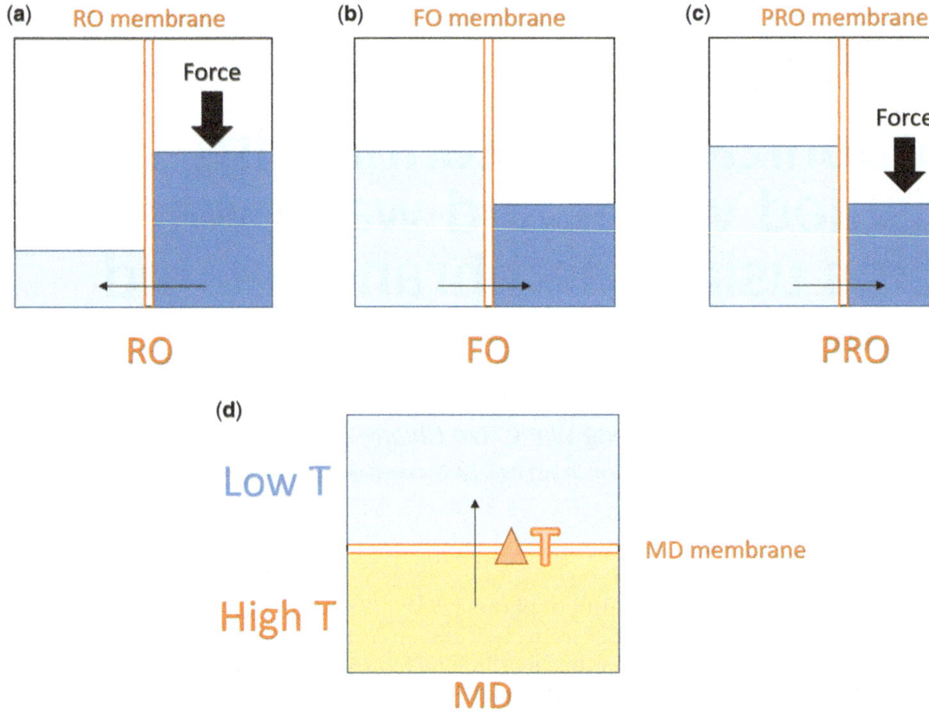

Figure 11.1 Configuration and principles of (a) RO, (b) FO, (c) PRO and (d) MD. Black arrows represent the water flow direction.

membrane technologies have been widely implemented in municipal water sectors (i.e., wastewater treatment and desalination plants).

According to the United Nations Food and Agriculture Organization (UN-FAO), more than 26 billion m^3 of domestic wastewater was treated in 2009. Moreover, wastewater treatment accounts for about 3% of the United States total electrical consumption annually (USEPA, 2006), and it is at average of 0.3% for China (Hao *et al.*, 2015). The generation of electricity required to drive the WWTPs operation in China could result in the production of more than 110 million tons of CO_2 annually (Hao *et al.*, 2015). Therefore, because of increasing energy cost, tremendous fossil fuel consumption and climate change, WWTPs including those using membrane technologies should be designed for improving energy efficiency and consider resource recovery as a key performance indicator. Considering that the membrane technologies used in wastewater treatment could be more energy intensive as compared to conventional wastewater treatment processes (e.g., MBRs versus CAS) (Mannina *et al.*, 2020), it could increase the carbon footprint and energy consumption in wastewater applications. Moreover, global desalination of seawater and brackish water via RO could contribute more than 50% of drinking water production (Judd, 2017). Although RO is the most economical technology for seawater desalination at the commercial scale as compared to the thermal technologies, and has achieved a decline in production cost from \$4.5 m^{-3} in 1997 to ~\$1.5 m^{-3} in 2000 for seawater reverse osmosis (SWRO) with the help of the pressure exchanger (PX) as energy recovery devices (ERDs) (Judd, 2017), it still requires a specific energy ranging between 3 and 4 kWh/m^3 which is more than double of the theoretical energy requirement (i.e., 1.06 kWh/m^3 for seawater with the salt concentration of 35 000 ppm and at a 50% recovery) (Ali *et al.*, 2018). Thus, without doubt, the current high energy demands from wastewater treatment and desalination need to be reduced to achieve both economically and environmentally sustainable water/wastewater treatment processes.

Efforts by the water industry and scientific community in research led to the development of some novel membrane processes, and an integration of membrane technologies to conventional systems could have much potential for energy saving, resource recovery and decarbonization. Therefore, it is of obvious interest, and it is necessary to summarize these latest developments in achieving energy self-sufficiency, and even carbon neutrality, by using novel membrane technologies for wastewater treatment/reclamation and desalination systems. Especially, novel membrane technologies that could reduce energy consumption with high decarbonization potential are discussed for wastewater treatment, including aerobic granular sludge membrane bioreactors (AGMBRs), algae membrane bioreactors (A-MBRs), anaerobic membrane bioreactors (AnMBRs), membrane biofilm reactors (MBfRs) and forward osmosis (FO) integrated processes. Moreover, this chapter also includes the membrane technologies for desalination with decarbonization potential, consisting of pressure retarded osmosis (PRO), forward osmosis-reverse osmosis (FO-RO) hybrid and forward osmosis-membrane distillation hybrid (FO-MD).

11.2 MEMBRANE STRATEGIES FOR DECARBONIZATION IN WASTEWATER TREATMENT AND RESOURCE RECOVERY

11.2.1 Aerobic granular sludge membrane bioreactors (AGMBRs)

The aerobic granular sludge membrane bioreactor (AGMBR) is a novel and promising technology for wastewater treatment/reclamation, which combines the aerobic granular sludge (AGS) and the membrane filtration to simultaneously remove organics and nutrients in wastewater (Chen *et al.*, 2017; Iorhemen *et al.*, 2019; Li *et al.*, 2019). Comparing to CAS in conventional MBRs, AGS has a denser structure, larger particle size and better settleability, which is beneficial to diminish membrane fouling in AGMBR (e.g., pore blocking and cake layer formation). Besides, Zhang *et al.* (2020) revealed that the scouring effect of AGS on the membrane surface also played a crucial role in alleviating membrane fouling due to its intrinsic particulate properties. Thus, compared to conventional MBRs, AGMBR could save energy for membrane fouling control to reduce the carbon footprint. Moreover, AGMBR coupled with the reverse osmosis (AGSMBR-RO) process was proposed for municipal wastewater reclamation, and the energy demand of AGSMBR-RO could be as low as 0.79 kWh/m^3, which is significantly lower than the range of 1.15–2.0 kWh/m^3 incurred in conventional municipal wastewater reclamation processes (Wang *et al.*, 2020). Meanwhile, membrane filtration of AGMBR is able to remove the all the suspended solid of AGS, and thus improves effluent quality (Liébana *et al.*, 2018).

To date, two types of AGMBRs have been widely studied according to operational modes, including batch mode and continuous mode. It is known that the sequencing batch reactor (SBR) is the ideal configuration for granules cultivation and long-term stability maintenance due to its alternant feast/famine condition, hydraulic selection pressure and strong hydraulic shear force. Therefore, the batch mode using SBR followed by membrane filtration has been verified as an effective configuration of AGMBR to mitigate membrane fouling and maintain long-term stability of granules. The transmembrane pressure (TMP) increasing rates were mostly lower than 0.3 kPa/d, which were 50–90% lower compared to those of the control MBRs (Li *et al.*, 2019; Thanh *et al.*, 2013; Truong *et al.*, 2018). Besides, the AGMBRs also showed good total nitrogen (TN) removal efficiency with higher than 55%, which was attributed to the simultaneous nitrification and denitrification (SND) due to the various microenvironments inside the granules (Li *et al.*, 2019; Thanh *et al.*, 2013; Vijayalayan *et al.*, 2014). However, it should be noted that AGMBR operated in the continuous mode is more attractive for application purposes, due to its reduced capital cost (CAPEX) and operational cost (OPEX) (Chen *et al.*, 2017).

However, cultivation and stability maintenance of granules under the continuous mode are still challenging due to its different hydraulic and operational condition, as compared to those of SBR, which limits the wide application of AGMBR operated under the continuous mode (Chen *et al.*, 2017; Corsino *et al.*, 2016; Li *et al.*, 2014). Recently, a few researchers have attempted to overcome this challenge by recreating or simulating similar hydraulic conditions as those in SBRs in a continuous configuration (Chen *et al.*, 2017; Corsino *et al.*, 2016). An internal circulation AGMBR, which was achieved by driving liquid downward in the anoxic zone and upward in the aerobic MBR zone, was designed to mimic the hydraulic condition in SBR to cultivate AGS in a continuous flow bioreactor by Chen *et al.* (2017). The granules were successfully cultivated after 35 days with an average particle size of 0.228 mm; however, the membrane filtration behavior was not considered in this study. Corsino *et al.* (2016) attempted to recreate the hydraulic conditions in SBR to cultivate granules in a novel hydrodynamic configuration of AGMBR. The results showed that the membrane filtration performance of the AGMBR was improved 90% as compared to a conventional MBR. However, the granules disintegrated in less than 20 days under the continuous operation mode. The information and clear mechanisms regarding the granules cultivation in continuous membrane reactors is still limited due to the complex mechanism of granulation, and therefore the cultivation and stability of granules is the key challenge for the scaling-up of the low-carbon-footprint AGMBRs.

11.2.2 Algae membrane bioreactors (A-MBRs)

Microalgae has been widely used in wastewater treatment/reclamation due to its distinctive capacity of transferring the nitrogen and phosphorus in wastewater to algal biomass. However, its poor settleability that leads to biomass loss and deteriorates effluent quality limits its application (Tang & Hu, 2016). Recently, algae membrane bioreactor (A-MBR), in which membrane filtration is used to overcome poor settleability of algae, has drawn much attention due to the increasingly stringent wastewater discharge standard.

In the wastewater treatment process, A-MBR is mostly used as a polishing step to remove the residual nutrients (i.e., nitrogen and phosphorus) from the effluent of the conventional biological treatment processes (Low *et al.*, 2016; Tang & Hu, 2016; Winkler & Straka, 2019). Meanwhile, high biomass production of algae can be achieved due to the complete retention of algae by membranes, in which value-added products such as biofuel, animal food and other bioproducts can be generated (Drexler & Yeh, 2014; Tang & Hu, 2016). Although membrane fouling resulting from the algogenic organic matters (AOMs) secreted by algae was a challenge for A-MBRs, the integration of wastewater treatment and biofuel production could be achieved simultaneously, and the algal biofuel was suggested to dominate 75% of the market share in the future (Nhat *et al.*, 2018). Of note, capture of CO_2, which is converted to algal biomass could be achieved, and the algal CO_2 capture rate could be as high as 91.59% for the treatment of terephthalic acid wastewater (Yang *et al.*, 2020). When compared to traditional algae cultivation systems (e.g., high-rate algal pond and photobioreactors), the A-MBR significantly improved the algae biomass yield and nutrient removal efficiencies (Bilad *et al.*, 2014; Drexler & Yeh, 2014; Tang & Hu, 2016).

Recently, a symbiotic system of algae-sludge MBR (AS-MBR) was proposed to improve the nutrients removal and mitigate membrane fouling (Sun *et al.*, 2018a, 2018b). The produced oxygen by algae during the photosynthesis process would be utilized by bacteria to degrade organic pollutants and oxide ammonium. In turn, the carbon dioxide generated by bacteria could be used by algae to synthesize algal biomass (Luo *et al.*, 2017b). Liang *et al.* (2013) reported that the presence of certain bacteria (e.g., *Bacillus licheniformis*) may create favorable conditions to promote the growth of *Chlorella.vulgaris* (i.e., a species of green microalga), further advancing the symbiotic relationship. Sun *et al.* (2018b) found that the removal efficiencies of total nitrogen and phosphate increased by 10 and 8% with the introduction of algae to MBR, respectively. Besides, the membrane filtration performance was improved by 50%, which was attributed to the inhibition of the filamentous bacteria and lower extracellular polymeric substances (EPS) concentration in the AS-MBR. Although AS-MBR is a promising technology for nutrient removal and membrane fouling mitigation to reduce energy consumption and carbon footprint, it should be noted that the performance of the AS-MBR would be affected by various factors such as the ratio of algae to biomass, influent wastewater characteristics and operation conditions (Sun *et al.*, 2018b).

11.2.3 Anaerobic membrane bioreactor (AnMBRs)

Mainstream anaerobic treatment has received considerable attention (Chernicharo *et al.*, 2015; Smith *et al.*, 2012) because it eliminates aeration for biomass, recovers energy in the form of methane-rich biogas and produces significantly less biomass than aerobic systems. Anaerobic membrane bioreactor (AnMBR) is an advanced wastewater treatment technology that combines the anaerobic process and membrane filtration technology. In comparison to conventional anaerobic processes, AnMBR, in which membranes are immersed into the anaerobic biomass or membranes are connected externally to the anaerobic bioreactors, has the advantages of better-quality effluent, lower biomass production and lower footprint (Lei *et al.*, 2018; Smith *et al.*, 2014; Xu *et al.*, 2020a). Moreover, due to decoupling of hydraulic retention time (HRT) and solid retention time (SRT), AnMBRs can maintain a longer SRT and a higher organic load in the bioreactors to effectively prevent methanogens loss and improve methane production efficiency. Therefore, AnMBR technology is currently recognized as a promising energy-positive technology to achieve the energy balance of wastewater treatment.

Pilot-scale AnMBR studies treating low-organic strength domestic wastewater in the past decade have reported that gaseous methane yield was around 0.1–0.3 L CH_4/g $COD_{removed}$ (Lim *et al.*, 2019; Robles *et al.*, 2020; Shin & Bae, 2018). Thus, net energy balance could theoretically be improved to make AnMBRs energy positive when applying waste-to-energy process, that is, converting biogas into electricity and heat as resource recovery (Figure 11.2) (Shin & Bae, 2018; Smith *et al.*, 2014).

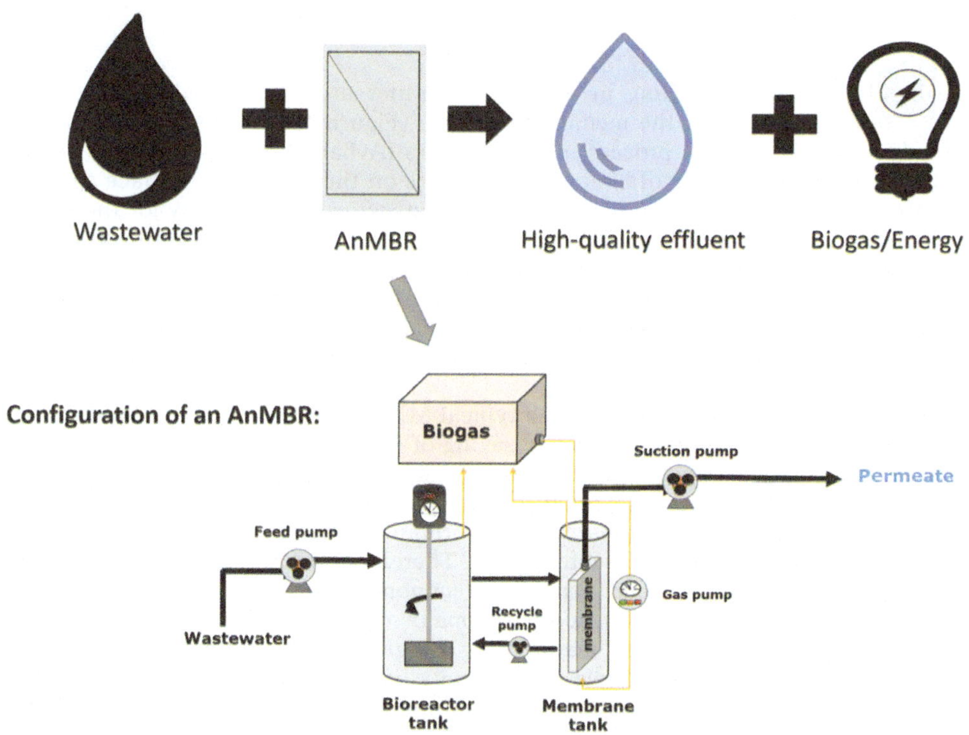

Figure 11.2 Schematic process of potential energy recovery from wastewater by using AnMBR (adopted from Shin & Bae, 2018) and a configuration of an AnMBR.

Yet, AnMBRs for treatment of domestic wastewater may not generate a high methane yield due to the low organic loadings and high portion of dissolution of methane in the effluents (Crone *et al.*, 2016). For high-strength wastewater treatment, AnMBR presented a high possibility to achieve energy self-sufficiency by converting waste to energy. For example, the methane content of biogas produced by an AnMBR treating landfill leachate ranged from 70 to 90% with a methane yield of 0.34 L/g $COD_{removed}$ (Xie *et al.*, 2014). Moreover, as for food waste digested by AnMBRs, net energy benefit could be potentially obtained, which could be due to the methane-rich biogas production of 0.21 ± 0.1 L CH_4/g $COD_{removed}$ (Galib *et al.*, 2016; Jeong *et al.*, 2017).

In general, AnMBRs have the potential to be an energy-positive technology for wastewater treatment/reclamation and water reclamation. However, serious issues such as serious membrane fouling (Robles *et al.*, 2012), lack of nutrients removal and high dissolved methane in the effluent (Crone *et al.*, 2016) have to be addressed in energy-efficient and cost-effective ways in order for the worldwide full-scale applications of AnMBRs. Instead of energy-intensive biogas sparging, more energy efficient AnMBRs may adopt proper strategies for fouling control, including physical (e.g., rotating membranes), chemical (e.g., NaClO) and biological methods (e.g., quorum quenching) (Shin & Bae, 2018; Xu *et al.*, 2020b; Yue *et al.*, 2018). Dissolved methane could be removed by using a hollow fiber membrane contactor which could achieve an average removal efficiency of more than 70% (Lim *et al.*, 2019). Novel hybrid technology like forward osmosis (FO)-AnMBR could help reduce dissolved methane in the effluent (Chen *et al.*, 2014). Noticeably, coupling anaerobic ammonium oxidation (anammox) with nitrite/nitrate-dependent anaerobic methane oxidation (n-DAMO) simultaneously removed up to 85% dissolved methane and more than 99% nitrogen from synthetic anaerobic effluent (Liu *et al.*, 2020), which would thus reduce greenhouse gas (i.e., CH_4) emission as well as the external carbon addition for the post-treatment of AnMBR's effluent.

11.2.4 Membrane biofilm reactors (MBfRs)

Membrane biofilm reactors (MBfRs), in which membranes are pressurized to supply a gaseous substrate to a biofilm formed on the membrane surface (Figure 11.3), have gained much attention as a sustainable water treatment process in recent years (Aybar *et al.*, 2014; Hasar, 2009; Martin & Nerenberg, 2012). The majority of biomass in MBfR is on the membrane surface as a functional layer of biofilm, in which unique microbial community structures could allow for the simultaneous removal of organics and nitrogen from wastewater. Moreover, in MBfR, high gas transfer rates could be achieved to save more energy as compared with the energy-intensive aeration, especially when high gas supply pressure is used, and thus a smaller size tank for biological process is required (i.e., low CAPEX) (Aybar *et al.*, 2014; Hasar, 2009; Martin & Nerenberg, 2012). Furthermore, MBfRs limits the release of volatile organic compounds (VOCs) and greenhouse gases from the bioreactors because it does not have intensive aeration bubbles.

Membrane-aerated biofilm reactor (MABR) is a typical MBfR, where oxygen is directly delivered via membrane to the functional biofilm. An energy saving of 40–75% could be achieved by MABRs due to the more efficient oxygen transfer, as compared to the conventional CAS (Martin & Nerenberg, 2012). For example, the energy consumption for biological operation by using MABR could be as low as 0.212 kWh/m^3 with excellent treatment performance (i.e., removal of 96.8% of TSS, 94.8% of NH_4 and 98.9% BOD) (Tirosh & Shechter, 2020). Therefore, MABRs showed a great potential to decrease the environmental impacts and increase the economic sustainability of wastewater treatment plants. However, membrane replacement cost is a major obstacle for its commercial development (Martin & Nerenberg, 2012). Moreover, maintaining a highly active biofilm in MABR could still be a challenge, as proper strategies to couple flow velocity or turbulence with moderate aeration rate should be implemented to keep the biofilm with optimum thicknesses (Martin & Nerenberg, 2012). Noticeably, quorum quenching was employed in an MABR, and it was demonstrated to control the EPS content as well as the thickness of biofilm via the degradation of acyl homoserine lactone (AHL) signal molecules (Taşkan *et al.*, 2020). To make MABR a cost-effective decarbonized technology for

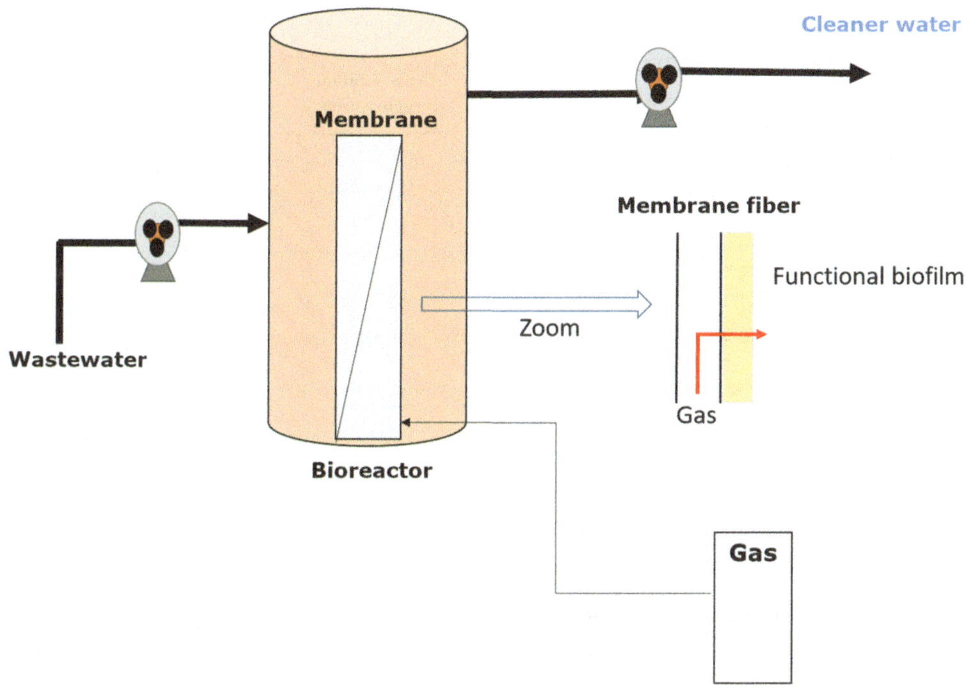

Figure 11.3 Membrane biofilm reactor (MBfR) for wastewater treatment.

advanced wastewater treatment, proper biofilm management as well as cost-effective membranes should be developed in the future.

11.2.5 Forward osmosis (FO) integrated processes for wastewater treatment/reclamation and resource recovery

11.2.5.1 Microfiltration forward osmosis membrane bioreactors (MF-FOMBRs)

Compared with hydraulic pressure-driven membrane technologies such as RO, the osmotic pressure-driven forward osmosis (FO) is featured with low fouling propensity and thus low energy input (Parida & Ng, 2013). In the FO process, the osmotic energy of the concentrated solution from concentrated solution side could draw water molecules from the dilute solution side across the FO membrane, while salts could be rejected (Figure 11.1b) (Parida & Ng, 2013). FO membranes could be used for resource recovery. For example, Bao *et al.* (2020) stated that the FO membrane modified with a moderate primary amine could achieve a high anti-fouling capability and recover ammonium with rejection above 94% for concentrating domestic wastewater. Phosphorus is also a non-renewable resource. However, due to the low concentration of phosphorus in domestic wastewater, there are limited technologies for direct phosphorus recovery from domestic wastewater. Microfiltration forward osmosis membrane bioreactor (MF-FOMBR), where a forward osmosis membrane bioreactor (FOMBRs) and a microfiltration (MF) membrane are operated in parallel, may achieve wastewater treatment/reclamation as well as more than 90% phosphorus recovery with a phosphorus content of 11.1–13.3% in recovered amorphous calcium phosphate precipitates (Figure 11.4a) (Qiu *et al.*, 2015). The FO membrane could reject the nutrients, while the MF membrane allows them to pass through, and the phosphorus is then recovered from the nutrient-rich MF permeate without addition of Fe^{3+}, Ca^{2+} and Mg^{2+}. As no biological activity is required in MF-FOMBR, obviating the need for the enrichment

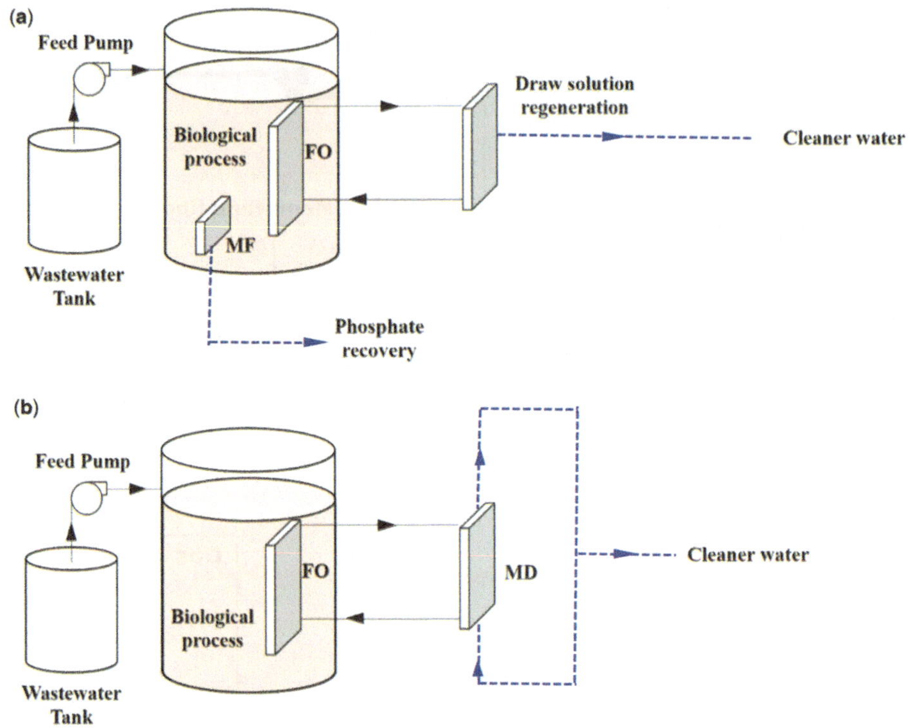

Figure 11.4 (a) System configuration of microfiltration forward osmosis membrane bioreactors (MF-FOMBRs) for wastewater reclamation as well as nutrient recovery and (b) forward osmotic membrane bioreactors (FOMBRs)-MD hybrid system (FOMBR-MD) for wastewater reclamation.

of phosphate accumulating organisms (PAOs) and the downstream disposal of phosphorus-rich sludge could lead to energy saving. Moreover, the FO membrane in the FOMBRs offers the advantage of low membrane fouling potential which would require lower hydraulic pressure, scouring intensity and frequency of backwashing (Achilli *et al.*, 2009; Qiu *et al.*, 2015), although FO membrane fouling could result from a reduction in the osmotic pressure driving force due to the increasing bioreactor salinity (Holloway *et al.*, 2015).

The energy consumption of sustainable MF-FOMBR should be competitive to the CAS systems, followed by the advanced treatment processes for the high-quality water reuse (Holloway *et al.*, 2015). Currently, the intensive energy input for recovery of draw solution (DS) significantly limits the scaling-up of FOMBR and other FO related technology including MF-FOMBR. Regeneration of DS for FOMBRs may be achieved by the RO process (Holloway *et al.*, 2015). According to the reverse osmosis system analysis system (ROSA) design software (Dow Filmtec, Edina, MN), the optimized approximate specific energy demand calculated to reconcentrate a draw solution of 40 g L^{-1}-NaCl is 1.6 kWh m^{-3} (Holloway *et al.*, 2015). It can drop to approximately 1.1 kWh m^{-3} if a pressure exchanger is to be incorporated into the RO system. However, this value would still exceed that of current advanced wastewater treatment processes (Holloway *et al.*, 2015). Thus, to overcome the major barrier for the implementation of full-scale MF-FOMBR, developing creative and efficient draw solution regeneration configurations is necessary to reduce the energy consumption associated with FOMBRs. Another pressure-driven filtration, namely nanofiltration (NF), which could achieve a high rejection of multivalent ions as well as a sufficiently low pressure for high water recovery rate, is a promising option for DS regeneration. High quality effluent was generated from a FOMBR and NF hybrid system for agricultural irrigation, and the DS replacement costs were reduced (Corzo *et al.*, 2018). Although seawater brine from SWRO process is considered a waste product, it can be used as an existing and easily accessible high osmotic pressure draw solution that will help increase the economic feasibility of existing FOMBR, as the DS reconcentration is not necessary (Qiu *et al.*, 2015). Moreover, it could create a more sustainable way for disposal of seawater brine, as the diluted brine from FOMBR systems could have a reduced influence on marine ecosystems. Thus, from the energy perspective, using seawater brine as DS for FOMBRs could be expected to make positive impacts on the sustainability of wastewater treatment and carbon footprint, though the diluted brine cannot be used as reusable water without further advanced treatment and the long-term operation of the system with regard to the accumulation of pollutants and membrane fouling needs to be further investigated (Qiu *et al.*, 2015).

11.2.5.2 Forward osmosis–membrane distillation hybrid (FO-MD) for wastewater treatment/reclamation

Compared to the pressure-driven filtration such as RO and NF to regenerate DS as well as reusable water, the FOMBRs with membrane distillation (MD) (i.e., FOMBR-MD) for wastewater treatment/reclamation has also been extensively studied (Morrow *et al.*, 2018). In an FOMBR-MD system, wastewater is fed into a bioreactor where aeration is supplied to the biomass and scours the FO membrane. Through osmosis, water diffuses from the bioreactor across the FO membrane into the draw solution. The diluted draw solution is sent to MD for reconcentration of draw solution and generation of product water (Figure 11.4b). MD is a thermally driven process to desalt water (Figure 11.1d). Water is transported as vapor from a high temperature solution to a low temperature solution through a microporous hydrophobic membrane due to a partial vapor pressure gradient, and MD can completely reject nonvolatile substances (Morrow *et al.*, 2018). When recovering FO draw solution, the requirement of energy of MD is relatively lower, as compared to the pressure-driven filtration technologies, since the former can use waste heat from industrial plants directly to thermally desalt draw solution. Furthermore, when feed solution salinity increases, the electrical energy requirement for RO would increase; however, MD is only marginally affected by the increase of salinity of feed solution (Amy *et al.*, 2017). Moreover, Luo *et al.* (2017a) stated that the FOMBR-MD system could effectively remove 30 significant trace organic contaminants of concern in wastewater. The rejections of all trace organics were more than 90%, which indicated that FOMBR-MD system could produce high-quality water. Thus, the FOMBR-MD system is considered a promising technology to produce high quality clean water with a low energy consumption and carbon footprint (Morrow *et al.*, 2018).

Alternatively, FO can be directly connected to MD (FO-MD) without biological process for industrial wastewater reclamation and/or resource recovery (Zhou *et al.*, 2017). For example, Nguyen *et al.* (2016) used the FO-MD systems to concentrate high-nutrient sludge, and the polytetrafluoroethylene membrane was the most effective of four types of MD membranes which could achieve a high-water flux of 10.28 LMH and approximately 100% of salt rejection. Ge *et al.* (2016) used a hydroacid complex (i.e., $Na_3[Cr(C_2O_4)_3]$) as draw solute to remove As(III) to a concentration below 10 μg/L (below WHO standard) by an FO-MD system. Dye wastewater treatment could also be achieved by this system using poly(acrylic acid) sodium (PAA-Na) salt (Ge *et al.*, 2012). Moreover, an FO-MD hybrid system could be also employed for the treatment of oily wastewater which not only included crude oil but also significant amounts of chemical additives (i.e., acetic acid) could be recovered (Zhang *et al.*, 2014).

Novel design of membrane could make FO-MD hybrid more competitive. For example, a symmetric FO membrane is endowed with a high-water flux as well as high salt rejection, which is due to the negatively charged sulfonate groups and the ultrathin symmetric architecture (Cheng *et al.*, 2019). Li *et al.* (2020) analyzed the economic feasibility of FO-MD for concentrating textile wastewater using a symmetric FO membrane. The results showed that by contrasting two commercial FO membranes, the symmetric FO membrane showed a superior performance in the FO-MD process. It could be due to the identical water transfer rate (WTR) between the FO and MD process, resulting in much lower energy consumption. Moreover, to achieve a concentration factor (CF) of 10 when treating 500 mL of textile wastewater using the symmetric FO membrane in the hybrid process, it only used the lowest total cost of 0.17 USD among the three. However, possible improvements with regards to the feasibility of various hard-to-remove compounds and other optimized draw solutes need to be further investigated to examine the promising potential of symmetric FO membranes for various types of wastewater treatment (Li *et al.*, 2020).

11.3 POTENTIAL MEMBRANE STRATEGIES FOR DECARBONIZATION IN DESALINATION

11.3.1 Pressure retarded osmosis (PRO) in desalination for power generation

Extracting energy from two solutions with different salinity gradients by transforming the osmotic pressure into mechanical work could be achieved by pressure retarded osmosis (PRO) technology (Figure 11.1c). In the PRO process, a solution of lower salinity such as river water is drawn by osmotic power of a solution with higher salinity such as seawater across a semipermeable membrane, and the work is harnessed by the turbine with an equal rate to the flow rate through the PRO membrane (O'Toole *et al.*, 2016).

However, not all PRO can make economically energy profits, and power density (W/m^2) which determines membrane productivity can be used to evaluate the performance of PRO systems. The Norwegian power company Statkraft suggested that the benchmark for commercial PRO should be more than 5 W/m^2 (O'Toole *et al.*, 2016). Although this PRO configuration can be used to extract renewable energy, the overall specific attainable energy may be quite low at approximately 0.1 kWh per m^3 when using seawater and river water in real application (O'Toole *et al.*, 2016). Specifically, when considering the pretreatment cost of the seawater, freshwater, and the employment of the turbine and other subsystems, the available net specific energy could be reduced to 0.124 kWh/m^3 in a river-to-sea PRO system (Figures 11.5a and 11.6) (O'Toole *et al.*, 2016). The main parasitic costs resulted from the pretreatment of freshwater and seawater for fouling mitigation (Figure 11.6). Thus, the major obstacle for the river-to-sea PRO system is the membrane fouling (Table 11.1), due to the fact that the fouling potential of the PRO mode (i.e., the porous layer of the membrane facing the feed solution) is higher as compared to that of FO mode (i.e., the active layer of the membrane facing the feed solution) (Figure 11.1b and c). For example, 50 ppm total organic carbon (TOC) and 5 mM calcium was found to lead to more severe fouling in the PRO mode, as compared to that of the FO mode in experiments (Parida & Ng, 2013). The largely declined water flow rate could have resulted from the serious fouling of the membrane porous support layer of PRO and increased concentrative internal concentration polarization (ICP) (Holloway *et al.*, 2015). Overall, the PRO systems for energy recovery using seawater and river water in real application is not economically feasible. Technological breakthroughs for mitigating PRO membrane fouling could allow for increasing the available net specific energy. Furthermore, during high-pressure operation, a support layer with high mechanical strength of PRO is also necessary (Sun & Chung, 2013), and it is a challenge for the efficient and economical osmotic power generation (Table 11.1).

Although RO is currently one of the most common and efficient desalination technologies for seawater desalination, much energy is still required to overcome the osmotic pressure of seawater (Ali *et al.*, 2018). Because of RO brine with a relatively higher concentration to, for example, seawater, it could be used to enable higher power production of PRO. Considering that more than 24 million m^3/d drinking water has been produced worldwide by desalination plants (Achilli *et al.*, 2014), there is tremendous osmotic power that could be recovered by PRO systems. Moreover, SWRO coupling with PRO (SWRO-PRO) could not only mitigate the energy input for the SWRO systems but could also reduce the discharge of RO brine which could reduce the adverse environmental impact on marine ecology (Figure 11.5b). According to Figure 11.6, the pretreatment cost could be more than 29% of the total osmosis power; however, brine entering the PRO subsystem could be considered to be free of

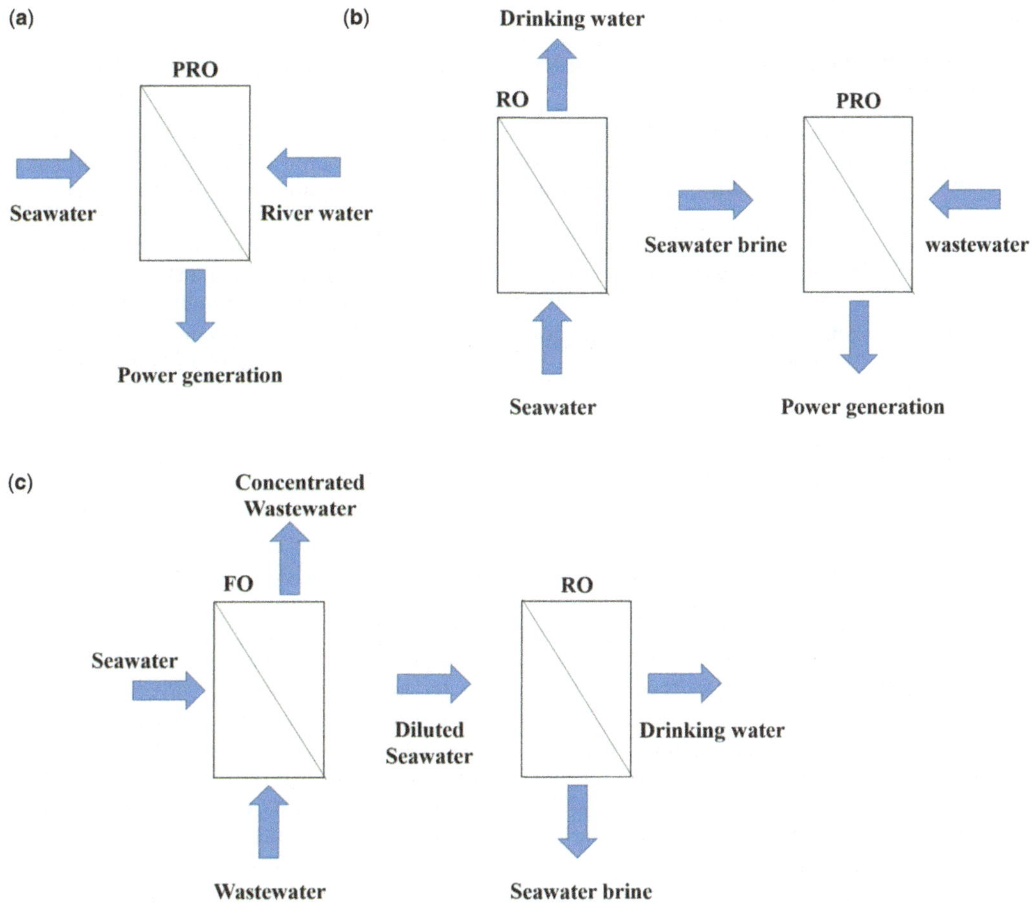

Figure 11.5 (a) System configuration of river-to-sea PRO system for power generation, (b) SWRO-PRO system for desalination and power generation and (c) FO and RO hybrid system for desalination and wastewater concentration.

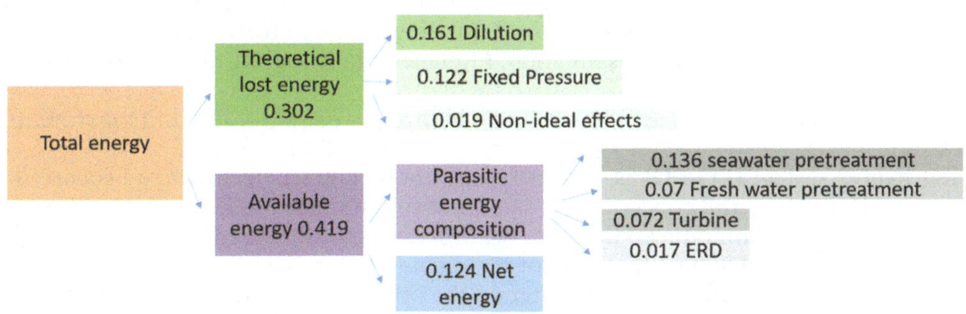

Figure 11.6 Sankey diagram of a river-to-sea PRO facility adopted from a previous study (O'Toole *et al.*, 2016). Total energy calculated based on the Gibbs free energy of mixing at 18°C. The unit is kWh/m^3.

Table 11.1 Hybrid membrane technologies for desalination with low energy consumption.

Process	Potential Hybrids	Driving Force	Potential Niches	Advantages	Disadvantages
RO		Applied pressure	Desalination	Low CAPEX	High energy consumption
RO	PRO	Applied pressure, osmotic pressure	Seawater desalination, energy saving	low energy consumption	High PRO membrane fouling, high CAPEX, need PRO membrane high mechanic tolerance
RO	FO	Applied pressure, osmotic pressure	Wastewater concentrate, desalination	Low energy consumption	High CAPEX, FO membrane fouling
FO	MD	Applied pressure, osmotic pressure	Seawater desalination	Low energy consumption	FO membrane fouling, low flux of MD, high CAPEX

foulants, as it has been already pretreated by the RO pretreatment system such as ultrafiltration (UF). It would thus eliminate the parasitic energy consumption, as compared to that of the river-to-sea PRO system (Achilli *et al.*, 2014). Moreover, the membrane power densities of the SWRO-PRO could be higher than 5 W/m^2 which largely exceeds the reported 1.5 W/m^2 for the river-to-sea PRO pilot system in Korea (Achilli *et al.*, 2014; Kim *et al.*, 2013). For the energy perspective, a SWRO-PRO pilot system at 50% RO recovery could yield approximately 1.1 kWh/m^3, which could thus reduce the specific energy for SWRO to around 1 kWh/m^3 (Prante *et al.*, 2014).

However, wastewater retentate, which has a salinity close to that of river water, was preferred as the feed water for the PRO system rather than river water due to the scarcity of freshwater in some regions such as Singapore (Wan & Chung, 2015). Therefore, membrane fouling due to the foulants from wastewater retentate could be still a major problem for the SWRO-PRO systems for power generation. The reduction in the power densities could largely result from the fouling on the porous substrate of the PRO membranes. Therefore, both UF and NF has been employed as pretreatment processes to boost the power densities to 6.6 and 8.9 W/m^2, respectively, as they could mitigate fouling potential of the wastewater retentate (Wan & Chung, 2015). Moreover, coagulation of feed wastewater could also be helpful (Wan *et al.*, 2019). As compared to the untreated wastewater which caused a 69.3% flux reduction, $AlCl_3$ and $NaAlO_2$ have been demonstrated to increase the normalized water flux to 66 and 64%, respectively. The initial water fluxes have also been increased to 25.5 and 24.8 LMH at 20 bars, respectively (Wan *et al.*, 2019). Rather than pretreatment, increasing the anti-fouling membrane properties could also be demonstrated to improve the SWRO-PRO, and a wide range of nanomaterials could be used to modify the membrane such as graphene-based materials, carbon nanotubes and zeolites, which endowed the PRO membranes with favorable membrane structures and enhanced desirable antifouling characteristics (Wan *et al.*, 2020). In general, to further scale up more PRO into RO systems for low-carbon-footprint desalination, PRO membrane fouling needs to be well-mitigated.

11.3.2 Forward osmosis-reverse osmosis (FO-RO) hybrid for desalination and wastewater concentration

FO and RO hybrid system (FO-RO) is considered as another green technology because it can be employed to desalt seawater and concentrate wastewater simultaneously (Figure 11.5c) (Linares *et al.*, 2016). Shaffer *et al.* (2015) announced that the 'FO process is not intended to replace RO, but rather is to be used to process feed waters that cannot be treated by RO'. The FO system can use seawater on one side of the FO membrane and wastewater on the other side, and it could lead to the reduction of osmotic pressure of seawater prior to RO desalination. Reducing the volume of wastewater could reduce the energy consumption for transportation as well as treatment processes (Linares *et al.*, 2016). Moreover, it could be more efficient to harvest energy (e.g., biogas) and nutrients (e.g., phosphates) from the resulting concentrated wastewater. The concentrated wastewater can potentially be treated by anaerobic treatment with enhanced biogas (CH_4) production (Amy *et al.*, 2017). For example, the methane yield of an AnMBR could progressively increase from 214 to 322 $mL_{\text{-}CH4}/g_{\text{-}COD}$ when the pre-concentration factor of domestic wastewater by FO increased from 1 to 10 (Vinardell *et al.*, 2021). The energy consumption of the overall FO-RO system can be offset by the enhanced biogas produced (Amy *et al.*, 2017).

Furthermore, potential electricity savings could be also achieved in the SWRO facilities by lowering the operating hydraulic pressure (Linares *et al.*, 2016). The brackish water RO membranes (BWRO) could also be used instead of SWRO membranes. Moreover, higher flux could be employed for the diluted seawater, which could thus increase the water recovery of the whole system. Furthermore, discharging brines with lower salinity would have less adverse impacts on the aquatic ecosystem. The specific energy consumption (SEC) associated to the FO-RO process for desalination ranged between 1.3 and 1.5 kWh m^{-3}, and it was calculated with the total capacity of 2400 m^3 d^{-1} based on a conservative estimate using a secondary wastewater effluent as the feed and seawater as the draw solution (Yangali-Quintanilla *et al.*, 2011). This range of SEC could be lower than that of the conventional RO process for seawater desalination.

However, the critical aspect in the CAPEX of FO-RO hybrid systems results from the FO membranes. The use of an FO-RO system could be more viable if commercial FO membrane modules can be produced with a reasonable price (Table 11.1) (Linares *et al.*, 2016). Overall, the FO-RO system could have a 56% lower OPEX due to the energy saving for diluted seawater desalination as compared to the conventional SWRO, although there is a 21% higher CAPEX due to the implementation of the FO systems (Linares *et al.*, 2016). To calculate the total cost per cubic meter of desalted water, the FO-RO hybrid desalination system could achieve a cost reduction of 16% as compared to the SWRO. Moreover, economic evaluation of the FO-RO hybrid process was evaluated comparing to a conventional two-stage SWRO. Spiral wound FO and plate-and-frame FO-RO hybrid processes can achieve cost reductions of $355.3 million and $310.2 million, respectively, over a period of 20 years (Im *et al.*, 2020).

11.3.3 Forward osmosis-membrane distillation hybrid (FO-MD) for desalination

A separation process must be employed for a sustainable FO system for draw solute regeneration as well as desalting water. Among RO, NF and MD, integration of forward osmosis with MD (FO-MD) have also been extensively investigated for desalination purpose (i.e., seawater and brackish desalination) due to the advantages of MD over pressure-driven processes (i.e., RO and NF). MD has a higher salt rejection than RO and NF because MD can completely reject nonvolatile substances (Morrow *et al.*, 2018). It could have a lower operation pressure among the three, rendering a promising process for desalinating highly saline streams, since a key attribute of the MD process is that flux and quality of produced water are not sensitive to salinity (less than 200 000 ppm) of the feed water (Amy *et al.*, 2017). Moreover, MD is usually operated at high temperatures, and it could use the waste heat and solar energy as thermal energy for desalination. Thus, the FO–MD process is especially favored when solar energy and waste heat is abundantly attainable near the MD plants (Wang *et al.*, 2015). Compared to the fouling encountered in RO and NF, the fouling in MD is significantly lower, and scale inhibitor and acid could be used to address the scaling in MD process (Amy *et al.*, 2017).

However, the MD process for desalination has difficulty in converting waste heat or solar energy to the overall MD system with high utilization efficiency (Amy *et al.*, 2017). Moreover, the reverse salt flux could be another hurdle for the FO-MD system, as it could reduce the osmotic driving force and increase the replenishment cost (Wang *et al.*, 2015). Furthermore, membrane performance with regarding to the flux, hydrophobicity and the wettability also need to be improved to reduce the CAPEX of the FO-MD systems (Table 11.1). In particular, low membrane flux could be a main disadvantage of MD, and modifications to different MD configurations could have different energetics and increase the transmembrane flux (González *et al.*, 2017). Lab-scale MD configuration includes direct contact MD (Francis *et al.*, 2014), air gap MD (Alsaadi *et al.*, 2015) and vacuum MD (Alsaadi *et al.*, 2014). Moreover, efficient internal heat recovery and satisfactory flux are also considered as main obstacles for MD module scale-up (Amy *et al.*, 2017). Overall, more research needs to be further conducted to improve the performance as well as the sustainability of FO-MD systems for low-carbon-footprint desalination.

11.4 CONCLUSIONS

In addition to meeting the tightened effluent discharge standards in many countries, membrane technologies are commonly used to address other objectives including protection of public health and ecological issues and producing reusable water. There is an increasing trend to consider energy consumption associated with various membrane technologies for water and wastewater treatments. To achieve truly sustainable treatment processes, the intensive energy consumption of membrane-based wastewater/water treatment plants is a key challenge and must be dealt with accordingly. Novel and modified membrane-based strategies are developed to reduce energy and carbon footprint and recover resources from water within the circular economy.

The way towards energy self-sufficient operation of the above summarized membrane-based processes for wastewater treatment/reclamation is aiming to directly capture energy and nutrients from wastewater (e.g., P, N and biogas) and minimize energy consumption such as aeration. As membrane fouling is still a main obstacle for membrane-based technologies, especially the AnMBRs and the direct membrane filtration, further optimization should be conducted to address it in a more cost-effective and holistic manner. Membrane-based desalination operations mainly rely on RO as the baseline conventional technology. However, it is challenged by the significant specific energy consumption as well as the adverse environmental impacts such as greenhouse gas emissions. Interesting membrane processes including FO, PRO and MD were proposed, and the hybrid technologies encompassing a mix of new and conventional technologies can generate clean water and sustainable electricity simultaneously. However, parasitic drawbacks are with these novel membrane-based strategies for desalination such as the serious fouling potential of PRO, the high membrane cost of FO and the low membrane flux of MD. We summarized the pains and gains of each hybrid technologies in this chapter and suggested that more research optimization and development is necessary for their next steps towards practical and worldwide implementation for energy saving and low carbon footprint.

REFERENCES

Achilli A., Cath T. Y., Marchand E. A. and Childress A. E. (2009). The forward osmosis membrane bioreactor: a low fouling alternative to MBR processes. *DesalinationI*, **239**(1–3), 10–21, https://doi.org/10.1016/j.desal.2008.02.022

Achilli A., Prante J. L., Hancock N. T., Maxwell E. B. and Childress A. E. (2014). Experimental results from RO-PRO: a next generation system for low-energy desalination. *Environmental Science and Technology*, **48**(11), 6437–6443, https://doi.org/10.1021/es405556s

Ali A., Tufa R. A., Macedonio F., Curcio E. and Drioli E. (2018). Membrane technology in renewable-energy-driven desalination. *Renewable and Sustainable Energy Reviews*, **81**, 1–21, https://doi.org/10.1016/j.rser.2017.07.047

Alsaadi A. S., Francis L., Amy G. L. and Ghaffour N. (2014). Experimental and theoretical analyses of temperature polarization effect in vacuum membrane distillation. *Journal of Membrane Science*, **471**, 138–148, https://doi.org/10.1016/j.memsci.2014.08.005

Alsaadi A. S., Francis L., Maab H., Amy G. L. and Ghaffour N. (2015). Evaluation of air gap membrane distillation process running under sub-atmospheric conditions: experimental and simulation studies. *Journal of Membrane Science*, **489**, 73–80, https://doi.org/10.1016/j.memsci.2015.04.008

Amy G., Ghaffour N., Li Z., Francis L., Linares R. V., Missimer T. and Lattemann S. (2017). Membrane-based seawater desalination: present and future prospects. *Desalination*, **401**, 16–21, https://doi.org/10.1016/j.desal.2016.10.002

Aybar M., Pizarro G., Boltz J., Downing L. and Nerenberg R. (2014). Energy-efficient wastewater treatment via the air-based, hybrid membrane biofilm reactor (hybrid MfBR). *Water Science and Technology*, **69**(8), 1735–1741, https://doi.org/10.2166/wst.2014.086

Bao X., She Q., Long W. and Wu Q. (2020). Ammonium ultra-selective membranes for wastewater treatment and nutrient enrichment: interplay of surface charge and hydrophilicity on fouling propensity and ammonium rejection. *Water Research*, **190**, 116678, https://doi.org/10.1016/j.watres.2020.116678

Bilad M., Arafat H. A. and Vankelecom I. F. (2014). Membrane technology in microalgae cultivation and harvesting: a review. *Biotechnology Advances*, **32**(7), 1283–1300, https://doi.org/10.1016/j.biotechadv.2014.07.008

Chen L., Gu Y., Cao C., Zhang J., Ng J.-W. and Tang C. (2014). Performance of a submerged anaerobic membrane bioreactor with forward osmosis membrane for low-strength wastewater treatment. *Water Research*, **50**, 114–123, https://doi.org/10.1016/j.watres.2013.12.009

Chen C., Bin L., Tang B., Huang S., Fu F., Chen Q., Wu L. and Wu C. (2017). Cultivating granular sludge directly in a continuous-flow membrane bioreactor with internal circulation. *Chemical Engineering Journal*, **309**, 108–117, https://doi.org/10.1016/j.cej.2016.10.034

Cheng W., Ma J., Zhang X. and Elimelech M. (2019). Sub-1 μm free-standing symmetric membrane for osmotic separations. *Environmental Science & Technology Letters*, **6**(8), 492–498, https://doi.org/10.1021/acs.estlett.9b00364

Chernicharo C., Van Lier J., Noyola A. and Ribeiro T. B. (2015). Anaerobic sewage treatment: state of the art, constraints and challenges. *Reviews in Environmental Science and Bio/Technology*, **14**(4), 649–679, https://doi.org/10.1007/s11157-015-9377-3

Corsino S., Campo R., Di Bella G., Torregrossa M. and Viviani G. (2016). Study of aerobic granular sludge stability in a continuous-flow membrane bioreactor. *Bioresource Technology*, **200**, 1055–1059, https://doi.org/10.1016/j.biortech.2015.10.065

Corzo B., de la Torre T., Sans C., Escorihuela R., Navea S. and Malfeito J. J. (2018). Long-term evaluation of a forward osmosis-nanofiltration demonstration plant for wastewater reuse in agriculture. *Chemical Engineering Journal*, **338**, 383–391, https://doi.org/10.1016/j.cej.2018.01.042

Crone B. C., Garland J. L., Sorial G. A. and Vane L. M. (2016). Significance of dissolved methane in effluents of anaerobically treated low strength wastewater and potential for recovery as an energy product: a review. *Water Research*, **104**, 520–531, https://doi.org/10.1016/j.watres.2016.08.019

Drexler I. L. C. and Yeh D. H. (2014). Membrane applications for microalgae cultivation and harvesting: a review. *Reviews in Environmental Science and Bio/Technology*, **13**(4), 487–504, https://doi.org/10.1007/s11157-014-9350-6

Francis L., Ghaffour N., Alsaadi A. S., Nunes S. P. and Amy G. L. (2014). Performance evaluation of the DCMD desalination process under bench scale and large scale module operating conditions. *Journal of Membrane Science*, **455**, 103–112, https://doi.org/10.1016/j.memsci.2013.12.033

Galib M., Elbeshbishy E., Reid R., Hussain A. and Lee H.-S. (2016). Energy-positive food wastewater treatment using an anaerobic membrane bioreactor (AnMBR). *Journal of Environmental Management*, **182**, 477–485, https://doi.org/10.1016/j.jenvman.2016.07.098

Ge Q., Wang P., Wan C. and Chung T.-S. (2012). Polyelectrolyte-promoted forward osmosis–membrane distillation (FO–MD) hybrid process for dye wastewater treatment. *Environmental Science and Technology*, **46**(11), 6236–6243, https://doi.org/10.1021/es300784h

Ge Q., Han G. and Chung T.-S. (2016). Effective As (III) removal by a multi-charged hydroacid complex draw solute facilitated forward osmosis-membrane distillation (FO-MD) processes. *Environmental Science and Technology*, **50**(5), 2363–2370, https://doi.org/10.1021/acs.est.5b05402

González D., Amigo J. and Suárez F. (2017). Membrane distillation: perspectives for sustainable and improved desalination. *Renewable and Sustainable Energy Reviews*, **80**, 238–259, https://doi.org/10.1016/j.rser.2017.05.078

Hao X., Liu R. and Huang X. (2015). Evaluation of the potential for operating carbon neutral WWTPs in China. *Water Research*, **87**, 424–431, https://doi.org/10.1016/j.watres.2015.05.050

Hasar H. (2009). Simultaneous removal of organic matter and nitrogen compounds by combining a membrane bioreactor and a membrane biofilm reactor. *Bioresource Technology*, **100**(10), 2699–2705, https://doi.org/10.1016/j.biortech.2008.12.024

Holloway R. W., Achilli A. and Cath T. Y. (2015). The osmotic membrane bioreactor: a critical review. *Environmental Science: Water Research and Technology*, **1**(5), 581–605, https://doi.org/10.1039/C5EW00103J

Im S. J., Jeong S., Jeong S. and Jang A. (2020). Techno-economic evaluation of an element-scale forward osmosis-reverse osmosis hybrid process for seawater desalination. *Desalination*, **476**, 114240, https://doi.org/10.1016/j.desal.2019.114240

Iorhemen O. T., Hamza R. A., Zaghloul M. S. and Tay J. H. (2019). Aerobic granular sludge membrane bioreactor (AGMBR): extracellular polymeric substances (EPS) analysis. *Water Research*, **156**, 305–314, https://doi.org/10.1016/j.watres.2019.03.020

Jeong Y., Hermanowicz S. W. and Park C. (2017). Treatment of food waste recycling wastewater using anaerobic ceramic membrane bioreactor for biogas production in mainstream treatment process of domestic wastewater. *Water Research*, **123**, 86–95, https://doi.org/10.1016/j.watres.2017.06.049

Judd S. J. (2017). Membrane technology costs and me. *Water Research*, **122**, 1–9, https://doi.org/10.1016/j.watres.2017.05.027

Kim Y. C., Kim Y., Oh D. and Lee K. H. (2013). Experimental investigation of a spiral-wound pressure-retarded osmosis membrane module for osmotic power generation. *Environmental Science and Technology*, **47**(6), 2966–2973, https://doi.org/10.1021/es304060d

Lei Z., Yang S., Li Y.-Y., Wen W., Wang X. C. and Chen R. (2018). Application of anaerobic membrane bioreactors to municipal wastewater treatment at ambient temperature: a review of achievements, challenges, and perspectives. *Bioresource Technology*, **267**, 756–768, https://doi.org/10.1016/j.biortech.2018.07.050

Li J., Cai A., Wang M., Ding L. and Ni Y. (2014). Aerobic granulation in a modified oxidation ditch with an adjustable volume intraclarifier. *Bioresource Technology*, **157**, 351–354, https://doi.org/10.1016/j.biortech.2014.01.130

Li J., Liu Y., Li X. and Cheng F. (2019). Reactor performance and membrane fouling of a novel submerged aerobic granular sludge membrane bioreactor during long-term operation. *Journal of Water Reuse and Desalination*, **9**(1), 1–9, https://doi.org/10.2166/wrd.2017.019

Li M., Li K., Wang L. and Zhang X. (2020). Feasibility of concentrating textile wastewater using a hybrid forward osmosis-membrane distillation (FO-MD) process: performance and economic evaluation. *Water Research*, **172**, 115488, https://doi.org/10.1016/j.watres.2020.115488

Liang Z., Liu Y., Ge F., Xu Y., Tao N., Peng F. and Wong M. (2013). Efficiency assessment and pH effect in removing nitrogen and phosphorus by algae-bacteria combined system of *Chlorella vulgaris* and *Bacillus licheniformis*. *Chemosphere*, **92**(10), 1383–1389, https://doi.org/10.1016/j.chemosphere.2013.05.014

Liébana R., Modin O., Persson F. and Wilén B.-M. (2018). Integration of aerobic granular sludge and membrane bioreactors for wastewater treatment. *Critical Reviews in Biotechnology*, **38**(6), 801–816, https://doi.org/10.1080/07388551.2017.1414140

Lim K., Evans P. J. and Parameswaran P. (2019). Long-term performance of a pilot-scale gas-sparged anaerobic membrane bioreactor under ambient temperatures for holistic wastewater treatment. *Environmental Science and Technology*, **53**(13), 7347–7354, https://doi.org/10.1021/acs.est.8b06198

Linares R. V., Li Z., Yangali-Quintanilla V., Ghaffour N., Amy G., Leiknes T. and Vrouwenvelder J. S. (2016). Life cycle cost of a hybrid forward osmosis–low pressure reverse osmosis system for seawater desalination and wastewater recovery. *Water Research*, **88**, 225–234, https://doi.org/10.1016/j.watres.2015.10.017

Liu T., Li J., Khai Lim Z., Chen H., Hu S., Yuan Z. and Guo J. (2020). Simultaneous removal of dissolved methane and nitrogen from synthetic mainstream anaerobic effluent. *Environmental Science and Technology*, **54**(12), 7629–7638, https://doi.org/10.1021/acs.est.0c00912

Low S. L., Ong S. L. and Ng H. Y. (2016). Characterization of membrane fouling in submerged ceramic membrane photobioreactors fed with effluent from membrane bioreactors. *Chemical Engineering Journal*, **290**, 91–102, https://doi.org/10.1016/j.cej.2016.01.005

Luo W., Phan H. V., Li G., Hai F. I., Price W. E., Elimelech M. and Nghiem L. D. (2017a). An osmotic membrane bioreactor–membrane distillation system for simultaneous wastewater reuse and seawater desalination: performance and implications. *Environmental Science and Technology*, **51**(24), 14311–14320, https://doi.org/10.1021/acs.est.7b02567

Luo Y., Le-Clech P. and Henderson R. K. (2017b). Simultaneous microalgae cultivation and wastewater treatment in submerged membrane photobioreactors: a review. *Algal Research*, **24**, 425–437, https://doi.org/10.1016/j.algal.2016.10.026

Mannina G., Cosenza A. and Rebouças T. F. (2020). A plant-wide modelling comparison between membrane bioreactors and conventional activated sludge. *Bioresource Technology*, **297**, 122401, https://doi.org/10.1016/j.biortech.2019.122401

Martin K. J. and Nerenberg R. (2012). The membrane biofilm reactor (MBfR) for water and wastewater treatment: principles, applications, and recent developments. *Bioresource Technology*, **122**, 83–94, https://doi.org/10.1016/j.biortech.2012.02.110

Melin T., Jefferson B., Bixio D., Thoeye C., De Wilde W., De Koning J., Van der Graaf J. and Wintgens T. (2006). Membrane bioreactor technology for wastewater treatment and reuse. *Desalination*, **187**(1–3), 271–282, https://doi.org/10.1016/j.desal.2005.04.086

Morrow C. P., Furtaw N. M., Murphy J. R., Achilli A., Marchand E. A., Hiibel S. R. and Childress A. E. (2018). Integrating an aerobic/anoxic osmotic membrane bioreactor with membrane distillation for potable reuse. *Desalination*, **432**, 46–54, https://doi.org/10.1016/j.desal.2017.12.047

Ng T. C. A. and Ng H. Y. (2010). Characterisation of initial fouling in aerobic submerged membrane bioreactors in relation to physico-chemical characteristics under different flux conditions. *Water Research*, **44**(7), 2336–2348, https://doi.org/10.1016/j.watres.2009.12.038

Nguyen N. C., Nguyen H. T., Ho S.-T., Chen S.-S., Ngo H. H., Guo W., Ray S. S. and Hsu H.-T. (2016). Exploring high charge of phosphate as new draw solute in a forward osmosis–membrane distillation hybrid system for concentrating high-nutrient sludge. *Science of the Total Environment*, **557**, 44–50, https://doi.org/10.1016/j.scitotenv.2016.03.025

Nhat P. V. H., Ngo H., Guo W., Chang S., Nguyen D. D., Nguyen P., Bui X.-T., Zhang X. and Guo J. (2018). Can algae-based technologies be an affordable Green process for biofuel production and wastewater remediation? *Bioresource Technology*, **256**, 491–501, https://doi.org/10.1016/j.biortech.2018.02.031

O'Toole G., Jones L., Coutinho C., Hayes C., Napoles M. and Achilli A. (2016). River-to-sea pressure retarded osmosis: resource utilization in a full-scale facility. *Desalination*, **389**, 39–51, https://doi.org/10.1016/j.desal.2016.01.012

Parida V. and Ng H. Y. (2013). Forward osmosis organic fouling: effects of organic loading, calcium and membrane orientation. *Desalination*, **312**, 88–98, https://doi.org/10.1016/j.desal.2012.04.029

Prante J. L., Ruskowitz J. A., Childress A. E. and Achilli A. (2014). RO-PRO desalination: an integrated low-energy approach to seawater desalination. *Applied Energy*, **120**, 104–114, https://doi.org/10.1016/j.apenergy.2014.01.013

Qiu G., Law Y.-M., Das S. and Ting Y.-P. (2015). Direct and complete phosphorus recovery from municipal wastewater using a hybrid microfiltration-forward osmosis membrane bioreactor process with seawater brine as draw solution. *Environmental Science and Technology*, **49**(10), 6156–6163, https://doi.org/10.1021/es504554f

Radjenović J., Petrović M. and Barceló D. (2009). Fate and distribution of pharmaceuticals in wastewater and sewage sludge of the conventional activated sludge (CAS) and advanced membrane bioreactor (MBR) treatment. *Water Research*, **43**(3), 831–841, https://doi.org/10.1016/j.watres.2008.11.043

Robles A., Durán F., Ruano M., Ribes J. and Ferrer J. (2012). Influence of total solids concentration on membrane permeability in a submerged hollow-fibre anaerobic membrane bioreactor. *Water Science and Technology*, **66**(2), 377–384, https://doi.org/10.2166/wst.2012.196

Robles Á., Durán F., Giménez J. B., Jiménez E., Ribes J., Serralta J., Seco A., Ferrer J. and Rogalla F. (2020). Anaerobic membrane bioreactors (AnMBR) treating urban wastewater in mild climates. *Bioresource Technology*, **314**, 123763, https://doi.org/10.1016/j.biortech.2020.123763

Shaffer D. L., Werber J. R., Jaramillo H., Lin S. and Elimelech M. (2015). Forward osmosis: where are we now? *Desalination*, **356**, 271–284, https://doi.org/10.1016/j.desal.2014.10.031

Shin C. and Bae J. (2018). Current status of the pilot-scale anaerobic membrane bioreactor treatments of domestic wastewaters: a critical review. *Bioresource Technology*, **247**, 1038–1046, https://doi.org/10.1016/j.biortech.2017.09.002

Smith A. L., Stadler L. B., Love N. G., Skerlos S. J. and Raskin L. (2012). Perspectives on anaerobic membrane bioreactor treatment of domestic wastewater: a critical review. *Bioresource Technology*, **122**, 149–159, https://doi.org/10.1016/j.biortech.2012.04.055

Smith A. L., Stadler L. B., Cao L., Love N. G., Raskin L. and Skerlos S. J. (2014). Navigating wastewater energy recovery strategies: a life cycle comparison of anaerobic membrane bioreactor and conventional treatment systems with anaerobic digestion. *Environmental Science and Technology*, **48**(10), 5972–5981, https://doi.org/10.1021/es5006169

Sun L., Tian Y., Zhang J., Cui H., Zuo W. and Li J. (2018a). A novel symbiotic system combining algae and sludge membrane bioreactor technology for wastewater treatment and membrane fouling mitigation: performance and mechanism. *Chemical Engineering Journal*, **344**, 246–253, https://doi.org/10.1016/j.cej.2018.03.090

Sun L., Tian Y., Zhang J., Li H., Tang C. and Li J. (2018b). Wastewater treatment and membrane fouling with algal-activated sludge culture in a novel membrane bioreactor: influence of inoculation ratios. *Chemical Engineering Journal*, **343**, 455–459, https://doi.org/10.1016/j.cej.2018.03.022

Sun S.-P. and Chung T.-S. (2013). Outer-selective pressure-retarded osmosis hollow fiber membranes from vacuum-assisted interfacial polymerization for osmotic power generation. *Environmental Science and Technology*, **47**(22), 13167–13174, https://doi.org/10.1021/es403270n

Tang T. and Hu Z. (2016). A comparison of algal productivity and nutrient removal capacity between algal CSTR and algal MBR at the same light level under practical and optimal conditions. *Ecological Engineering*, **93**, 66–72, https://doi.org/10.1016/j.ecoleng.2016.04.008

Taşkan B., Hasar H. and Lee C. H. (2020). Effective biofilm control in a membrane biofilm reactor using a quenching bacterium (Rhodococcus sp. BH4). *Biotechnology and Bioengineering*, **117**(4), 1012–1023, https://doi.org/10.1002/bit.27259

Thanh B. X., Visvanathan C. and Aim R. B. (2013). Fouling characterization and nitrogen removal in a batch granulation membrane bioreactor. *International Biodeterioration and Biodegradation*, **85**, 491–498, https://doi.org/10.1016/j.ibiod.2013.02.005

Tirosh U. and Shechter R. (2020). Membrane aerated biofilm reactor (MABR)–distributed treatment of wastewater at low energy consumption. Frontiers in Water-Energy-Nexus – Nature-Based Solutions, Advanced Technologies and Best Practices for Environmental Sustainability, 513–515, https://doi.org/10.1007/978-3-030-13068-8_128

Truong H. T. B., Nguyen T., Thi P. and Bui H. M. (2018). Integration of aerobic granular sludge and membrane filtration for tapioca processing wastewater treatment: fouling mechanism and granular stability. *Journal of Water Supply: Research and Technology – Aqua*, **67**(8), 846–857, https://doi.org/10.2166/aqua.2018.104

US EPA. (2006). Wastewater Management Fact Sheet: Energy Conservation. EPA 832-F-06-024.

Vijayalayan P., Thanh B. X. and Visvanathan C. (2014). Simultaneous nitrification denitrification in a batch granulation membrane airlift bioreactor. *International Biodeterioration and Biodegradation*, **95**, 139–143, https://doi.org/10.1016/j.ibiod.2014.05.020

Vinardell S., Astals S., Jaramillo M., Mata-Alvarez J. and Dosta J. (2021). Anaerobic membrane bioreactor performance at different wastewater pre-concentration factors: An experimental and economic study. *Science of the Total Environment*, **750**, 141625, https://doi.org/10.1016/j.scitotenv.2020.141625

Wan C. F. and Chung T.-S. (2015). Osmotic power generation by pressure retarded osmosis using seawater brine as the draw solution and wastewater retentate as the feed. *Journal of Membrane Science*, **479**, 148–158, https://doi.org/10.1016/j.memsci.2014.12.036

Wan C. F., Cui Y., Gai W. X., Cheng Z. L. and Chung T.-S. (2020). Nanostructured membranes for enhanced forward osmosis and pressure-retarded osmosis. In Sustainable Nanoscale Engineering. Elsevier, pp. 373–394.

Wan C. F., Jin S. and Chung T.-S. (2019). Mitigation of inorganic fouling on pressure retarded osmosis (PRO) membranes by coagulation pretreatment of the wastewater concentrate feed. *Journal of Membrane Science*, **572**, 658–667, https://doi.org/10.1016/j.memsci.2018.11.051

Wang P., Cui Y., Ge Q., Tew T. F. and Chung T.-S. (2015). Evaluation of hydroacid complex in the forward osmosis–membrane distillation (FO–MD) system for desalination. *Journal of Membrane Science*, **494**, 1–7, https://doi.org/10.1016/j.memsci.2015.07.022

Wang S., Chew J. W. and Liu Y. (2020). Development of an integrated aerobic granular sludge MBR and reverse osmosis process for municipal wastewater reclamation. *Science of the Total Environment*, **748**, 141309, https://doi.org/10.1016/j.scitotenv.2020.141309

Winkler M. K. and Straka L. (2019). New directions in biological nitrogen removal and recovery from wastewater. *Current Opinion in Biotechnology*, **57**, 50–55, https://doi.org/10.1016/j.copbio.2018.12.007

Xie Z., Wang Z., Wang Q., Zhu C. and Wu Z. (2014). An anaerobic dynamic membrane bioreactor (AnDMBR) for landfill leachate treatment: performance and microbial community identification. *Bioresource Technology*, **161**, 29–39, https://doi.org/10.1016/j.biortech.2014.03.014

Xu B., Albert Ng T. C., Huang S. and Ng H. Y. (2020a). Effect of quorum quenching on EPS and size-fractioned particles and organics in anaerobic membrane bioreactor for domestic wastewater treatment. *Water Research*, **179**, 115850, https://doi.org/10.1016/j.watres.2020.115850

Xu B., Albert Ng T. C., Huang S., Shi X. and Ng H. Y. (2020b). Feasibility of isolated novel facultative quorum quenching consortiums for fouling control in an AnMBR. *Water Research*, **169**, 115251, https://doi.org/10.1016/j.watres.2019.115251

Yang Q., Li H., Wang D., Zhang X., Guo X., Pu S., Guo R. and Chen J. (2020). Utilization of chemical wastewater for CO_2 emission reduction: purified terephthalic acid (PTA) wastewater-mediated culture of microalgae for CO_2 bio-capture. *Applied Energy*, **276**, 115502, https://doi.org/10.1016/j.apenergy.2020.115502

Yangali-Quintanilla V., Li Z., Valladares R., Li Q. and Amy G. (2011). Indirect desalination of Red Sea water with forward osmosis and low pressure reverse osmosis for water reuse. *Desalination*, **280**(1–3), 160–166, https://doi.org/10.1016/j.desal.2011.06.066

Yue X., Koh Y. K. K. and Ng H. Y. (2018). Membrane fouling mitigation by NaClO-assisted backwash in anaerobic ceramic membrane bioreactors for the treatment of domestic wastewater. *Bioresource Technology*, **268**, 622–632, https://doi.org/10.1016/j.biortech.2018.08.003

Zhang S., Wang P., Fu X. and Chung T.-S. (2014). Sustainable water recovery from oily wastewater via forward osmosis-membrane distillation (FO-MD). *Water Research*, **52**, 112–121, https://doi.org/10.1016/j.watres.2013.12.044

Zhang W., Liang W., Zhang Z. and Hao T. (2020). Aerobic granular sludge (AGS) scouring to mitigate membrane fouling: performance, hydrodynamic mechanism and contribution quantification model. *Water Research*, **188**, 116518, https://doi.org/10.1016/j.watres.2020.116518

Zhou Y., Huang M., Deng Q. and Cai T. (2017). Combination and performance of forward osmosis and membrane distillation (FO-MD) for treatment of high salinity landfill leachate. *Desalination*, **420**, 99–105, https://doi.org/10.1016/j.desal.2017.06.027

doi: 10.2166/9781789061796

Chapter 12

Natural treatment systems and integrated watershed management for decarbonization

*Hannah R. Molitor and Jerald L. Schnoor**

Department of Civil and Environmental Engineering, University of Iowa, Iowa City, IA, USA
**Correspondence: jerald-schnoor@uiowa.edu*

12.1 INTRODUCTION

Water and wastewater utilities use tremendous amounts of energy and emit copious quantities of greenhouse gases (GHGs) through direct emissions from the facilities themselves (Scope 1 emissions), purchased electricity and energy from outside suppliers (Scope 2 emissions), and emissions related to the usage of water by customers (e.g., hot water) and energy/transportation costs associated with moving equipment and personnel to the site each day (Scope 3 emissions).

For common emission estimates that cover Scopes 1 and 2, water and wastewater utilities are estimated to emit 3–7% of all global greenhouse gas emissions (Trommsdorff, 2015). California is the most extreme example and uses 20% of all the state's electricity just in supplying water (Loge, 2016). Water is a heavy commodity and moving it around via pumping uses tremendous amounts of electricity. If the electricity is supplied by highly polluting coal-fired power plants, this especially increases the carbon footprint of utilities. Aeration of wastewater by blowers and sparged air represents another huge energy investment and a concomitant emission of greenhouse gases. Finally, treatment of carbonaceous biochemical oxygen demand (BOD) and nitrogen in wastewater results in direct emissions of carbon dioxide (CO_2), methane (CH_4) and nitrous oxide (N_2O) that are potent biogenic greenhouse gases. Flaring, the purposeful burning of digester gas, poses another highly-emitting CO_2 operation by some utilities.

As alternatives to or subsequent processes to engineered systems, natural treatment systems (NTS) have great potential to decarbonize the wastewater treatment by leveraging microbial and/or plant communities. NTS can be used as tertiary (polishing) operations at wastewater treatment plants; or they can be utilized separately to sequester carbon from the atmosphere (negative emissions), treat stormwater runoff for removal of metals and organic contaminants (green infrastructure), or prevent erosion and runoff of nutrients and pesticides in agricultural applications. NTS are engineered to use a minimal amount of energy and mechanization. Instead, they utilize soil, plants/algae, bacteria, and fungi to achieve sequestration of metals, carbon storage in wood and soils, or biodegradation of organic contaminants to innocuous end-products (e.g., H_2O, CO_2, HCl). They tend to be low cost with low energy consumption, low emissions, and aesthetically pleasing to the public. Consequently, judicious incorporation of NTS into the water and wastewater treatment sectors can mitigate GHGs with low-energy, economical technologies that are adaptable to most regions of the world.

Johnston and Karanfil (2013) estimated the greenhouse gas emissions associated with seven utilities in the southeastern US and determined an average of 1240 kg carbon dioxide equivalents (CO_2-eq)/million gallons (MG) when Scopes 1 and 2 emissions were evaluated. On an energetic basis, they found energy use in the water cycle to vary from 1250 to 6500 kWh/MG with wastewater treatment processes using the most energy. Energy required to treat and distribute drinking water ranged from 250 to 3500 kWh/MG.

Indicative of operations which may cause greenhouse gas emissions in the utility industry, Figure 12.1 shows the conveyance of water in drinking water and wastewater processes. Total energy utilization for all the operations ranges from 1030 to 36 200 kWh/MG (Griffiths-Sattenspiel & Wilson, 2009). Pumping, customer usage (Scope 3 emissions) and wastewater treatment represent large energy inputs and GHG emissions, thus, opportunities for GHG reductions. Meanwhile sludge handling and natural treatment systems offer the potential for large energy savings – even net zero emissions by virtue of utility-generated electricity, combined heat and power (CHP), and carbon sequestration (negative emissions). Obviously, one of the first considerations should be to power the entire system with renewable energy to the maximum extent possible (solar, hydro, biomass and wind) or low-carbon electricity provided by nuclear power. If space is available onsite, solar panels for powering the facility avoids Scope 2 emissions, a highly attractive investment. Water conservation, fewer chemicals, less pumping, and more gravity flow offer other options to reduce GHG emissions (Erickson *et al.*, 2008). In wastewater processes, anaerobic operations result in less sludge production and more methane production which can be used for process heating or microturbine electricity generation. Process modifications such as anaerobic membrane bioreactors can be run in colder climates than previously thought, which offers an alternative to achieve net zero emissions (McCarty *et al.*, 2011). Burning biosolids (biomass) to make district heat and electricity is another possibility in addition to the application of biosolids onto nearby agricultural land for co-benefits of soil conditioning and carbon sequestration.

One of the keys to reducing greenhouse gas emissions in the water industry is to view the entire water cycle holistically – the One Water movement (*One Water Hub/US Water Alliance*). When we pump-up groundwater or surface water as the source for drinking water, we begin the cycle. Drinking water becomes wastewater (used water), and then used water is returned to streams or groundwater after appropriate treatment, thus completing the cycle. Drinking water becomes wastewater, and wastewater becomes drinking water.

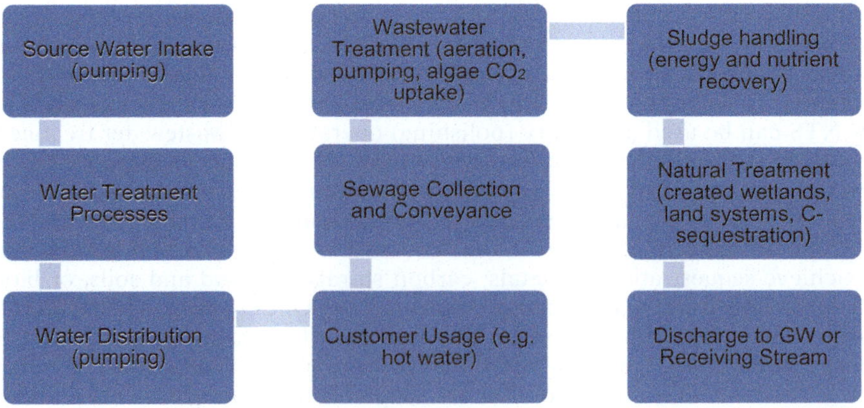

Figure 12.1 Water and wastewater utilities general flowscheme. Energy is required for each step in the process with concomitant greenhouse gas emissions.

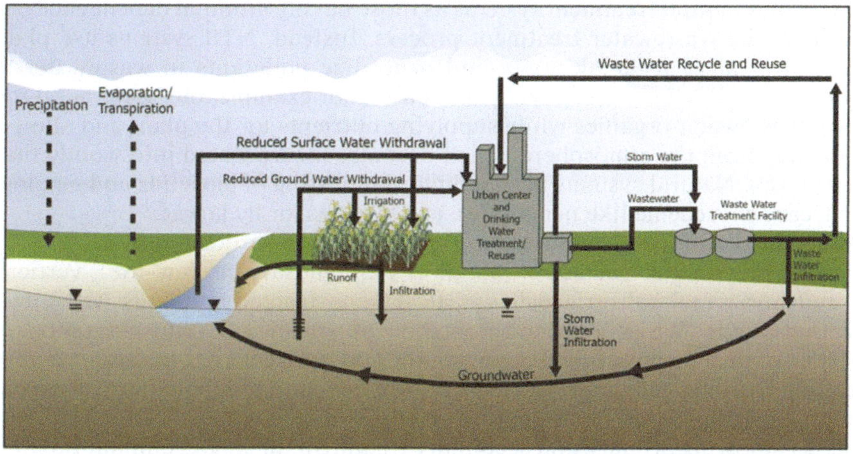

Figure 12.2 Water and wastewater treatment as a part of the One water cycle (withdrawal, drinking water treatment, wastewater treatment, recovery of nutrients/water/energy, fertigation onto crops, and recharge of groundwater and surface water).

In Integrated Watershed Management, drinking water treatment plants and wastewater treatment and recovery are designed within the context of the entire system, the One Water Cycle. We can create a circular economy around the use of water. Figure 12.2 depicts the water cycle, illustrating some opportunities for conservation, integrated management, and reduction of carbon emissions. Under integrated watershed management, wastewater treatment is more properly termed 'water reclamation and reuse' where nutrients, energy, and water are recovered, and greenhouse gas emissions are mitigated. By reusing water, the withdrawals of source water are decreased, thus reducing the energy and carbon footprint of the entire water cycle. Water is reused for graywater applications, irrigation/ fertigation, aquifer recharge, and even for direct potable reuse. By treating and infiltrating stormwater, aquifers are replenished and more water is available for reuse. Some treated wastewater is also available for aquifer storage and recovery via infiltration basins. Excess nutrients from wastewater can be applied as irrigation water onto crops (fertigation) as illustrated in Figure 12.2.

In this chapter, we describe natural treatment system technologies, their advantages and disadvantages, and their potential to decarbonize the water sector when incorporated in integrated watershed management. We emphasize phytoremediation and microalgal cultivation in two engineering research examples.

12.2 NATURAL TREATMENT SYSTEMS

Natural treatment systems can be as effective as engineered treatment systems but their reliance on treatment mechanisms original to nature is unique. The processes we use to improve our anthropogenically-altered water quality and soil health have been purifying water and soil systems for eons. NTS depend primarily on ecosystems of microorganisms and/or plants for pollutant sequestration or biodegradation to treat wastewater, contaminated soils, and contaminated waters. These microbial- and phyto-based treatment systems will remove carbon, nutrients, some pathogens, and are particularly favorable for the removal of trace contaminants of emerging concern at low concentrations including pharmaceuticals, consumer care products, and pesticides. When incorporated into traditional water and wastewater operations, NTS can greatly reduce GHG emissions with the ultimate goal of 'net zero emissions'.

The US EPA defines natural treatment systems as those having minimal dependence on mechanical elements to support the wastewater treatment process. Instead, NTS systems use plants, bacteria, archaea, fungi and/or algae to break down and neutralize pollutants in wastewater. Often, these natural components work symbiotically with each other. For example, the bacteria on roots of plants may break down wastewater organics while supplying nutrients for the plant and allowing the plant to fix carbon dioxide from the atmosphere which becomes incorporated into woody tissue and soils (carbon sequestration). Natural systems may include composting of biosolids and employ other small animals in the treatment scheme like nematodes, earthworms, or fly larvae.

Through site remediation and wastewater treatment, NTS protect public and environmental health. NTS include free water surface wetlands (FWS), subsurface flow (SF), vertical subsurface flow (VSSF), and horizontal subsurface flow (HSSF) constructed wetlands (CW), biofilters (BF), waste stabilization ponds (WSP), and land irrigation of wastewater onto plantations of trees and grasses (PHYTO). With the goal of conserving water and nutrients, groundwater can be recharged through infiltration basins, wastewater nutrients are applied onto cropland (fertigation), and biosolids applications onto land provide soil conditioning and carbon sequestration. Such NTS systems can be employed in-series following primary and secondary treatment for small communities and developing countries, or they can be used as 'polishing' or tertiary treatment on the back-side of conventional wastewater treatment facilities. As a tertiary treatment or effluent polishing step, NTS can remove antimicrobial drugs that may otherwise be released by conventional wastewater treatment systems in microdoses that can confer antibiotic-resistance to pathogenic bacteria (Ryan *et al.*, 2011). In the same way, NTS can also sequester or degrade anthropogenic compounds that are toxic to aquatic systems. Where NTS use photosynthetic organisms, waters or polluted soils can be simultaneously treated and oxygenated, thereby improving water quality or supporting aerobic environments for further pollutant degradation.

Beyond the value of low-cost, low-energy, decarbonizing treatment, certain NTS provide marketable products that offset treatment costs or energy requirements. Atmospheric carbon dioxide is first fixed as biomass, and then can be used for biofuels, fertilizer, feed, biochar feedstock, fiber for pulp or paper, or burned directly to generate power. However, the expense of biomass transport and access to quality storage options can be prohibitive. Biomass can also be converted to biochar, a stable product for storage, through pyrolysis. As a soil additive, biochar increases the health of soil through increased water and nutrient capacity and subsequently improves the carbon sequestration abilities of the soil (Saeid & Chojnacka, 2019).

Natural treatment systems can be simpler, more cost-effective, efficient, and reliable than conventional treatment infrastructure. NTS are particularly attractive as a means to meet wastewater treatment standards, remove nutrients and micropollutants, and sequester or utilize carbon because they require less capital and operational investment than conventional methods (Mahmood *et al.*, 2013). The systems are relatively inexpensive to install and rarely rely upon the chemical inputs or mechanical parts necessary for 'gray infrastructure,' making them effective for areas that have limited access to power, specialized equipment, or skilled workers. These qualities make NTS appealing to small communities or those in developing countries, as it facilitates effective wastewater treatment and avoids a substantial investment that many communities cannot afford.

Of course, limitations exist such as the high degree of treatment specified in some effluent discharge permits that may be difficult to achieve. Cold weather, temperate climates, and seasonal events (floods, droughts) may also limit the application of NTS. However, NTS have great potential and adaptability to various environments. The wide range of technologies included in the NTS umbrella naturally cover a wide range of treatment scenarios; NTS are used to treat wastewater, stormwater, agricultural runoff, and contaminated sites. The ability of NTS biomass to utilize nitrogen and phosphorous makes it extremely useful for remediating high nutrient concentrations in wastewater and agricultural runoff. Additionally, specialized NTS can sequester or transform certain hazardous organic species and heavy metals, which makes them suitable to treat contaminated sites. NTS can also act as a stormwater filter and remove pollutants from stormwater runoff before it enters rivers, lakes, or groundwater. For each of these treatment scenarios, NTS employ mechanisms long-established in the natural world that today's engineers now channel into designed systems.

12.2.1 Constructed wetlands

Constructed wetlands are natural biological treatment systems with a variety of applications and design parameters. They are traditionally used for conventional wastewater treatment of BOD and nutrients, but they are also useful for degradation of emerging organic contaminants (consumer product chemicals, pharmaceuticals, pesticides) in tertiary treatment or polishing applications. They are planted with rooted vegetation (e.g., reeds, rushes, sedges, cattails) and with slow flow filtration paths configured as horizontal (surface and subsurface flow) (Figure 12.3), or vertical flow for deeper penetration into the root zone where the highest concentrations of degrading microorganisms reside. Slow filtration through the root zone is the key to the process. Crites *et al.* (2014) list design specifications for various configurations of constructed wetlands. Constructed wetlands have shallow waters and provide the opportunity for photons from the sun to degrade susceptible pollutants through direct or indirect photolysis. In addition to photodegradation of chemical contaminants, photolysis can inactivate pathogens through natural photosensitizers (i.e., DOM) (Wenk *et al.*, 2019).

Generally, researchers agree that created wetlands have the potential to serve as substantial carbon sinks, especially relative to conventional wastewater treatment methodologies (Rosli *et al.*, 2017). The growth of vegetation in constructed wetlands fixes carbon dioxide from the atmosphere into biomass, and organic carbon accumulates in sediments from senescence and root turnover (Nahlik & Fennessy, 2016). However, methane release from anaerobic sediments and nitrous oxide emissions must be considered in the overall greenhouse gas balance, as well as the CO_2 released when microbes degrade organic materials in wastewater.

On a mass basis, each mg of CH_4 has 28 times the global warming potential (GWP) of a mg of CO_2, calculated over a timescale of 100 years. A mg of nitrous oxide (N_2O) has 265 times the GWP of CO_2 on that same mass basis. N_2O emitted today remains in the atmosphere for an average of 121 years, while CH_4 resides for an average of about 12.4 years (Myhre *et al.*, 2013).

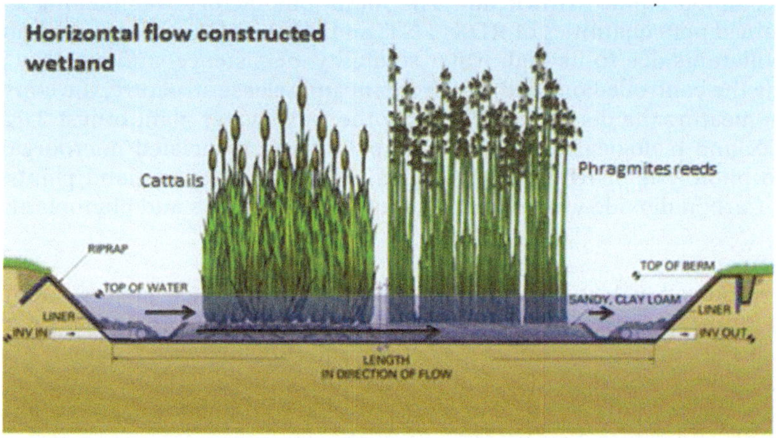

Figure 12.3 A typical horizontal flow constructed wetland treats conventional pollutants like BOD and nutrients following primary treatment in rural settings or as a tertiary/polishing step following secondary treatment. Toxic trace organics from pharmaceuticals, consumer-care products, industrial chemicals, and pesticides may be degraded in such systems by biological processes and photolysis. Adapted from https://waterpurificationengineering.weebly.com/constructed-wetlands.html.wetlands.html

In a wide literature analysis, Mander *et al.* (2014) found that free water surface (FWS) constructed wetlands had substantially lower CO_2 emissions than subsurface flow CWs, 95.8–137.0 mg m^{-2} h^{-1}, respectively. Methane emissions ranged from 3.0 to 6.4 mg m^{-2} h^{-1}, while N_2O release rates were small but significant at 0.09–0.13 mg m^{-2} h^{-1}. From this reference, it can be estimated that constructed wetlands have lower greenhouse gas emissions than conventional forms of wastewater treatment like activated sludge. They contribute a smaller net source of total greenhouse gases on a CO_2-equivalent basis: 191–332 mg CO_2-eq m^{-2} h^{-1}. Roughly 41–50% of global warming potential (GWP) was due to CO_2, 44–54% due to CH_4, and 5–6% due to N_2O emissions.

Overall, created wetlands may constitute a source or a sink of greenhouse gas emissions. It depends on the wastewater influent concentrations (BOD, NH_4^+, NO_3^-), flow design and size of unit operations, and the timescale of the analysis. In one particularly careful study, measured methane emissions were highly variable but indicated a significant relationship with temperature and the density of vegetation. Average methane emissions in the vegetation were 7.8 at 15°C and 24.5 mg m^{-2} h^{-1} at 24°C, respectively. Nitrous oxide emissions ranged from 0.5 to 1.9 g m^{-2} y^{-1}. Net carbon dioxide sequestration was measured as 0.27–2.4 kg m^{-2} y^{-1} which represented 12–67% of the CO_2 photosynthesized into biomass. N_2O emissions were a significant fraction of total GHG emissions (12–29%) according to de Klein and van der Werf (2014). For this example, the constructed wetland was a net sink for greenhouse gases expressed on a CO_2-equivalent basis of 30.8–274 mg CO_2-eq m^{-2} h^{-1}. CO_2 was photosynthesized into plant biomass while N_2O and CH_4 represented GHG emissions, which were more than offset by carbon sequestration. Such a study demonstrates the potential for created wetlands to serve as a net sink for greenhouse gases resulting in 'negative emissions'.

Floating mats of plants can also be used to create superior removal of pollutants from free water surface wetlands and ponds. According to Pavlineri *et al.* (2017), floating wetlands removed 58% of total nitrogen, 48.75% of total phosphorus, 72.8% total NH_4-H, and 57.8% of chemical oxygen demand (COD). Floating wetlands may also sequester CO_2 into plant biomass, but a net greenhouse gas analysis was not performed in this study.

An example of a horizontal created wetland is shown at the Iowa Army Ammunition Plant in Middletown, Iowa, in Figure 12.4. Contaminated groundwater and effluent from the nearby munitions factory were treated by a 2-acre free-surface wetland planted with native vegetation, especially *Sagittaria* spp. (common name, arrowhead). The munitions factory was making C4-explosive and wastewater contained ppm quantities of RDX, TNT, and HMX. RDX was the most problematic of the toxic chemical pollutants due to its high water solubility, persistence, and mobility in groundwater. After constructing the controlled outlet through a dam and release structure, the constructed wetland was successful in meeting the discharge permit for the wastewater plant of just 2 μg/L (ppb) due to photolysis of RDX and biological degradation by plants and associated microorganisms. Removal was attributed to photolysis of RDX and biological degradation by wetland plants and associated microorganisms. Carbon dioxide was sequestered into plant biomass and phytoplankton.

12.2.2 Treatment lagoons

Lagoons (waste stabilization lagoons or ponds) are shallow (1.2–2.4 meters) manmade structures designed to hold and treat wastewater with bacteria and microorganisms that break down various contaminants over the designed hydraulic residence time (Bowman *et al.*, 2002). Like natural lakes, lagoons can stratify into anaerobic, facultative and aerobic layers. These systems are intended to take in wastewater, remove nutrients and decrease chemical and biological oxygen demand (without mixing or aeration), and return the water back to the environment. Lining, usually made of clay or geosynthetics, prevents leakage of contaminated water into the groundwater. The simplicity and affordability of treatment lagoons make them popular for small, rural communities or agricultural operations with lax effluent discharge regulations. Independently, lagoon systems are rarely able to treat to the stringent effluent limits year-round nor do the systems offer much operator control. Common operational problems include overgrowth of algae (which may be managed through biomass

Figure 12.4 Aerial view (left) and photo of the inlet to the horizontal flow, free surface pond at the iowa army ammunition plant in middletown, iowa. It is planted with arrowhead (native vegetation) and resulted in the treatment of RDX contaminant to meet required effluent discharge permit.

harvesting) (Steinmann *et al.*, 2003), sludge buildup, uncontrolled effluent ammonia concentrations, strong odors during spring or fall inversions, and the creation of habitat for insect vectors (e.g., mosquitos). Uncovered anaerobic lagoons for animal manure treatment are particularly problematic as a major source of CH_4 and a small contributor of N_2O; in fact, manure management accounted for 9.7% of all United States anthropogenic CH_4 emissions in 2018 (Desai & Camobreco, 2020).

In consideration of the shortcomings of lagoon systems for wastewater treatment, research is ongoing to optimize alternative wastewater treatment strategies that will meet stringent nutrient discharge regulations yet still be affordable for small communities and farms. This is an opportunity to implement technologies that will reduce GHG emissions, produce salable goods, and better protect receiving waters.

12.2.3 Bioremediation and biofiltration

The goal of bioremediation is to harness specialized microbial enzymes to completely mineralize pollutants or decrease their concentrations to levels below regulatory limits. During the process, environmental conditions are altered to encourage the growth of microbes that degrade target pollutants. This is often achieved through bioaugmentation, the addition of microbial cultures, or biostimulation, the addition of rate limiting nutrients or electron acceptors, to increase the rate of biodegradation. Given suitable conditions for high rates of microbial activity, there are a wide range of compounds for which this NTS is applicable: municipal wastewater, pesticides, industrial chemicals, crude oil components, chlorinated solvents, and so on. Indigenous or introduced microbes can treat a variety of contaminated sites but have significant limitations. Bioremediation of contaminated soils typically requires relatively long treatment times and extensive monitoring. Additionally, microbes often function best under optimized laboratory conditions and their treatment efficiency can decrease with variations of pH, nutrient content, temperature, or the presence of toxic compounds (Karigar & Rao, 2011).

Biological filtration NTS, often used for stormwater treatment, generally function as a combination of physical and biological treatment methods. Filter media, typically sand or activated carbon, physically traps contaminants through sorption or slows contaminant flow rates and provides surface area for microbial growth. The microorganisms form biofilms on the media and consume or sequester contaminants, as in bioremediation. Sorption is especially significant where highly porous activated carbon serves as the filter medium and can remove organic compounds and low concentrations of heavy metals. Activated carbon's removal efficacy is often predicted by a compound's hydrophobicity but can also depend on pore diffusivity (molecular volume), electrostatic/pi-pi interactions, and hydrogen donor/acceptor interactions with specific surface groups (Webb *et al.*, 2020). Biofiltration is also inadvertently used in drinking water treatment where pollutant-degrading microbes colonize granular or powdered activated carbon for contaminant removal.

12.2.4 Microalgal cultivation

Phototrophs, including microalgae, can use solar energy to sequester CO_2 and take up nutrients from wastewater. CO_2 may be sourced from the atmosphere or waste streams, especially for microalgae that can tolerate high concentrations of potentially toxic flue gas components, including NO_x, SO_x, and CO. Cyanobacteria and eukaryotic microalgae can fix CO_2 rapidly; at a rate 10–50 times greater than that of terrestrial plants (Iasimone *et al.*, 2017). Through photosynthesis, microalgae fix carbon from CO_2 and release O_2 as a by-product. If grown in symbiotic microalgae-bacterial communities, microalgae produce the O_2 which serves as an electron acceptor for aerobic bacteria and fixes the CO_2 produced by the same bacteria. Oxygen production can also save aeration costs and energy in subsequent wastewater treatment processes; 50% or more of total energy consumption can be attributed to aeration in some wastewater treatment plants (Lemar & de Fontaine, 2017). Microalgae can remove nitrogen and phosphorus from wastewater, as well as more challenging contaminants at low concentrations, including pesticides, heavy metals, and other inorganic and organic contaminants.

Microalgae is typically cultivated in one of two general systems, open or closed. Open cultivation systems include ponds, lagoons, or raceways. Closed systems contain microalgal culture in transparent vessels, generally termed 'photobioreactors', which may be flat panels, tubes, or plastic bags. Open systems are less expensive and easier to operate than photobioreactors but provide little control over CO_2 mass transfer, culture contamination, and evaporation rates. Without separation of HRT and SRT, both system types require large land areas. Cultivated microalgal biomass can be harvested for revenue to offset other process costs; harvesting cells from dilute solution is especially energy- and cost-intensive, making up 20–30% of the biomass production cost (Fasaei *et al.*, 2018; Van Den Hende *et al.*, 2011).

Waste streams of otherwise emitted CO_2 (from bioethanol plants, cement manufacturers, and flaring operations) have great potential to be used as substrate. Microalgal cultivation operates through similar mechanisms to other biomass-accumulating NTS but does so at an accelerated rate, which ties directly to an accelerated rate of CO_2 sequestration and greater potential for the decarbonization of wastewater treatment.

A detailed description of the decarbonization potential of microalgal wastewater treatment can be found in Chapter 9.

12.2.5 Land treatment systems

Land application of wastewater from the water cycle can take many forms. If land is available to the utility, it is advantageous to use the land for treatment of waste, nutrients and to sequester carbon dioxide from the atmosphere into woody biomass and organic carbon in the soil. Utilities are wise to utilize public/private partners within their sewershed and watershed for investments in green infrastructure. Such partnerships offer the possibility of avoiding Scope 2 emissions while sequestering carbon into soils and woody biomass. Negative emissions can be the first step towards the utility goal of achieving 'net-zero emissions'.

Phytoremediation is one of the most advantageous land treatment systems for wastewater. Traditionally, phytoremediation has been used to clean-up groundwater and contaminated soils containing legacy pollutants, often at brownfield industrial sites. However, wastewater from the utility industry may also be applied continuously as a land treatment onto vegetation. In this application, polishing or tertiary treatment of wastewater can be accomplished by spray- or drip-irrigation (surface or sub-surface) onto plantations of trees inter-planted with a mixture of cool season and warm season

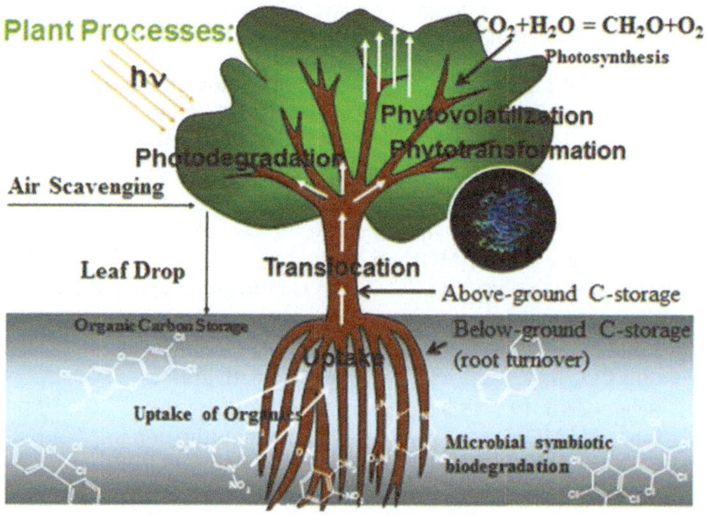

Figure 12.5 Phytoremediation schematic for the processes of photosynthesis and storage of carbon from the atmosphere into woody biomass and soils via root turnover. Other processes include rhizosphere biodegradation, plant uptake, translocation, plant phytotransformation, phytovolatilization, and photodegradation of organic chemicals.

grasses. Dense, shallow roots of the grasses and deep penetrating roots of the trees can facilitate effective treatment of nutrients, fine particles, and organic chemicals to very low target concentrations. Organic chemicals are degraded by rhizosphere bacteria and by the plant itself; nutrients (N,P) are taken-up and removed by the plant; and carbon dioxide is sequestered from the atmosphere into woody biomass and as organic carbon in soils (Figure 12.5). Volatilization may also play a treatment role in any NTS that employs plants, but it is especially prominent in those applications for treatment of volatile organic contaminants with high rates of transpiration regulated by climate, soil and plant types. Under these conditions, contaminants are absorbed through the roots, along with water and nutrients, and carried up the xylem to the plant's stomata where volatile organic compounds can change from the liquid to gas phase. Upon volatilization, some species are photochemically degraded while others may persist for hours or more in the atmosphere.

Regenerative agriculture refers to the use of best management practices (BMPs) to restore and preserve biodiversity and soil quality – it is a form of sustainable farming that values the long-term productivity of soil organic carbon and fertility. BMPs include no tillage, diverse cover crops, multiple crop rotations and legumes, intercropping, recycling of manure through beneficial grazing, minimizing synthetic fertilizers and pesticides, use of perennial crops such as energy crops, silvopasturing (integrating trees, forage, and grazing livestock), and restoring and creating wetlands. Such practices have the potential to reduce erosion and add soil organic carbon where it has been badly depleted in the past. Due to the large area of arable land in the world, it is possible to sequester highly significant amounts of carbon dioxide from the atmosphere each year through regenerative agriculture. Minx *et al.* (2018) estimate 2.3 Gigatons CO_2/yr (2.3 billion metric tons/yr $= 2.3 \times 10^{15}$ g CO_2/yr) could be removed from the atmosphere by soil carbon sequestration which represents roughly 6% of total annual anthropogenic emissions currently.

Water utilities could partner with farmers in their watershed or urban landowners in the sewershed to create negative emissions and sequester carbon dioxide from the atmosphere. Such partnerships could involve utilities supplying irrigation water and fertigation of crops resulting in the removal of nitrogen and phosphorus to meet stringent discharge permits. This form of 'water quality trading'

Table 12.1 Potential to reduce CO_2 and other greenhouse gas (GHG) emissions from the water and wastewater utility industry by negative emissions from natural treatment systems (or solar panels).

Activity	Total GHG Emissions Emitted or Sequestered (CO_2-eq)	Required Resources to off-set GHGs from 1.0 MG of Utility Water and Wastewater (Estimated, this Work)	Reference
Water and wastewater utility industry	Emitted: 1240 kg CO_2-eq/MG (550–2190 g CO_2-eq/MG)		Johnston and Karanfil (2013)
Created wetlands	Sequestered:[a] 0.27–2.4 kg CO_2-eq/m^2-yr	517–4590 m^2 of wetlands	de Klein and van der Werf (2014)
Hybrid poplar buffer strip with grasses	Sequestered: 27.5–29.3 metric tons CO_2/ha-yr	423–451 m^2 of hybrid poplar with grasses	Ney *et al.* (2005)
Microalgal cultivation	Sequestered: 12.7–15.5 metric tons CO_2/MG	2300–2800 m^2 of *S. obliquus* cultivation in vertical manifold PBRs	Molitor and Schnoor (2020); Zhu *et al.* (2018)
Native forests in US and Iowa	Sequestered: 3.7–7.0 metric tons CO_2/ha-yr	1770–3350 m^2 of forest land	Ney *et al.* (2002, 2005)
Regenerative agriculture, soil cover crops and carbon farming	Sequestered: 0.2–7.4 metric tons CO_2-eq/ha-yr	1680–62 000 m^2 of farmland (in partnership)	Lal (2004, 2015); Minx *et al.* (2018); Tellatin and Myers (2018)
Solar panels	Power generated: 6.6 kW solar panels avoid 10.6 metric tons of CO_2/yr from electricity production (US avg)	0.77 kW of solar power or three solar panels 4.8 m^2 (52 ft^2) required to off-set 1.0 MG of utility CO_2-eq emissions	Authors assumed 300-watt solar panels and United States average electricity mix to calculate GHG emissions avoided

[a]Created wetlands can serve as a net source or sink for GHGs depending on conditions.

within the watershed is also possible whereby farmers sequester carbon into agricultural soils much more cheaply than utilities can remove or reduce it. A direct payment to farmers for the service of negative emissions is also possible. The goal of the utility is to recover as much resource from their wastewater as possible including dollars, water, nutrients, and energy while avoiding greenhouse gas emissions or creating negative emissions as off-sets.

Typical CO_2 emissions from one million gallons (1.0 MG) of water withdrawn, used, treated, and discharged by utilities (as in Figure 12.1) could be off-set by 1680–62 000 m^2 of farmland practicing regenerative agriculture for the sequestration of carbon into soils (Lal, 2004, 2015; Minx *et al.*, 2018; Tellatin & Myers, 2018). Only a relatively small portion of the CO_2 emitted by utilities would be considered biogenic emissions, and the majority is from energy requirements to pump the water, run the plants and treat the water. Based on the results in Table 12.1, it is clear that considerable land is required to create enough negative emissions to off-set greenhouse gas emissions from water and wastewater utilities. The most land would be required for regenerative agriculture and the least for hybrid poplar buffer strips. Created wetlands show promise also, but the literature is mixed as to their net benefits (carbon sinks or sources).

One million gallons of water produced by the water utility is a rather small volume of water by utility standards. Even a small community of 5000 people using a total of 200 gallons per capita-day would generate 1.0 million gallons per day (MGD) or 365 MG per year. Multiplying the areas in Table 12.1

(third column for hybrid poplar buffers) by 365 yields a required spatial area of 15.4–2260 ha – quite a lot of land for a small utility to control. It is not practical to off-set 100% of the greenhouse gas emissions from water utilities by NTS; rather it suggests one option for utilities to consider reducing their carbon footprint.

To illustrate, the last row in Table 12.1 supposes that the utility purchases or partners to obtain solar power for electrifying its operations or, alternatively, to off-set their Scope 2 greenhouse gas emissions. The relatively small area (4.8 m^2) of solar panels required to off-set emissions from 1 MG of water suggests a viable option for the utility to reduce its carbon footprint. For the 1 MGD water utility (365 MG per year), only 1750 m^2 (0.175 ha) of solar panels would be required to off-set 100% of the carbon footprint. Utilizing NTS would require much more land. Thus, most water utilities which have reduced their carbon footprints have done so through: (1) electrifying pumping and other operations with solar power and battery storage; (2) upgrading anaerobic digestion of wastewater solids; (3) producing power from methane digester gas by microturbines; (4) producing combined heat and power (CHP) by internal combustion engines using digester gas; (5) improving sludge stabilization and biosolids applications or reducing landfilling; (6) implementing demand side management with water conservation by smart water metering and variable pricing campaigns.

Indeed, Wong and Law-Flood (2011) have provided examples of several water utilities which have successfully reduced their carbon footprints, mainly by anaerobic digestion of wastewater solids and utilization of the gas (Wong & Law-Flood, 2011). Examples include: Sheboygan, Wisconsin achieved 35–50% reduction in energy; Nashua, New Hampshire saved $750 000 in annual electricity and landfill costs from digester and biosolids improvements and 20% reduction in energy; Gloversville-Johnstown, New York combined heat and power (CHP) system generated 100% of total energy needs on-site; Essex Junction, Vermont used microturbines and CHP to provide 37–39% of its total energy needs; Pittsfield, MA employed digester gas for CHP internal combustion engines and microturbines to save 29% of energy needs; East Bay Municipal Utility District, California used internal combustion engines for CHP and saved 90%; and Fairhaven, MA increased their organic solids loading to augment volume of digester gas resulting in CHP to save 73% of energy.

12.3 CASE STUDY: CARBON SEQUESTRATION AND AGRICULTURAL RUNOFF TREATMENT THROUGH PHYTOREMEDIATION

An example of phytoremediation of runoff from agricultural lands is illustrated in Figure 12.6 at Amana, Iowa. The setting consists of row crop agriculture of corn and soybeans in rotation with herbicides and fertilizers applied liberally. Approximately 0.5–1 kg per acre of active ingredient herbicides (atrazine, alachlor) and 68 kg N per acre (as anhydrous ammonia) are applied each spring. To protect the stream and groundwater quality, three rows of hybrid poplar trees (*Populus deltoides x nigra*, DN-34) were spaced 2.4 m apart with 1 m between trees. Trees were planted in a trenched row with bare cuttings (poles) protruding 20 cm out of the soil surface initially. The cuttings leafed-out and rapidly grew roots down to the surface of the groundwater table at a depth of 1.2–2.4 m, thus taking up runoff chemicals and surface groundwater nitrates simultaneously. The stand is shown in Figure 12.6 after seven years of growth. About 200 trees were lost or thinned during the seven-year period for a total of 1478 trees over each acre (0.4 ha). Delivery of eroded sediment from the row crops to the stream was cut to nearly zero, nitrate and pesticides were uptaken, (Paterson & Schnoor, 1992, 1993) and nitrate (NO_3^-) in groundwater was cut from 100 mg L^{-1} to less than 5.0 mg L^{-1}. However, the stream received nitrate during large storm events when NO_3^- was delivered by tile drainage lines directly to the surface water without being filtered through the root zone of the trees.

An analysis of the carbon sequestration by hybrid poplar into above-ground woody biomass at the Amana site after years was reported by Ney *et al.* (2005). It amounted to 22.05–22.92 metric tons C per acre (i.e., 54.5–56.6 metric tons C per hectare). To sequester that much carbon into the woody

Figure 12.6 Hybrid poplar plantation after seven years of growth at Amana, Iowa, to treat runoff of pesticides and nutrients from agriculture row crops while sequestering carbon dioxide into woody biomass and soils.

biomass would require a CO_2-sequestration rate from the atmosphere of 200–208 metric tons CO_2 per hectare over seven years. On average annually, the CO_2-sequestration into woody biomass was therefore equivalent to 27.5–29.3 metric tons CO_2 ha^{-1} y^{-1} as shown in Table 12.1. Forests in the US sequester 3.57–5.03 metric tons carbon per hectare per year in above-ground biomass. (Ney *et al.*, 2005) Unmanaged forests in Iowa averaged 5.06 metric tons C ha^{-1} y^{-1}, (Ney *et al.*, 2002) with a range of of 3.7–7.0 metric tons CO_2 ha^{-1} y^{-1} (Table 12.1). However, most of the carbon stock in the riparian zone buffer strip shown in Figure 12.6 is stored in the soils (below ground). Average soil organic carbon in the top 30 cm of soil was 5.35%. (Ney *et al.*, 2005) However, the amount sequestered into below-ground soils on an annual basis is relatively small compared to the above-ground woody biomass sequestration.

In native forests of Iowa, carbon stocks are estimated to be 137.3 metric tons C per ha, of which 60.8% is in soils (top 30 cm), 24.5% stored in above-ground biomass in trees, 3.8% resides in trees below-ground (roots), 10.4% lies in carbon on the forest floor, and 0.5% is comprised of understory vegetation (Ney *et al.*, 2002). By far the most carbon in native Iowa forests resides in below-ground soils and roots, 83.6–90.7 metric tons C ha^{-1}. Above-ground carbon content (in trees, understory, and the forest floor) ranges from 35.2 to 61.9 metric tons C ha^{-1} with the greatest amount in native oak-hickory forests. Soil carbon is a large pool which builds-up and oxidizes very slowly but can be lost by erosion relatively quickly over decades to centuries. Net benefits of reforestation of native forests in Iowa include the possibility to remove 7.0 metric tons CO_2 ha^{-1} y^{-1} of which 6.21 metric tons CO_2 ha^{-1} y^{-1} goes into above-ground forest biomass and 0.79 metric tons CO_2 ha^{-1} y^{-1} sequesters into below-ground soil and roots (Ney *et al.*, 2002).

12.4 CASE STUDY: POWER PLANT FLUE GAS AND FERTILIZER WASTEWATER TREATMENT BY NUTRITIOUS MICROALGAE

Microalgae are a promising alternative livestock feed source, with greater resource-to-biomass conversion efficiency than conventional agricultural crops (corn, soy, wheat, etc.). High-protein microalgae are of particular promise because they are able to remove nitrogen and phosphorus from wastewater, fix CO_2 and other pollutants from flue gas, tolerate saline water, and occupy less land footprint per unit of produced biomass. This NTS resource recovery scheme simultaneously treats waste, produces valuable biomass, and decarbonizes agriculture and waste treatment systems.

Table 12.2 Nutrient and flue gas component utilization rates by *S. obliquus* grown with simulated coal-fired power plant flue gas and triple-nitrogen Bold's Basal Medium during the exponential growth phase.

Flue Gas and Nutrient Components	Maximum Microalgal Utilization Rate (mg L^{-1} d^{-1})	Comments
CO_2	1300±80	Either fixed as biomass or escaped the PBR, less than 0.1% CO_2 went into solution
SO_2	6.9±0.4	Majority accumulated in culture medium as SO_4^{2-} at a rate of 130±20 mg L^{-1} d^{-1}
NO_2	–	Non-detectable relative to NO_3^-
CO	–	Not assessed, possibly oxidized to a negligible quantity of CO_2
NO_3^-	200±10	Majority from culture medium
PO_4^{3-}	22±3	Exclusively from culture medium

In our research, we seek to inform technology scale-up for power plant and industrial flue gas treatment, and wastewater treatment, by microalgae. At bench scale, we modeled substrate inhibition of microalgal biomass productivity by CO_2, stimulated microalgal growth with simulated coal-fired power plant emissions, and enhanced microalgal settleability with simulated emissions through increased production of extracellular polymeric substances (EPS) (Table 12.2).

12.4.1 Scaling-up microalgal cultivation

From experimental microalgal data, characteristics of fertilizer plant wastewater (pH 6.8, 47.3 mg/L N, 7.5 mg/L P, negligible organic carbon concentration), and data from the University of Iowa power plant, we present a full-scale scheme to treat 1 MGD wastewater and simultaneously utilize emitted flue gas carbon through microalgal cultivation (Figure 12.7).

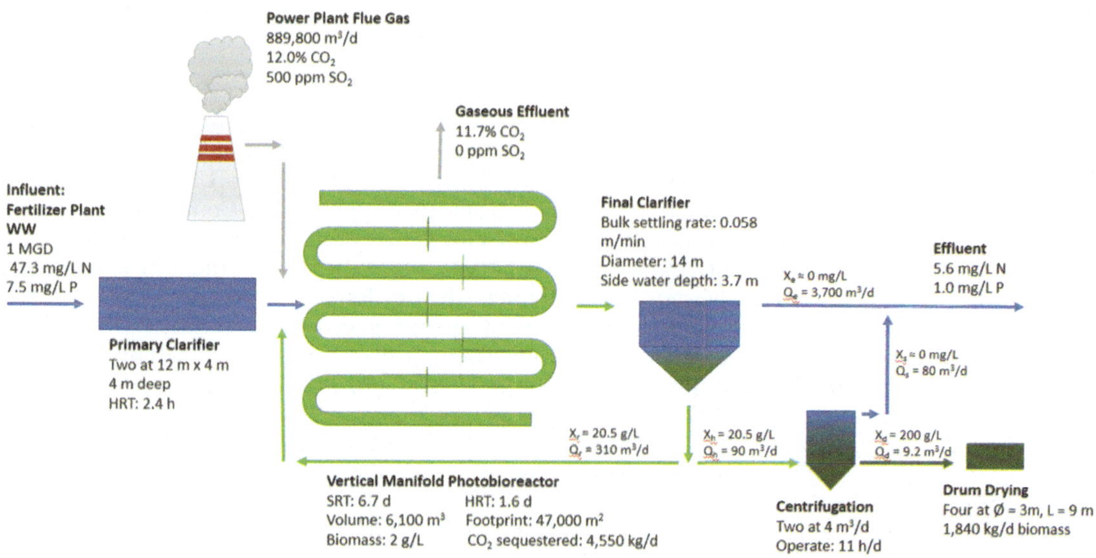

Figure 12.7 Schematic of treatment of 1 MGD fertilizer plant wastewater and CO_2 sequestration from power plant flue gas by microalgal photobioreactors. Treatment processes (clarification, PBRs, and final clarification) and followed by biomass harvesting.

The wastewater is treated through primary clarification, a vertical manifold PBR with biomass recycle, and final clarification while a co-located power plant (889 800 m^3/d; 12% CO_2) supplies CO_2 to the PBR. Biomass dewatering and drying is achieved through centrifugation and drum drying. The influent solids concentration is first reduced using two rectangular clarifiers, each 12 × 4 m and 4 m deep with 2.4 h hydraulic residence time (Metcalf & Eddy, 2003). The clarifier effluent is then fed to the PBR where nitrogen and phosphorus are removed by *S. obliquus* (dry biomass composition: 50.7 ± 0.1% C, 6.44 ± 0.04% N, 1.0 ± 0.1% P) (Molitor & Schnoor, 2020). The PBR system was designed to meet generic wastewater P effluent limits of 1.0 mg/L, as the system is barely N-limited, with maximum culture density of 2 g/L (Acién *et al.*, 2012). The designed solids retention time and hydraulic retention time are 6.7 and 1.6 d, respectively. Consequently, the PBR requires 6100 m^3 of working volume and a recycle rate of 310 m^3/d of 20.5 g/L biomass from the final clarifier.

To achieve a surface overflow rate of 27 m^3/m^2/d, a final clarifier is 14 m in diameter with side water depth of 3.7 m because the microalgae have a relatively rapid bulk settling rate of 0.058 m/min (measured settling rate of 1 m/min adjusted to account for a fraction of the biomass settling at that rate). The possibility of bulk settling in this scenario, facilitated by high concentrations of EPS, is unique and advantageous to efficient biomass harvesting. The settled biomass that is harvested (90 m^3/d), rather than recycled to the PBRs, is dewatered through centrifugation in two centrifuges with capacities of 4 m^3/d and operating at 11 h/d (Tredici *et al.*, 2016). The centrifuges produce microalgal paste of approximately 20% solids, which is then dried to approximately 5% solids through four drum dryers (Tang *et al.*, 2003).

The treatment scheme has effluent concentrations of 5.6 mg/L N and 1.0 mg/L P, produces 1840 kg/d dried microalgal biomass, and sequesters 4550 kg/d CO_2.

Microalgae are well suited to treat high nitrate wastewaters, without carbon source supplementation, which is generally considered challenging for other microbes (Pinar *et al.*, 1997). Fortunately, there are thousands of sources of domestic secondary effluent, some of which are co-located with CO_2-emitting power plants. Beyond municipal wastewater, industrial and agricultural wastewater streams with low organic carbon and high nitrogen concentrations (such as explosives factory wastewater, fertilizer plant wastewater, agricultural run-off, and irrigation return waters) would be strategic options for microalgal substrate (Ji *et al.*, 2018). However, heavy metal and pathogen contamination from certain wastewaters could impede production of microalgae for animal feed.

12.4.2 GHG and land footprints

Based on the carbon-sequestration rate of the microalgal biomass of the reactor, 4550 kg/d CO_2 is utilized. However, this rate of CO_2 sequestration only represents 2.4% of the influent CO_2 because the stack emissions rate and CO_2 concentration exceeds the capacity of the PBRs. Utilization could be improved by optimizing biomass production or increasing cultivation volumes. Comparing the results in Table 12.1 to those in Figure 12.7, it is notable that the cultivation area required to simply sequester the GHG emissions from the conventional treatment of 1 MG wastewater is significantly less than the resources required to *treat* 1 MGD wastewater with microalgal PBRs.

The estimated GHG emissions associated with the proposed microalgal treatment scheme were calculated from literature values for the energy requirements for unit processes including: influent wastewater pumping (140.2 kWh/d); primary settling (15.5 kWh/d); PBR mixing, module pumping, gas delivery, and circulation (934.4 kWh/ha/d); biomass return pumping (42.3 kWh/d); final settling (15.5 kWh/d); centrifugation (9.5 kWh/h); and drum drying (5.1 kWh/kg algae) (Goldstein & Smith, 2002; Tang *et al.*, 2003; Tredici *et al.*, 2016) Under the assumption that electricity would be provided from the University of Iowa power plant, the expected GHG emission rate is 0.244 kg/kWh CO_2-eq (primary fuel: bituminous coal, secondary fuels: oil, gas, and biomass) (eGRID, 2018). Accounting for electricity use and CO_2-sequestration, the projected net GHG emissions from the proposed scheme are −1080 kg/d CO_2-eq (−1160 kg/d CO_2, −1150 kg/d SO_2, 5.8 kg/d NO_x, 1.2 kg/d CH_4, and 0.16 kg/d N_2O). Though only 24 kg/d SO_2 (an indirect GHG) will be accumulated in the microalgal biomass, the remainder of the SO_2 (1140 kg/d) will be rapidly oxidized and accumulate at a rate of approximately 200 mg/L/d in the wastewater to a non-inhibitory concentration of 280 mg/L SO_4^{2-}. If electricity were to be provided by wind turbines, at a GHG emission rate of 1.8×10^{-2} kg/kWh CO_2 (Alsaleh & Sattler, 2019), the projected net GHG emissions from the proposed scheme would be −4300 kg/d CO_2-eq.

The corresponding land footprint is 11.6 acres based on a literature-sourced land-footprint-to-cultivation-volume ratio of 7.7 m^2/m^3, the value for a full-scale 1300 m^3-system of vertical manifold PBRs designed by A4F-AlgaFuel, S.A. to treat cement plant flue gases (Torzillo & Chini Zittelli, 2015). Systems external to the PBRs are included in the area-to-cultivation-volume ratio, though the PBRs account for the vast majority of the treatment facility footprint. As technologies for wastewater treatment with microalgae progress, it can be expected that areal productivity will increase significantly through improved understanding of cultivation conditions for increased nutrient uptake rates and optimal HRT and SRT.

12.4.3 Microalgal end products

Since fossil fuel use and CO_2-generating industrial processes are unlikely to cease in the near future, emissions must be mitigated through other means. Recovery of CO_2 as a resource may be enabled through microalgal uptake and the produced biomass may be used for biofuels, fertilizer, commodity chemicals, or feed (Khan *et al.*, 2018; Rebolloso-Fuentes *et al.*, 2001; Silkina *et al.*, 2019). Some autotrophic microorganisms reduce CO_2 for biosynthesis, including microalgae which is also a source of protein (Matassa *et al.*, 2016). Additionally, waste heat from power plants or industrial processes may be recovered to maintain favorable culture temperatures (between 15 and 35°C) (Gassan *et al.*, 2010).

In this scenario, the *S. obliquus* biomass characteristics inform options for beneficial use within an array of possible microalgae-derived end products, the sale of which may offset the cost of this decarbonization technology. The biomass energy content was comparable between microalgae grown with control gases and simulated emissions, as indicated by the H:C ratios of 0.15:1 and 0.14:1, respectively. These values indicated that the whole biomass had relatively low energy content and, if unprocessed, would be better suited to animal feed than as fuel/feedstock. If the biomass were to be combusted for energy, the biomass grown under control conditions and with simulated emissions would produce 1.76 and 1.92 g CO_2/g biomass, respectively. The relatively high N:C values suggested that the biomass would be slow to compost as fertilizer. While the *S. obliquus* control culture protein content ($46.6 \pm 0.8\%$) exceeded that of soy ($40.3 \pm 0.6\%$), that of *S. obliquus* grown with simulated emissions was ($31 \pm 0.8\%$), significantly lower. However, the *S. obliquus* grown with simulated emissions was sufficiently rich in lysine and methionine to have value as a ruminant livestock feed additive.

This work is motivated by the need to overcome cultivation and harvesting barriers to producing marketable microalgal biomass and removing pollutants from flue gas (CO_2, SO_x, and NO_x) in full-scale operations. Thus, use of energy-intensive fertilizer and freshwater resources will decrease, wastewater treatment costs will be offset, and GHG emissions will be lessened while simultaneously producing sustainable biomass products.

12.5 CONCLUSIONS AND OUTLOOK

Natural treatment systems are effective in the uptake of harmful pollutants and nutrients from wastewater systems or contaminated ecosystems, while simultaneously providing opportunities for decarbonization (Table 12.3). As previously mentioned, NTS require little mechanical or technological input to function, making them preferable to chemical or energy-intensive treatments in terms of economic access and greenhouse gas emissions. Furthermore, phytoremediation, microalgae cultivation, and constructed wetlands have potentially salable biomass end products. Microalgal cultivation operates at an accelerated rate, relative to terrestrial plants, and therefore may hold the greatest potential for rapid decarbonization in the wastewater treatment sector, when managed similarly to conventional microbial treatment processes (separate HRT and SRT; tailored conditions to

Table 12.3 Summary of natural treatment system mechanisms, advantages, disadvantages, and means of decarbonization.

Natural Treatment System	Mechanisms	Advantages	Disadvantages	Means of Decarbonization
Constructed wetlands	• Plant/algal uptake • Volatilization • Photolysis • Microbial biodegradation	• Many suitable plant species • Micropollutants removal	• Not suitable for large municipalities • Land area intensive • Potential to introduce invasive species	Plant/algal uptake and storage in biomass and sediments
Lagoons	• Microbial biodegradation • Algal uptake of N,P	• Low cost • Simple design and installation • Low operational skill	• Possible groundwater contamination • Sludge accumulation • Odor • Limited treatment efficacy	Algal uptake and sedimentation Sometimes counterproductive: CH_4 release
Microalgal cultivation	• Phycoremediation • Sequestration	• Rapid carbon sequestration and nutrient consumption • Salable biomass	• Susceptible to pollutants • Dewatering inefficiencies	Algal uptake and utilization
Biofiltration	• Microbial biodegradation • Filtration	• Micropollutant removal and degradation • Possible groundwater impacts • Microbes become specialized	• Overwhelmed by high pollutant concentrations • Expensive O&M for industrial waste • Bioclogging	No carbon sequestration: GHG emissions may be similar to conventional wastewater treatment
Phytoremediation/ Land application	• Rhizosphere bioremediation • Plant uptake	• Low cost, green • Aesthetically pleasing • Carbon sequestration • Increased soil fertility • Decreased toxicity • GW monitoring but no discharge permit required	• Large land requirement • Long treatment period • Potential toxicity of contaminants to trees, grasses	Plant uptake into woody biomass and root turnover sequestering organic carbon into soils

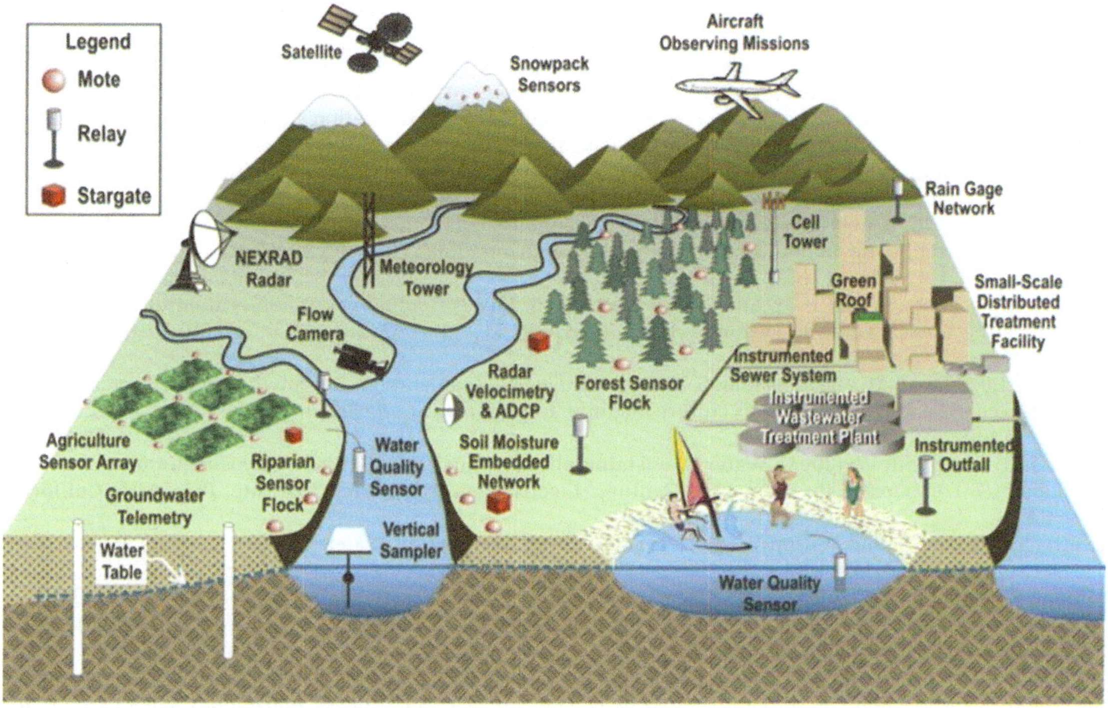

Figure 12.8 Integrated watershed approach.

increase nutrient uptake rates). Combining NTS with conventional wastewater treatment is an attractive method for improving treatment, meeting more stringent permit requirements for N and P, and at the same time capturing carbon via photosynthesis.

Natural treatment systems incorporated into traditional water and wastewater operations can be further integrated with watershed management to reduce the carbon footprint of water utilities with the ultimate goal of 'net zero emissions'. Figure 12.8 illustrates an integrated watershed approach where the water cycle is monitored to reduce costs and greenhouse gas emissions, restore soil carbon, enhance water quality, and conserve water while replenishing aquifers. Small-scale distributed water treatment facilities and wastewater treatment plants are instrumented with sensors to continuously monitor the state of the system in concert with the One-Water concept. NTS are integrated into the water and wastewater operations to provide for carbon sequestration, the uptake of excess nutrients and biodegradation of xenobiotic organic chemicals.

REFERENCES

Acién F. G., Fernández J. M., Magán J. J. and Molina E. (2012). Production cost of a real microalgae production plant and strategies to reduce it. *Biotechnology Advances*, **30**(6), 1344–1353, https://doi.org/10.1016/j.biotechadv.2012.02.005

Alsaleh A. and Sattler M. (2019). Comprehensive life cycle assessment of large wind turbines in the US. *Clean Technologies and Environmental Policy*, **21**(4), 887–903, https://doi.org/10.1007/s10098-019-01678-0

Bowman R. H., Gloyna E. F., Middlebrooks E. J., Pearson G. F., Reed S. and Reid L. C. (2002). Wastewater Technology Fact Sheet: Facultative Lagoons. Office of Water, United States Environmental Protection Agency, Washington, DC, USA.

Crites R. W., Middlebrooks E. J., Bastian R. K. and Reed S. C. (2014). Natural Wastewater Treatment Systems. CRC Press, Boca Raton, FL, USA.

de Klein J. J. M. and van der Werf A. K. (2014). Balancing carbon sequestration and GHG emissions in a constructed wetland. *Ecological Engineering*, **66**, 36–42, https://doi.org/10.1016/j.ecoleng.2013.04.060

eGRID. (2018). University of Iowa Main Power Plant. Emissions & Generation Resource Integrated Database (eGRID). Office of Atmospheric Programs, Clean Air Markets Division, Washington, DC, USA.

Desai M. and Camobreco V. (2020). Inventory of US Greenhouse Gas Emissions and Sinks: 1990–2018. United States Environmental Protection Agency, Washington, DC, USA.

Erickson D., Orrett E. B. and Rosenblum J. (2008). Greenhouse Gas Emissions Related to Water and Wastewater Services: Baseline, Reduction Strategies, and Recommendations. City of Santa Rosa Utilities Department, Santa Rosa, CA, USA, p. 167.

Fasaei F., Bitter J. H., Slegers P. M. and van Boxtel A. J. B. (2018). Techno-economic evaluation of microalgae harvesting and dewatering systems. *Algal Research*, **31**, 347–362, https://doi.org/10.1016/j.algal.2017.11.038

Gassan H., Eugenia M. M. and Sebastián S. (2010). Influence of temperature on growth of scenedesmus obliquus in diluted olive mill wastewater as culture medium. *Engineering in Life Sciences*, **10**(3), 257–264, https://doi.org/10.1002/elsc.201000005

Goldstein R. and Smith W. (2002). Water & Sustainability (Volume 4): U.S. Electricity Consumption for Water Supply & Treatment – the Next Half Century. Electric Power Research Institute, Palo Alto, California, USA.

Griffiths-Sattenspiel B. and Wilson W. (2009). The Carbon Footprint of Water. A River Network Report. River Network, Portland, OR, USA.

Iasimone F., De Felice V., Panico A. and Pirozzi F. (2017). Experimental study for the reduction of CO_2 emissions in wastewater treatment plant using microalgal cultivation. *Journal of CO2 Utilization*, **22**, 1–8, https://doi.org/10.1016/j.jcou.2017.09.004

Ji F., Yin H., Zhang H., Zhang Y. and Lai B. (2018). Treatment of military primary explosives wastewater containing lead styphnate (LS) and lead azide (LA) by mFe0-PS-O_3 process. *Journal of Cleaner Production*, **188**, 860–870, https://doi.org/10.1016/j.jclepro.2018.04.029

Johnston A. H. and Karanfil T. (2013). Calculating the greenhouse gas emissions of water utilities. *Journal AWWA*, **105**(7), E363–E371, https://doi.org/10.5942/jawwa.2013.105.0073

Karigar C. S. and Rao S. S. (2011). Role of microbial enzymes in the bioremediation of pollutants: A review. *Enzyme Research*, **2011**, 11, https://doi.org/10.4061/2011/805187

Khan M. I., Shin J. H. and Kim J. D. (2018). The promising future of microalgae: current status, challenges, and optimization of a sustainable and renewable industry for biofuels, feed, and other products. *Microbial Cell Factories*, **17**(1), 36, https://doi.org/10.1186/s12934-018-0879-x

Lal R. (2004). Soil carbon sequestration to mitigate climate change. *Geoderma*, **123**(1), 1–22, https://doi.org/10.1016/j.geoderma.2004.01.032

Lal R. (2015). Cover cropping and the 4 per thousand proposal. *Journal of Soil and Water Conservation*, **70**, 141A–141A, https://doi.org/10.2489/jswc.70.6.141A

Lemar P. and de Fontaine A. (2017). Energy Data Management Manual for the Wastewater Treatment Sector. Better Buildings. Office of Energy Efficiency and Renewable Energy, United States Department of Energy, Washington, DC, USA.

Loge F. (2016). Is Using Less Water the Secret to Cutting our Greenhouse gas Emissions? The Guardian. Guardian Media Group, London, UK.

Mahmood Q., Pervez A., Zeb B. S., Zaffar H., Yaqoob H., Waseem M., Zahidullah and Afsheen S. (2013). Natural treatment systems as sustainable ecotechnologies for the developing countries. *BioMed Research International*, **19**, 796373.

Mander Ü., Dotro G., Ebie Y., Towprayoon S., Chiemchaisri C., Nogueira S. F., Jamsranjav B., Kasak K., Truu J., Tournebize J. and Mitsch W. J. (2014). Greenhouse gas emission in constructed wetlands for wastewater treatment: a review. *Ecological Engineering*, **66**, 19–35, https://doi.org/10.1016/j.ecoleng.2013.12.006

Matassa S., Verstraete W., Pikaar I. and Boon N. (2016). Autotrophic nitrogen assimilation and carbon capture for microbial protein production by a novel enrichment of hydrogen-oxidizing bacteria. *Water Research*, **101**, 137–146, https://doi.org/10.1016/j.watres.2016.05.077

McCarty P. L., Bae J. and Kim J. (2011). Domestic wastewater treatment as a net energy producer–can this be achieved? *Environmental Science and Technology*, **45**(17), 7100–7106, https://doi.org/10.1021/es2014264

Metcalf & Eddy. (2003). Physical Unit Processes. In: Wastewater Engineering: Treatment and Reuse, 4th edn, G. Tchobanoglous, F. L. Burton and H. D. Stensel (eds), McGraw-Hill, Boston, pp. 393–394.

Minx J. C., Lamb W. F., Callaghan M. W., Fuss S., Hilaire J., Creutzig F., Amann T., Beringer T., de Oliveira Garcia W., Hartmann J., Khanna T., Lenzi D., Luderer G., Nemet G. F., Rogelj J., Smith P., Vicente J. L., Wilcox J. and del Mar Zamora Dominguez M. (2018). Negative emissions – part 1: research landscape and synthesis. *Environmental Research Letters*, **13**(6), 063001, https://doi.org/10.1088/1748-9326/aabf9b

Molitor H. R. and Schnoor J. L. (2020). Using simulated flue gas to rapidly grow nutritious microalgae with enhanced settleability. *ACS Omega*, **5**(42), 27269–27277, https://doi.org/10.1021/acsomega.0c03492

Myhre G., Shindell D., Bréon F. M., Collins W., Fuglestvedt J., Huang J., Koch D., Lamarque J. F., Lee D., Mendoza B., Nakajima T., Robock A., Stephens G., Takemura T. and Zhang H. (eds) (2013). Anthropogenic and Natural Radiative Forcing. Cambridge University Press, Cambridge, UK.

Nahlik A. M. and Fennessy M. S. (2016). Carbon storage in US wetlands. *Nature Communications*, **7**(1), 13835, https://doi.org/10.1038/ncomms13835

Ney R. A., Schnoor J. L. and Mancuso M. A. (2002). A methodology to estimate carbon storage and flux in forestland using existing forest and soils databases. *Environmental Monitoring and Assessment*, **78**(3), 291–307, https://doi.org/10.1023/A:1019939003210

Ney R. A., Meyers T. A., Espina A. and Schnoor J. L. (2005). How many samples are required? Evaluating a model for verification of carbon sequestration in a hybrid poplar buffer strip. *Environmental Monitoring and Assessment*, **102**(1–3), 375–388, https://doi.org/10.1007/s10661-005-6393-8

Paterson K. G. and Schnoor J. L. (1992). Fate of alachlor and atrazine in a riparian zone field site. *Water Environment Research*, **64**(3), 274–283, https://doi.org/10.2175/WER.64.3.13

Paterson K. G. and Schnoor J. L. (1993). Vegetative alteration of nitrate fate in unsaturated zone. *Journal of Environmental Engineering*, **119**(5), 986–993, https://doi.org/10.1061/(ASCE)0733-9372(1993)119:5(986)

Pavlineri N., Skoulikidis N. T. and Tsihrintzis V. A. (2017). Constructed floating wetlands: a review of research, design, operation and management aspects, and data meta-analysis. *Chemical Engineering Journal*, **308**, 1120–1132, https://doi.org/10.1016/j.cej.2016.09.140

Pinar G., Duque E., Haidour A., Oliva J., Sanchez-Barbero L., Calvo V. and Ramos J. L. (1997). Removal of high concentrations of nitrate from industrial wastewaters by bacteria. *Applied and Environmental Microbiology*, **63**(5), 2071–2073, https://doi.org/10.1128/aem.63.5.2071-2073.1997

Rebolloso-Fuentes M. M., Navarro-Pérez A., García-Camacho F., Ramos-Miras J. J. and Guil-Guerrero J. L. (2001). Biomass nutrient profiles of the microalga nannochloropsis. *Journal of Agricultural and Food Chemistry*, **49**(6), 2966–2972, https://doi.org/10.1021/jf0010376

Rosli F. A., Lee K. E., Choo Ta G., Mokhtar M., Latif M. T., Goh T. and Simon N. (2017). The use of constructed wetlands in sequestrating carbon: an overview. *Nature Environment and Pollution Technology*, **16**, 813–819.

Ryan C. C., Tan D. T. and Arnold W. A. (2011). Direct and indirect photolysis of sulfamethoxazole and trimethoprim in wastewater treatment plant effluent. *Water Research*, **45**, 1280–1286, https://doi.org/10.1016/j.watres.2010.10.005

Saeid A. and Chojnacka K. (2019). Organic Farming. Elsevier B.V., The Netherlands.

Silkina A., Ginnever N. E., Fernandes F. and Fuentes-Grünewald C. (2019). Large-scale waste bio-remediation using microalgae cultivation as a platform. *Energies*, **12**(14), 2772–2788, https://doi.org/10.3390/en12142772

Steinmann C. R., Weinhart S. and Melzer A. (2003). A combined system of lagoon and constructed wetland for an effective wastewater treatment. *Water Research*, **37**, 2035–2042, https://doi.org/10.1016/S0043-1354(02)00441-4

Tang J., Feng H. and Shen G.-Q. (2003). Drum dryers. In: Encyclopedia of Agricultural, Food, and Biological Engineering, D. R. Heldman and C. I. Moraru (eds), Marcel Dekker Inc., New York, NY, USA, pp. 211–214.

Tellatin S. and Myers R. (2018). Cover Crops and Carbon Sequestration. United States Department of Agriculture, National Institute for Food and Agriculture, Sustainable Agriculture Research & Education, College Park, MD, USA.

Torzillo G. and Chini Zittelli G. (2015). 'Tubular photobioreactors.' Volume 2: products and refinery design. In: Algal Biorefineries, A. Prokop, R. K. Bajpai and M. E. Zappi (eds), Springer International Publishing AG Switzerland, Cham, Switzerland, pp. 187–212.

Tredici M. R., Rodolfi L., Biondi N., Bassi N. and Sampietro G. (2016). Techno-economic analysis of microalgal biomass production in a 1-ha Green wall panel (GWP®) plant. *Algal Research*, **19**, 253–263, https://doi.org/10.1016/j.algal.2016.09.005

Trommsdorff C. (2015). Can the Water Sector Deliver on Carbon Reduction? The Source. International Water Association, London, UK.

Van Den Hende S., Vervaeren H., Desmet S. and Boon N. (2011). Bioflocculation of microalgae and bacteria combined with flue gas to improve sewage treatment. *New Biotechnology*, **29**(1), 23–31, https://doi.org/10.1016/j.nbt.2011.04.009

Webb D. T., Nagorzanski M. R., Powers M. M., Cwiertny D. M., Hladik M. L. and LeFevre G. H. (2020). Differences in neonicotinoid and metabolite sorption to activated carbon are driven by alterations to the insecticidal pharmacophore. *Environmental Science and Technology*, **54**(22), 14694–14705, https://doi.org/10.1021/acs.est.0c04187

Wenk J., Nguyen M. T. and Nelson K. L. (2019). Natural photosensitizers in constructed unit process wetlands: photochemical characterization and inactivation of pathogen indicator organisms. *Environmental Science and Technology*, **53**(13), 7724–7735, https://doi.org/10.1021/acs.est.9b01180

Wong S. C. and Law-Flood A. (2011). Tapping the Energy Potential of Municipal Wastewater Treatment: Anaerobic Digestion and Combined Heat and Power in Massachusetts. Massachusetts Department of Environmental Protection, Boston, MA, USA.

Zhu Y., Jones S. B. and Anderson D. B. (2018). Algae Farm Cost Model: Considerations for Photobioreactors. Pacific Northwest National Laboratory, United States.

doi: 10.2166/9781789061796

Chapter 13

Microbial electrochemical communication in carbon and electron flow for CO_2 methanation

Aijie Wang[1], Bo Wang[2], Zechong Guo[3], Weiwei Cai[4] and Wenzong Liu[1]*

[1]State Key Lab of Urban Water Resource and Environment, School of Civil and Environmental Engineering, Harbin Institute of Technology Shenzhen, Shenzhen 518055, China
[2]Center for Electromicrobiology, Section of Microbiology, Department of Biology, Aarhus University, Aarhus C 8000, Denmark
[3]School of Environmental and Chemical Engineering, Jiangsu University of Science and Technology, Zhenjiang 212100, China
[4]School of Civil Engineering, Beijing Jiaotong University, Beijing 100044, China
**Correspondence: waj0578@hit.edu.cn*

13.1 INTRODUCTION

Currently, sustainable technologies have been widely explored and developed to capture/fix or utilize CO_2 via chemical, electrochemical, photochemical, and biotechnological approaches accompanied by less or green energy input (Centi & Perathoner, 2009; Kondratenko *et al.*, 2013). Anaerobic digestion (AD) involves bacterial fermentation of organic wastes in the absence of free oxygen, which consequently supports methanogens to produce CH_4 as fuel (Yu *et al.*, 2021). Due to various biological consortia and different by-products involved in complex substrates biodegradation, CH_4 production is determined by two main pathways of acetoclastic and hydrogenotrophic methanogensis. Importantly, hydrogenotrophic methanogens can convert CO_2 to CH_4 by utilizing H_2 produced by acetogenesis. However, H_2 is usually not accumulated much in the system because of the sensitive feedback of H_2 to acetogens.

Recently, electrochemical CO_2 reduction is developed into an effective method for the activation and transformation of stable CO_2 molecules, which is driven by electron flow from anode to cathode (Jhong *et al.*, 2013). Importantly, the cathode surface chemistry of CO_2 reduction always requires efficient and special catalysts. The current technique of microbial electrocatalysis CO_2 reduction offers promising prospects for reducing carbon levels in a sustainable manner, taking full advantage of CO_2-derived chemical commodities (Bajracharya *et al.*, 2017). Bioelectrocatalytic reduction of CO_2 employs versatile microbes to achieve carbon reduction. On the anode side, biological electrons can be collected by exoelectrogens from waste organics in wastewater, waste biomass, and so on., and on the cathode side, the cathode can supply H_2 and electrons for communities, which can directly utilize CO_2 as a final electron acceptor for their metabolism, to produce high-value-added chemicals, containing one or more carbons, like CH_4, acetate, and so on. Therefore, carbon cycling driven by various functional bacteria has been investigated on the acceleration and regulation process via electron transfer.

The carbon cycle, as one of the crucial natural processes on the earth, includes a large number of organisms, in which diverse species of microorganisms play joint efforts in organic degradation, energy utilization, and biosynthesis. Different types of interspecies interactions work and connect for mutualism, in which two or more distinct species living in close proximity rely on each other for nutrients, protection, and/or other life functions (Kouzuma *et al.*, 2015). Efficient interspecies communication for electromethanogenesis (EM) has been disclosed through direct interspecies electron transfer (DIET) in the form of electric currents in syntrophic consortia (Kouzuma *et al.*, 2015; Lovley, 2017b). Typically, a representative example of such syntrophy is found in an integrated system of electrochemically enhanced AD for accelerating CH_4 production, where the pathway of CO_2 reduction by reducing equivalents, for example H_2 and electrons, can be dominant between syntrophic partners of exoelectrogens and hydrogenotrophic methanogens (Rotaru *et al.*, 2014b). Different types of electrochemically enhanced AD systems have been developed under external applied voltages (Huang *et al.*, 2022), indicating improved properties on microbial system stability (Liu *et al.*, 2016c; Cai *et al.*, 2019), gas production (Cai *et al.*, 2016b; Liu *et al.*, 2016b), and low-temperature adaption (Liu *et al.*, 2016a). The direct biological conversion of electrical current and CO_2 into CH_4 by electromethanogenesis has been observed (Cheng *et al.*, 2009).

It has been ascertained that external energy can power microbes and microbial productions (Lovley, 2011; Lovley & Nevin, 2013). EM requires a relatively low energy input (0.2–0.8 V) to provoke CO_2 reduction and CH_4 recovery from wastewaters or wastes and simultaneously enhance organics degradation when couples of an AD system (Cai *et al.*, 2016b). Nonetheless, the energy consumption is always evaluated through electricity input, moreover, the electricity delivery is restricted over the long-term operation in rural or remote areas (Wang *et al.*, 2020c). Therefore, renewable energy powering electromethanogenesis is crucial to widespread application. Solar energy is regarded as one of the most feasible options to conform to the energy demand worldwide because of its availability and enormous solar energy (120,000 TW) striking the earth daily (Kamat, 2007). However, natural solar energy is a day-night intermittent power, which is supposed to be a limitation aspect to overcome. The utilization of electricity from renewable sources for CO_2 reduction can also alleviate the existing challenges associated with the intermittent output of renewable energy by storing the electricity in chemical forms.

13.2 CARBON CONVERSION AND ELECTRON FLOW TO ELECTROMETHANOGENESIS

The energy recovery in the form of CH_4 has been considered as the promising way to achieve energy- and carbon-neutral in wastewater treatment plants (WWTPs) through AD (Silvestre *et al.*, 2015). In WWTPs, most organics in the wastewater will be converted into biomass deposited in the sludge phase. The waste sludge digestion will be the main or even the sole way to CH_4 generation in WWTPs. Biogenic CH_4 is usually produced from the anaerobic conversion of organic substrates by methanogens in anaerobic bioreactors.

Commonly, two general pathways for CH_4 production in AD systems are formed as hydrogenotrophs and methylotrophs based on whether CO_2 or methyl compounds are utilized as main carbon sources (Thauer *et al.*, 2008). Complex organics are hydrolyzed and fermented into small molecular matters, like long-chain fatty acids, monosaccharides, and amino acids, and then to short-chain fatty acids, which are suitable for acetoclastic methanogens. Meanwhile, H_2 generates in the acidogenic fermentation and acetogenic processes. To prevent inhibition of acetogenic bacteria, very low H_2 pressure is commonly required in the AD system. H_2 scavengers, including hydrogenotrophic methanogens, homo-acetogens, and so on, have to consume H_2 quickly to make sure continuous digestion (Appels *et al.*, 2008), which is one of the key limiting processes to achieve high efficiency of methanogenesis. EM offers a new option for CH_4 production enhancement from organic waste streams. Besides anaerobic methanogenesis, EM (bioelectrochemical reduction of CO_2 to CH_4) has also been considered as a novel biogenic pathway for CH_4 production (Cai *et al.*, 2021).

EM takes place in a niche with a plentiful supply of H_2 or electrons, like H_2-producing fermenters, cathodic surface, or conductive carriers, and CO_2 can be reduced directly or indirectly to CH_4 by methanogens through various pathways. Firstly, hydrogenotrophic methanogens can be enhanced through mediated electron transfer with molecular H_2 as an electron carrier. Besides fermentative H_2, which usually has very low pressure in an AD system, more H_2 can be further supplied on cathode via electrochemical reactions (H_2 evolution under cathodic catalysts) or bioelectrochemical processes (produced by exoelectrogens which are capable of accepting electrons from solid cathode) (Equation (13.1)) and consumed by hydrogenotrophic methanogens to produce CH_4 (Equation (13.2)) (Villano *et al.*, 2010). H_2 evolution theoretically requires a cathode potential of −0.41 V vs. SHE, but the more negative potential is practically needed due to various electrochemical losses (Wang *et al.*, 2020b):

$$2H^+ + 2e^- \rightarrow H_2 \tag{13.1}$$

$$4H_2 + CO_2 \rightarrow CH_4 + 2H_2O \tag{13.2}$$

Secondly, EM can also directly be conducted without an electron carrier. Some hydrogenotrophic methanogens (e.g. *Methanobacterium* species) can accept electrons directly from the cathode and reduce CO_2 to CH_4 with the participation of protons produced by the anode (Equation (13.3)) (Lohner *et al.*, 2014). This process requires a cathode potential of −0.24 V vs. SHE:

$$CO_2 + 8H^+ + 8e^- \rightarrow CH_4 + 2H_2O \tag{13.3}$$

Thirdly, except for H_2, some other organic compounds (such as acetate, ethanol, butyrate, succinate etc.) derived from electric current and CO_2 via cathodic electrosynthesis also provide more available substrates for methanogens (Steinbusch *et al.*, 2010). Acetate synthesized on the cathode with acetogenic microorganism (e.g., *Sporomusa ovata*) as catalysts (Equation (13.4)) (Nevin *et al.*, 2011) can be subsequently converted into CH_4 by acetotrophic methanogens (Equation (13.5)) (Jiang *et al.*, 2013), while other compounds require syntrophic metabolism via interspecies electron transfer:

$$H_2 + CO_2 \rightarrow \text{Acetate} \tag{13.4}$$

$$\text{Acetate} \rightarrow CH_4 + CO_2 \tag{13.5}$$

Compared with anaerobic methanogenesis, EM has its own particularities in carbon conversion and electron flow routes. Carbon conversion routes in electromethanogenesis are more diverse as a result of the involvement of electrochemical processes, which is generally beneficial for the stability of the system. In addition, organic substrate degradation and CH_4 production are separated in different regions, making it easier to create a favorable microenvironment for sensitive methanogens.

In AD, the electron transfer between different microbial populations is usually through intermediate products (electron carriers), which is implicit and spontaneous. In the EM, the external voltage is like a pump, extracting electrons from the source (organic substrates) and delivering them into the product (CH_4). Due to the existence of external voltage and circuit, a portion of the electron flow becomes explicit and mandatory, which substantially improves the monitorability and controllability of the system. In light of the above reasons, EM has great potentials to overcome the unfavorable factors that plague the conventional methanogenic pathway in anaerobic bioreactors and improve CH_4 production efficiency.

13.2.1 Functional communities and genes involved in carbon conversion

Electrochemically enhanced ADs mainly depend on the ability of certain microorganisms capable of extracellular electron transport (Yu *et al.*, 2021). Primarily, well-established biofilm on anode plays a role of efficient electron collection from organics biodegradation (Wang *et al.*, 2020a). Well-known common exoelectrogenic bacteria (e.g., *Geobacter, Shewanella*, and *Desulfovibrio*) are found related to c-type cytochrome genes in these systems, moreover, a relatively high functional and phylogenetic diversity of microorganisms can be developed despite the feedback of a single substrate, like acetate-alone. For example, functional genes related to substrate degradation accounted for 15–25% of carbon degradation and fixation in all gene categories, while functional genes related to complex carbon utilization accounted for ca. 10% in all detected genes (Liu *et al.*, 2010). A variety of carbon degradation genes, including amylase, xylanase, and endochitinase, were varied considerably among reactor operations. Accordingly, bioreactors with high coulombic efficiency and energy harvest (H_2 yield) have the greatest capability for using a variety of complex carbon sources. A significant correlation is reported between coulombic efficiency and community composition ($r = 0.84$, $P = 0.025$), and COD removal and carbon degradation ($r = 0.84$, $P = 0.035$) with community structure (Liu *et al.*, 2010). However, carbon degradation communities on anode are not significantly related to the cathode terminal products.

When fermentative substrates (glucose) are used for reactors, functional genes with high diversities involve in complex carbon degradation, including carbon degradation genes of cellulose, hemicellulose, lignin, starch, pectin, and chitin (Varrone *et al.*, 2014). More microorganisms in anode biofilm are detected from cytochrome genes, such as cytochromes derived from *Geobacter metallireducens* for metal reduction, *Bradyrhizobium* sp. involved in the oxidation of organic contaminants, *OmcA/MtrC* from *Geobacter sulfurreducens, Shewanella sediminsis, S. oneidensis, S. amazonensis, S. loihica*, and *S. pealeana*. It indicates that the reactors with the highest energy recovery showed a higher (total) amount of cytochrome genes. Meanwhile, microorganisms related to carbon fixation genes for rubisco, carbon monoxide dehydrogenase, and propionyl-CoA carboxylase are detected. For example, propionyl-CoA carboxylase genes derived from *Roseiflexus* sp., *Nitrobacter hamburgensis, Chloroflexus aggregans*. Also, a part of detected bacteria are uncultured, like carbon monoxide dehydrogenase (CODH) from an uncultured bacterium (lab clone), and rubisco derived from an uncultured bacterium.

A comprehensive community structure is supposed to be well developed from various substrates, though higher functional diversity can be detected when utilizing complex carbon than simple sole carbons from acetate to glucose. Genes from all major functional categories indicate that microbial communities are able to perform a large variety of functions. Necessarily, the variability of community functions is not (only) related to the presence of exoelectrogens, as a fraction of carbon degradation functions can be played by non-exoelectrogens of fermentation and or symbiotic relationships with other bacteria. Accordingly, the extracellular electron transfer pathway has an important impact on the methanogenic community structure in the reactors. Applied voltages increased the relative abundance of hydrogenotrophic methanogens over acetoclastic methanogens. In control reactors with open circuit, a relatively high abundance of acetoclastic methanogens are presented. However, six genera were detected with the ability of H_2 utilization and CH_4 production in closed circuit, including *Methanobacterium, Methanococcus, Methanoculleus, Methanocorpusculum, Methanospirillum*, and *Candidatus Methanoregula*, but only one genus of *Methanosarcina* was defined as acetate utilization methanogens.

13.2.2 Organic conversion for CH_4 production under electrochemistry regulation

Due to the attractive prospects of EM, some researchers recently have paid attention to promoting the CH_4 production capacity of anaerobic bioreactors with electrochemical intervention as auxiliary strengthening approaches. Anaerobic digestion plays a leading role in the CO_2 fixation market, which is highly mature in both technical and theoretical directions. However, AD is easily vulnerable to unfavorable factors, such as unsuitable temperature, non-neutral pH, or toxic inhibitors. The efficiency of CH_4 production may fluctuate drastically when encountering environmental changes. In an AD bioreactor, energy flow and carbon conversion happen spontaneously in a closed system, lacking direct external intervention measures. That leads to the difficulty in quickly constructing a normal order in the system (such as the startup or recovery from the crash) or effectively adjusting the CH_4-producing processes when the system is unstable. The microbial electrocatalytic process, in contrast, is able to exert a directional intervention to organic degradation pathways through external electric energy input, which can be an effective complementary approach for AD. Therefore, an electrochemically enhanced AD is proposed in the expectation of integrating the advantages of these two technologies and providing an ideal choice for carbon fixation.

Since the concept of the electrochemically enhanced AD is proposed by Willy Verstraete in 2006 (Pham *et al.*, 2006), many research works have been conducted to verify the feasibility of coupling technology. It was reported that CH_4 production performance and the stability of AD systems can be promoted by introducing a microbial electrocatalytic process (Malaeb *et al.*, 2013; Wang *et al.*, 2022b). The potential mechanism for these promoting effects has been analyzed in recent studies, and it is believed that microbial community distribution and carbon conversion pathways in the anaerobic bioreactor are positively regulated by bioelectrochemical processes (Guo *et al.*, 2017a). Applying voltage in a complicated anaerobic environment is beneficial for the creation of suitable habitat for anaerobic and facultative populations by reducing electrode potential and boosting hydrogenotrophic methanogens by providing additional H_2 (Wang *et al.*, 2009). The influence of electrochemistry includes both instant direct contribution and long-term indirect effect (such as affecting the distribution of anaerobic community during the start-up stage) (Liu *et al.*, 2016c; Zhao *et al.*, 2015). It is disclosed by sequencing techniques in some researches that external voltage, as a positive growth condition or selective pressure, has substantial regulatory effects on microbial community structures. In particular, external voltage leads to an increase in electrogenic microorganisms and hydrogenotrophic methanogens (Liu *et al.*, 2016b).

Besides microorganism community distribution, the external voltage can also influence the organic degradation pathway positively in an anaerobic system. According to an electron-balance analysis of the glucose digestion process in our case (Figure 13.1) (Guo *et al.*, 2017a), more electrons

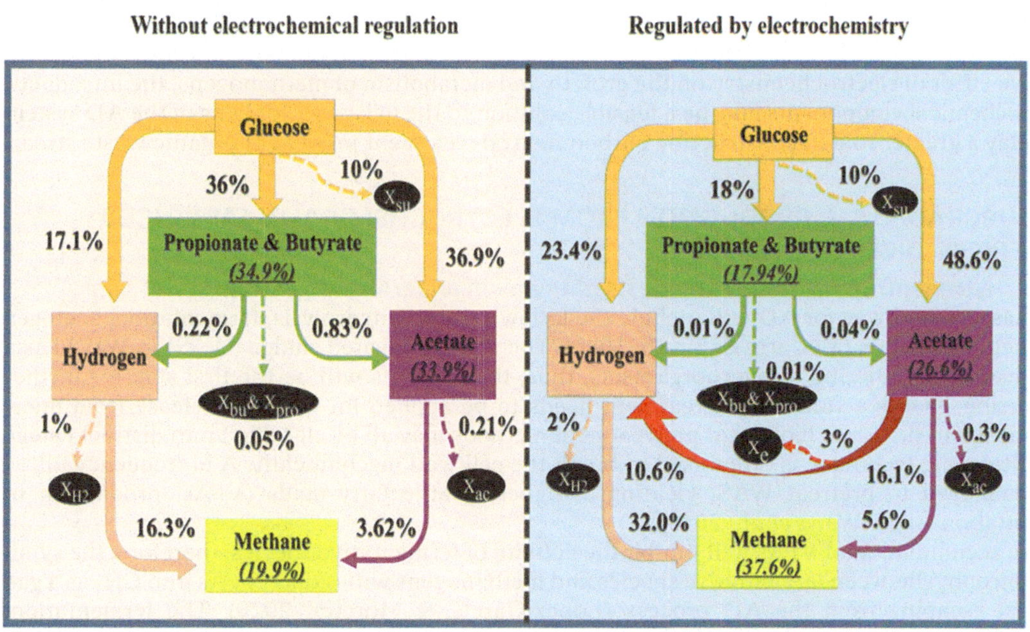

Figure 13.1 Electron balance analysis for glucose digestion in absence and presence of electrochemical process (Guo *et al.*, 2017a).

obtained from the oxidation of glucose were transferred and finally stored in propionate and butyrate (34.9%) in an AD system without electrochemical regulation. Hydrogenotrophic methanogenesis was unrestricted, but acetotrophic methanogenesis was inadequate due to relatively low amounts of acetotrophic methanogens, and only a small amount of acetate was converted into CH_4 (3.6%). The waste of electron flow in fermentation products (propionate, butyrate, and even acetate) resulted in unsatisfactory CH_4-producing performance. With the regulation of bioelectrochemistry, however, more electrons were transferred from glucose to H_2 (23.4%) and acetate (48.6%), instead of butyrate and propionate, and the bioelectrolysis reaction created an additional pathway between acetate and H_2, about 10.6% of the total electrons were transferred from acetate into H_2 through circuit current, which alleviated the limitation of acetotrophic methanogens and led to a significant enhancement of CH_4 production.

The influence of electrochemistry on the anaerobic degradation process is largely determined by the electrochemical efficiency of electrodes (e.g., current density, cathodic catalysis efficiency, etc.) and the complexity of the system. Generally, the ratio of bioelectrochemical reactions in the total electron flow and the direct contribution of electrochemistry on biogas production will be promoted by increasing the electrochemical efficiency of electrodes (current density) through various methods (such as increasing electrode areas, using efficient electrode materials, optimizing reactor configuration, etc.) (Wang *et al.*, 2017). Supplying simple substrates (such as acetate and ethanol) is beneficial for establishing an efficient anode biofilm and obtaining good electrochemical performances. With complex compounds (such as wastewater or waste sludge) as carbon substrates, however, the biofilm communities and anaerobic digestion processes became more complex (Zhang *et al.*, 2011). The relative abundance of electrode-respiring bacteria and the electron transfer efficiency were generally lower than those in the systems using acetate, which may weaken the effects of electrochemistry on the AD process (Zhang *et al.*, 2011).

Methanogenesis is generally the rate-limiting step of AD with soluble substrates, and methanogens are the most important but also the most vulnerable members in the CH_4-producing functional microbial community. In fact, many bottlenecks in AD technology (such as long lag time, poor stability) can be ultimately attributed to the high sensitivity and slow growth of methanogens. Considering the positive effect of electrochemistry on the growth and metabolism of methanogens, the introduction of electrochemical components may be a feasible solution to the existing problems of the AD system and may play a greater role in constructing carbon-neutral treatment process of organic waste streams.

13.3 UPGRADED CH_4 PRODUCTION FROM ELECTROCHEMICALLY ENHANCED ANAERPBIC DIGESTION

13.3.1 Hytrogentrophic methanogenesis pathway in anaerobic digestion

The classical pathway for AD will include the following steps: hydrolysis, fermentation, acetogenesis, and methanogenesis (McCarty & Smith, 1986). The waste activated sludge (WAS) in WWTPs can be perceived as a gathering of microorganisms, thus, the cell lysis will be the first step for further AD conversion, that is a suitable pretreatment needs to be applied for organics release from microbes. Chemical, physical, and biological pretreatment methods may all be classified into distinct categories. The ultrasonic technique is often used to break the cell wall mechanically. A bi-frequency ultrasonic was employed to pretreat WAS, yielding a higher volatile fatty acids (VFAs) production in the fermentation stage (Wang *et al.*, 2018).

The accumulation of VFAs will inhibit the activity of fermentative species, therefore, the symbiosis (or syntrophy) between fermentative species and methanogens will reduce VFAs into CH_4 as a gaseous product escaping from the AD process (Lopez-Garcia & Moreira, 2020). The fermentation will produce VFAs as end products, which will be utilized by acetate producers and methanogens (Ziels *et al.*, 2019). Currently, there are eight orders of methanogens, belonging to three phyla (Lyu *et al.*, 2018). The novel methanogens were continuously discovered, implying the diversity of methanogens.

However, the metabolism type of methanogens was rarely monotonous in AD due to the substrate limitation for methanogens. Commonly, the H_2/CO_2 and acetate will be utilized as the substrate for CH_4 production, which are termed hydrogenotroph and acetotroph, respectively. Although there are methylotrophs in natural ecosystems for methanogenesis, the lower abundances of methylotrophs result in less attention in AD reactors.

Among the nutrient type of methanogens in AD of WAS, the hydrogenotrophic methanogens, utilizing H_2 as an electron donor and CO_2 as an electron acceptor, are an important adjuster of H_2. As the fermentative species need to reduce NADH into NAD^+, and the regeneration of NAD^+ ensures the completion of glycolysis, H_2 will be sensitive as its accumulation will thermodynamically affect the NAD^+ regeneration (Stams & Plugge, 2009). The H_2 formation in the fermentation stage also provides an insight into biological H_2 production in dark fermentation. However, recent progress in bioelectrochemistry which combined dark fermentation and microbial electrolysis led to a breakthrough in H_2 production from glucose (Varanasi *et al.*, 2019; Wang *et al.*, 2020b). Naturally, the hydrogenotrophic methanogens will capture H_2 as a precursor for CH_4 production, therefore, the CH_4 is considered as an inevitable product in a hybrid system (Wang *et al.*, 2009).

Recently, a hybrid model for electrochemically enhanced AD was proposed to enable an accelerated CH_4 production rate from waste activated sludge (WAS) (Liu *et al.*, 2016b). Electrode H_2 evolution provides a unique niche for hydrogenotrophic methanogens as a continuous H_2 supplier. A novel electrochemically enhanced AD reactor was carried out to test its performances in CH_4 production which was fed with WAS. The CH_4 production rate was enhanced by the introduction of electrodes, approaching approximately 3.2 times the control AD. Based on the electron balance, the sum of control CH_4 production and current contribution is highly aligned with the enhanced CH_4 production. The current value can reach as high as 12 mA (the maximum), which is higher than that of lab-scale bioelectrochemical reactors. A further dynamic model of methanogenesis was built in electrochemically enhanced AD, indicating the CH_4 production rate would be improved by 1.4 times feeding with glucose (Guo *et al.*, 2017a). The configuration of electrochemically enhanced AD will affect the performance in the CH_4 production rate. The higher ratio of cathode/anode enhanced the CH_4 production rate (increased by 56–180%) in our previous study (Guo *et al.*, 2017b). Furthermore, the independent cathode also caused an enhancement in CH_4 production in a continuous model with glucose as substrate for cathode and acetate for anode (Cai *et al.*, 2016a). The electrochemically enhanced AD provided a promising way to accelerate the CH_4 production from WAS, which is valuable for AD as the long sludge retention time is the bottleneck in its implementation. The improvement of the CH_4 production rate potentially enables the AD to shorten the sludge retention time with the similar performance of CH_4 yield.

13.3.2 Microbial community evolution in cathode biofilm for methanation

In the electrochemically enhanced AD model, the hydrogenotrophic methanogens will be dominant in the cathodic biofilm, which has been verified by our studies (Gao *et al.*, 2021; Perona-Vico *et al.*, 2019; Siegert *et al.*, 2015). This finding is consistent with our initial hypothesis and previous studies, the growth of hydrogenotrophic methanogens will be stimulated in the presence of H_2 evolution. The *Methanobacterium* was the enriched genus, working as hydrogenotrophic methanogens at the cathode. Additionally, the facultative acetoclastic methanogens were also found to be present in the cathodic biofilm, as well as acetogens. Therefore, a postulation of cathodic CH_4 generation was proposed, on the one hand, the hydrogenotrophic methanogens will accept H_2/electrons from the cathode to produce CH_4; on the other hand, acetogens will consume H_2 from the cathode to produce acetate for acetoclastic methanogens (Figure 13.2).

Indeed, according to the previous studies related to the pure culture of methanogens on an electrode, the cathodic CH_4 generation theory can be modified with a direct pathway of CH_4 production from electrons flow to methanogens. Initially, the first report was provided by Derek R. Lovley (Rotaru *et al.*, 2014a), the labeled C^{14} CH_4 can be produced from the co-culture of *Methanosaeta* and *Geobacter*,

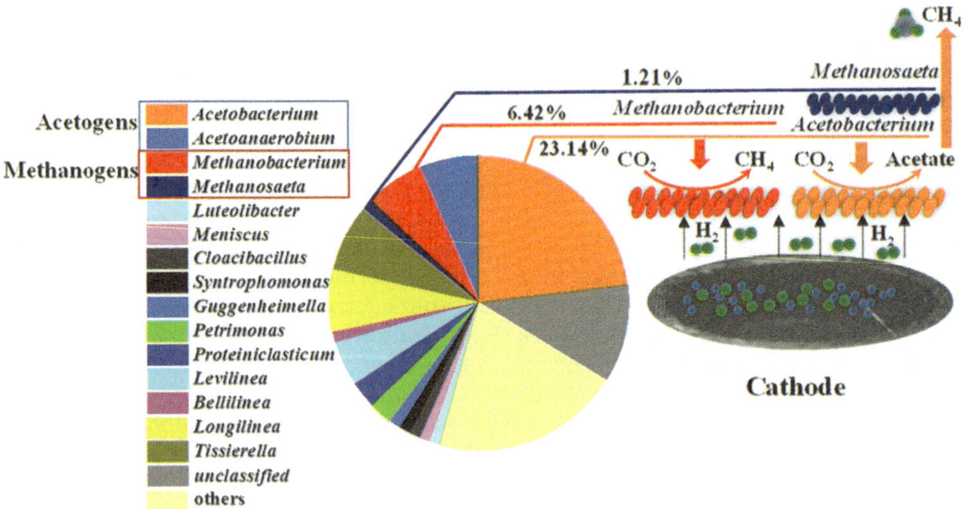

Figure 13.2 Putative pathways of cathodic methanogenesis (Cai *et al.*, 2016b).

proving the *Methanosaeta* was capable of accepting electrons directly from *Geobacter*. This finding induced considerable attention on the DIET (Rotaru *et al.*, 2014a). A mutant of *Methanococcus* without hydrogenase was subsequently cultured as working species, confirming the direct electron transfer without H_2 dependence (Lohner *et al.*, 2014). Moreover, *Methanosarcinales* was verified to perform multiple modes (hydrogenase-mediated and free extracellular enzyme-independent modes) of electrode interactions on cathodes (Rowe *et al.*, 2019), which expands the underlying electron transfer mechanisms at the cathode. Recently, a defined coculture of *Methanobacterium* strain YSL and *Geobacter metallireducens* can grow via DIET (Zheng *et al.*, 2020). The main classifications of methanogens were putative to have the ability to achieve extracellular electrons to produce CH_4 (Gao & Lu, 2021). Therefore, the DIET in methanogens is more broadly distributed than we expected.

Although the pure culture was found to be able to directly acquire electrons from the cathode, the electron transfer in microbial electrolysis assisted anaerobic digestion normally occurred within the biofilm which consists of diverse species. There are two questions that should be noticed: firstly, why the *Methanobacterium* or *Methanobrevibacter* always dominated in the biofilm rather than *Methanosaeta* or *Methanosarcina* that both can utilize H_2/electrons from the cathode; secondly, whether the DIET will support the holistic biofilm performances. A further study revealed the microenvironment of the cathode is differing from the bulk solution as the rapid consumption of proton (Cai *et al.*, 2020). The *mcrA* sequencing technology provided a higher resolution into the classification of methanogens at the species level, the basophilic methanogens was confirmed as the enriched *Methanobacterium* genus at the cathode with the extreme alkaline microenvironment (at the μm-scale) (Cai *et al.*, 2018a). Therefore, the extreme condition may induce the enrichment of special methanogens, such as the *Methanobacterium* genus. In the mixed culture, H_2 was firstly confirmed as the electron carriers in the syntrophy between fermentative species and methanogens, then the formate was found to be an alternative to H_2 with extremely high efficiency (Stams & Plugge, 2009). The novel DIET was proved as one kind of electron carrier, a recent study suggested it can be an option for diverse syntrophs as the presence of e-pili in *Syntrophus* resembled the type IV pili of *Geobacter* (Walker *et al.*, 2020). A modeling confirmed that the formate can reach 317×10^3 e^- cp^{-1} s^{-1} for electron transfer in syntrophy, whose the DIET ability will be 44.9×10^3 e^- cp^{-1} s^{-1} that is higher than that of H_2-mediate (5.24×10^3 e^- cp^{-1} s^{-1}) (Storck *et al.*, 2016). Clearly, the formate will be the ideal mediates for electron transfer, which normally is not the main pathway for syntrophy as the formate is thermodynamically reversible to form H_2 (Cai *et al.*, 2020). Besides, the growth of methanogens without cytochrome was restricted on formate, as the H_2 is still an intermediate inside of methanogens through coenzyme F_{420}-dependent formate dehydrogenase, the formate-dependent will lose an advantage to compete with others who own lower H_2 thresholds in the natural environments (Thauer *et al.*, 2008). In typical AD reactors, H_2 is still the main intermediate for electron transfer within syntrophic bacteria and methanogens.

H_2 was detected in our subsequent test in cathodic biofilm (Cai *et al.*, 2020). A clear gradient of H_2 concentration implied it will be the electron carriers at the cathode for methanogenesis. Meanwhile, the cyclic voltammetry (CV) test was used to couple with the microsensor, which provided semi-quantitative results of H_2 contribution in total electron transfer. It was surprising to discover only less than 50% of electrons would be transported via H_2, thus, the DIET may contribute significantly to cathodic methanogenesis. However, the extracellular electron transfer pathways have been extensively expanded according to recent progress, not only the pili, flavin, cytochrome, but also exDNA, vesicle, extracellular polymer substances (EPS) have been involved (Liu *et al.*, 2020; Lovley, 2017a; Saunders *et al.*, 2020; Xiao *et al.*, 2017). The H_2-free electron pathway also may include multi-mediates rather than only direct electron transfer, which still needs further study.

13.3.3 Microbial network of electrochemically enhanced AD

In electrochemically enhanced AD reactors, microbial communities with the ability of extracellular electron transfer play a role of a bridge connection to fermentative bacteria (FB) and methanogens (Liu *et al.*, 2016b), that is, multiple pathways are developed via an electron transfer pathway from carbon degradation to CH_4 (Figure 13.3). When using high applied voltage (e.g., >0.5 V (Liu *et al.*, 2010) and an efficient metal catalyzer on the cathode (Liu *et al.*, 2019), for example Pt/C, modified Ni (Cai *et al.*, 2018b; Wang *et al.*, 2019), and so on, H_2 can be detected noticeably as one of the important cathodic products. The H_2 production rate, which can be determined by applied voltage (Wang *et al.*, 2009), indirectly has great effects on methanogens. Moreover, a gradual increasing CH_4 production over operation time illustrates a substantial development of methanogenic communities in the system. The communities utilized H_2 can fully develop in the cathode biofilm on acetate and CH_4 production, and then change suspended solution and anode communities via diffused H_2 as electron carrier (Cai *et al.*, 2016b).

The original electrode biofilm interacted with fermentative communities to establish a new electroactive community system, which was soon joined by methanogens, resulting in a complex community network (Figure 13.4) (Liu *et al.*, 2016c). It has been reported that bioelectrochemical

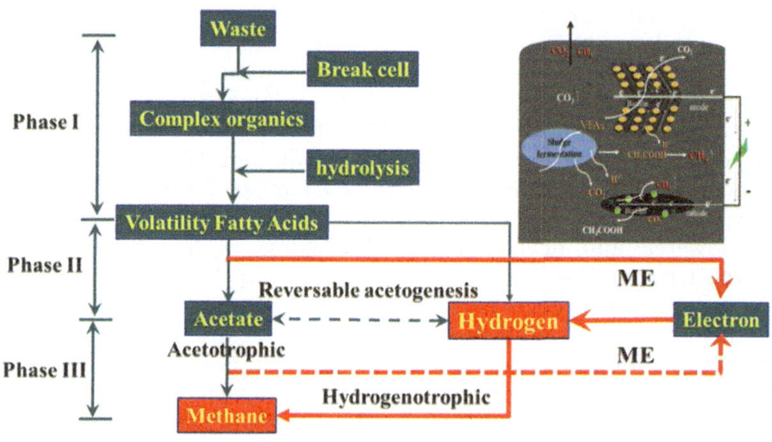

Figure 13.3 Enhanced hydrogenotrophic methanogenesis via the extracellular electron transfer pathway.

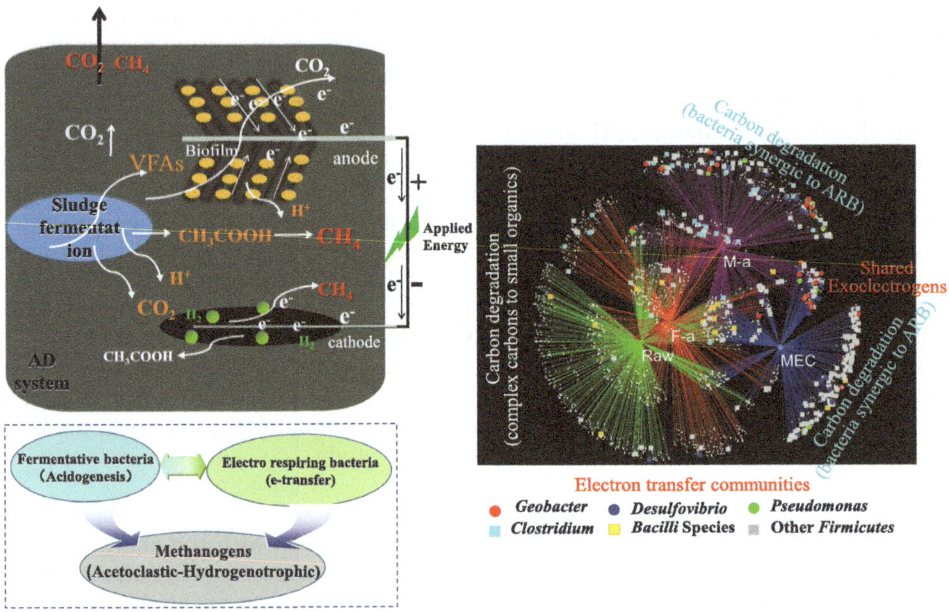

Figure 13.4 Microbial network of acidogenic bacteria, electro-respiring bacteria and methanogens.

communities are highly enriched in dominant functional groups related to *Proteobacteria* (~60%) and *Firmicutes* (20–30%), which conduct substrate degradation and electron transfer with significant potential on complicated carbon utilization, as core communities in the context of extracellular electron transfer. They are supposed to play a very important role in carbon recycling as it has been reported that *Firmicutes* may have a symbiotic relationship with anode respiring bacteria (Zhang *et al.*, 2011). Representative anaerobic fermentative bacteria account for 1–10% overall communities, consisting of *Citrobacter* (class *Gammaproteobacteria*), *Macellibacteroides* (class *Bacteroidia*) and two genera of class *Clostridia* (*Proteiniclasticum* and *Sedimentibacter*) were enriched after prefermentation and became the predominant bacteria. *Proteiniclasticum*, responsible for the degradation of proteins to produce HAc, propionic acid (HPr) and iso-butyric acid (iso-HBu) (Zhou *et al.*, 2018), had an abundance of 6%. *Macellibacteroides* with abundance of 5–7% can metabolize various carbohydrates for HAc, HBu, and iso-HBu generation (Jabari *et al.*, 2012). The genera *Citrobacter* and *Sedimentibacter* can degrade organics to produce VFAs and H_2 during acidogenesis and acetogenesis. Compared to the original anode biofilm communities, some bacteria also shift during the connection to sludge fermentation process. Four genera of the class *Clostridia* (namely *Acetoanaerobium*, *Acetobacterium*, *Anaerovorax* and *Fusibacter*) decreased in all integrated reactors feeding sludge fermentation liquid.

Methanogens are powered and developed in all biofilms in integrated systems. Meanwhile, the abundance of the fermentation bacteria in the sludge can be significantly improved, but the acetotrophic methanogen shows a decrease (Wang *et al.*, 2019). The acetotrophic methanogens can be enriched in anode biofilm in all conditions with enough substrate carbons, though the substrate competition exists between anode microorganisms and methanogens on acetate utilization, while hydrogenotrophic methanogens will be further enriched with higher abundance in cathode biofilm with continuous applied voltages (Cai *et al.*, 2019), including *Methanocorpusculum*, *Methanosphaerula*, *Methanoregula*, *Methanospirillum*, *Methanobacterium* and *Methanobrevibacter*, which closely related to the H_2 evolution of cathodic reaction, supplying favor substrates for hydrogenotrophic methanogens in anaerobic condition. Community structure and the methanogen production pathway will be directed by the biocathode materials with high H_2 evolution reaction (HER) activity. A high HER tends to select H_2 instead of electrons for methanogenesis. As a result, the dominating microorganisms at the cathode with HER capacity can be selected similarly, usually including hydrogenotrophic *Methanobacterium* and *Methanospirillum* (Ferry *et al.*, 1974; Rotaru *et al.*, 2014a), and the *Methanobacterium* could not only directly convert H_2 into CH_4 but also electrons (Cheng *et al.*, 2009).

13.4 CO_2 METHANATION DRIVEN BY SOLAR-POWERED BIOELECTROCHEMICAL SYSTEM

13.4.1 Solar intermittent driven-power accelerates bioelectrochemical performances

There has been a flurry of studies that reported that continuous direct electric current powered microbes would affect the growth of cells and microbial metabolic behavior. Thus, a majority of research gathered increasing attention in the regulation of applied voltages or potentials (Ding *et al.*, 2016). For example, the production rate of CH_4 has achieved the highest with 0.052 m^3 $CH_4 \cdot m^{-3}$ reactor·d^{-1} when 0.8 V applied voltage was employed in a dual-chambered electrochemically enhanced AD reactor (Ding *et al.*, 2016). In another study, it was observed that the microbial community of electrode biofilms shifted in the absence or presence of electrode potentials, whereas higher proportions of functional genes were upregulated under the scenario of an open circuit (Liu *et al.*, 2010). Together, these results gave a clue that an excess of energy over microbial metabolism may be supplied via continuous electric electricity and thus jeopardized the overall performances of bioelectrochemical systems.

Recently, another electro-driven mode of intermittent electrical electricity to facilitate bioenergy recovery or wastewater conversion exhibited unexpectedly fabulous outcomes. It was noted that higher and more stabilized removal efficiencies of phenol were accomplished in the single-chambered microbial electrolysis cell via the intermittent power with 1 day-on/1 day-off (Ailijiang *et al.*, 2016). Long-term operation of a periodic disconnection power supply could significantly lower the internal resistance of BES, improve the COD removal efficiency, and enhance H_2 recovery efficiency (Cho *et al.*, 2019; Hussain *et al.*, 2018). When further considering the inherent property of solar light, as one of the most credible discretions, that endows the day-night intermittency but also as the potential limitation, solar intermittent driven-powered energy was also revealed that could effectively facilitate bioenergy recovery and simultaneous wastewater treatment. Wang *et al.* (2020c, 2020d) demonstrated that natural solar light, as a day/night intermittent power, could effectively enhance CH_4 production from wastewater and regulate electron transfer protein of cytochrome c, and this study further displayed the underlying impact of solar energy to electrochemically enhanced AD. Wang *et al.* (2022a) further developed natural solar-powered microbes to reduce CO_2 into CH_4 for efficient carbon capture. Comprehensively comparing other reported research involved in EM with CO_2 reduction to CH_4 via a biocathode (Table 13.1), it could be easily concluded that the presence of membrane has a prevalence in such systems and higher applied voltages were provided due to the occurrence of water splitting at the anode rather than the oxidation of organic matters. In addition, there is the out of tune between the efficiency of current (i.e. coulombs) to carbon-containing product (i.e. CH_4) and the production rate of the reductive product, implying energy investment does not completely convert into CH_4 and thus causing another nominal squandering of resources. Further considering the other exterior operation parameters control, for example, electrode modification, temperature and pH control, and so on, admittedly, natural solar-powered EM still exhibited a considerably competitive competence in bioenergy recovery via green intermittent driving force than via continuous direct electricity.

Table 13.1 The comparisons of carbon capture via microbial electromethanogenesis in CH_4 production rates.

Inoculum[a]	Reactor Configuration and Operation Mode	Temperature (°C)	Cathode Material	Applied Voltage (V)	Inorganic Carbon Source/ conc. ($mol \cdot L^{-1}$)	Current Density ($A \cdot m^{-2}$)	CH_4 Production Rate (mol $CH_4 \cdot m^{-3}$ reactor$\cdot d^{-1}$)	Current to CH_4 Efficiency (%)	Reference
Anaerobic sludge from a packed bed biofilm reactor fed with a synthetic mixture of fatty acids and alcohols	DC Batch	35	Carbon paper	Cathode potential −0.75 V vs. SHE	CO_2/ Quantity sufficient	Cathode current density −0.69	0.173±0.026	76±7	Villano *et al.* (2010)
Activated sludge	DC Continuous	30	Graphite felt	Cathode potential −0.55 V vs. NHE[b]	HCO_3^-/0.06	3.8	0.268	51.3	Van Eerten-Jansen *et al.* (2012)
Enriched activated sludge	DC Batch	30±1	Carbon felt	Cathode potential −0.95 V vs. SHE[b]	CO_2/ Quantity sufficient	3.37	0.024	89.2	Jiang *et al.* (2013)
Enriched crushed anaerobic granular sludge	DC Continuous	37±2	Graphite felt	Cathode potential −0.70 V vs. SHE	HCO_3^-/0.024	3.0	0.35	84.3 (Coulombic efficiency)	Xu *et al.* (2014)
The effluent of the anode chamber of a thermophilic microbial fuel cell fed by formation water and sodium acetate	DC Batch	55	Carbon cloth	Cathode potential −0.5 V vs. SHE	HCO_3^-/0.03	0.175	0.233	93	Fu *et al.* (2015)
Anaerobic sludge	DC Continuous	31±1	Graphite felt	Cathode potential −0.7 V vs. SHE	HCO_3^-/0.06	2.9	0.464	73	(Van Eerten-Jansen *et al.* 2015)
A marine lithoautotrophic *Methanobacterium*-like archaeon strain IM1	DC Batch	21	Graphite rod	Cathode potential −0.6 V vs. SHE	CO_2/ Quantity sufficient	Cathode current density −0.5	0.0134	80 (Coulombic efficiency)	(Patil *et al.* 2015)
An enriched hydrogenotrophic methanogenic culture	DC Batch	22±2	Porous carbon felt	Cathode potential −0.8 V vs. SHE	CO_2 (1.65 atm)	Ca.15	0.03±0.015	39 (Coulombic efficiency)	Dykstra and Pavlostathis (2017)

Anaerobic sludge	TC[c] Batch	30	Heat-treated stainless steel felt (HSSF), stainless steel felt (SSF), and graphite felt (GF)	3.5±0.3	HCO_3^-/0.06	7.1	HSSF: 12.86 SSF: 9.19 GF: 15.72	HSSF: 60.8 SSF: 56.9 GF: 69.4	Sangeetha *et al.* (2017)
Anaerobic mixed sludge including granular sludge from the paper industry WWTP[b] and sludge from the municipal WWTP[b]	DCBatch	30	Granular activated carbon	Cathode potential −0.58 V	CO_2/ Quantity sufficient	35	4.3	66	Liu *et al.* (2018)
Anaerobic sludge	DC Continuous	30	Graphite granules	2.8±0.1		35	4.1	67	
Activated sludge from a full-size, enlarged granular sludge-bed reactor for starch wastewater treatment	DC Continuous	30±1	Carbon felt	Cathode potential −1.0 V vs. Ag/AgCl	HCO_3^-/0.06	2.442±0.484	0.24±0.05	44.27±4.01	Yang *et al.* (2020)
			Neutral red-modified carbon felt[e]			7.622±1.436	1.39±0.17	58.90±11.47	
			Anthraquinone-2,6-disulfonate-modified carbon felt[e]			4.377±1.230	0.84±0.24	60.88±4.01	
The effluent of a CH_4-producing single-chamber microbial electrolysis cell	DC Batch	30	Nickel foam	Potentiostatic: cathode potential −0.9 V vs. Ag/AgCl	HCO_3^-/0.048 and CO_2/ Quantity sufficient	1.47	2.68	25	Mao *et al.* (2021)
				Galvanostatic: current density 2.14 A·m^{-2}		2.14	4.91	36	
				Applying constant voltage: 1.98 V		3.23	12.05	90	
The effluent of a CH_4-producing single-chamber microbial electrolysis cell	SC Batch No pH control	25±2	Pt-coated carbon cloth	1 V (powered by solar light)	HCO_3^-/0.164	17.3	5.47±0.29	211.13±3.89	Wang *et al.* (2022a)

[a]For the double-chamber system, the inoculum specifically refers to the cathode.
[b]Cathode potential vs. SHE (standard hydrogen electrode) or vs. NHE (normal hydrogen electrode).
[c]TC refers to a three-chamber composed of one anode chamber in the middle and facing two cathode chambers.

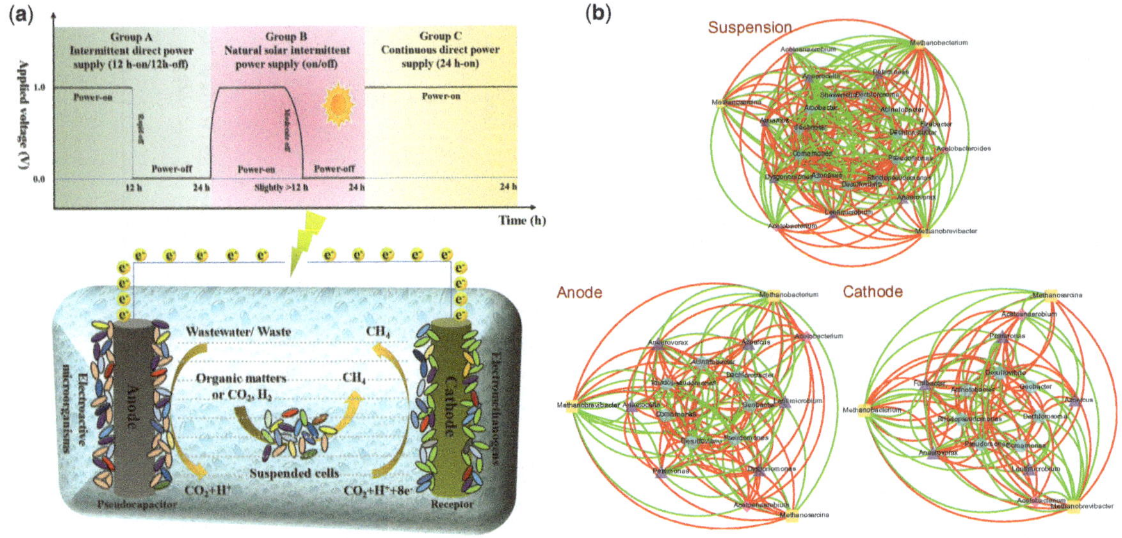

Figure 13.5 (a) schematic diagram of the single-chambered membrane-less bioelectrochemical system driven by the intermittent electric field applied by manual on-off (group A) or natural solar power (group B) and continuous electrical field (group C). (b) molecular ecological networks (MEN) visualization of OTUs from functional microbial consortia in the anode biofilm, cathode biofilm, and bulk solution, respectively. Each node represents an OTU (species), and different colors and shapes of nodes signifier categories of specific functional genera: electroactive microorganisms (sky blue ellipse), methanogens (yellow round rectangle), acetogens (pink diamond) and anaerobic fermentative bacteria (purple triangle). A red edge indicates a positive interaction between two individual nodes, while a Green edge indicates a negative interaction (reproduced from Wang *et al.*, 2020d).

13.4.2 Intermittent electro field mediates mutualistic interspecies electron transfer

The mechanisms of electron transfer involved in bioelectrochemical systems have been widely untangled, and mainly occurred at the bioelectrode, however, the correlation and contribution of planktonic microbial consortia to CO_2 reduction into CH_4 are usually ignored and have not been well revealed yet. The primary pathways of electron transfer for EM in the bulk solution could be classified into two categories: (i) indirect interspecies electron transfer (IIET), which is further subdivided into the intermediate-mediated IIET (i-IIET), such as H_2 and formate (Cai *et al.*, 2020); the electron shuttle-mediated IIET (e-IIET), such as the diffusive exchange of electrons between species through soluble electron shuttles such as H_2 (Lovley, 2017b); (iii) and the DIET.

Wang *et al.* (2020c) developed natural solar-powered EM to recover CH_4 from the wastewater and demonstrated that the solar-intermittent driving mode displayed excellent higher CH_4 production and the efficiencies of electron transfer and energy recovery (Figure 13.5a). In addition, the result of molecular ecological networks (MEN) analysis disclosed that electroactive microorganisms (EAMs) played a pivotal role in three positions (bioanode, biocathode and suspension), and greatly electrochemically communicated with methanogens (Figure 13.5b) (Wang *et al.*, 2020d). Also, hydrogenotrophic methanogens, mainly the genera of *Methanobacterium* and *Methanobrevibacter*, endowed more links and exhibited a more positive connection in bioelectrodes. In contrast, acetotrophic methanogens, mainly the genus of *Methanosarcina*, showed more negative relationships in the bioelectrodes. Lower positive connections in the bulk solution and higher positive relationships in the bioelectrodes were presented in the acetogens. Fermentative bacteria (FB), with the ability to degrade organic matters, showed tight affiliations with methanogens and acetogens attached to the biocathode. More complex and diverse connectivity of EAMs, methanogens, acetogens and FB appeared in the bulk solution and suggested obvious mutualistic symbiosis, cooperation, and competition. Together, electrode biofilms had a more positive association and the planktonic microbial community showed the opposite connectivity. Hence, either microbial assembly in the anode or the bulk solution were closely linked to carbon source conversion, whereas the cathode to biosynthesis CH_4 was relatively independent, basically relying on electron transfer to bridge the communication with the anode and suspension.

13.5 CHALLENGES AND PERSPECTIVES

CO_2 capture and utilization for the production of gaseous/liquid energy carriers is a promising way to obtain value-added commodities and moderate the rising CO_2 in the atmosphere. Bioelectrochemical CO_2 reduction is an effective avenue to the reduction of stable CO_2 molecule via extracellular electron transfer, which can be accelerated by external energy input. The interspecies electron transfer provides multiple ways to enhance more hydrogenotrophic methanogens for CO_2 reduction. Also, mutualistic relationships are often established between specific partners that are able to sense each other through mechanisms that have evolved to make their interactions more efficient and robust.

There are still promising potentials to promote microbial growth and biochemical reactions of microorganisms in a specific niche. Hydrogenotrophic methanogens can grow faster (doubling time 4–8 hours) than acetoclastic methanogens (*Methanosaeta*, doubling time 5–7 days) (Wu *et al.*, 1992). Assisted by DIET, higher growth rates using different substrates (acetate, propionate, butyrate, long-chain fatty acids, glycerol, protein, glucose, and starch) can support hydrogenotrophic methanogens (Tang *et al.*, 2015). Some species, for example, *Methanothrix*, are only known to reduce CO_2 to CH_4 with electrons derived from DIET. Low potential electrons derived from DIET may improve the growth of *Methanothrix* species beyond that the possibility with acetate as the sole substrate (Wang *et al.*, 2016), further contributing to *Methanothrix* activity in soils and sediments (Lovley, 2017b).

With the development of high efficient (bio)materials, the CO_2 can also be reused for chemical, biological or physical purposes (bioenergy with carbon capture and utilization, BECCU). A new hybrid microbial photoelectrochemical system can perform improved microbial anode capability of oxidizing waste organics in wastewater (Lu *et al.*, 2020). The CO_2 reduction to CO has been achieved on the nanowire silicon photocathode integrated with a selective single-atom nickel catalyst (Si NW/Ni SA). Similar to H_2, CO is also one of the suitable substrates for microbial products (CH_4, acetate, etc.). In the future, efficient (bio)materials with conductive biofilms derived from DIET may be developed for efficient biological CO_2 capture and utilization technologies. Also, such technologies can reduce carbon emission in the WWTPs to contribute achievement to the carbon emission restriction (Mallapaty, 2020).

ACKNOWLEDGEMENTS

The authors would like to acknowledge the support from the Shenzhen Overseas High-level Talents Research Startup Program from Harbin Institute of Technology, Shenzhen. This research was financially supported by the Natural Science Foundation of Guangdong Province for Distinguished Young Scientists (No. 2021B1515020084), the National Natural Science Foundation of China (No. 51778607), the National Natural Science Foundation of China NSFC-ISF Joint Program (No. 41961144024), and the Shenzhen Science and Technology Innovation Program (No. KQTD2019092917 2630447).

REFERENCES

Ailijiang N., Chang J., Liang P., Li P., Wu Q., Zhang X. and Huang X. (2016). Electrical stimulation on biodegradation of phenol and responses of microbial communities in conductive carriers supported biofilms of the bioelectrochemical reactor. *Bioresource Technology*, **201**, 1–7, https://doi.org/10.1016/j.biortech.2015.11.026

Appels L., Baeyens J., Degrève J. and Dewil R. (2008). Principles and potential of the anaerobic digestion of waste-activated sludge. *Progress in Energy and Combustion Science*, **34**(6), 755–781, https://doi.org/10.1016/j.pecs.2008.06.002

Bajracharya S., Srikanth S., Mohanakrishna G., Zacharia R., Strik D. P. B. T. B. and Pant D. (2017). Biotransformation of carbon dioxide in bioelectrochemical systems: state of the art and future prospects. *Journal of Power Sources*, **356**, 256–273, https://doi.org/10.1016/j.jpowsour.2017.04.024

Cai W., Han T., Guo Z., Varrone C., Wang A. and Liu W. (2016a). Methane production enhancement by an independent cathode in integrated anaerobic reactor with microbial electrolysis. *Bioresource Technology*, **208**, 13–18, https://doi.org/10.1016/j.biortech.2016.02.028

Cai W. W., Liu W. Z., Yang C. X., Wang L., Liang B., Thangavel S., Guo Z. C. and Wang A. J. (2016b). Biocathodic methanogenic community in an integrated anaerobic digestion and microbial electrolysis system for enhancement of methane production from waste sludge. *ACS Sustainable Chemistry and Engineering*, **4**(9), 4913–4921, https://doi.org/10.1021/acssuschemeng.6b01221

Cai W., Liu W., Zhang Z., Feng K., Ren G., Pu C., Sun H., Li J., Deng Y. and Wang A. (2018a). mcrA sequencing reveals the role of basophilic methanogens in a cathodic methanogenic community. *Water Research*, **136**, 192–199, https://doi.org/10.1016/j.watres.2018.02.062

Cai W. W., Liu W. Z., Sun H. S., Li J. Q., Yang L. M., Liu M. J., Zhao S. L. and Wang A. J. (2018b). Ni5P4-NiP2 nanosheet matrix enhances electron-transfer kinetics for hydrogen recovery in microbial electrolysis cells. *Applied Energy*, **209**, 56–64, https://doi.org/10.1016/j.apenergy.2017.10.082

Cai W. W., Liu W. Z., Zhang Z. J., Feng K., Ren G., Pu C. L., Li J. Q., Deng Y. and Wang A. J. (2019). Electro-driven methanogenic microbial community diversity and variability in the electron abundant niche. *Science of the Total Environment*, **661**, 178–186, https://doi.org/10.1016/j.scitotenv.2019.01.131

Cai W., Liu W., Wang B., Yao H., Guadie A. and Wang A. (2020). Semiquantitative detection of hydrogen-associated or hydrogen-free electron transfer within methanogenic biofilm of microbial electrosynthesis. *Applied and Environmental Microbiology*, **86**(17), e01056-20.

Cai W., Wang B., Liu W., Yao H., Deng Y. and Wang A. (2021). Sessile methanogens dominated cathodic biofilm: distribution and network in physiological transitions. *Science of the Total Environment*, **795**, 148724, https://doi.org/10.1016/j.scitotenv.2021.148724

Centi G. and Perathoner S. (2009). Opportunities and prospects in the chemical recycling of carbon dioxide to fuels. *Catalysis Today*, **148**(3–4), 191–205, https://doi.org/10.1016/j.cattod.2009.07.075

Cheng S., Xing D., Call D. F. and Logan B. E. (2009). Direct biological conversion of electrical current into methane by electromethanogenesis. *Environmental Science and Technology*, **43**(10), 3953–3958, https://doi.org/10.1021/es803531g

Cho S. K., Lee M. E., Lee W. and Ahn Y. (2019). Improved hydrogen recovery in microbial electrolysis cells using intermittent energy input. *International Journal of Hydrogen Energy*, **44**(4), 2253–2257, https://doi.org/10.1016/j.ijhydene.2018.07.025

Ding A., Yang Y., Sun G. and Wu D. (2016). Impact of applied voltage on methane generation and microbial activities in an anaerobic microbial electrolysis cell (MEC). *Chemical Engineering Journal*, **283**, 260–265, https://doi.org/10.1016/j.cej.2015.07.054

Dykstra C. M. and Pavlostathis S. G. (2017). Methanogenic biocathode microbial community development and the role of bacteria. *Environmental Science and Technology*, **51**(9), 5306–5316, https://doi.org/10.1021/acs.est.6b04112

Ferry J. G., Smith P. H. and Wolfe R. (1974). Methanospirillum, a new genus of methanogenic bacteria, and characterization of Methanospirillum hungatii sp. nov. *International Journal of Systematic and Evolutionary Microbiology*, **24**(4), 465–469.

Fu Q., Kuramochi Y., Fukushima N., Maeda H., Sato K. and Kobayashi H. (2015). Bioelectrochemical analyses of the development of a thermophilic biocathode catalyzing electromethanogenesis. *Environmental Science and Technology*, **49**(2), 1225–1232, https://doi.org/10.1021/es5052233

Gao K. and Lu Y. (2021). Putative extracellular electron transfer in methanogenic Archaea. *Frontiers in Microbiology*, **12**, 611739, https://doi.org/10.3389/fmicb.2021.611739

Gao T., Zhang H., Xu X. and Teng J. (2021). Integrating microbial electrolysis cell based on electrochemical carbon dioxide reduction into anaerobic osmosis membrane reactor for biogas upgrading. *Water Research*, **190**, 116679, https://doi.org/10.1016/j.watres.2020.116679

Guo Z., Liu W., Yang C., Gao L., Thangavel S., Wang L., He Z., Cai W. and Wang A. (2017a). Computational and experimental analysis of organic degradation positively regulated by bioelectrochemistry in an anaerobic bioreactor system. *Water Research*, **125**, 170–179, https://doi.org/10.1016/j.watres.2017.08.039

Guo Z., Thangavel S., Wang L., He Z., Cai W., Wang A. and Liu W. (2017b). Efficient methane production from beer wastewater in a membraneless microbial electrolysis cell with a stacked cathode: the effect of the cathode/anode ratio on bioenergy recovery. *Energy and Fuels*, **31**(1), 615–620, https://doi.org/10.1021/acs.energyfuels.6b02375

Huang Q., Liu Y. and Dhar B. R. (2022). A critical review of microbial electrolysis cells coupled with anaerobic digester for enhanced biomethane recovery from high-strength feedstocks. *Critical Reviews in Environmental Science and Technology*, **52**(1), 50–89, https://doi.org/10.1080/10643389.2020.1813065

Hussain S. A., Perrier M. and Tartakovsky B. (2018). Long-term performance of a microbial electrolysis cell operated with periodic disconnection of power supply. *RSC Advances*, **8**(30), 16842–16849, https://doi.org/10.1039/C8RA01863D

Jabari L., Gannoun H., Cayol J. L., Hedi A., Sakamoto M., Falsen E., Ohkuma M., Hamdi M., Fauque G., Ollivier B. and Fardeau M. L. (2012). Macellibacteroides fermentans gen. nov., sp. nov., a member of the family Porphyromonadaceae isolated from an upflow anaerobic filter treating abattoir wastewaters. *International Journal of Systematic and Evolutionary Microbiology*, **62**(Pt 10), 2522–2527, https://doi.org/10.1099/ijs.0.032508-0

Jhong H. R., Ma S. C. and Kenis P. J. A. (2013). Electrochemical conversion of CO_2 to useful chemicals: current status, remaining challenges, and future opportunities. *Current Opinion in Chemical Engineering*, **2**(2), 191–199, https://doi.org/10.1016/j.coche.2013.03.005

Jiang Y., Su M., Zhang Y., Zhan G., Tao Y. and Li D. (2013). Bioelectrochemical systems for simultaneously production of methane and acetate from carbon dioxide at relatively high rate. *International Journal of Hydrogen Energy*, **38**(8), 3497–3502, https://doi.org/10.1016/j.ijhydene.2012.12.107

Kamat P. V. (2007). Meeting the clean energy demand: nanostructure architectures for solar energy conversion. *Journal of Physical Chemistry C*, **111**(7), 2834–2860, https://doi.org/10.1021/jp066952u

Kondratenko E. V., Mul G., Baltrusaitis J., Larrazabal G. O. and Perez-Ramirez J. (2013). Status and perspectives of CO_2 conversion into fuels and chemicals by catalytic, photocatalytic and electrocatalytic processes. *Energy and Environmental Science*, **6**(11), 3112–3135, https://doi.org/10.1039/c3ee41272e

Kouzuma A., Kato S. and Watanabe K. (2015). Microbial interspecies interactions: recent findings in syntrophic consortia. *Frontiers in Microbiology*, **6**, 477.

Liu W., Wang A., Cheng S., Logan B. E., Yu H., Deng Y., Nostrand J. D., Wu L., He Z. and Zhou J. (2010). Geochip-based functional gene analysis of anodophilic communities in microbial electrolysis cells under different operational modes. *Environmental Science and Technology*, **44**(19), 7729–7735, https://doi.org/10.1021/es100608a

Liu D. D., Zhang L., Chen S., Buisman C. and ter Heijne A. (2016a). Bioelectrochemical enhancement of methane production in low temperature anaerobic digestion at 10 degrees C. *Water Research*, **99**, 281–287, https://doi.org/10.1016/j.watres.2016.04.020

Liu W., Cai W., Guo Z., Wang L., Yang C., Varrone C. and Wang A. (2016b). Microbial electrolysis contribution to anaerobic digestion of waste activated sludge, leading to accelerated methane production. *Renewable Energy*, **91**, 334–339, https://doi.org/10.1016/j.renene.2016.01.082

Liu W., He Z., Yang C., Zhou A., Guo Z., Liang B., Varrone C. and Wang A.-J. (2016c). Microbial network for waste activated sludge cascade utilization in an integrated system of microbial electrolysis and anaerobic fermentation. *Biotechnology for Biofuels*, **9**(1), 1–15, https://doi.org/10.1186/s13068-015-0423-8

Liu D., Roca-Puigros M., Geppert F., Caizán-Juanarena L., Na Ayudthaya S. P., Buisman C. and Ter Heijne A. (2018). Granular carbon-based electrodes as cathodes in methane-producing bioelectrochemical systems. *Frontiers in Bioengineering and Biotechnology*, **6**, 78, https://doi.org/10.3389/fbioe.2018.00078

Liu W., Wang L., Gao L. and Wang A.-J. (2019). Hydrogen and methane production in bioelectrochemical system: biocathode structure and material upgrading, Chapter 5.10. In: Microbial Electrochemical Technology, S. V. Mohan, S. Varjani and A. Pandey (eds), Elsevier, The Netherlands, pp. 921–953.

Liu X., Jing X., Ye Y., Zhan J., Ye J. and Zhou S. (2020). Bacterial vesicles mediate extracellular electron transfer. *Environmental Science and Technology Letters*, **7**(1), 27–34, https://doi.org/10.1021/acs.estlett.9b00707

Lohner S. T., Deutzmann J. S., Logan B. E., Leigh J. and Spormann A. M. (2014). Hydrogenase-independent uptake and metabolism of electrons by the archaeon *Methanococcus Maripaludis*. *ISME Journal*, **8**(8), 1673–1681, https://doi.org/10.1038/ismej.2014.82

Lopez-Garcia P. and Moreira D. (2020). The syntrophy hypothesis for the origin of eukaryotes revisited. *Nature Microbiology*, **5**(5), 655–667, https://doi.org/10.1038/s41564-020-0710-4

Lovley D. R. (2011). Powering microbes with electricity: direct electron transfer from electrodes to microbes. *Environmental Microbiology Reports*, **3**(1), 27–35, https://doi.org/10.1111/j.1758-2229.2010.00211.x

Lovley D. R. (2017a). Happy together: microbial communities that hook up to swap electrons. *ISME Journal*, **11**(2), 327–336, https://doi.org/10.1038/ismej.2016.136

Lovley D. R. (2017b). Syntrophy goes electric: direct interspecies electron transfer. *Annual Review of Microbiology*, **71**, 643–664, https://doi.org/10.1146/annurev-micro-030117-020420

Lovley D. R. and Nevin K. P. (2013). Electrobiocommodities: powering microbial production of fuels and commodity chemicals from carbon dioxide with electricity. *Current Opinion in Biotechnology*, **24**(3), 385–390, https://doi.org/10.1016/j.copbio.2013.02.012

Lu L., Li Z. D., Chen X., Wang H., Dai S., Pan X. Q., Ren Z. Y. J. and Gu J. (2020). Spontaneous solar syngas production from CO_2 driven by energetically favorable wastewater microbial anodes. *Joule*, **4**(10), 2149–2161, https://doi.org/10.1016/j.joule.2020.08.014

Lyu Z., Shao N., Akinyemi T. and Whitman W. B. (2018). Methanogenesis. *Current Biology*, **28**(13), R727–R732, https://doi.org/10.1016/j.cub.2018.05.021

Malaeb L., Katuri K. P., Logan B. E., Maab H., Nunes S. P. and Saikaly P. E. (2013). A hybrid microbial fuel cell membrane bioreactor with a conductive ultrafiltration membrane biocathode for wastewater treatment. *Environmental Science and Technology*, **47**(20), 11821–11828, https://doi.org/10.1021/es4030113

Mallapaty S. (2020). How China could be carbon neutral by mid-century. *Nature*, **586**(7830), 482–483, https://doi.org/10.1038/d41586-020-02927-9

Mao Z., Sun Y., Zhang Y., Ren X., Lin Z. and Cheng S. (2021). Effect of start-up process using different electrochemical methods on the performance of CO_2-reducing methanogenic biocathodes. *International Journal of Hydrogen Energy*, **46**(4), 3045–3055, https://doi.org/10.1016/j.ijhydene.2020.02.002

McCarty P. L. and Smith D. P. (1986). Anaerobic wastewater treatment. *Environmental Science & Technology*, **20**(12), 1200–1206, https://doi.org/10.1021/es00154a002

Nevin K. P., Hensley S. A., Franks A. E., Summers Z. M., Ou J. H., Woodard T. L., Snoeyenbos-West O. L. and Lovley D. R. (2011). Electrosynthesis of organic compounds from carbon dioxide is catalyzed by a diversity of acetogenic microorganisms. *Applied and Environmental Microbiology*, **77**(9), 2882–2886, https://doi.org/10.1128/AEM.02642-10

Patil S. A., Arends J. B. A., Vanwonterghem I., van Meerbergen J., Guo K., Tyson G. W. and Rabaey K. (2015). Selective enrichment establishes a stable performing community for microbial electrosynthesis of acetate from CO_2. *Environmental Science and Technology*, **49**(14), 8833–8843, https://doi.org/10.1021/es506149d

Perona-Vico E., Blasco-Gomez R., Colprim J., Puig S. and Baneras L. (2019). NiFe -hydrogenases are constitutively expressed in an enriched Methanobacterium sp. population during electromethanogenesis. *Plos One*, **14**(4), e0215029, https://doi.org/10.1371/journal.pone.0215029

Pham T. H., Rabaey K., Aelterman P., Clauwaert P., De Schamphelaire L., Boon N. and Verstraete W. (2006). Microbial fuel cells in relation to conventional anaerobic digestion technology. *Engineering in Life Sciences*, **6**(3), 285–292, https://doi.org/10.1002/elsc.200620121

Rotaru A.-E., Shrestha P. M., Liu F., Shrestha M., Shrestha D., Embree M., Zengler K., Wardman C., Nevin K. P. and Lovley D. R. (2014a). A new model for electron flow during anaerobic digestion: direct interspecies electron transfer to Methanosaeta for the reduction of carbon dioxide to methane. *Energy and Environmental Science*, **7**(1), 408–415, https://doi.org/10.1039/C3EE42189A

Rotaru A. E., Shrestha P. M., Liu F., Markovaite B., Chen S., Nevin K. and Lovley D. (2014b). Direct interspecies electron transfer between Geobacter metallireducens and Methanosarcina barkeri. *Applied and Environmental Microbiology*, **80**(15), 4599–4605, https://doi.org/10.1128/AEM.00895-14

Rowe A. R., Xu S., Gardel E., Bose A., Girguis P., Amend J. P. and El-Naggar M. Y. (2019). Methane-linked mechanisms of electron uptake from cathodes by *Methanosarcina barkeri*. *Mbio*, **10**(2), e02448-18.

Sangeetha T., Guo Z. C., Liu W. Z., Gao L., Wang L., Cui M. H., Chen C. and Wang A. J. (2017). Energy recovery evaluation in an up flow microbial electrolysis coupled anaerobic digestion (ME-AD) reactor: role of electrode positions and hydraulic retention times. *Applied Energy*, **206**, 1214–1224, https://doi.org/10.1016/j.apenergy.2017.10.026

Saunders S. H., Tse E. C. M., Yates M. D., Otero F. J., Trammell S. A., Stemp E. D. A., Barton J. K., Tender L. M. and Newman D. K. (2020). Extracellular DNA promotes efficient extracellular electron transfer by pyocyanin in *Pseudomonas aeruginosa* biofilms. *Cell*, **182**(4), 919–932, https://doi.org/10.1016/j.cell.2020.07.006

Siegert M., Yates M. D., Spormann A. M. and Logan B. E. (2015). Methanobacterium dominates biocathodic archaeal communities in methanogenic microbial electrolysis cells. *ACS Sustainable Chemistry and Engineering*, **3**(7), 1668–1676, https://doi.org/10.1021/acssuschemeng.5b00367

Silvestre G., Fernandez B. and Bonmati A. (2015). Significance of anaerobic digestion as a source of clean energy in wastewater treatment plants. *Energy Conversion and Management*, **101**, 255–262, https://doi.org/10.1016/j.enconman.2015.05.033

Stams A. J. M. and Plugge C. M. (2009). Electron transfer in syntrophic communities of anaerobic bacteria and archaea. *Nature Reviews Microbiology*, **7**(8), 568–577, https://doi.org/10.1038/nrmicro2166

Steinbusch K. J. J., Hamelers H. V. M., Schaap J. D., Kampman C. and Buisman C. J. N. (2010). Bioelectrochemical ethanol production through mediated acetate reduction by mixed cultures. *Environmental Science and Technology*, **44**(1), 513–517, https://doi.org/10.1021/es902371e

Storck T., Virdis B. and Batstone D. J. (2016). Modelling extracellular limitations for mediated versus direct interspecies electron transfer. *Isme Journal*, **10**(3), 621–631, https://doi.org/10.1038/ismej.2015.139

Tang Y.-Q., Shigematsu T., Morimura S. and Kida K. (2015). Dynamics of the microbial community during continuous methane fermentation in continuously stirred tank reactors. *Journal of Bioscience and Bioengineering*, **119**(4), 375–383, https://doi.org/10.1016/j.jbiosc.2014.09.014

Thauer R. K., Kaster A.-K., Seedorf H., Buckel W. and Hedderich R. (2008). Methanogenic archaea: ecologically relevant differences in energy conservation. *Nature Reviews Microbiology*, **6**(8), 579–591, https://doi.org/10.1038/nrmicro1931

Van Eerten-Jansen M. C. A. A., Heijne A. T., Buisman C. J. N. and Hamelers H. V. M. (2012). Microbial electrolysis cells for production of methane from CO_2: long-term performance and perspectives. *International Journal of Energ Research*, **36**(6), 809–819, https://doi.org/10.1002/er.1954

Van Eerten-Jansen M. C. A. A., Jansen N. C., Plugge C. M., de Wilde V., Buisman C. J. N. and ter Heijne A. (2015). Analysis of the mechanisms of bioelectrochemical methane production by mixed cultures. *Journal of Chemical Technology and Biotechnology*, **90**(5), 963–970, https://doi.org/10.1002/jctb.4413

Varanasi J. L., Veerubhotla R., Pandit S. and Das D. (2019). Biohydrogen production using microbial electrolysis cell: recent advances and future prospects, chapter 5.7. In: Microbial Electrochemical Technology, S. V. Mohan, S. Varjani and A. Pandey (eds), Elsevier, The Netherlands, pp. 843–869.

Varrone C., Van Nostrand J. D., Liu W., Zhou B., Wang Z., Liu F., He Z., Wu L., Zhou J. and Wang A. (2014). Metagenomic-based analysis of biofilm communities for electrohydrogenesis: from wastewater to hydrogen. *International Journal of Hydrogen Energy*, **39**(9), 4222–4233, https://doi.org/10.1016/j.ijhydene.2014.01.001

Villano M., Aulenta F., Ciucci C., Ferri T., Giuliano A. and Majone M. (2010). Bioelectrochemical reduction of CO(2) to CH(4) via direct and indirect extracellular electron transfer by a hydrogenophilic methanogenic culture. *Bioresource Technology*, **101**(9), 3085–3090, https://doi.org/10.1016/j.biortech.2009.12.077

Walker D. J. F., Nevin K. P., Holmes D. E., Rotaru A.-E., Ward J. E., Woodard T. L., Zhu J., Ueki T., Nonnenmann S. S., McInerney M. J. and Lovley D. R. (2020). Syntrophus conductive pili demonstrate that common hydrogen-donating syntrophs can have a direct electron transfer option. *ISME Journal*, **14**(3), 837–846, https://doi.org/10.1038/s41396-019-0575-9

Wang A., Liu W., Cheng S., Xing D., Zhou J. and Logan B. E. (2009). Source of methane and methods to control its formation in single chamber microbial electrolysis cells. *International Journal of Hydrogen Energy*, **34**(9), 3653–3658, https://doi.org/10.1016/j.ijhydene.2009.03.005

Wang L. Y., Nevin K. P., Woodard T. L., Mu B. Z. and Lovley D. R. (2016). Expanding the eiet for DIET: electron donors supporting direct interspecies electron transfer (DIET) in defined co-cultures. *Frontiers in Microbiology*, **7**, 236.

Wang L., Liu W., He Z., Guo Z., Zhou A. and Wang A. (2017). Cathodic hydrogen recovery and methane conversion using Pt coating 3D nickel foam instead of Pt-carbon cloth in microbial electrolysis cells. *International Journal of Hydrogen Energy*, **42**(31), 19604–19610, https://doi.org/10.1016/j.ijhydene.2017.06.019

Wang H., Cai W.-W., Liu W.-Z., Li J.-Q., Wang B., Yang S.-C. and Wang A.-J. (2018). Application of sulfate radicals from ultrasonic activation: disintegration of extracellular polymeric substances for enhanced anaerobic fermentation of sulfate-containing waste-activated sludge. *Chemical Engineering Journal*, **352**, 380–388, https://doi.org/10.1016/j.cej.2018.07.029

Wang L., He Z. W., Guo Z. C., Sangeetha T., Yang C. X., Gao L., Wang A. J. and Liu W. Z. (2019). Microbial community development on different cathode metals in a bioelectrolysis enhanced methane production system. *Journal of Power Sources*, **444**, 227306, https://doi.org/10.1016/j.jpowsour.2019.227306

Wang B., Liu W., Cai W., Li J., Yang L., Li X., Wang H., Zhu T. and Wang A. (2020a). Reinjection oilfield wastewater treatment using bioelectrochemical system and consequent corrosive community evolution on pipe material. *Journal of Bioscience and Bioengineering*, **129**(2), 199–205, https://doi.org/10.1016/j.jbiosc.2019.09.001

Wang B., Liu W., Varrone C., Yu Z. and Wang A. (2020b). Hydrogen and methane generation from biowaste: enhancement and upgrading via bioelectrochemical systems. In: Bioelectrochemical Systems: Vol. 2 Current and Emerging Applications, P. Kumar and C. Kuppam (eds), Springer Singapore, Singapore, pp. 83–130. https://doi.org/10.1007/978-981-15-6868-8_5

Wang B., Liu W., Zhang Y. and Wang A. (2020c). Bioenergy recovery from wastewater accelerated by solar power: intermittent electro-driving regulation and capacitive storage in biomass. *Water Research*, **175**, 115696, https://doi.org/10.1016/j.watres.2020.115696

Wang B., Liu W., Zhang Y. and Wang A. (2020d). Intermittent electro field regulated mutualistic interspecies electron transfer away from the electrodes for bioenergy recovery from wastewater. *Water Research*, **185**, 116238, https://doi.org/10.1016/j.watres.2020.116238

Wang B., Liu W., Zhang Y. and Wang A. (2022a). Natural solar intermittent-powered electromethanogenesis towards green carbon reduction. *Chemical Engineering Journal*, **432**, 134369, https://doi.org/10.1016/j.cej.2021.134369

Wang X. T., Zhang Y. F., Wang B., Wang S., Xing X., Xu X. J., Liu W. Z., Ren N. Q., Lee D. J. and Chen C. (2022b). Enhancement of methane production from waste activated sludge using hybrid microbial electrolysis cells-anaerobic digestion (MEC-AD) process-a review. *Bioresource Technology*, **346**, 126641, https://doi.org/10.1016/j.biortech.2021.126641

Wu W.-M., Jain M. K., De Macario E. C., Thiele J. H. and Zeikus J. G. (1992). Microbial composition and characterization of prevalent methanogens and acetogens isolated from syntrophic methanogenic granules. *Applied Microbiology and Biotechnology*, **38**(2), 282–290, https://doi.org/10.1007/BF00174484

Xiao Y., Zhang E., Zhang J., Dai Y., Yang Z., Christensen H. E. M., Ulstrup J. and Zhao F. (2017). Extracellular polymeric substances are transient media for microbial extracellular electron transfer. *Science Advances*, **3**(7), e1700623, https://doi.org/10.1126/sciadv.1700623

Xu H., Wang K. and Holmes D. E. (2014). Bioelectrochemical removal of carbon dioxide (CO_2): an innovative method for biogas upgrading. *Bioresource Technology*, **173**, 392–398, https://doi.org/10.1016/j.biortech.2014.09.127

Yang H.-Y., Wang Y.-X., He C.-S., Qin Y., Li W.-Q., Li W.-H. and Mu Y. (2020). Redox mediator-modified biocathode enables highly efficient microbial electro-synthesis of methane from carbon dioxide. *Applied Energy*, **274**, 115292, https://doi.org/10.1016/j.apenergy.2020.115292

Yu Z., Liu W., Shi Y., Wang B., Huang C., Liu C. and Wang A. (2021). Microbial electrolysis enhanced bioconversion of waste sludge lysate for hydrogen production compared with anaerobic digestion. *Science of the Total Environment*, **767**, 144344, https://doi.org/10.1016/j.scitotenv.2020.144344

Zhang Y. F., Min B., Huang L. P. and Angelidaki I. (2011). Electricity generation and microbial community response to substrate changes in microbial fuel cell. *Bioresource Technology*, **102**(2), 1166–1173, https://doi.org/10.1016/j.biortech.2010.09.044

Zhao Z., Zhang Y., Quan X. and Zhao H. J. B. T. (2015). Evaluation on direct interspecies electron transfer in anaerobic sludge digestion of microbial electrolysis cell. *Bioresource Technology*, **200**, 235–244, https://doi.org/10.1016/j.biortech.2015.10.021

Zheng S., Liu F., Wang B., Zhang Y. and Lovley D. R. (2020). Methanobacterium capable of direct interspecies electron transfer. *Environmental Science and Technology*, **54**(23), 15347–15354, https://doi.org/10.1021/acs.est.0c05525

Zhou A., Liu Z., Varrone C., Luan Y., Liu W., Wang A. and Yue X. (2018). Efficient biorefinery of waste activated sludge and vinegar residue into volatile fatty acids: effect of feedstock conditioning on performance and microbiology. *Environmental Science Water Research and Technology*, **4**, 1819, https://doi.org/10.1039/C8EW00266E

Ziels R. M., Nobu M. K. and Sousa D. Z. (2019). Elucidating syntrophic butyrate-degrading populations in anaerobic digesters using stable-isotope-informed genome-resolved metagenomics. *Msystems*, **4**(4), e00159-19, https://doi.org/10.1128/mSystems.00159-19

doi: 10.2166/9781789061796

Chapter 14
Thermal energy from wastewater

*James McQuarrie**

Tetra Tech Engineering Services Company, Denver, CO, USA
**Correspondence: jim.mcquarrie@tetratech.com*

14.1 INTRODUCTION

Governments, power utilities, public institutions, and private corporations are beginning to establish goals and putting plans into action to decarbonize their footprints. Thankfully, costs for renewable electricity generation and grid-scale battery energy storage are now competitive with the cost of fossil-based thermoelectricity and forecasted to trend lower through 2030 (See Chapter 2). This trend in grid electrical power generation along with other innovations in batteries is allowing other historically energy intensive societal needs like personal automobile transportation to piggyback onto the low-carbon energy trend via electrification. Unlike electricity, there is no trend of similar scale to replace natural gas and other fossil-based combustion fuels for indoor heating and hot water with low-carbon energy. Yet, energy consumption for indoor heating and hot water is substantial. For scale, consider Figure 14.1 below that shows in most of the United States, more than half of total residential energy consumption is for indoor heating and hot water. In temperate climates like the northeastern United States, more than two-thirds of total household energy consumption is expressly for indoor heat and hot water. Similarly, most residential consumption for natural gas is for indoor heating and hot water.

Electrical resistance heating and application of heat pumps are two transition alternatives to the current heavy reliance on combustion-based heating systems discussed above. With electrical resistance heating, 100% of the electrical energy input is converted to heat output giving the local system a coefficient of performance (CoP) of 1. Heat-pump based systems can perform at a CoP commonly in the range of 2–6, depending largely on the temperature and mass flow characteristics of the connected thermal source. Heat pumps achieve these levels of efficiency since they are designed to transfer thermal energy from a local source to a sink (i.e., residential or office space) rather than generate the heat itself from a primary energy source. Figure 14.2 illustrates the range of emission reduction that can be achieved through application of heat pumps compared with electrical resistance heating and how water-source heat pumps can provide the most favorable CoP in terms of unit heat (or cooling) provided per unit of primary energy put into the system. All of the systems summarized in Figure 14.2 are electrical-based heating systems. Therefore, if the electricity is generated from 100% renewables, all types of electrical heating systems would be low carbon. However, the figure shows

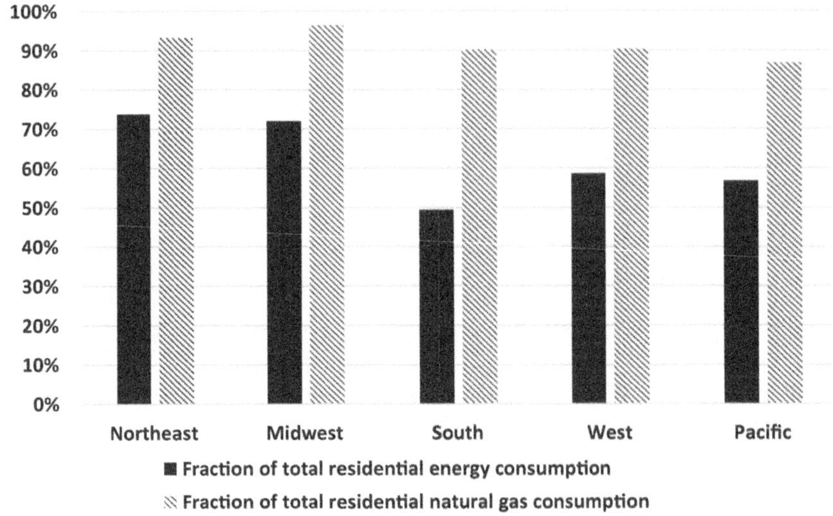

Figure 14.1 Fraction of total residential energy consumption and natural gas consumption used for indoor heating and hot water (adapted from US Energy Information Association, 2018).

that heat pump-based systems reduce the burden of electrical energy consumption compared with electrical resistance heating.

To achieve deep decarbonization, transition of indoor heating and hot water systems away from combustion of fossil-based fuels is necessary. In addition, much greater system-level efficiencies are needed within the built environment that go beyond individual building systems. Rather, over the next few decades, a new generation of modern district heating (DH) and district energy systems (DES) are necessary in the major cities throughout the world that connect and couple buildings (i.e., demands or loads) with non-primary thermal energy (i.e., sources). Lily Riahi, Policy Unit, Climate,

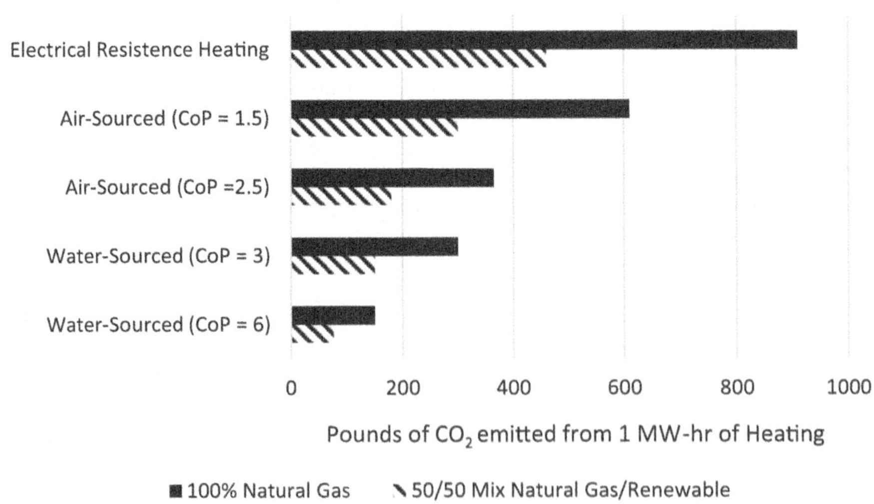

Figure 14.2 Range of CO_2 emission reductions achieved through application of heat pump systems based on the portfolio makeup of the electrical generating plant that supplies the grid.

Energy and Environmentally Sound Technologies UNEP (2015) identifies modern district energy as the most effective approach for many cities to transition to sustainable heating and cooling, by improving energy efficiency and enabling higher sharing of renewable thermal energy. Countries such as Denmark have made modern district energy the cornerstone of their energy policy to reach their goal of 100% renewable energy, and, similarly, other countries, such as China, are exploring synergies between high levels of wind production and district heating. Like the modern electrical grid that consists of more (but smaller) distributed electrical generation stations that put renewable electricity into the grid, modern DH and DES systems link thermal energy sources to a loop that distributes thermal energy to meet connector energy load requirements. Cities are responsible for 70% of the global energy demand. At the same time, cities provide the population density and potential for economic growth necessary to support and sustain modern DH and DES systems. For obvious reasons where there is population density, there is also wastewater. Figure 14.3 shows that in addition to other sources, public investments in wastewater infrastructure can provide value-added service to society in providing a local source for thermal energy to a campus or district scale DH or DES system.

14.2 WASTEWATER AS A THERMAL ENERGY SOURCE

Wastewater can be quite warm. Thermal energy from commercial and industrial discharges, residential hot showers, dishwashing, clothes washers, and other appliances results in the embedding of substantial quantities of thermal energy in wastewater. Along the distance of travel within the built environment, the mass volume of wastewater increases as lateral connections add to the interceptor. In high-density urban areas where campus and district-scale energy systems are most likely to be contemplated, the rate of thermal energy discharged to the sewer can outpace the rate of heat dissipation, providing a quite favorable initial temperature (T_i) condition for heat recovery. Over longer distances, the wastewater in the sewer approaches the ground temperature of the surrounding soil (geo-exchange) which is still warmer than ambient wintertime air temperature conditions. Figure 14.4 helps to illustrate that during winter the wastewater temperature is quite warm compared with ambient air conditions, making it attractive from a thermal energy transfer perspective. In Figure 14.4, the moving average of the air temperature over the seasons is shown in this climate (Philadelphia, PA) to range from -7 to 4°C (20–40 F) during winter. During this same period, the moving average temperature of wastewater in the interceptor is 16–21°C (61–70 F). Conversely, during summer, ambient air temperature trends above the wastewater temperature making the sewer an option for sinking thermal energy (i.e., cooling) in certain DES system designs.

By some means of thermal energy exchange (i.e., exchanger or direct heat pump), the quantity of heat transferred from wastewater or treated effluent for beneficial use can be estimated using the following equation:

$$Q = \dot{m} \times c \times \Delta T$$

where Q is heat transfer (watts); m is the mass flow rate of water (g/sec); c is the specific capacity of water (4.186 J/g-C); ΔT is the temperature drop incurred during heat transfer, T_f minus T_i (C).

As an example, the quantity of heat transferred from a wastewater flow rate of 3.785 million liters per day (mld) is calculated below and assumes a temperature differential of 4°C across the heat exchanger:

$$Q = \frac{43.8\,L}{\text{sec}} \times \frac{1000\,g}{L} \times \frac{4.186\,J}{g - C} \times (17C - 13C) = 733\,\text{kilowatts}$$

In this example, 733 kilowatts (KW) of heat are drawn from the system. Figure 14.5 helps to depict the range in quantity of heat that could be recovered from an interceptor, at a lift station or from effluent at a treatment plant. Due to the specific capacity of water and the relatively favorable initial temperature of wastewater, a considerable quantity of heat can be transferred to meet indoor heating and hot water needs.

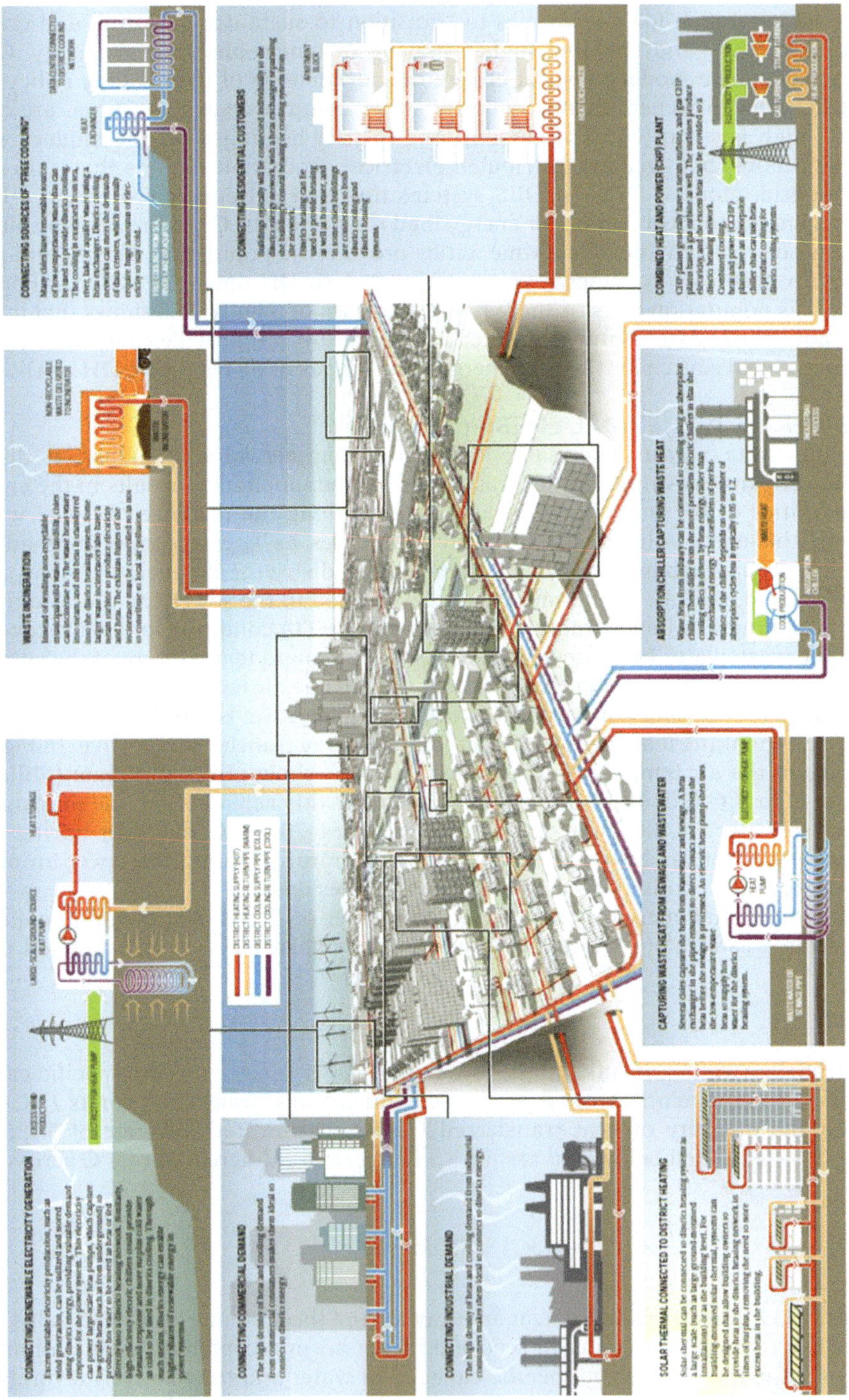

Figure 14.3 Illustration of a modern district energy system where a variety of non-primary thermal energy sources are used in conjunction with a common system to provide low-carbon heating (and cooling) to end-users (from UNEP, 2016).

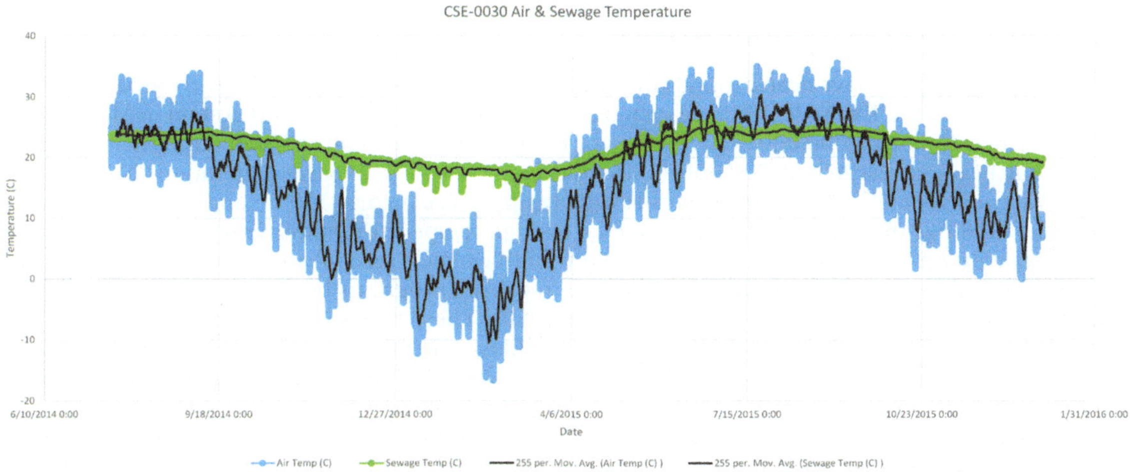

Figure 14.4 Temperature of sewer water and of the ambient air for two years, sewer is the Upper Schuylkill River East Side Interceptor Sewer, Philadelphia Water Department (Kohl, 2019).

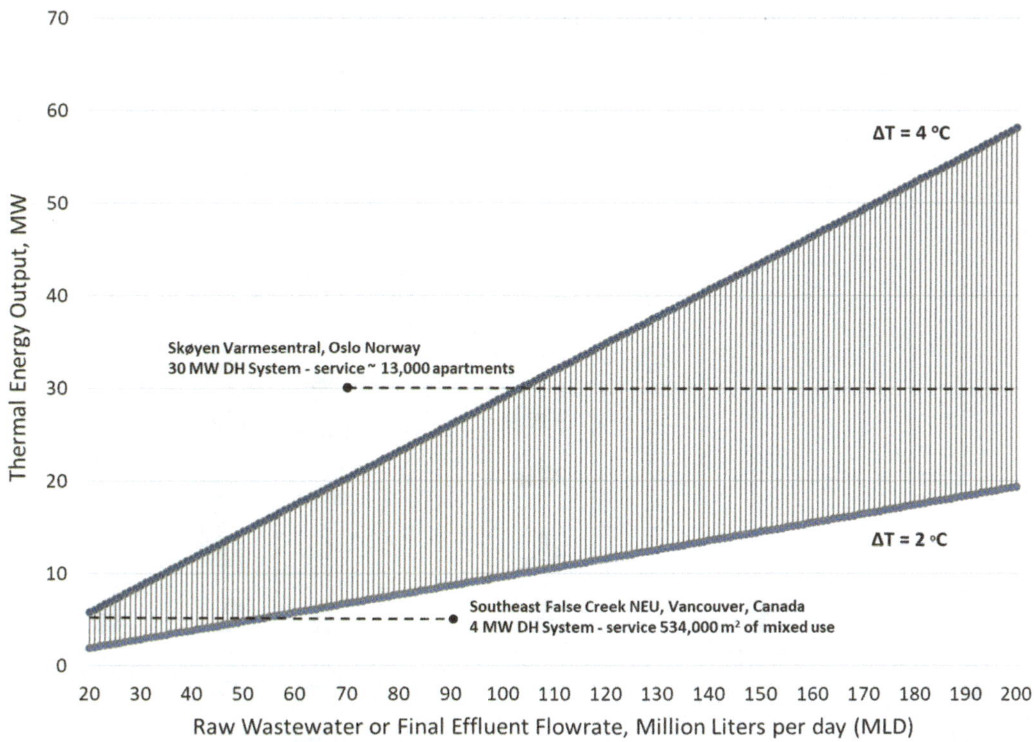

Figure 14.5 Range of recoverable heat (MW) as a function of flow rate at DH/DES access point and dT drop during contact with the heat exchanger or heat pump. A ΔT of 2°C is reasonable for an ambient loop DES type system and ΔT of 4°C is a reasonable upper range for a DH system using a direct heat pump. Two existing DH systems are shown for reference.

14.3 INTEGRATION OF THERMAL ENERGY RECOVERY FROM WASTEWATER WITH MODERN DISTRICT ENERGY SYSTEMS

District heating systems have been around since the 19th century, with some of the early urban systems still in operation today. Examples of these systems include the Steam Operations system in Manhattan that serves an estimated 3,000,000 New Yorkers or the Paris Urban Heating Company system in Paris France that heats an estimated 500,000 household equivalents. The early generation DH systems (first and second generation) produce steam or high temperature (nearly 100°C) and distribute heat to connected buildings through a network of pipes. The high temperature requirement of these systems means they rely on combustion of primary energy sources such as coal in the past or natural gas today. Combined heat and power (CHP) and waste-to-energy programs began to integrate into these district and campus-scale systems in the 1970s and 1980s. These systems are more efficient than the original steam systems with lower GHG emission profiles. The advent of fourth generation district and campus heating systems, which operate at much lower temperature hot water than previous generation systems, has allowed for low-grade thermal energy sources to enter the portfolio of sources for contribution into DH loops. The installation of this generation of DH system is primarily motivated by curbing CO_2 emissions. Heat pump transfer and amplification of thermal energy enables municipal wastewater (raw wastewater or treated effluent) to be used as a source for thermal energy recovery. An illustration of a DH system utilizing a wastewater interceptor is provided in Figure 14.6. In this figure, a campus utility plant interfaces with the interceptor. A heat pump located in the campus utility plant extracts heat from the passing sewage and transfers the heat into a secondary hot water loop that, with pumping, circulates the hot water to transfer heat to its end-use systems.

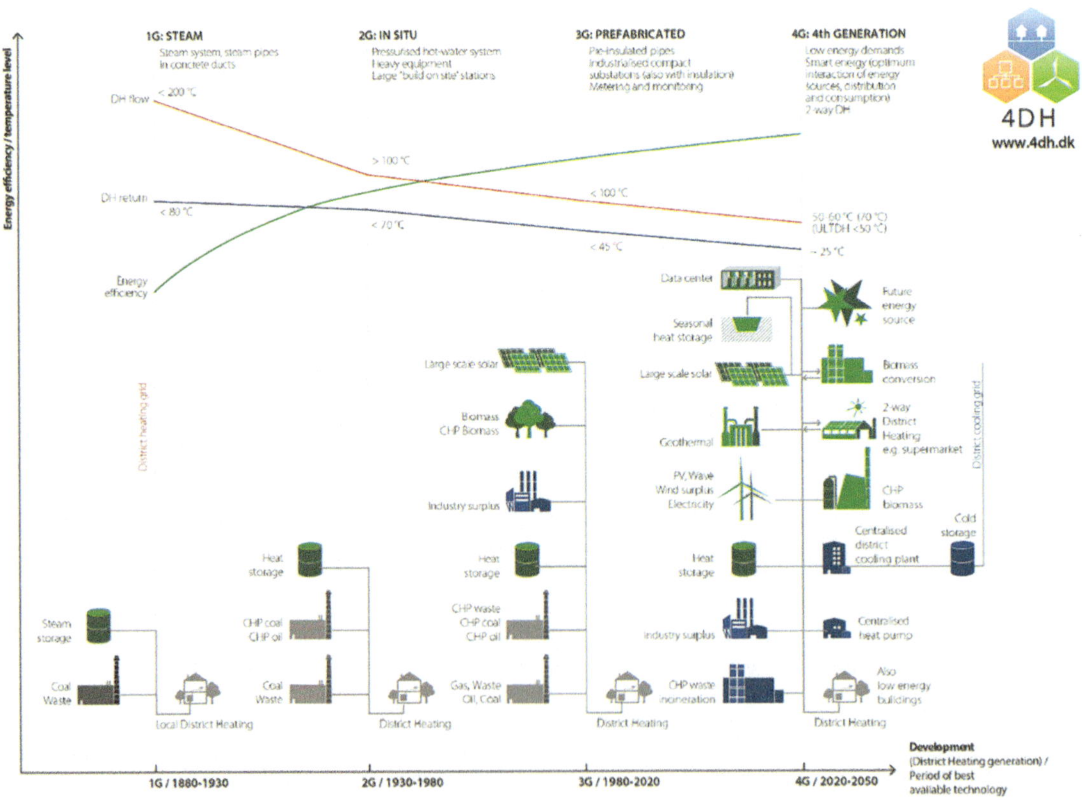

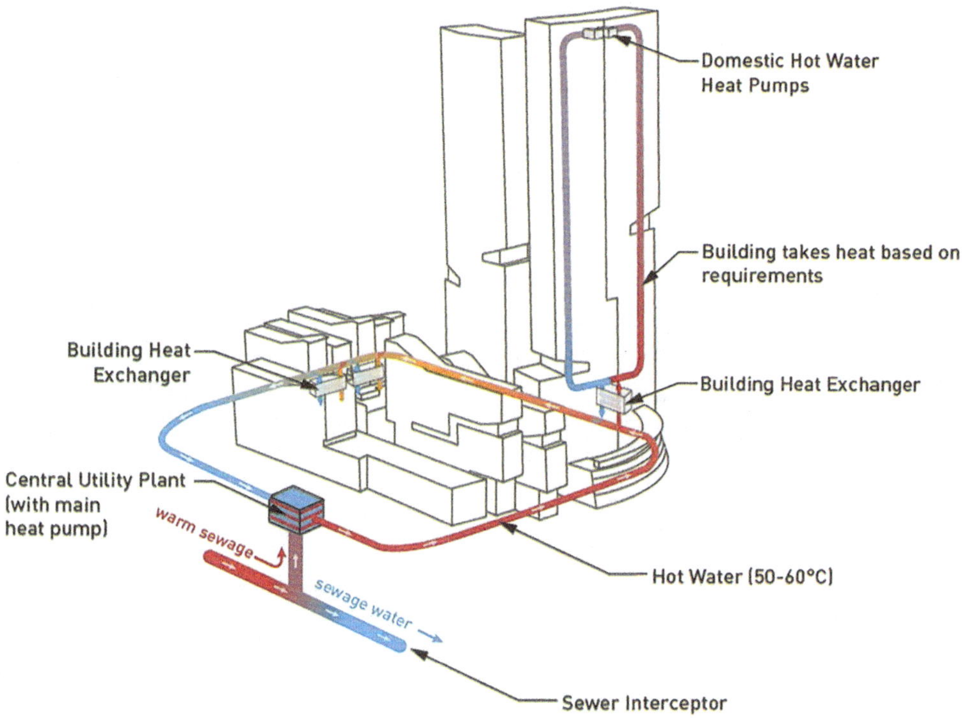

Figure 14.6 Schematic illustration of a fourth generation DH system sourced to a wastewater interceptor and serving district buildings with building heat and hot water.

The advent of fourth generation district and campus heating systems which operate at much lower temperature hot water than previous generation systems has allowed for low-grade thermal energy sources like wastewater to enter the portfolio of sources to supply DH and DES loops (Lund *et al.*, 2018)

Building on the low carbon performance capabilities of fourth generation systems, fifth generation district energy systems (DES) operate at even lower temperatures and seek to provide a campus with year-round low carbon heating and cooling (also known as eco-loops). These systems operate on the premise of an ambient temperature loop that circulates tepid water year-round through a network of buildings. For example, during winter the loop may circulate 17°C water in the main loop to each building, a heat pump extracts and amplifies heat from the loop. In mixed building use environments, some buildings may be doing the opposite where the heat pump is rejecting heat back (e.g., a data center) into the loop for other buildings to pick up and utilize. During summer, the loop may circulate at 22°C and within each building the heat pump is rejecting heat into the loop. Throughout, the wastewater in the interceptor provides the system-level sink or source as needed to balance the overall needs of the system. Figure 14.7 illustrates an example DES using a wastewater interceptor as heating source or sink. Within the campus utility plant, wastewater is passed through a heat exchanger where the heat is exchanged with the ambient circulation loop. A secondary loop in each building utilizes a local heat pump to provide heating or cooling as required by the building. In the built environment, DES systems such as these can provide groups of buildings with extremely efficient and low carbon heating and cooling. An air sourced or ground sourced heat pump (refer to Figure 14.2) can be installed at the campus utility plant if needed to provide peaking capacity during extreme cold or warm periods of the year.

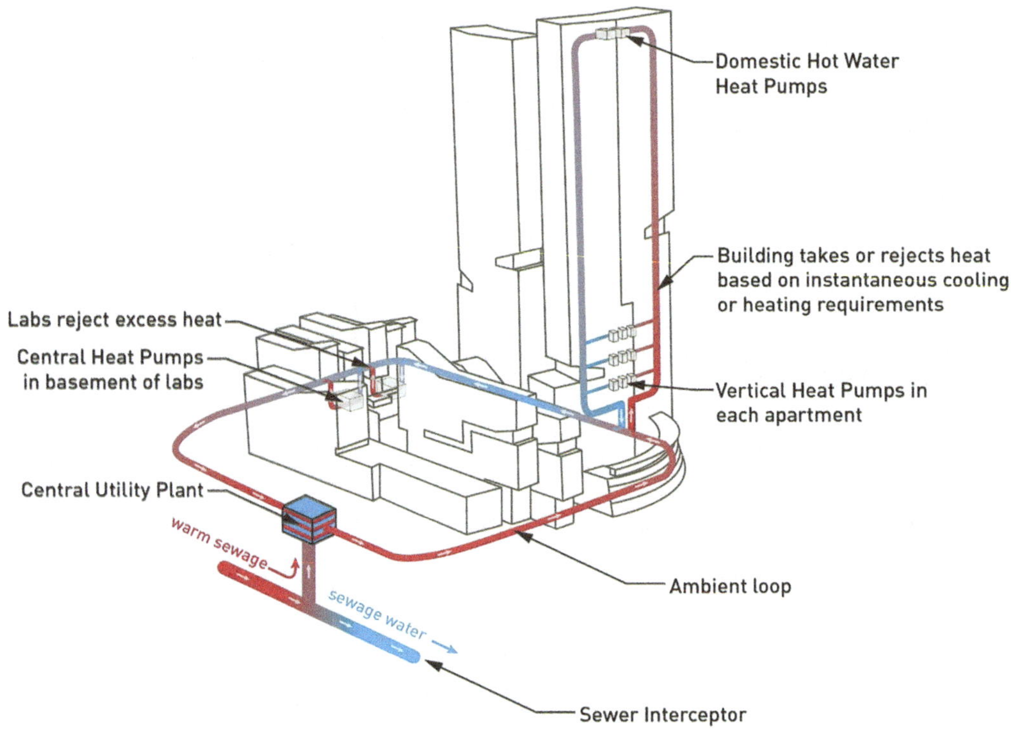

Figure 14.7 Schematic illustration of a fifth generation DES system sourced to a wastewater interceptor and serving district buildings with heating and cooling. Heat is exchanged from sewage to separated ambient loop in the Central Utility Plant.

Importantly, if cooling is required no additional central plant or pipework is needed as the apartment heat pump works in reverse cycle and instead of absorbing energy from the loop rejects energy to the loop. This can deliver cost savings compared with having a central plant heating system and a central plant cooling system installed in parallel.

14.4 ASSESSING THE TECHNICAL FEASIBILITY OF THERMAL ENERGY RECOVERY FROM WASTEWATER

The built urban environment in cities includes decades of public investment in wastewater conveyance and treatment infrastructure. Of course, the prime function of this infrastructure investment is to manage and treat wastewater to protect public health and the environment. However, from a high-level technical perspective, these public investments in a subterranean urban network which convey substantial quantities of low-grade thermal energy have the potential to bring further environmental benefit to the ratepayer in providing a local source of heat and contribute towards broader city or regional goals in decarbonization. At scale, raw wastewater and final effluent can provide a substantial amount of urban indoor heating requirements when coupled with fourth or fifth generation district or campus scale systems. A comprehensive national survey of the recoverable heat from sewage and treated effluent in the UK estimated that roughly 18.3 terawatt-hours of heat could be recovered annually and satisfy 3.6% of the country's heating demand (Wilson & Worall, 2021).

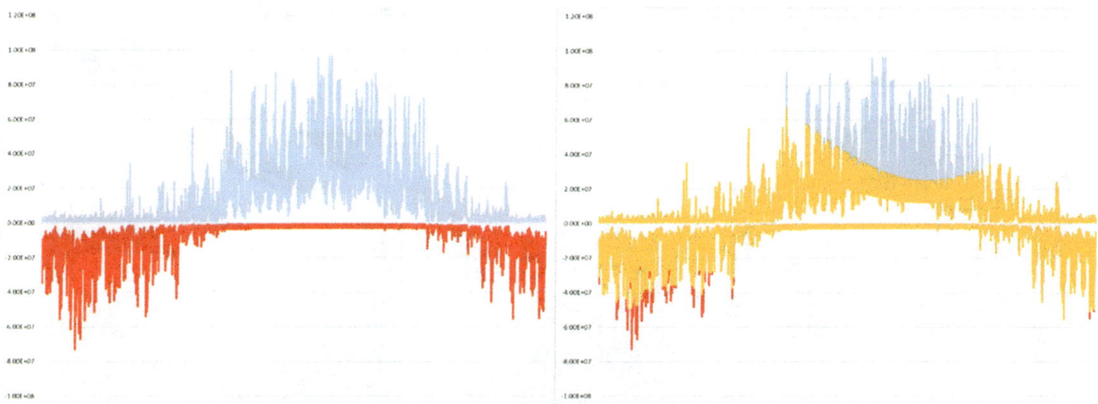

Figure 14.8 Energy model output of building campus energy demands (a) and overlay of thermal energy recovered from a nearby wastewater interceptor (b). The red shows seasonal heating demands, and the blue shows seasonal cooling demands. The yellow shows the coverage of heating and cooling requirements provided through the interceptor.

For a given application or site, it is necessary to understand the potential for thermal energy from wastewater to meet or match building requirements. The assessment must consider the building energy demand profile and compare it to the diurnal and seasonal profile of the thermal energy capabilities of the wastewater (or effluent). An overlay of the building energy model demands with the recoverable thermal energy in the wastewater flow help to determine fit. Figure 14.8(a) provides an example of an hour-by-hour one-year output of a building energy model (eQuest software, Department of Energy). The model incorporates local climate data, building details and architectural treatments for energy efficiency. Figure 14.8(b) provides an overlay of the heat transfer capacity of a nearby interceptor based on the initial wastewater temperature (T_i) and the estimated dry weather flow pattern in the interceptor. In this example, a campus DES coupled to the nearby interceptor can satisfy nearly all the heating requirements and about two-thirds of its summertime cooling requirements. The electrical input for the pumps can be satisfied with grid or on-site PV. The CoP of the water-sourced heat pumps helps reduce the overall electrical requirement of the system. The need for natural gas for heating in winter has been minimized.

14.5 ADOPTION OF THERMAL ENERGY RECOVERY FROM WASTEWATER

It is estimated that globally there are over 500 applications of thermal energy recovery from wastewater (Schmid, 2008). The scale of these installations varies with many at building scale (e.g., ~500 KW) while others like the Skøyen heating plant in Oslo, Norway, at 30 MW of heat recovery capacity, help to illustrate the potential for significant contribution by a local sewerage authority towards decarbonization. The Skøyen heating plant contributes to the larger city district energy and is estimated to provide 130 million kWh of heat annually which corresponds to satisfying the heating and hot water needs for 13,000 apartments.

The Southeast False Creek Neighborhood Energy Utility in Vancouver, British Columbia, provides another example of a fourth generation DH system that supplies indoor heat and hot water to a mix of residential and commercial building space. The heating plant is located under an overpass near a city sewage pump station. A large heat pump (see Figure 14.9) extracts heat directly from screened sewage and exchanges this heat into a second loop that circulates the heat through the Southeast False Creek District with low carbon indoor heating and hot water. The 4 MW phase of the system has been in

Figure 14.9 A large scale heat pump (~4 MW) extracts heat directly from screened raw wastewater to serve the Southeast False Creek District (courtesy city of Cancouver).

operation since 2010 and as of 2019 provided space heating and hot water to roughly 5,750,000 ft^2 of residential and commercial space.

Figure 14.10 provides a schematic illustration of how a heat pump is applied in this instance to extract heat and boost the temperature in the supply loop. In the source loop shown on the left, warm wastewater is passed through the low-pressure evaporator side of the heat pump. In this side of the heat pump cycle, the refrigerant in the pump draws heat out of the passing fluid. The cooled

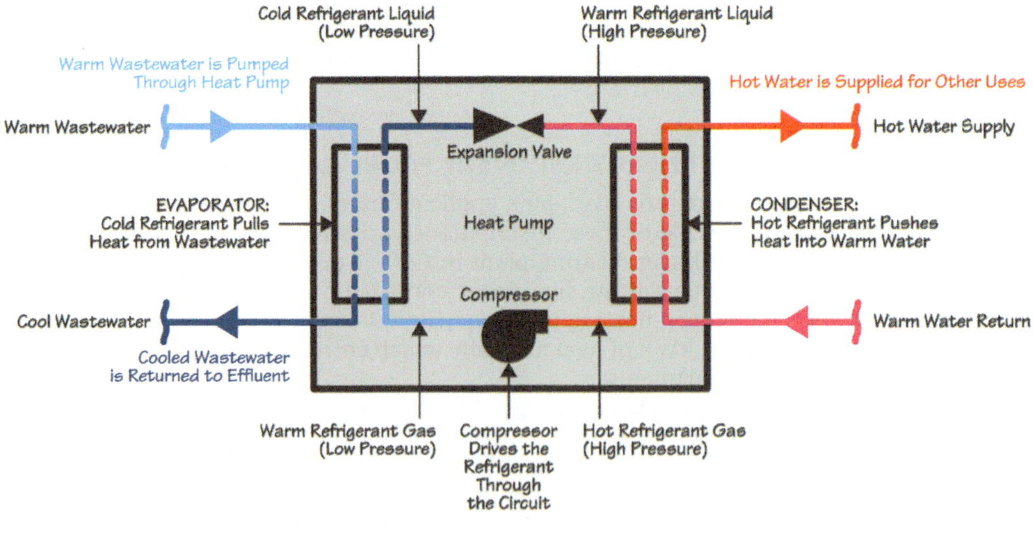

Figure 14.10 Schematic illustration of a heat pump applied to wastewater in a fourth generation DH system.

wastewater exits the heat pump evaporator and is returned to the sewer. The warmed refrigerant gas is then compressed and transferred to the supply loop where under high pressure, the now hot refrigerant gas in the condenser exchanges the heat into the hot water supply line where incoming return flow from the neighborhood takes heat from the hot refrigerant and exits as hot water in the supply line. An expansion valve releases the refrigerant back to the low-pressure evaporator side to begin the cycle again and absorb heat.

The National Western Center (NWC) in Denver, Colorado is a mixed-use redevelopment that includes event space, arenas, research, and education spaces. The thermal energy from the wastewater system at the NWC provides an example of a 5th generation ambient loop DES system that supplies indoor heating during winter and some cooling in summer. The campus utility plant for the site has a nominal capacity of 4 MW and draws heat (or rejects) from an interceptor that runs under the site. A two-pipe loop extends across the site and provides a source of indoor heating and cooling to 119,000 square meters of event space and research space.

14.6 OPPORTUNITIES AND BARRIERS TO THERMAL ENERGY FROM WASTEWATER

The 2021 Intergovernmental Panel on Climate Change (IPCC) report punctuated the point that the world's climate is changing. A significant reduction in global CO_2 emissions requires revisiting the traditional silos of infrastructure service that will challenge the status quo. The incorporation of low-grade thermal energy sources into connected systems is one viable means to reduce indoor heating related emissions but will require open leadership, new policy, unique partnerships, and collaborative business model arrangements. Continued population growth through the remainder of the 21st century will result in many new cities with older cities undergoing infill development, redevelopment, and further densification. These high-density living environs present opportunities for fourth and fifth generation DH and DES systems to be developed that incorporate low-grade sources like wastewater and thus enable low-carbon heating. Twentieth-century public investment in sewer and treatment infrastructure were often co-located along rivers and in what were at the time industrialized areas. Today, these former industrial sites tend to be prime areas for redevelopment and densification within cities to accommodate population and economic growth. These brownfield redevelopment projects coupled with local climate action and energy plans set the stage for potential thermal energy from wastewater systems. Figure 14.11 illustrates a general framework for enabling the partnerships, policy, and techno-economic assessments necessary to determine whether a potential thermal energy from wastewater system is viable.

14.6.1 Defining strategic planning

Given their nature, recovery and reuse of low-temperature or low-grade thermal energy sources requires unique partnerships and buy-in from a variety of stakeholders. Cities and local government that have implemented climate action to reduce carbon footprint help 'set the table' for these initial discussions and eventual partnerships. As an example, in 2014 New York City committed to reducing its GHG emissions by 80% by 2050 compared to 2005 levels (New York City Mayor's Office of Sustainability, 2016). Subsequent efforts by the city resulted in the *Building Emissions Law* that will require large buildings in the city to monitor emissions and invest in systems that reduce buildings emissions and especially from the systems that provide indoor heating and cooling (Local Law 97, 2019). This is an example directive under which innovative approaches and partnerships can develop to tap local sources of low-temperature heat to create low-carbon heating systems. Another example is the Netherlands efforts to transition off fossil natural gas. The country's heating policies, under its 2019 Climate Act, target substantial reductions including addressing that over a third of the energy consumed in the Netherlands is used to heat buildings and homes. New housing is no longer allowed to connect to the gas grid and existing homes and buildings must find low-carbon alternatives to fossil fuels by 2050. Decarbonization policy that recognizes the need to address indoor heating, as

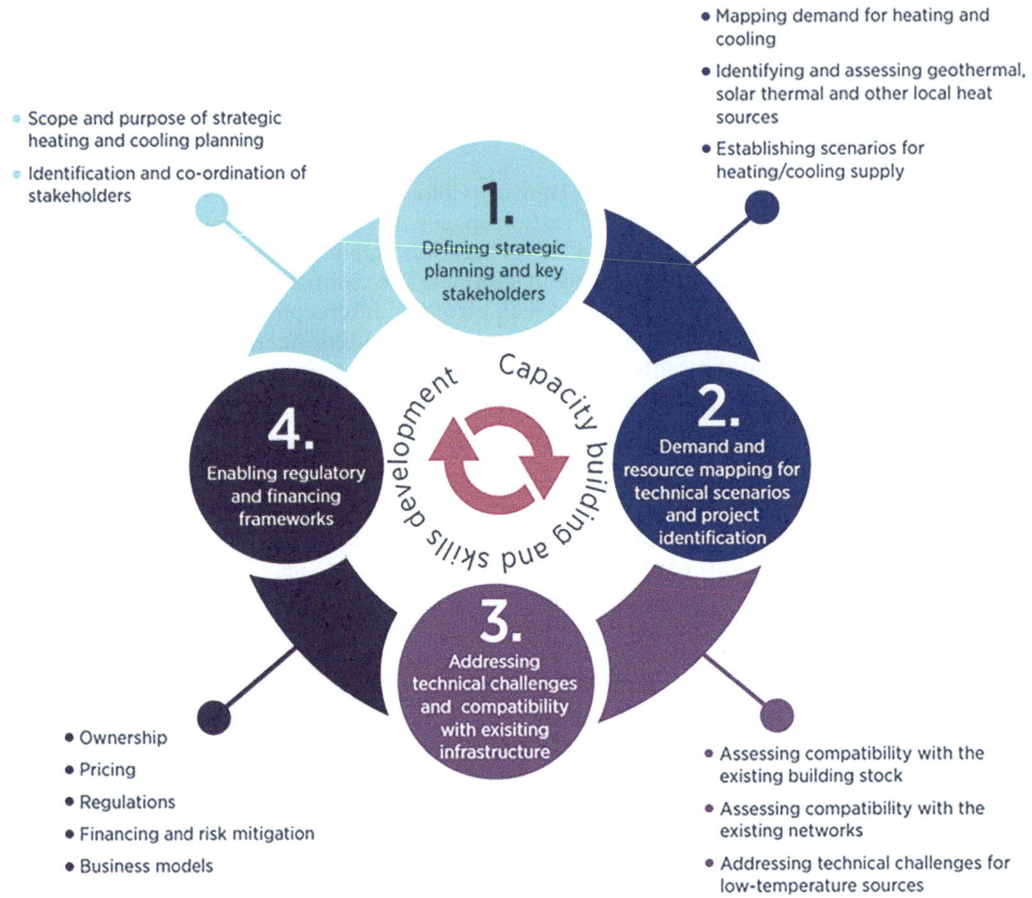

Figure 14.11 Schematic framework for enabling the integration of low-temperature sources into district energy system (Bertelsen *et al.*, 2021).

illustrated with these two examples, help to align end-user needs (e.g., building owner) with potential local low-grade thermal energy source owners (e.g., data center or wastewater sanitation district). In support of these broader goals, wastewater utilities can establish sewer heat recovery policies which establish the utility's position on thermal energy recovery from its infrastructure as well as other technical, administrative and/or financial requirements.

14.6.2 Demand and resource mapping

A challenge with matching and integrating low-grade thermal energy source opportunities with indoor heating demands is awareness and coordination of a potential end user with the thermal energy source. Local governments can help close this gap through their planning departments by developing spatial information that shows the locale and scale of thermal sources and, if applicable, the boundaries of designated heating districts under development or in planning. These maps can include the array of potential low-grade thermal energy sources typically found in an urban landscape. Wastewater treatment facilities and collection system interceptors can be included on these maps showing the estimated available heat in the sewer networks. For a prospective developer or development site, the resource mapping allows the developer to contemplate campus energy options at the early master planning stages of design. For cities with defined strategic plans in place (e.g., Defining Strategic Planning), a wastewater utility can provide mapping layers of interceptor systems with wastewater flows to provide a high-level understanding of available thermal energy available. Under mandates that require alternatives to fossil natural gas, resource mapping and site proximity to available heating sources could become a factor in the overall benefit, attraction and appraised value of a build site to a developer.

14.6.3 Technical feasibility

The next step in a potential thermal energy from wastewater application is to conduct a technical assessment of the specific site conditions and infrastructure. Designation of enterprise zones in cities which include master planning provisions for DH or DES can help address many of the technical matters that could otherwise be challenging to address. A technical assessment for TEW needs to consider access to the interceptor, pump station or other wastewater asset and the requirements of the wastewater utility in terms of how a potential system would interface with the wastewater infrastructure in a manner that would not interfere with normal operations. Other site considerations are the relative complexities of the right of ways and easements that would be required to install, operate and maintain the system. Partnering with the local wastewater utility on a potential thermal energy from wastewater agreement can also generate cost-sharing opportunities to improve the wastewater system itself. As a site undergoes redevelopment and thermal energy from wastewater is installed, the partners can work in opportunities to incorporate necessary improvements to the wastewater infrastructure such as interceptor re-alignment, replace aging infrastructure, or install odor control. The process of going through a technical feasibility assessment allows all the stakeholders to contribute to the overall concept, have their needs addressed and develop buy-in within their respective organizations for the project.

14.6.4 Regulatory and financing frameworks

Energy systems and the sale of energy, even if low-grade, may fall under the jurisdiction and authority of an energy commission or public utility oversight committee. If so, there may be the need for an exemption or require a change in statute to allow for this type of modern heating system as opposed to the governance structures and policy built around fossil-based energy companies and services. Given its potential significance, it is important to understand and, if necessary, address any regulatory barriers that may prevent DES and/or DH systems (both from a technical perspective as well as governance).

Generally, there are three different business models that have been used for thermal energy from wastewater systems. These business models need to address the overall financing of the system, the partnering requirements of the wastewater utility, plus the authority of entity to own, operate and bill customers for service. The following are brief examples of three different business models that have been utilized for thermal energy from wastewater-based heating systems.

14.6.4.1 Public special purpose utility

The model for the Southeast False Creek Neighborhood Energy Utility (NEU) in Vancouver, British Columbia is based on a small special purpose energy district formed within or under an already established public utility. The NEU was formed as a special purpose small utility, owned, and operated by the City of Vancouver. The City of Vancouver also owns and operates the sewer and nearby pump station. The NEU is responsible for providing indoor heating and hot water for the specially designated Southeast False Creek NEU. Development within this district is required through covenant to participate in the NEU. The billing rates for NEU service to its customers are reviewed and set annually by Vancouver City Council along with the other rate charge reviews for other city services (e.g., water and sewer).

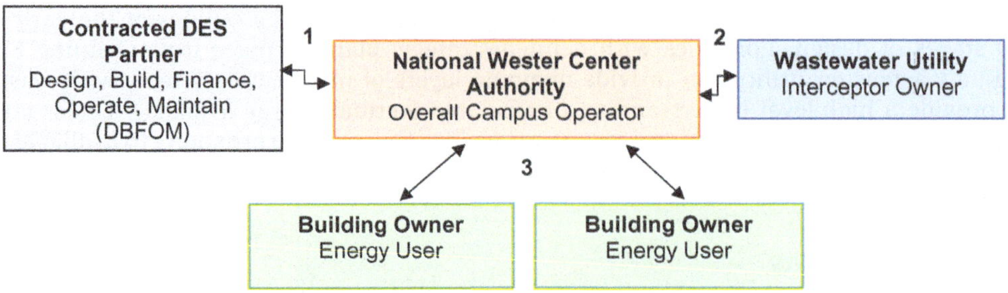

Figure 14.12 General structure of relationships that form the basis of the thermal energy recovery system at NWC.

14.6.4.2 Public-private energy service agreement

The model for the National Western Center (NWC) is based on a public-private partnership where the campus entity agrees to commit and pay a private partner for services including recovery of the private partner's initial capital outlay to build the system. The NWC Authority is responsible for making energy payments to the contracted district energy partner who owns and operates the system. In turn, through operating agreements with the prime building owners on the campus the NWC Authority invoices these entities on a routine basis. The NWC Authority has another operating agreement with the regional wastewater utility that owns the interceptor that serves the campus utility plant that allows and commits access to the interceptor beneath the site, for thermal energy recovery. The term for the campus energy agreement is 40 years.

Figure 14.12 illustrates the coordination of relationships that come together to partner on thermal energy recovery from wastewater. In the example of the NWC thermal energy recovery system, three agreements form the business structure:

(1) **Campus Energy Agreement** – 40-yr fixed price contract between the NWC Authority and the DES Partner to provide DBFOM services for the ambient loop system. The DES Partner is a consortium of engineering, construction, finance, and O&M services companies. The NWC Authority is responsible for the full monthly energy payment that includes capital repayment, O&M, and renewal.
(2) **Operating Agreement** – Operating agreement between NWC Authority and wastewater utility to allow access to interceptor for heating and cooling. The agreement covers the operating plan between the organizations to coordinate performance expectations and lines of communication.
(3) **End User Operating Agreements** – Agreements between NWC Authority and end users on campus. The NWC Authority invoices users on a monthly basis for their portion of the energy payment. The payment is based on the end user's portion of the total connected capacity plus an administrative fee.

14.6.4.3 Joint operator/owner partnership

The Skøyen heating plant is a large TEW facility in connection with the sewage drainage tunnel owned and operated by the Water and Sewerage Authority in Oslo municipality. Two heat pumps with a total power of 30 MW that collect heat from the sewage in the tunnel. In addition, the plant has an electric boiler with a capacity of 12 MW. The plant extracts low-carbon, renewable energy from the wastewater that is converted to 130 million kWh of heat, which corresponds to the need for heat and hot water for 13 000 apartments. The heating plant is jointly operated and owned by Fortum Oslo Farme and the municipality. In joint ownership business models take advantage of the joint sharing of responsibility in terms of the economics, performance and sustainability benefits of the system. The Skøyen heating plant is part of a much larger district energy DH system.

REFERENCES

Bertelsen N., Mathiesen B. V., Djørup S. R., Schneider N. C. A., Paardekooper S., Sánchez García L., Thellufsen J. Z., Kapetanakis J., Angelino L. and Kiruja J. (2021). Integrating Low-Temperature Renewables in District Energy Systems: Guidelines for Policy Makers. International Renewable Energy Agency, Masdar City P.O. Box 236, Abu Dhabi, United Arab Emirates, Available at https://vbn.aau.dk/en/publications/d68fe575-1109-40a3-b758-b48af9939f80 (Accessed September, 2021).

Kohl P. (2019). State of the Science and Issues Related to Heat Recovery from Wastewater, Project Number ENER10C13/4788, Water Research Foundation, Denver, CO, USA.

Lily Riahi, Policy Unit, Climate, Energy and Environmentally Sound Technologies UNEP (2015). District Energy in Cities: Unlocking the Potential of Energy Efficiency and Renewable Energy, UNEP, Copenhagen Centre for Energy Efficiency (C2E2), ICLEI – Local Governments for Sustainability and UN-Habitat, United Nations Environment Programme, Gigiri Nairobi, Kenya.

Local Law 97 (2019). Local Laws for the City of New York. Available at https://www1.nyc.gov/assets/buildings/local_laws/ll97of2019.pdf

Lund H., Ostergaard P. A., Chang M., Werner S., Svendsen S., Sorknaes P., Thorsen J. E., Hvelplund F., Ole Gram Mortensen B., Vad Mathiesen B., Bojesen C., Duic N., Zhang X. and Moller B. (2018). The status of 4th generation district heating: research and results. *Energy*, **164**, 147–159. https://doi.org/10.1016/j.energy.2018.08.206 (Accessed September 2021).

New York City Mayor's Office of Sustainability (2016). New York City's Roadmap to 80×50. Available at https://www1.nyc.gov/assets/sustainability/downloads/pdf/publications/New%20York%20City's%20Roadmap%20to%2080%20x%2050_Final.pdf (Accessed October 2021).

Schmid F. (2008). Sewage Water: Interesting Heat Source for Heat Pumps and Chillers. Proceedings of the 9th International IEA Heat Pump Conference, Zürich, Switzerland, 20–22 May 2008, pp. 1–12.

US Energy Information Association (2018). Annual household site end-use consumption by fuel in the U.S. – totals, 2015. Available at https://www.eia.gov/consumption/residential/data/2015/index.php?view=consumption (accessed 29 December 2020)

Wilson M. P. and Worall F. (2021). The heat recovery potential of 'wastewater': a national analysis of sewage effluent discharge temperatures. *Environmental Science: Water Research and Technology*, **7**, 1760–1777. https://doi.org/10.1039/D1EW00411E, https://doi.org/10.1039/D1EW00411E

doi: 10.2166/9781789061796

Chapter 15

Concept wastewater treatment plants in China

Jiuhui Qu[1,2]*, Hongqiang Ren[2,3], Hongchen Wang[2,4], Kaijun Wang[1,2], Gang Yu[1,2], Bing Ke[1,5], Han-Qing Yu[1,6], Xingcan Zheng[1,7] and Ji Li[1,8]

[1]School of Environment, Tsinghua University, Beijing 100084, China
[2]Expert Committee for China's Concept WWTPs, Beijing 100044, China
[3]School of the Environment, Nanjing University, Nanjing 212013, China
[4]School of Environment and Nature Resources, Renmin University of China, Beijing 100872, China
[5]Administrative Centre for China's Agenda 21, Ministry of Science and Technology, Beijing 100038, China
[6]CAS Key Laboratory of Urban Pollutant Conversion, University of Science and Technology of China, Hefei 230026, China
[7]North China Municipal Engineering Design & Research Institute, Tianjin 300074, China
[8]School of Environment and Civil Engineering, Jiangnan University, Wuxi 214122, China
*Correspondence: jhqu@rcees.ac.cn

15.1 INTRODUCTION

China has witnessed fast growth in public infrastructure over the past several decades. In order to maintain the rapid development of cities and industries, the government and experts are seeking to solve the environmental problems that are far behind its economic development. The total number of wastewater treatment plants in Chinese cities increased by 10-fold, from 481 to 5640, during the period from 2000 to 2018. In the meantime, the construction of thousands more wastewater treatment plants (WWTPs) in the near future is being planned. To explore the suitable wastewater treatment paradigm for China, the China Concept WWTP Committee (CCWC), initiated by several academic leaders in 2014, was formed. The committee laid out a grand vision to ponder the goals of wastewater management in 21st century China. They proposed a new concept for these future WWTPs: sustainable water quality, resource recovery, energy neutrality, and environmental friendliness, which they called the 'Concept Plant'. China's Concept WWTP is expected to lead a national paradigm shift of the wastewater industry.

Water pollution control is currently one of the most pressing challenges faced by China (Lu *et al.*, 2015; Yu *et al.*, 2019). In the battle against environmental pollution, wastewater treatment plays a pivotal role. Although China has the largest wastewater treatment capacity and market of the world, the development history of its wastewater industry is actually very short. Wastewater management in China was almost in blank until 40 years ago when several public environmental incidents raised the urgency of water environment protection. In the 1980s, the National Environmental Protection Bureau was set up, and the first large-scale WWTP with a treatment capacity of 260,000 m^3/d was constructed in Tianjin. Since then, accompanied with the rapid economic development and urbanization, the

Figure 15.1 Growth of municipal WWTPs number in China during 2007–2017 (Qu *et al.*, 2019).

amount of municipal wastewater increased drastically, and the wastewater composition became increasingly complicated because of the entering of more industrial wastewater into the sewers. To address these challenges, China started to build more centralized WWTPs and supplementary facilities. The construction speed and WWTPs scale have been increasing continuously over the years till the end of the 'Twelfth Five-Year Plan' period (Figure 15.1).

The water environmental pollution in China has also aggravated the shortage of water resources, especially in North China regions which face more severe water deficiency. To overcome this limitation, wastewater reclamation and reuse presents a key pathway. Beijing has been pioneering in this direction and has achieved great progress in constructing water reclamation infrastructures. In 2016, the Beijing Gaobeidian WWTP, with the treatment capacity of 1 million m^3/d, was successfully upgraded into a reclaimed water plant, announcing a transition of wastewater management in China from simply treatment to reclamation. However, the overall water reclamation ratio in China is still very low, and the reclaimed water was mainly reused as landscape water due to relatively low quality. Currently, the reclaimed wastewater still cannot compete in price with conventional water supply, and establishment of water reuse infrastructures and program is at slow pace.

After nearly 40 years of remarkable development, China's wastewater industry has grown into the largest one in the world. It now possesses more than 5000 municipal WWTPs with a daily treatment capacity of nearly 200 million m^3/d (Figure 15.1). Accordingly, the wastewater treatment ratio has increased substantially, reaching 90% by 2018. These WWTPs play key roles in water environmental protection by reducing the environmental release of pollutants. The wastewater management mode has also changed from the unitary government-dominated construction and operation into a multiple system that involves both the government and enterprises. Such a transition not only lessened the financial burden of the government to a certain extent, but also improved the construction and operation efficiencies of wastewater facilities.

15.2 THE CURRENT CHALLENGES OF CHINA'S URBAN WASTEWATER TREATMENT FACILITIES

Despite the impressive progress achieved, the government-dominated, pursuant-type development of the wastewater industry in China also left behind many problems. In particular, there are still

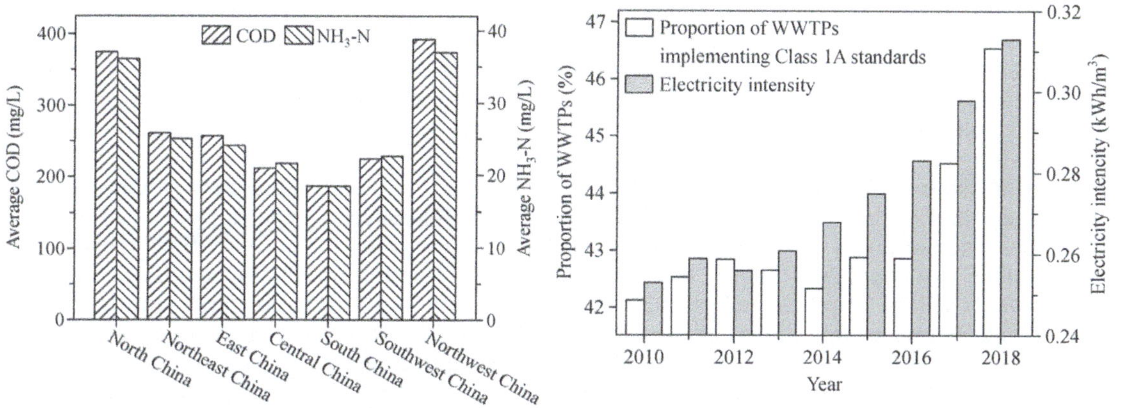

Figure 15.2 (left) The geographic distribution of influent COD and NH_3-N concentrations of WWTPs in China; (right) proportion of WWTPs implementing class 1A effluent standards and the energy consumption intensity of WWTPs in China (Qu *et al.*, 2019).

considerable gaps in the design and operation performance of the treatment facilities compared to those in the developed countries. For example, most plant designs and operations did not consider the sustainability development demand, instead they overstress the pollutants abatement in order to meet the stringent national Class 1A effluent standards. Therefore, in most WWTPs primary sedimentation tanks were omitted while delayed aeration and additional biofiltration were widely implemented, resulting in overtreatment and significantly increased energy/chemical consumption (Figure 15.2). This was aggravated by the lagged development of a sewer system, especially at county level. As a consequence, China's wastewater management suffer from insufficient wastewater collection on one hand and a lower operating ratio of the WWTPs on the other. Such insufficient municipal wastewater collection, plus the dilution by stormwater, significantly lowers the organic strength of wastewater while complicates operation (Figure 15.2).

Consequently, the organic loading in wastewater is typically too low to support efficient denitrification, and to solve this problem external electron donors such as methanol have to be applied. In addition, the low organic content and high sand composition of wastewater sludge also prohibits its anaerobic digestion, a common practice for bioenergy recovery worldwide. It is estimated that less than 3% of WWTPs in China are equipped with anaerobic digestion, among which a large fraction are in poor operation (Jin *et al.*, 2014; Zhang *et al.*, 2016). Therefore, there is almost no energy recovery in China's WWTPs currently, not to mention the recovery of the nutrient resources. How to improve the sustainability of wastewater treatment in China remains a critical issue to be addressed.

15.3 WASTEWATER CONCEPT PLANT PROVIDES A VISION AND EXAMPLE FOR FUTURE DEVELOPMENT

Remarkable progress has been made by China's wastewater industry in infrastructure construction and technology innovation. However, with continued population growth and urbanization in the future, the water shortage will become more severe and urban ecology may become more vulnerable. Thus, the target of wastewater management is shifting from solely pollutant abatement to water reuse, resource recovery and water ecology restoration. This has been reflected by the recent policy changes in China (Wang & Gong, 2018).

For many years, China has been enforcing an end-of-pipe pollution control strategy, that is emphasizing wastewater treatment and water environmental remediation. However, the overall environmental quality has shown no obvious improvement. In 2015, the Chinese government issued the Action Plan on Water Pollution Control (The State Council, 2016), opening a new age of water environmental protection aimed at improving the quality of the overall water ecological environment instead of simple water quality control (Hansen *et al.*, 2018; Holdgate, 1987). This means that the frontier of pollution control will extend from WWTPs to the upstream sewer networks and the downstream rivers and wetlands.

With the vision of turning a WWTP from a site of pollutant removal into a plant of energy, water and fertilizer and an integrated part of urban ecology, in 2014 several experts from top institutes, universities and authority in China jointly proposed the program for constructing a brand-new 'concept plant' (Jin *et al.*, 2014). This China Concept WWTP Committee (CCWC) envisioned that the concept plants will be implemented in 2030–2040, practicing low-carbon concepts, and intensively applying and demonstrating global advanced technologies that have been and will be engineered, so as to fully meet China's sustainable development goal and hopes to become the benchmark of municipal WWTPs in the world. Over the past few years, the CCWC has gathered global insights and established cooperation with many domestic institutions. Discussion and exchanges, visits, collaborative research, formulation of plans, work on engineering practice, and gathering of feedback have been carried out. In 2015, the CCWC initiated and hosted the 'Urban Sewage Treatment Concept Plant' campus creative design competition under the theme of 'Concept Plant – Water Future – My Heart'. The competition was attended by nearly 1000 students from more than 100 universities across the country, effectively enlightening the thinking of the wastewater industry and conveying the ideas to society, especially the younger generation (Figure 15.3).

The CCWC has successfully promoted the implementation of the concept WWTP into practice. The first attempt was completed in Suixian, Henan, and another plant is under construction in Yixing, Jiangsu. In the near future, the Yixing concept plant will be the most instructive plant, leading the WWTPs upgrading to a large-scale, sustainable wastewater treatment plant. In 2018, the first concept plant was built in Suixian County, Henan Province, via a public-private-partnership (PPP) model (Figure 15.4).

Figure 15.3 International competition of campus creative design 'the concept WWTP for the future of water – concept WWTP in My mind'.

Figure 15.4 Sui county NO.3 WWTP aerial view and scenery.

The completed Sui County No. 3 WWTP serves a population of ~900,000 and treats wastewater at an average flowrate of 40,000 cubic meters per day (CMD). In the first phase, the average design flowrate is 20,000 CMD. The plant includes a liquid treatment area, an organic waste processing area, a constructed wetland, agriculture and sponge city demonstration areas, and an office building and education center. Wastewater is treated by preliminary treatment (screens and aerated grit chamber), primary clarification and fermentation, and a step feed activated sludge process with biological nutrient removal. Secondary effluent is polished by denitrification filters and disinfected by ozonation, which is also effective for the destruction of trace levels of emerging contaminants. Treated effluent passes through a constructed wetland, replenishing local surface water bodies. The good effluent quality makes it possible for potential reuse in industrial applications.

The organic waste processing system is designed to treat 100 tons/day of organic waste. Sludge produced from wastewater treatment is co-digested with manure collected from livestock and poultry farms and straw from agricultural operations throughout the county. This uses the DANAS (Dry ANAerobic System) process, a dry anaerobic digestion technology developed by CSDWS. Design capacity for the first phase project is 50 tons/day. The central load rate of organic matter is more than 85%. Co-digestion not only mitigates the non-point source pollution problems in the county, but also produced 510,000 m^3 of biogas, 438,765 kWh of electricity and 4500 tons of fertilizer in 2020. The constructed wetland, agricultural demonstration area (using the organic fertilizer produced onsite),

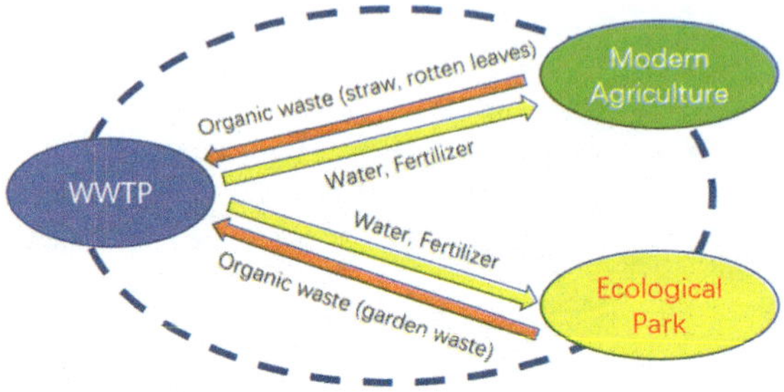

Figure 15.5 Material cycle in the Sui County No.3 WWTP.

and sponge city demonstration area together constitute the ecological park, which creates a synergy between wastewater treatment and the surrounding environment. The office building houses a modern control center and an exhibition hall that displays the treatment technologies employed at the plant. It also serves as an education center to demonstrate the importance of environmental protection and how various resources can be recovered from wastewater and beneficially reused.

The Sui County Concept Factory is combined with the local organic fertilizer factory, adopting the method of 'bartering things, leaving biogas in the middle', that is, the organic fertilizer factory is responsible for the collection, storage and transportation of livestock and poultry manure, and the concept factory produces organic fertilizer raw materials to supply the organic fertilizer plant, which produces biogas for power generation. This method maintains the healthy operation of the organic matter center and realizes the resource recovery and harmless utilization of sludge. The products comply with the 'China Organic Fertilizer Standard (NY525-2012)' (Figure 15.5).

Having adopted the goals set out by the CCWC, the Sui County No. 3 WWTP project has received national recognition. Achievements include an energy self-sufficiency of 50% (Figure 15.6).

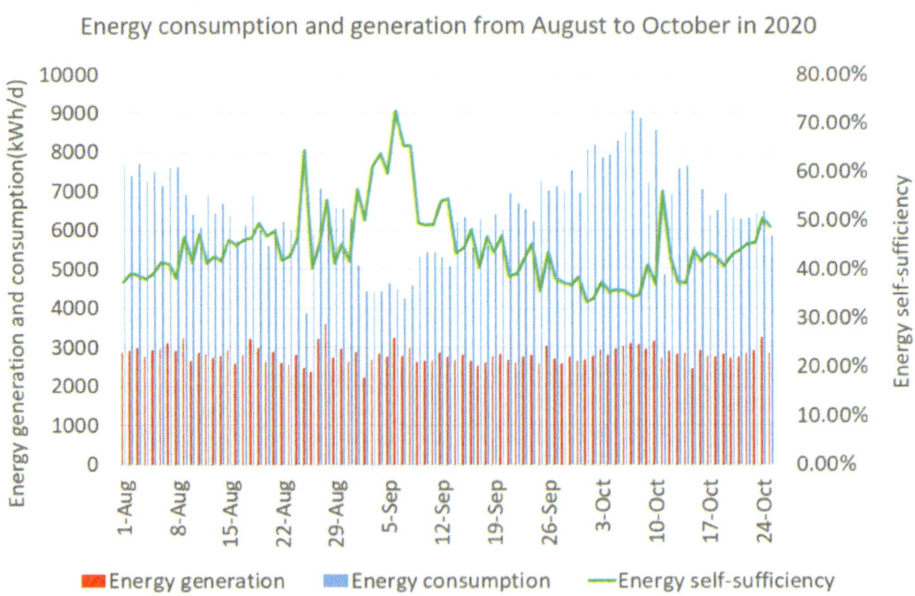

Figure 15.6 Energy self-sufficiency in Sui County No.3 WWTP.

Table 15.1 Yixing wastewater resource concept plant.

Parameters	BOD_5	COD	SS	TN	NH_3-N	TP	pH
Influent quality, mg/L	150–200	480	250	65	55	5–8	6–9
Effluent standard, mg/L	<5	<40	<10	<3	<1	<0.1	6–9

Another example is Yixing Concept Water Resource Recovery Facility (WRRF). From the beginning of 2017 to the end of 2019, after five changes in the draft, it was finally completed by the Beijing Municipal Institute, SUP Atelier and THUPDI Architectural Design Branch. The construction of this plant will be completed in 2021. Yixing concept plant will not only become a water resource recovery facility, but also serve as a full-scale R&D center aiming at comprehensive research and verification of emerging technologies. This brand-new demonstration integrates a pollutant reduction factory with energy, water, and fertilizer factories, which will become a new type of environmental infrastructure that integrates with the surrounding neighborhoods. It is hoped that through the Yixing concept plant, the concept of 'sewage is a resource and sewage treatment plant is a resource factory' will be clearly conveyed to the whole society, and the public's inherent perception of sewage treatment plants will be changed.

The urban sewage resource concept plant that was built in Yixing adopts 'three-in-one' construction, which consists of a water purification center with a capacity of 20,000 tons/day, a collaborative processing center for organic matter with a capacity of 100 tons/day, and a production-oriented R&D center. The sewage treatment part has achieved superior nitrogen and phosphorus removal (TN <3 mg/L, TP <0.1 mg/L, Table 15.1), and its cost performance is significantly better than the current domestic sewage plants. The organic matter co-processing center can treat sludge and cyanobacteria, kitchen waste and straw to produce energy (energy self-sufficiency rate >60%) and fertilizer, and the production-oriented R&D center is composed of two state-of-the-art pilot facilities, displaying the world's most advanced sewage treatment technology in real time (Figure 15.7).

Figure 15.7 Yixing wastewater resource concept plant.

In the Yixing concept WWTP, there will be an innovation center for demonstration and commercialization of the leading-edge technologies with great engineering potential. Those technologies will be selected and demonstrated in the plant with a wastewater treatment capacity of ~1000 t/d for technology showcases. Yixing concept WWTP will serve as a great platform for those innovative technologies to be applied and promoted in this industry. Another 4–5 concepts plants will be designed and built in the next few years as well.

15.4 THE NEXT PARADIGM OF WASTEWATER TREATMENT

Looking back at the 40-years rapid development history of China's wastewater sector, there are both remarkable achievements and numerous failures. Although China has almost accomplished the wastewater infrastructure construction, many problems are left behind, including the under-developed sewers and sludge disposal facilities, remaining high energy consumption and insufficient operating performance, poor linkage between the WWTP effluent discharge standards and the local conditions and environmental protection demand, and lack of synergistic planning between humans and nature.

China has entered the era of environmental stewardship and ecological civilization. Under this background, the wastewater treatment concept plants were conceived and built, showing far-reaching significance for the future. Many water companies have shown great interest and strong desire for building concept plants. China National Development and Reform Commission has also recently issued guidelines on promoting the utilization of wastewater as a resource, which will further promote concept WWTP developments.

New challenges and opportunities like the global pandemic, the commitment on carbon neutrality by mid-century, and the general trend of global environmental governance, are driving the continued evolution of the concept plants. In response to these new demands, the concept plant committee will have a broader vision, carry out more extensive cooperation, continue to 'think-practice-innovate', and optimize the construction model. They strive to build approximately 100 concept plants in the coming 5–8 years, applying advanced technologies and covering various local conditions, different capacities, distinctive features, and diverse models. These will promote substantial changes and upgrades in industry construction forms, technology, and standards. The concept plants are expected to reshape the wastewater industry and lead the paradigm shift in China and the world.

REFERENCES

Hansen M. H., Li H. and Svarverud R. (2018). Ecological civilization: interpreting the Chinese past, projecting the global future. *Global Environmental Change*, **53**, 195–203, https://doi.org/10.1016/j.gloenvcha.2018.09.014

Holdgate M. W. (1987). Our Common Future: The Report of the World Commission on Environment and Development. Oxford University Press, Oxford & New York.

Jin L., Zhang G. and Tian H. (2014). Current state of sewage treatment in China. *Water Research*, **66**, 85–98, https://doi.org/10.1016/j.watres.2014.08.014

Lu Y., Song S., Wang R., Liu Z., Meng J., Sweetman A. J. and Wang T. (2015). Impacts of soil and water pollution on food safety and health risks in China. *Environment International*, **77**, 5–15, https://doi.org/10.1016/j.envint.2014.12.010

Qu J., Wang H., Wang K., Yu G., Ke B., Yu H. Q. and Gong H. (2019). Municipal wastewater treatment in China: development history and future perspectives. *Frontiers of Environmental Science and Engineering*, **13**(6), 1–7.

The State Council. (2016). The 13th Five-Year Plan for the Construction of Urban Sewage Treatment and Recycling Facilities. The State Council of the People's Republic of China, Beijing (in Chinese).

Wang M. and Gong H. (2018). Imbalanced development and economic burden for urban and rural wastewater treatment in China – discharge limit legislation. *Sustainability*, **10**(8), 2597, https://doi.org/10.3390/su10082597

Yu C., Huang X., Chen H., Godfray H. C. J., Wright J. S., Hall J. W. and Taylor J. (2019). Managing nitrogen to restore water quality in China. *Nature*, **567**(7749), 516–520, https://doi.org/10.1038/s41586-019-1001-1

Zhang Q. H., Yang W. N., Ngo H. H., Guo W. S., Jin P. K., Dzakpasu M. and Ao D. (2016). Current status of urban wastewater treatment plants in China. *Environment International*, **92**, 11–22, https://doi.org/10.1016/j.envint.2016.03.024

doi: 10.2166/9781789061796

Chapter 16

Data science tools to enable decarbonized water and wastewater treatment systems

Kathryn B. Newhart[1], Amanda S. Hering[2] and Tzahi Y. Cath[3]*

[1]United States Military Academy, West Point, NY, USA
[2]Baylor University, Waco, TX, USA
[3]Colorado School of Mines, Golden, CO, USA
**Correspondence: kathryn.newhart@westpoint.edu*

16.1 INTRODUCTION

Data science tools can leverage historical and currently generated data to inform and impact how water and wastewater distribution and treatment systems are monitored and controlled. Despite decades of advancement in data-driven modeling of engineered environmental systems, water and wastewater treatment facilities continue to use basic monitoring, analysis approaches, and control schema. Conventionally, models in water and wastewater treatment are derived from the fundamental understanding of the phenomena responsible for the removal of contaminants (e.g., gravity separation and settling, chemical and microbial kinetics). Due to the large size and complexity of full-scale treatment facilities, these models are rarely sufficiently accurate for process monitoring and control. Rather, control thresholds such as upper and lower limits for individual process variables are set based on historical performance and on operators' understanding of a specific system to define normal operating conditions. Such values are static and include a large factor of safety to account for all possible water quality, environmental, and operational conditions; ultimately substantially reducing the efficiency of a system. An alternative to these static, physical, mathematical models are empirical, data-driven models. These 'intelligent' models rely on the relationships between variables identified within a data set without explicitly defining the relationship based on pre-existing knowledge. Data-driven modeling (DDM) has intensified in recent years as the expense of data collection and storage has decreased while processing speeds have exponentially increased. However, the water and wastewater treatment sectors have not fully realized these technological advancements. Manesis *et al.* (1998) presumed that the limiting factors to adopting DDM in the treatment industry were: (1) the underdeveloped field of intelligent control; and (2) the lack of familiarity with DDM by engineers. Despite increased interest in the scientific literature, full-scale application of DDM in water and wastewater treatment systems is still limited due to the second of Manesis' factors. The purpose of this chapter is to familiarize water treatment engineers with DDM methods to achieve decarbonization objectives in the water and wastewater treatment sectors.

DDM includes both statistical and machine learning (ML) methods, and while they may seem similar, there is no single approach that is universally 'better' than the other due to differences in their purpose and requirements. Statistical models are inherently probabilistic models, meaning that a measure of uncertainty comes automatically with the model. Thus, when used to analyze, summarize, and draw conclusions from data, statistical models include a margin of error that depends on the noise in the data. Assumptions about the shape of the distribution of the noise in the data or the functional form of the relationship between variables (e.g., linear, exponential, polynomial) are needed for these models to be valid. On the other hand, ML models are incredibly flexible and can handle modeling nonlinear and complex relationships among variables. They do not require any assumptions regarding the sampling distribution or shape of relationships between variables. However, uncertainty quantification cannot be as readily derived from ML models, a large number of internal, tuning parameters must be selected, and they often require very large sample sizes to fit them. Both approaches can be used to achieve the same goal and are agnostic to specific processes or systems, but they differ philosophically with statistical models taking a stochastic approach and ML models taking an algorithmic approach. A more in-depth discussion of the distinction between statistical and ML models can be found in Boulesteix and Schmid (2014). Some may argue that ML models are 'black boxes' compared to statistical models; indeed, both types of models are black boxes relative to physics-based models. The importance of each variable in the model is automatically provided by a statistical model and takes more work to obtain from a ML model, but both types require some interpretation to understand each variable's influence on the response of interest (Ljung, 2010).

The objective of DDM for decarbonization is to minimize energy consumption and inefficiencies, and maximize resource and energy recovery to ultimately reduce direct and indirect greenhouse gas (GHG) emissions. Sources of direct GHG emissions include oxidation of organic matter, byproducts of the biological nitrogen removal processes, and biogas production from anaerobic digestion (AD) and combustion. Indirect GHG emissions include emissions associated with electrical energy consumption, external carbon for denitrification, and sludge disposal and recovery (Flores-Alsina *et al.*, 2011). Simulation studies have theorized the impact of different control schemes on GHG footprint for water and wastewater utilities (Barbu *et al.*, 2017), but are based on qualitative assumptions regarding operating costs and effluent quality indexes. This is because simulation studies approximate, but do not always accurately represent, the true multivariate relationships among variables at full-scale water treatment plants (WTP) and wastewater treatment plants (WWTP). For example, Oppong *et al.* (2013) compared the Pearson correlation coefficients between inputs and outputs of a full-scale AD and the most popular simulation model (benchmark simulation model no. 2 or BSM2) (Jeppsson *et al.*, 2007). They found large differences in magnitude and/or direction for all variable pairs. Additionally, the full-scale outputs did not match the BSM2 outputs. This discrepancy could be caused by many factors, including the influence of other variables and process disturbances such as change to influent composition, infrequent sampling of certain variables, and a different operating range at full-scale than is possible in simulation. Ultimately, the simulation model for this case was overly simplistic and unable to capture the true behavior of the AD process at full-scale. Dellana and West (2009) compared the prediction performances of a statistical model and an ML model on simulated and real WWTP data, showing conflicting results as to which model can 'best' predict effluent nitrogen and phosphorus. While the statistical model had a lower prediction error for some simulated cases, the ML model had the lowest prediction error for all cases using the real WWTP data. Ultimately, DDM at full-scale for decarbonization must explicitly target features known to impact energy consumption and GHG emissions; however, the actual impact may be difficult to extrapolate for an individual facility.

The work presented in this chapter is intended to introduce the reader to DDM methods that can be used within a larger decarbonization strategy and considerations to apply them appropriately. This chapter is organized into five sections. In Section 16.2, data preparation, common DDM methods, and metrics for comparing model performance are presented. In Section 16.3, unit processes in WTP and

WWTP are discussed in which the Section 16.2 methods may be used to maximize decarbonization. In Section 16.4, recommendations for full-scale implementation of DDM are presented, and Section 16.5 includes concluding remarks.

16.2 PRINCIPLE DATA SCIENCE TOOLS

DDM is largely divided into statistical learning and machine learning (a subset of artificial intelligence); however, there is no single method that performs 'best' for decarbonization systems. Depending on the application (i.e., system), consumer (i.e., plant operator, engineer, data science consultant), the quality and quantity of data available, and the objective of the analyses (i.e., forecasting, prediction, optimization), then a statistical, machine, or hybrid statistical-ML approach may be the most appropriate. Hybrid models are particularly advantageous in prediction and forecasting models due to the ability of statistical and ML models to effectively capture lower- and higher-dimensional relationships, respectively. Examples of hybrid configurations for prediction include using a statistical model as an input to a ML model (Newhart *et al.*, 2020), or a linear combination of a statistical model of a predictor and ML model of the statistical model residuals (Zheng & Zhong, 2011). In this section, we discuss important components of developing an intelligent water system: data preparation, dimension reduction for feature selection, prediction of important process control variables, optimization of machine learning models, and metrics for determining the 'best' model or approach. The process of developing a DDM is illustrated in Figure 16.1.

16.2.1 Data preparation

When aggregating data for use in DDM, there are a variety of considerations to ensure that the data are representative of the process being modeled:

(1) *Water quality sampling*: In-line sensors or analyzers are not available for all water quality variables that are monitored due to cost, a time-consuming or complex analytical method, or a lack of developed sensing technology. However, many of these variables are required for regulatory reporting (e.g., *E. coli*) or performance evaluations (e.g., volatile fatty acids). The sampling method will determine the timestamp assigned to the sample and how online data will be aggregated to best represent the environmental and operational conditions under which the sample was collected. Samples of the representative water or solids matrix are collected in one of three ways for laboratory analysis.

   (a) *Grab*: Analysis results are representative of the conditions only at the time that the sample was collected. These are often used for water quality variables that change with time if the sample is stored in ambient conditions (e.g., biological decay).

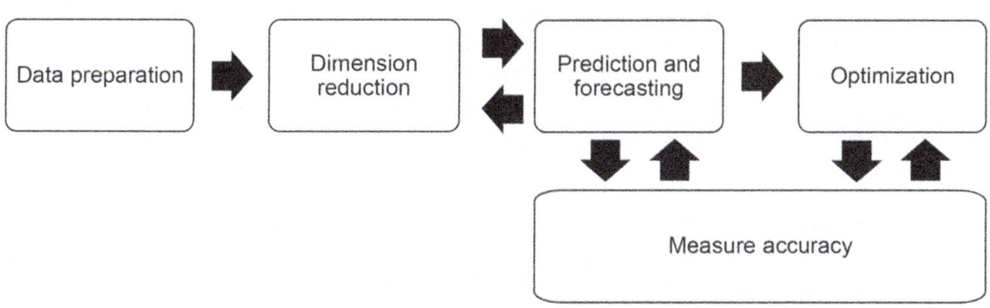

Figure 16.1 Stages of DDM workflow.

   (b) *Time-composite*: Analysis results are representative of a time-weighted, arithmetic average, which is independent of flow. An autosampler draws a set volume of sample at a desired temporal frequency. The aggregated sample is assumed to represent conditions over time.
   (c) *Flow-composite*: Analysis results are representative of an event-weighted, arithmetic average, which is dependent on flow. An autosampler draws a volume of sample proportional to the flow of water. At high flows, a larger volume of sample is taken relative to that taken at low flows. The aggregated sample is the best representation of actual contaminant loading because it accounts for both flow and time.
   (d) *Spatial-composite*: When there is spatial variability in water quality (e.g., poorly mixed rectangular tank), samples from various locations may be taken and combined. The aggregated sample represents the average conditions throughout the space.

(2) **Frequency**: Data are collected by WTP and WWTP at different time intervals depending on the availability of measuring devices (e.g., sensors, analyzers) and staff to maintain the devices, laboratory equipment and staff, and regulatory requirements for specific variables. When all water quality and operational variables are to be aggregated, a wide range of intervals must be considered for interpolation (e.g., seconds, 10 minutes, daily, 2–3 days per week). In DDM, there are two major implications for variables with different frequencies:
   (a) A single interval should be determined that is sufficiently granular for the application (e.g., daily) but is still sufficiently large to avoid inappropriately imputing variables that are not collected as frequently. Aggregation or interpolation approaches must be appropriate approximations of the true environmental and operating conditions for variables that are more or less frequent, respectively, than the interval determined above. Aggregation is most often performed by using arithmetic, time-weighted, or flow-weighted averages. Interpolation can be done linearly or by carrying forward the last measured value; however, interpolation should be done with caution. For example, linear interpolation between observations to 'fill in' missing data is frequently used in practice but is not necessarily an accurate representation of how the majority of water quality variables change with time (Newhart *et al.*, 2021).
   (b) Real-time applications must consider the veracity of instantaneous data and the physical location of the sensor. Many in-line sensors require at least 5–20 minutes to stabilize for a reliable measurement. Therefore, a frequency should be selected based on the time required to achieve a stable moving average for critical predictor variables. Additionally, the time between different sensor measurements is non-linearly related to the flow rate (i.e., hydraulic retention time), and observations may need to be lagged to accurately represent treatment performance for a given influent water quality.

(3) *Normalization*: Normalization of data is generally required of most DDM to make changes in individual variables independent of their magnitudes, although there are exceptions with some ML models (Maleki *et al.*, 2018). For example, a large change in concentration may be 1 mg/L for one constituent, but this may be considered a small change for a second constituent. Conventionally, data for DDM are normalized using either: (i) mean-center (subtract the mean from each value and divide by standard deviation); or (ii) max–min (subtract the minimum from each value and divide by the maximum minus the minimum) techniques.

(4) *Autocorrelation*: Many water quality measurements are highly correlated with the previous measurement for the sample variable, also known as *autocorrelation*. Autocorrelation and partial autocorrelation function plots can help determine the number of previous observations (i.e., lagged observations) that should be considered as predictors in prediction models (Maleki *et al.*, 2018; Perendeci *et al.*, 2009; Wu & Lo, 2010).

16.2.2 Measuring accuracy

To measure the accuracy of prediction methods for a given real-world application, there are a variety of metrics to consider. Fundamentally, there are two types of DDM error: training and testing. Training data are used to fit the models, whereas testing data are not used to fit the model and instead provide insight as to how well the model will perform in real-time or with unknown conditions. There is no standard in the literature as to which specific metrics are used, but frequently one measure of both training and testing error is used to assess model fit and performance. Therefore, it is important to understand the advantages or limitations of an accuracy metric for different applications.

The coefficient of determination (R^2) is the most well-known measure of model accuracy in environmental engineering. The most common application of R^2 is the comparison between training data (y_i or $\mathbf{y} = (y_1,\dots,y_n)$) and model predicted ($\hat{y}_i$) values using the total sum of squares (SST) and error sum of squares (SSE):

$$R^2 = 1 - \frac{SSE}{SST} \tag{16.1}$$

where $SST = \Sigma_{i=1}^{n}(y_i - \bar{y})^2$, $SSE = \Sigma_{i=1}^{n}(y_i - \hat{y}_i)^2$, and $\bar{y} = \frac{1}{n}\Sigma_{i=1}^{n} y_i$. In the case where the modeled values exactly match the actual values, $R^2 = 1$. A linear regression plot of $\hat{y}_i$ vs y_i is helpful to understand the distribution of error and differences in R^2 across multiple models. These plots can answer diagnostic questions such if there are outliers, particular groups of observations that are over or underestimated, or a range of y_i values with greater variability. R^2 does have some limitations, such as being sensitive to outliers, not providing a good measure of the magnitude of the differences, and not penalizing more complex models that have more parameters to estimate. Thus, it is important to understand the underlying differences between two models and their predictions before comparing their R^2 values, or use a different metric for evaluation.

It is expected for a model to have a higher R^2 on the training data compared to the testing data. However, a large difference in training and testing R^2 values can indicate that the model is overfit to the data. When a model is overfit, there are more parameters in the model than necessary to capture the overall pattern. Figure 16.2c is an example of overfitting where the model has too many parameters, and consequently fits to error in the data. A model can also be underfit (Figure 16.2a), where there are an insufficient number of model parameters to adequately capture changes in the dependent variable. Given that R^2 will be higher for an overfit model than a balanced model, R^2 should always be complemented with other measures of error. For example, model fitting criteria such as Akaike Information Criterion (AIC, Equation (16.2)) (Akaike, 1974) and Bayesian Information Criterion

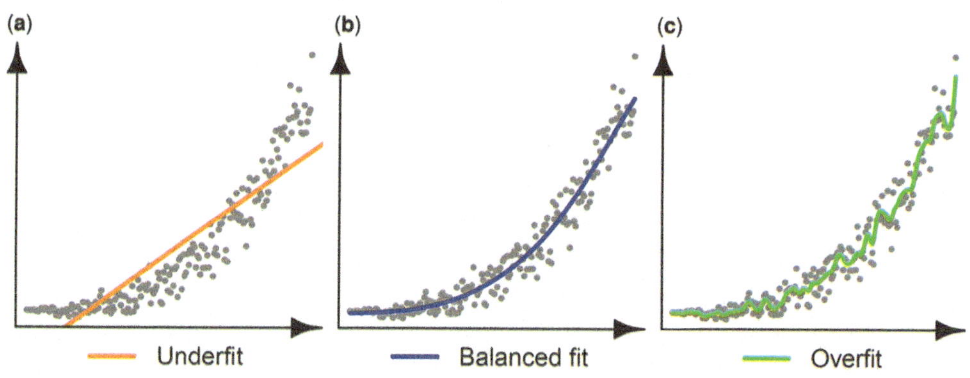

Figure 16.2 Example of an underfit model (orange), a robust/balanced model (blue), and an overfit model (Green).

(BIC, Equation (16.3)) (Schwarz, 1978) are metrics that balance model error against the number of parameters in a model, as follows:

$$AIC = n \cdot \ln\left(\frac{SSE}{n}\right) + 2 \cdot p \tag{16.2}$$

$$BIC = n \cdot \ln\left(\frac{SSE}{n}\right) + \ln(n) \cdot p \tag{16.3}$$

where n is the number of observations, and p is the number of model parameters. The difference between AIC and BIC is the penalty for the number of parameters. When comparing different models (e.g., number of parameters for a given model type), the model with the lowest error and fewest parameters will have the lowest AIC or BIC. When compared for the same model, AIC will choose a model with more inputs than BIC. In this case, it is suggested that models are selected that are generally favored by both AIC and BIC (Burnham & Anderson, 2004; Kuha, 2004; Vrieze, 2012). For example, if AIC is minimized with five parameters and BIC is minimized with three parameters, then a model with four parameters may be best.

To better understand the magnitude of model error, metrics that are not normalized (R^2) or penalized (AIC, BIC) can be used to measure the difference (or squared difference) between the actual and predicted observations. Mean absolute error (MAE, Equation (16.4)), mean squared error (MSE, Equation (16.5)), and root mean squared error (RMSE, Equation (16.6)) are examples of such metrics:

$$MAE = \frac{\sum_{i=1}^{n} |y_i - \hat{y}_i|}{n} \tag{16.4}$$

$$MSE = \frac{SSE}{n} \tag{16.5}$$

$$RMSE = \sqrt{\frac{SSE}{n}} \tag{16.6}$$

The individual metric selected depends on the desired sensitivity to large errors. For example, MAE depends on absolute errors as opposed to the squared errors of MSE and RMSE, so it is less influenced by large differences between actual and modeled values. An additional consideration is whether the metric is applied to the training or testing data. When MAE, MSE, or RMSE are used on both training and testing data in a single publication, some authors will use MAPE, MSPE, and RMSPE to indicate the prediction or testing metric. However, it is also common to see AIC, BIC, and/or R^2 used as the training metrics and MAE, RMSE, and/or RMSE as the testing metrics.

16.2.3 Dimension reduction

In many real-world DDM scenarios, the relationships between input and output variables are not well understood or defined; thus, irrelevant variables are often unintentionally included in DDM. The selection of a subset of variables necessary and sufficient to achieve the model's objective is called feature selection (Kira & Rendell, 1992). Including input variables that are highly correlated to each other or are simply noisy can reduce the effectiveness of predictive models. First, the inclusion of redundant information will increase the time and computing requirements without significantly improving prediction accuracy. Second, many statistical models can become numerically unstable in the presence of multicollinearity, wherein multiple variables supply overlapping information. Third, the interpretability of a model is reduced with additional nonessential input variables. Finally, in detecting faults, including noisy variables that do not change substantially over the monitoring period, makes it more difficult to detect faults (Harrou *et al.*, 2021).

Several methods to approach the problem of dimension reduction by feature selection are described here. Statistical dimension reduction methods such as correlation coefficients and principal component analysis (PCA) can be used prior to model building. Stepwise variable selection and lasso modeling approaches are both oftentimes used in the modeling step to reduce the number of variables in a model. A comparable stepwise approach for variable selection in ML models is also described.

16.2.3.1 Correlation statistics

Correlation coefficients take on values between −1 and +1. The sign indicates the direction of the relationship between two variables, X and Y, and the magnitude indicates the strength of the relationship. A value of one in absolute value indicates that the two variables are perfectly related, and zero indicates that they are completely uncorrelated. There are multiple statistical correlation metrics, although only Pearson's and Spearman's will be presented here to illustrate the difference between magnitude-based and rank-based correlation coefficients. A further discussion on correlation coefficients for water-related data can be found in Helsel and Hirsch (2002).

Pearson's correlation coefficient (r, Equation (16.7)) is the most popular and measures the strength and direction of a linear relationship for a set of n independent observations, $\{(x_1, y_1), (x_2, y_2), \ldots, (x_n, y_n)\}$. It is defined as:

$$r = \frac{\sum(x_i - \bar{x})(y_i - \bar{y})}{\sqrt{\sum(x_i - \bar{x})^2 \sum(y_i - \bar{y})^2}} \tag{16.7}$$

where $\bar{x}$ and $\bar{y}$ are the sample mean of the x_i and y_i values, respectively. Variables that are not linearly related may still be related but have a relatively low r. Spearman's rank-order correlation coefficient (r_s, Equation (16.8)) is a nonparametric variation of r that is able to measure the strength and direction of a monotonic relationship between two variables. For example, as X increases, Y increases for all X and Y, but not necessarily linearly. The Spearman coefficient achieves this by comparing the ranks (i.e., position when observations are ranked from smallest to largest) of each pair of observations, as opposed to the values themselves, as follows:

$$r_s = 1 - \frac{6\sum(x_{i_{\mathrm{rank}}} - y_{i_{\mathrm{rank}}})^2}{n(n^2 - 1)} \tag{16.8}$$

where $x_{i_{\mathrm{rank}}}$ is the rank of x_i, and similarly for $y_{i_{\mathrm{rank}}}$. Depending on the direction and magnitude of the expected correlation as well as the sample size, different correlation coefficients may be more appropriate. Figure 16.3 illustrates how Spearman is less sensitive to small sample size and nonlinear behavior than Pearson. However, both are unable to quantify relationships that both increase and decrease in direction. It is important to note that different correlation coefficient values should not be directly compared. For example, Pearson's r values should only be compared to other Pearson's r values to determine if one pair of features is more highly linearly correlated than another.

16.2.4 Principal component analysis

PCA is a dimension reduction method that creates linear combinations of existing variables to sequentially capture the most variation in the data (Jackson, 1991; Wise and Gallagher, 1996). Each linear combination is a principal component (PC) and is orthogonal to other components; thus, each component represents a different source or direction of variation and is linearly uncorrelated to the others. Figure 16.4 illustrates how a three-variable system can be reduced to two independent PCs using PCA, as the sum of the variance captured by PC1 and PC2 is 100%. In a hypothetical water

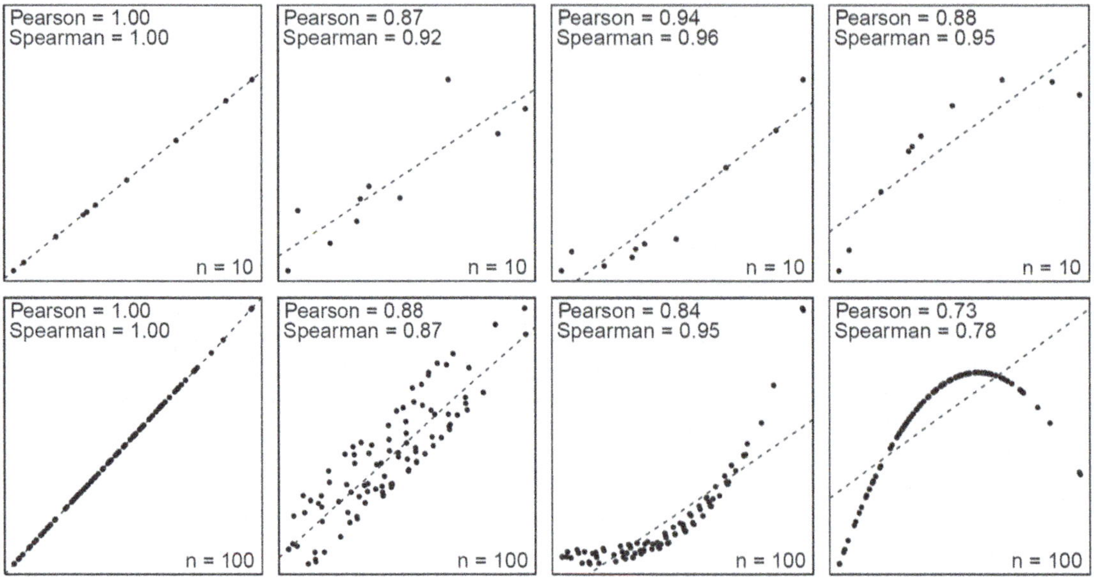

Figure 16.3 Examples of Pearson's r and Spearman's r_s for different sample sizes, $n = 10$ (top) and $n = 100$ (bottom), and data with and without noise (black dots) compared to a fitted linear regression (dashed line).

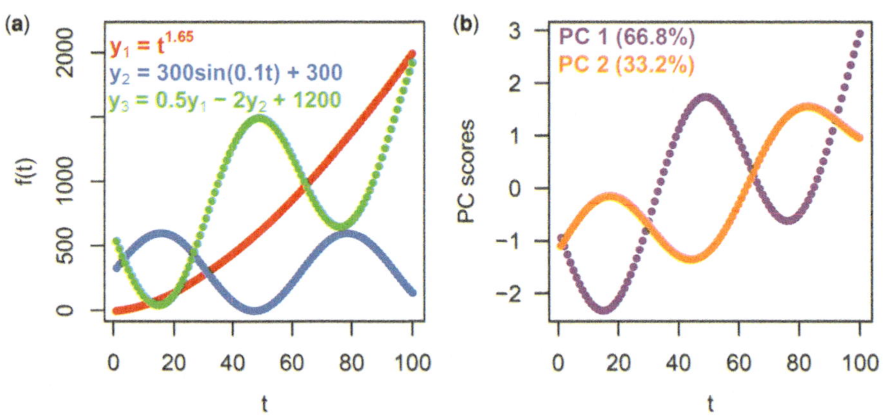

Figure 16.4 (a) Two nonlinear functions (y_1 and y_2) and a linear combination of the nonlinear functions (y_3) are plotted as functions of t; (b) PC1 scores and PC2 scores are the scaled observations from (a) multiplied by the variable loadings (i.e., rotation matrix) for PC1 and PC2, respectively. The percent of total variation captured by each PC is in parenthesis.

treatment system, y_1 could be taken as the ideal growth rate of algae, y_2 as solar irradiation, and y_3 as the actual growth rate. Even though y_1 and y_2 are nonlinear functions, Figure 16.4b shows how their linear combination (y_3) is captured in the first component. The remaining variation in the three-variable system not explained by PC1 is captured by PC2.

The most popular applications of PCA are dimension reduction for multiple regression models (Section 16.2.4.1) (Wallace *et al.*, 2016; Wang *et al.*, 2017) and general process insight (Corominas *et al.*, 2018). For modeling, the PCs with the largest loadings (i.e., magnitude of variance in a particular direction) are retained for model building. The specific number of components retained for model building depends on the percentage of variation described by the PCA subspace, and usually a value in the range of 90–99% is used as a threshold for the number of components retained.

Some limiting factors of conventional PCA, as well as the majority of standard statistical methods for water and wastewater applications, are the assumptions of stationarity (constant mean and variance), linearity, and independence over time. Modifications such as rolling training windows, nonlinear dimension reduction methods, and lagging observations can help approximate the conditions required for methods such as PCA (Kazor *et al.*, 2016; Odom *et al.*, 2018). Newhart *et al.* (2019) describes these adaptations in detail for municipal wastewater treatment.

16.2.4.1 Stepwise variable selection

Oftentimes, the fitted parameters of prediction models are used as indicators of corresponding variable importance; however, the applicability of this approach is DDM-method dependent. For example, the complex, non-linear relationships between predictors in many ML models are not easily summarized by the individual predictor weights in all cases. Therefore, if greater process understanding is the objective rather than prediction accuracy, then methods should be selected that are more interpretable as opposed to ML methods. If prediction accuracy is the primary driver, and ML models are used that do not reveal mechanistic information, then stepwise variable selection is an option for feature selection to reduce dimensionality and remove irrelevant predictors.

Stepwise variable selection methods, such as forward and backward selection algorithms, iteratively add or remove predictors, respectively, for a given model using information criteria (e.g., RMSE, AIC, BIC) to determine which variables enter or exit the model. However, the stepwise approach can bias the parameters estimates, and performing a complete search of all subsets of the variables can be computationally infeasible even if the number of variables is moderate. Both forward and backward stepwise selection can be compared to ensure the similarity of results (John *et al.*, 1994).

16.2.4.2 Lasso

Statistical models such as multiple regression are most often fit using ordinary least squares (OLS), which estimates the parameters such that the SSE between the actual and predicted values are minimized. While the OLS model fitting approach results in unbiased estimators, measurement error in the training dataset can produce high variance, which makes the model difficult to generalize (James *et al.*, 2013). Another approach to statistical model fitting that introduces a small amount of bias but that reduces variability and model complexity is *lasso* (Least Absolute Shrinkage and Selection Operator) (Tibshirani, 1996). Lasso performs variable selection and parameter estimation simultaneously, as opposed to performing these tasks in two separate steps.

Lasso shrinks unimportant predictors' coefficients to zero, thereby selecting only those variables with nonzero coefficients to remain in the model; however, conventional lasso does not always select the correct subset of variables. For example, if two variables are highly correlated, they will both be included in the final variable selection. To address this, Zou (2006) proposed *adaptive lasso*, which has been used for fault detection in a sequencing-batch membrane bioreactor for municipal wastewater treatment (Newhart *et al.*, 2020). *Fused lasso* is another variation of lasso that can address timeseries data where sequential time-varying coefficients are expected to be similar. In fused lasso, the difference between adjacent coefficients are penalized as opposed to the coefficients themselves (Hastie *et al.*, 2015; Tibshirani *et al.*, 2005). An example of fused lasso in WWTP for fault isolation can be found in Klanderman *et al.* (2020). A variation of lasso that is useful in the biological sciences is *group lasso*. Group lasso identifies groups of variables that are jointly in or out of the model (Yuan & Lin, 2006). An example of group lasso in WWTP can be found in Bai *et al.* (2019).

16.2.4.3 SHapley Additive exPlanations

SHapley Additive exPlanations (SHAP) is similar to stepwise variable selection in that input variables are sequentially added to a model, but it differs in how output values are used to determine the best subset. First, a baseline expected value is assumed based on average training data, and input variables are added incrementally to calculate new expected values. The difference between expected values of sequential input features indicates the magnitude and direction of an input variable's influence on the output value. However, all possible permutations of input variables must be tested to account for interaction effects between input variables. The *Shapley value* is the average contribution of an input variable based on all possible combinations of variables compared to the baseline (Shapley, 1951) and is a method-agnostic approach to variable selection in ML (Lundberg & Lee, 2017).

16.2.5 Prediction and forecasting

Predictive models use a mathematical representation of a process such that a set of predictors can be used to approximate a response variable over a given range of operating conditions. When the response variable is the future value of a predictor, the model is said to be a *forecast*. The major difference between forecasting and predictive models is in how they are used. Prediction models are often used to explore the relationships among predictor and response variables and to estimate *in-sample* values. Conversely, forecasting models are used to forecast *future* values of the response variable and should also account for the temporal dependence from one observation to the next. Predictive and forecasting models can be easily incorporated into an existing distributed control system (DCS) at WTP and WWTP such that difficult-to-measure variables can be approximated in real-time for control purposes (Newhart *et al.*, 2020). The use of predictive models in this case are frequently referred to as *soft-sensors*, whose name is derived from the distinction between 'hardware-based' sensors that include conventional in-line instrumentation and 'software-based' predictive and/or forecasting models.

16.2.5.1 Multivariate regression

The simplest modeling method is using a linear combination of predictor variables (X) to calculate a response variable (Y). While the predictors may take on different transformations (e.g., log normal, exponential), assumptions of linearity, no multicollinearity, and homoscedasticity are made with this model. Generally, the model error is taken to be normally distributed with mean zero and a given variance, and then the assumption of a normal distribution also applies to Y, conditional on the predictors. To validate if the data meet these requirements, scatterplots, histograms, and correlation coefficients of predictor variables and the residuals of the fitted model are used. Given the complexity of the water and wastewater treatment process, multiple regression models rarely provide the most accurate prediction, but they are almost always a good starting point. One benefit of multiple regression models is that when all of the predictor variables are standardized, then their estimated coefficients can be directly compared in terms of the strength of the relationship of each predictor variable on the response variable. These models can provide operational guidance and an initial understanding of the phenomena driving treatment performance, whereas ML prediction methods do not automatically provide this information.

Often, linear relationships hold for a narrow range of operating conditions, so multiple models may be needed to approximate a wider, realistic range. A multiple linear model that includes conditional coefficients for different ranges of predictors is a *spline*-based model. Given the limitations of the linear form of the multiple regression model, the individual terms can also be replaced with nonlinear functions, such as basis functions in the case of multivariate adaptive regression spline (MARS) (Friedman, 1991) or polynomials in the case of generalized additive models (GAM) (Hastie & Tibshirani, 1999).

When cyclic patterns are evident in variables observed over time and their corresponding autocorrelation plots, a multivariate regression with sine and cosine terms (i.e., Fourier series) can be used:

$$f(x) = \sum_{k=1}^{K} \alpha_k \cos\left(\frac{2\pi kx}{T}\right) + \sum_{k=1}^{K} \beta_k \sin\left(\frac{2\pi kx}{T}\right) + \varepsilon \tag{16.9}$$

where K is the number of cosine and sine pairs; x is the time in some period T; and α_k and β_k are estimated model coefficients. For example, Newhart *et al.* (2020) used a linear combination of sine, cosine, and process variables to model ammonia concentration in an activated sludge system, and the sine and cosine terms captured the diurnal variation in ammonia concentration. In this case, T was 1440 minutes (equating to 1 day for the length of a single cycle), and x was the minute of a day. Variable selection methods can be used to choose K.

16.2.5.2 Neural networks

Neural networks are one of the most widely studied ML predictive approaches in municipal water and wastewater treatment (Khataee & Kasiri, 2011). Neural networks are a form of computational intelligence that map input variables to output variables (i.e., predictor and response variables, respectively) in a way that mimics how a biological neural pathway is formed (Beale & Jackson, 1990; Bishop, 1995; Kasabov, 1996). Artificial neural networks (ANN) are the simplest form of neural network containing three layers of computational nodes (i.e., neurons): an input layer, a single hidden layer, and an output layer. If multiple hidden layers are used, the structure is said to be a deep neural network (DNN) and can be used to solve highly complex problems, but DNN require a large amount of data and time to train (Schmidhuber, 2015).

When an ANN is trained, weights (w) and biases (b) are adjusted to minimize the error between the actual and predicted output. The most popular training algorithm for feedforward ANN (in which the output from one layer is the input to the next layer) is backpropagation. The backpropagation algorithm works by computing the gradient of the loss function (also known as the cost or objective function) with respect to each weight, computing the gradient one layer at a time, and iterating backward from the last layer (Nielsen, 2015). Measures of error to use as a loss function to compare different model structures (e.g., number of nodes in a hidden layer, types of activation functions) can be found in Section 16.2.2.

Each node consists of an activation function (step, linear, or non-linear) that takes normalized inputs from the preceding layer, adjusting each input using weights and biases. A summary of different activation functions in ANN can be found in Sharma *et al.* (2020). The most widely used ANN activation function in environmental engineering is the sigmoid function where x is a vector of inputs to a node; w is a corresponding vector of weights to a node; and b is the node's bias:

$$\text{output} = \sigma(w \cdot x + b) \tag{16.10}$$

where $\sigma(z) \equiv (1/1 + e^{-z})$.

A neural network that uses sigmoid functions in the hidden layer and a linear function in the output layer is more commonly referred to as a multilayer perceptron (MLP) network. Another ANN that is gaining popularity is the radial basis function (RBF) neural network. In an RBF network, nonlinear radial distance functions are used in the hidden layer with a linear output layer. All ANN discussed to this point assume that the observations used in training and prediction are independent of each other, but a type of neural network for autocorrelated data called recurrent neural network (RNN) is gaining popularity in environmental data settings (Newhart *et al.*, 2021). In RNN, the output from nodes is used as an input for the next observation. This internal memory feature of RNN allows observations to be considered in an ordered sequence. In summary, there are a myriad of neural network configurations defined by the use of different activation functions in different layers. The literature has not yet established the 'best' neural network for water and wastewater treatment; thus, it is important to trial a range of options when developing a predictive model for a specific application.

16.2.5.3 Fuzzy logic

Fuzzy set theory (also known as 'fuzzy logic' or FL) allows for the general categorization of data without definitive boundaries by assigning partial membership (Zadeh, 1973). Put simply, FL allows for observations to be placed in multiple categories (assigning some categories as more probable than others) to account for uncertainty. Figure 16.5 illustrates the functions used to apply FL to a heating system. Input data (e.g., sensor measurements, labels) are 'fuzzified' by applying a membership function to assign linguistic variables, such as the triangle membership function assigning temperature classifications in Figure 16.5a or the Gaussian membership function assigning a heater power adjustment in Figure 16.5b. The membership function assigns multiple values between 0 and 1 for each linguistic label, where 0 indicates the observation does not belong to the given fuzzy set and 1 indicates that the observation belongs completely within the fuzzy set. For example, if the temperature in Figure 16.5 is between two values, then the observation has partial membership in two labels such as 0.7 Cold and 0.3 Cool; although it is not essential that the degrees of membership for all linguistic variables sum to 1. 'If-then' rules are then applied to each fuzzy set ('inference'). A rule set may establish that 'If the temperature is cold, increase the power to the heater substantially,' and 'If the temperature is cool, increase the power to the heater slightly.' In Figure 16.5b, the inferred values correspond to a large approximately 0.8 increase and a slight 0.2 increase. A center value can be calculated by using a weighted average of the inferred values to produce a single, numerical output ('defuzzify'), but alternative approaches also exist. In the heater example, the center value approach results in a power adjustment to the heater between substantial and slight.

Because if-then rules are explicitly defined, the method is considered an *expert system* as opposed to a data-driven system. However, fuzzy inference rule weights can also be identified using DDM methods such as neural networks for more complex problems, but they may lose the true interpretability of the 'if-then' expert structure (Hüllermeier, 2015; Jang, 1993). FL controllers have been proposed for use in process industries with time-varying and non-linear systems for decades, including in water and wastewater treatment (Ferrer *et al.*, 1998; Fiter *et al.*, 2005).

The two most common FL approaches for writing conditional statements are the Mamdani and the Takagi-Sugeno (TS). Mamdani fuzzy rules follow straightforward 'if-then' logic. In the example, 'if the acid flow is low, then the pH is high,' *acid flow* and *pH* are linguistic variables, and *low* and *high*

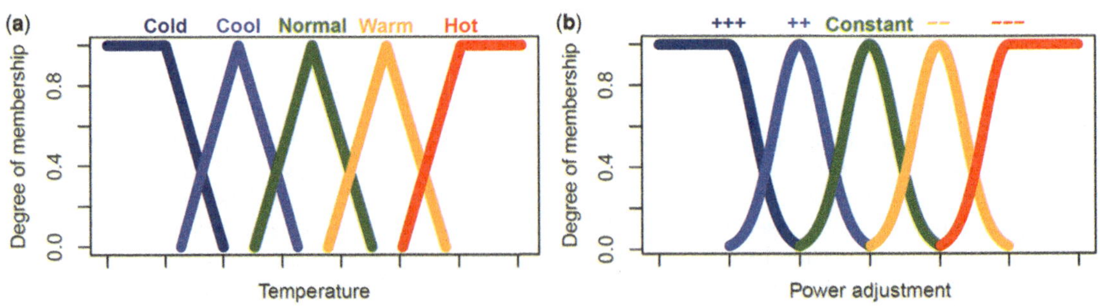

Figure 16.5 Example of (a) triangular membership functions for fuzzification and (b) a Gaussian membership function for defuzzification. A '++' or '−−' indicates a slight change to power and a '+++' or '−−−' indicates a substantial change to power in the corresponding direction (increase or decrease).

are linguistic values of the membership functions. In contrast, TS fuzzy rules use similar 'if' logic and a mathematical equation (e.g., constant, linear, nonlinear combination of input variables) (Takagi & Sugeno, 1983). For example, 'if the acid flow is low, then $pH = k \cdot \text{flow}z + c$,' where k, z, and c are fitted model parameters.

The steps to develop an intelligent FL controller are (Manesis *et al.*, 1998):

(1) Divide variables into manipulated and controlled. A controlled variable quantifies a characteristic of a system (e.g., performance, water quality). A manipulated variable is adjusted to keep a controlled variable at its set point. For example, a recirculation pump flow rate is an example of a manipulated variable, while the concentration of suspended solids, which varies with the recirculation pump flow rate, is a controlled variable.

(2) Establish a set of linguistic descriptors for each manipulated variable (e.g., high, normal, low), which are understood by plant operators. The granularity is directly related to the number of descriptors, although between three and five is appropriate for most control applications. For each set, determine a membership function (Ross, 2010), but the individual function is less important than the number of linguistic descriptors in a set (Sadollah, 2018).

(3) Define if-then rules to form the knowledge base using the linguistic descriptors for manipulated and controlled variables. The form of if-then rules depends on whether the system is Mamdani or TS.

(4) Select a method of weighted averaging for the membership functions of the manipulated and controlled variables (e.g., max-min, center of gravity (COG)).

Adaptive neuro-fuzzy inference systems (ANFIS) are five-layer networks that combine the advantages of ANN and TS FL by fuzzifying and defuzzifying the inputs and outputs to an ANN, respectively, to improve prediction accuracy for noisy data (Abraham, 2005). Due to this hybrid structure, ANFIS is considered to be a universal estimator (Jang *et al.*, 1997). The same method of training ANN (e.g., back-propagation) is used to the tune the FL parameters; however, the membership function itself must be defined. Additionally, the TS rules established in the first layer of ANFIS no longer have the advantage of interpretability compared to conventional FL models, and alternative variable importance approaches must be used to understand the input-output relationships. Applications of ANFIS in water and wastewater treatment are described in Sections 16.3.2 and 16.3.4.

16.2.5.4 Decision trees

A 'decision tree' in ML is a heuristic modeling technique based on a series of binary classifications (e.g., $x > 1$) to classify or predict a variable. There are many potential quantitative (regression) and qualitative (classification) questions that can be answered with decision trees, such as which base to add given a specific water quality as illustrated in Figure 16.6. There are many advantages of tree-based models, including the lack of assumptions regarding the distribution of individual variables or type of relationship (e.g., linear or nonlinear) between predictors and response variables. Instead of a single global model to describe the entire dataspace, multiple models are effectively created through branching and can handle a large number of unique cases. Finally, tree-based models are resistant to outliers (Steinberg & Colla, 1995), but they can be sensitive to changes in the branch splits and in which variables the splits are specified. A good introduction to and mathematical description of classification and regression trees can be found in Sutton (2005).

The oldest and most prevalent technique of fitting tree-based models is bootstrap aggregation, also known as *bagging* (Breiman, 1996). In bagging, many artificial samples are created by sampling with replacement from the data, a procedure termed *bootstrapping*. Then, each of these bootstrapped samples is used to train a prediction method, each obtains a prediction or classification, and then the results are combined by either averaging for regression or voting for classification tasks. *Boosting* is a

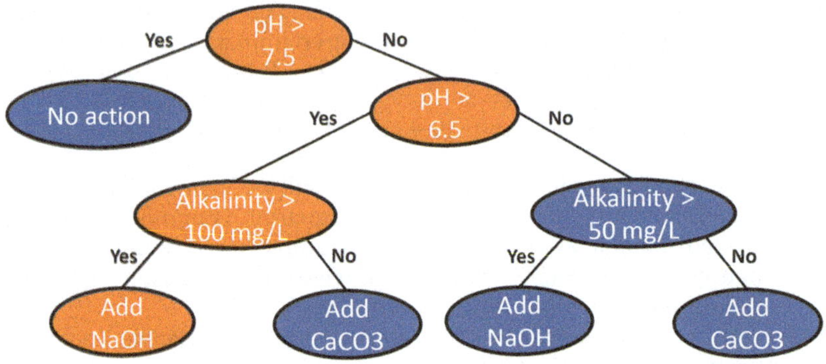

Figure 16.6 Example of a decision tree for determining which base (sodium hydroxide, NaOH, or calcium hydroxide, CaOH) to add based on water quality characteristics. Each node (circle) represents a binary classifier. The orange nodes represent the path taken to reach an action of 'add NaOH' for water with a pH of 7 and alkalinity of 150 mg/L.

variant of bagging in which a weighted average is used to aggregate the results for regression models, and the re-sampling of a bootstrap sample changes with each model fitting iteration. By including more incorrectly predicted observations in subsequent training steps, models are created that can handle unique cases. Additional classifier algorithms have been developed since bagging and boosting with the most popular being the *AdaBoost* algorithm, which uses a weighted average of a series of single binary classifiers that are determined by more heavily weighting incorrectly classified samples of previous classifiers (Freund & Schapire, 1997).

The most popular decision tree approach in water and wastewater treatment are *random forests*. A random forest (RF) is the average of a large number of decision trees created by recursively subsetting input variables and random resampling of training observations (i.e., bootstraping) (Breiman, 2001). The weights of each binary node from the fitting algorithm (specifically boosting or AdaBoost) can be used to determine the importance of a specific predictor in a model. However, input variable weights will vary depending on how nodes are split and therefore can be an inconsistent indicator of variable importance. Finally, to fit an RF model, the number of trees, the maximum depth of a tree, the minimum number of samples to form a split at a node, and the maximum number of variables to evaluate the best split are important hyperparameters that require tuning, resulting in a direct tradeoff between computational burden and accuracy.

16.2.6 Optimization

Optimization algorithms have two major practical applications: optimization of a predictive model (i.e., lowest error by adjusting internal model parameters or hyperparameters) (Le *et al.*, 2019) or finding the optimum set of inputs for an existing predictive model. This chapter focuses on the latter application of optimization as it is more relevant to the objectives of decarbonization. It is the combination of the data-driven predictive model and optimization that constitute model predictive control.

Metaheuristic algorithms are non-exact frameworks designed to search a solution space for the global optimum without calculating every possible solution. Given that there are very few exact mechanistic models for water and wastewater processes that are sufficiently accurate for monitoring and control at full-scale (Newhart *et al.*, 2019), metaheuristic algorithms are used to identify the optimum solution for the predictive models described in Section 16.2.4. The three most popular algorithms used in the water and wastewater distribution and treatment literature are genetic algorithms, particle swarm optimization, and simulated annealing, which will be presented in this section. For additional exact and heuristic methods not listed here, Beheshti and Shamsuddin (2013) provide a more complete list of optimization methods.

16.2.6.1 Genetic algorithms

A relatively quick method of searching for an optimum point in a predictive model is the genetic algorithm (GA) (Reeves & Rowe, 2002). Sections 16.3.1 and 16.3.4 describe applications for pump scheduling and AD operation, respectively. Based on the idea of Darwinian evolution, the phases of identifying the optimum solution using GA are:

(1) Initial population: each 'individual' is characterized by a set of variables ('genes') and demonstrates a potential solution. These individual observations can be randomly generated to allow for the entire range of possible solutions or based on original data.
(2) Fitness function: individuals are assigned a 'fitness score' based on a fitness function. The function quantifies the quality of an individual solution for a variety of criteria, such as minimizing energy consumption.
(3) Selection: The individuals with the highest fitness scores are selected.
(4) Crossover: Pairs of the fittest individuals swap genes, and the resulting 'offspring' have new sets of variables and are added to the population.
(5) Mutation: A low percentage of offsprings' genes experience random changes to maintain diversity within the population.
(6) Termination: To maintain a constant population size, the least fit individuals are removed. The sequence of fitness, selection, crossover, and mutation continue until the population converges.

The advantage of a GA is its ability to handle problems where the solution space is large and the boundaries of the solution space are difficult to identify. This is achieved by using a small number of individuals distributed throughout the solution space. While the population size can be made larger, it significantly increases computation time. The number of computational steps is generally equal to the number of generations (steps 2–4) multiplied by the size of the population. Because the fitness of the individual is used to determine the optimum solution rather than a derivative or gradient (as in traditional optimization), a GA tends to identify global as opposed to local optimums in a solution space (Kurek & Ostfeld, 2013). A drawback of GAs is that the process of generating individuals could produce technically infeasible solutions. In this case, a model of the solution space must be developed using an alternative method, explicitly defining plausible boundaries, or using discrete rather than continuous individuals (Sadatiyan Abkenar *et al.*, 2015).

16.2.6.2 Particle swarm optimization

Particle swarm optimization (PSO) is a robust stochastic optimization procedure based on the natural movement and intelligence of swarms (e.g., birds, fish) (Eberhart & Kennedy, 1995). Like GA, PSO is a population-based search method. Once a model space is defined, a population of particles (i.e., observations) are randomly initialized with a position in each dimension of the model space and a velocity vector. Each particle will iteratively search for a minimum value (i.e., fitness value) in the model space. Similar to swarms in nature, particles will use the knowledge of local minimum from other particles to inform the next search direction. However, when there are too few initialized particles, the search can become trapped in a local minimum, which is a problem that GAs are better at avoiding (Beheshti & Shamsuddin, 2013). When there are a large number of particles, the global minimum can be found but is more computationally intensive. When the number of particles is reasonable, PSO can be a computationally efficient alternative to GA (Hassan *et al.*, 2005).

Each particle contains three vectors: the current position (x), the location of the best solution that the particle has encountered so far ($\boldsymbol{p}$), and the direction (i.e., gradient) of particle travel ($\boldsymbol{v}$). Particles will travel in a direction that is a combination of the local best ($\boldsymbol{p}$, based on the current position) and the global best ($\boldsymbol{g}$, based on the $\boldsymbol{p}$ of all particles) (Equation (16.11)):

$$\boldsymbol{v}_{i+1} = W\boldsymbol{v}_i + \left(c_1 r_1 \left(\boldsymbol{p} - \boldsymbol{x}_i\right)\right) + \left(c_2 r_2 \left(\boldsymbol{g} - \boldsymbol{x}_i\right)\right), \quad \boldsymbol{v}_{\min} \leq \boldsymbol{v}_{i+1} \leq \boldsymbol{v}_{\max} \tag{16.11}$$

where W is the inertial weight; c_1 is the influence of the local minimum on the velocity vector (i.e., self-confidence factor); c_2 is the influence of the global minimum on the velocity vector (swarm confidence factor); and r_1 and r_2 are randomly generated numbers between 0 and 1.

16.2.6.3 Simulated annealing

Simulated annealing (SA) is a search technique that is based on a common thermodynamic principal that the probability distribution of a collection of atoms in equilibrium at a given temperature (Equation (16.12), Metropolis *et al.*, 1953) to identify the global optimum of a model space (Kirkpatrick *et al.*, 1983):

$$e^{-\frac{\Delta D}{T}} > R(0,1) \tag{16.12}$$

where ΔD is the change of distance between states, T is a synthetic temperature that represents the range of solution space considered for a different state, and $R(0, 1)$ is a random number between 0 and 1. The probabilistic approach is important to avoid being stuck at local minima by exploring a reasonable space of solutions to find a global minimum. At each step of SA, a neighboring state (s^*) is compared to the current state (s) and probabilistically decides between moving the system to state s^* or staying in-state s. These probabilities ultimately lead the system to move to states of lower energy. Typically, this step is repeated until the system reaches a state that is sufficient for the application, or until a given computation budget (e.g., number of iterations) has been exhausted. To strategically achieve a global optimum when the objective function is declining slowly, an 'annealing schedule' can be used in which T is iteratively reduced when the objective function plateaus. However, if the initial step size between states is not sufficiently small, then there is no guarantee that the global minimum will be found. In practice, the computational requirements of such granularity generally exceed the improvements in performance (Trosset, 2001). In conventional applications, SA continues to iterate until no change in the objective function is found for 300 iterations (Prakash *et al.*, 2008). If the search space is generally smooth or if there are multiple local minimums, SA could terminate early or be stuck in a local minimum. In these cases, PSO may be a better alternative. Section 16.3.3 describes how SA can be used for aeration control in WWTP.

16.3 DATA SCIENCE APPLICATIONS TO SELECT TREATMENT SYSTEMS

In the following sections, we provide illustrations of the diversity of DDM methods and frameworks used to achieve similar objectives (i.e., energy optimization) for common energy-intensive processes in WTP and WWTP: pumping, chemical addition (coagulation), aeration (nitrification), and biogas generation (anaerobic digestion).

16.3.1 Pump optimization

Pumping of water, wastewater, and biosolids can consume a significant fraction of a utility's energy demand and maintenance costs (Shi, 2011); upwards of 90% for many drinking water utilities (Cherchi *et al.*, 2015). In water treatment distribution systems, pumping schedules are used to reduce energy consumption; however, the optimum solution is difficult to identify with traditional modeling approaches due to distribution systems representing a highly non-linear system with multiple constraints. Examples of real-world constraints of pumps, although these principles can be generally applied to most mechanical equipment at WTP and WWTP, include design inefficiencies, minimum and maximum run times, maximum starts per hour, minimum rest times, minimum and maximum flow rates, maximum discharge pressures, minimum and maximum plant production rates, and lead-lag order of start-up or shutdown (Cherchi *et al.*, 2015). Current 'state-of-the-art' control approaches for 'cost-free' energy optimization include ON-OFF scheduling based on tank level (Nybo *et al.*, 2017), variable-frequency drives (VFD) that allow an individual pump's speed to be adjusted, load shifting (i.e., upstream pump scheduling), and process optimization (Shankar *et al.*, 2016). Examples in the

Table 16.1 Examples of data science water applications for pump optimization.

Author(s)	Objective	Method	Configuration	Results
Torregrossa *et al.* (2017)	Efficiency monitoring	FL	Mandami	Reduced pump energy consumption by 18.5%
Sadatiyan Abkenar *et al.* (2015)	Pump scheduling	GA	Discrete	Identified lowest energy strategies with minimal switches
Kebir *et al.* (2014)	Real-time VFD adjustment	FL	Mandami	40% energy savings versus ON/OFF (theoretical)
Zhang *et al.* (2012)	Pump scheduling	ANN	PSO	8–24% energy savings versus ON/OFF

scientific literature are discussed below and summarized in Table 16.1. Commercial optimization software for water distribution systems, the infrastructure (digital and physical) to support the software, labor, and training for operators typically have a 2–5 year payback period, with energy cost reductions ranging from 5–15% (Badruzzaman *et al.*, 2014).

Torregrossa *et al.* (2017) developed an FL pump performance metric that monitors efficiency and recommends preventative or immediate maintenance accounting for flow conditions. To do this, an efficiency index based on the mass of water lifted and energy consumed is calculated, and a rolling median is used to distinguish the long-term trend from fluctuations attributed to changing conditions. Multiple consecutive days with negative short-term fluctuations is indicative of needed maintenance. The immediacy of the maintenance response is determined by an FL system that weighs the long- and short-term efficiency, and the economic consequences of maintenance versus replacement are evaluated assuming maintenance is able to restore the pump to a baseline efficiency compared to a new, more efficient pump.

Sadatiyan Abkenar *et al.* (2015) used a GA approach to optimize the pumping schedule for two pumps in a hydraulic model of a moderate size water distribution system in Monroe, MI, USA, simultaneously minimizing energy while including an additional penalty for high pressures. A continuous approach that used pairs of start and stop times as genes produced infeasible solutions (i.e., conflicting ON or OFF times). To mitigate this, any mutation that produced an infeasible solution was 'repaired' prior to calculating the fitness of the solution. An alternative discrete approach that used a binary ON or OFF indicator for 1-hour intervals, where each interval is a gene, only produced feasible solutions.

Kebir *et al.* (2014) modeled a full-scale WWTP that relied on a sequential ON/OFF influent pumping strategy, which is inherently inefficient, and proposed a new FL controller that adjusts a pump's VFD by the deviation from average height of an upstream reservoir; reporting a hypothetical 40% reduction in energy. Zhang *et al.* (2012) used an ANN to develop an energy consumption model for a given flowrate, pump configuration for parallel operation, and upstream reservoir levels. They then determined the optimum pump schedule for a given flowrate, desired reservoir level, and physical constraints of the system using PSO; reporting a hypothetical 8–24% reduction in energy.

An important consideration for large WTP or WWTP is the cost of energy, especially if the cost of energy changes throughout the day or if utilities are billed based on monthly energy consumption peaks. Authors in scientific literature largely neglect changing energy costs with time. Rather, energy consumption models are developed based on the proxy of a VFD frequency or the energy rating of individual pieces of equipment. In most cases, minimizing energy consumption will result in the lowest costs; however, forecasting models may need to incorporate a variation of a cost function that describes true cost instead of using consumption as a proxy. For example, initiating pumping may not be the most energy efficient action at a given time but may reduce pumping costs over the course of a day if the immediate demand (when the tank levels *must* be lowered) coincides with increased energy cost.

16.3.1.1 Coagulation

Coagulation is the process in WTP (and in some instances WWTP) in which chemical (i.e., coagulant) is added to destabilize colloidal and suspended particulate matter, allowing the particles to aggregate (i.e., floc) and be more easily removed by gravity due to the larger, neutrally-charged aggregate mass. The generation and transportation of chemicals for coagulation and flocculation can account for 5–20% of a WTP's carbon footprint (Biswas & Yek, 2016); therefore, precision chemical treatment could account for significant cost and carbon savings, depending on the size of the treatment facility and initial water quality. To reduce the amount of chemical used to treat water, dose control strategies must be designed that can adjust for non-ideal physiochemical reaction kinetics due to poor mixing and changing water quality. However, this is rarely the case in full-scale treatment. In WTP, chemical dosing is primarily flow-paced in which a concentration of chemical per unit volume of water is maintained by adjusting the flow rate of a chemical dosing pump proportional to the flow of water. The concentration setpoint is usually only adjusted when a major water quality change or process upset occurs because the identification of an 'ideal' dose in a laboratory bench-scale experiment (i.e., jar tests) is time-consuming and labor-intensive, and results can greatly differ from full-scale. Therefore, the use of data-driven methods of chemical dosing could significantly improve treatment stability and reduce the carbon footprint of treatment facilities. Examples in the scientific literature are discussed below.

The application of ANN to predict coagulant dosing is not a novel concept. Van Leeuwen *et al.* (1999) were able to predict alum dose for a given water quality using historical jar test data and an ANN; although a multiple linear regression model achieved similar results. Ten years later, Maier *et al.* (2009) used the same data as Van Leeuwen to predict treated water quality (turbidity, color, pH, UV-254, residual alum) and optimal alum dose using a DNN (two-layer ANN) and was able to reduce the standard deviation of the prediction error by 37%. Zangooei *et al.* (2016) used historical jar test data to predict turbidity using pH, initial turbidity, temperature, type of coagulant (e.g., solid or liquid poly aluminum chloride from different vendors), and concentration of coagulant. An MLP with two hidden layers outperformed an RBF ANN and FL regression model and required less time to train. Similarly, Wu and Lo (2008) found that an ANN outperformed an ANFIS prediction model for treated water quality when influent water quality data were available. In the absence of real-time water quality, the ANFIS model was able to more accurately predict treated water quality based on historical trends and the present-day dose. When historical dosing data were available, Wu and Lo (2010) found that the inclusion of the previous timestep's coagulant dose (output variable of DNN model) reduced testing error.

Chen and Hou (2006) observed that multiple regression models were able to predict coagulant dose and pH adjustment dose for surface water using historical data. However, two models were developed separately for low and high influent turbidity conditions. Chen and Hou furthered their work to adjust feedback control parameters using Mamdani FL in order to minimize coagulant dosing while simultaneously achieving effluent turbidity and pH goals. Bello *et al.* (2014) proposed a linearized TS fuzzy model predictive control strategy to improve coagulant dose control stability by maintaining the surface charge and pH of the treated water. Depending on the quality of the available data and online instrumentation, FL controllers can improve precision and stability over conventional cascade control.

In order to minimize coagulant dosing, influent water quality, final water quality, and the coagulant dose need to be aggregated to train a predictive model for treated water quality. Predictive model options include multiple regression, ANN, or DNN. To bring the predictive power of the models to a full-scale utility, the best predictive model can then be incorporated into a control strategy in a variety of ways. The most basic control option is a standard cascade control in which the coagulant dose is increased when a treated water quality variable such as turbidity exceeds a threshold. However, this requires a well-understood dose-response relationship. A strict rule to increase the dose when treated water quality goals are not met could increase the concentration of coagulant beyond the need to satisfy electroneutrality, thereby causing effluent turbidity to worsen as particles re-stabilize in suspension (Tchobanoglous *et al.*, 2014). The proposed FL controllers could prevent such overdosing by including rule sets that account for the worsening water quality, but this approach would require more complex programming and a method for tuning consequent statement parameters. Adjustment can be done manually within the existing structure of the existing DCS at WTP, but will need to be done externally if an ANFIS is used. The same concern of programmatic complexity is raised if additional ANN or DNN are developed to identify the required coagulant dose for a given effluent water quality. In this case, the controller must operate on a separate server system, which can provide outputs to an existing cascade control strategy.

16.3.2 Nitrification

Biological nutrient removal (BNR) is the most expensive, variable, and difficult-to-model process in WWTP; yet it is required at the majority of modern facilities around the world in order to achieve the required nitrogen and phosphorus removal. The difficulty in modeling and control is due to two factors common to most WWTP processes: lack of reliable instrumentation and non-ideal process conditions at full scale. The microbial solid–liquid matrix where the treatment takes place (i.e., activated sludge or AS) interferes with common in-line instrumentation measurements due to biofilm growth on the instrument itself and competing ions or solids interference. Instrumentation that utilizes light rather than ion transfer, such as dissolved oxygen (DO) concentration, are sufficiently robust to provide reliable measurements with less frequent cleaning and maintenance. DO is a critical water quality parameter to measure in activated sludge systems because the availability of specific forms of oxygen determines the active microorganisms and, consequently, specific contaminant transformation. Aqueous oxygen available as free oxygen (O_2) will increase the DO concentration and is an indicator of *aerobic* conditions. When oxygen is only available in the form of nitrate (NO_3), conditions are *anoxic*. When no oxygen is available, conditions are *anaerobic*. It is the strategic alternation of these oxidation conditions that transforms contaminants of concern (namely carbon, nitrogen, phosphorus, and, to a lesser extent, sulfur) into the gas or solid phase, and thus reduces the aqueous concentration. While DO sensors can ensure aeration conditions are met, the measurement itself is a proxy for the completion of the contaminant transformation. For example, low strength wastewater (e.g., low concentration of organic materials) will not require as much oxygen in order to achieve treatment goals; however, aeration will continue to be provided to maintain a DO setpoint in the majority of systems regardless of demand.

A sequencing-batch reactor (SBR) is a commonly used wastewater treatment technology in which a single biologically-active, completely-mixed reactor undergoes a sequence of different operating conditions to achieve contaminant removal. The most common control strategy for SBRs uses timed sequences with distinct DO concentration setpoints for each phase in the treatment cycle. The DO setpoints are determined by operator experience and a general knowledge of the environmental conditions required at each phase, each of which activates a unique set of microorganisms. This control strategy requires only one in-line instrument (the DO sensor) and can ensure the desired treatment is achieved under stable influent conditions. However, DO is a *surrogate* for the actual contaminants removed in the process and cannot guarantee that effluent water quality standards are met. Historically, this uncertainty has been addressed by increasing DO setpoints to fully oxidize chemical contaminants and ensure microbial processes are not substrate-limited. This approach increases the energy consumption of the treatment process, accounting for 35–50% of a wastewater utility's total energy (Newhart *et al.*, 2020) and the second largest operational cost behind labor (Lindtner *et al.*, 2008). To reduce the energy consumption associated with aeration in SBRs and other secondary biological treatment systems (i.e., conventional and novel activated sludge configurations), new intelligent monitoring and control strategies are needed. Examples from literature are discussed below and summarized in Table 16.2.

Traoré *et al.* (2005) proposed an FL aeration control strategy for a step-fed, pilot-scale SBR treating municipal wastewater. The rules for the FL DO controller determined the air flow to maintain a

Table 16.2 Examples of data science water applications for aeration.

Author(s)	Objective	Method	Configuration	Results
Traoré *et al.* (2005)	DO control	FL	Mamdani	Improved stability over a wider range of conditions
Ferrer *et al.* (1998)	DO control	FL	Mamdani	40% energy reduction versus ON/OFF
Fiter *et al.* (2005)	DO control	FL	Mamdani	10% energy reduction versus ON/OFF
Du *et al.* (2018)	DO control	NN	RBF	Reduced aeration energy by 100 kWh/d
Asadi *et al.* (2017)	DO optimization	SA	MARS	30% reduction in airflow

DO setpoint from the measured DO and the cycle phase. Compared to an ON/OFF DO control approach (when DO measurement exceeds setpoint, turn off air; when DO measurement is less than the setpoint, turn on air) and conventional proportional-integral-derivative (PID) control, the fuzzy controller was able to maintain the DO setpoint with greater precision over a wider range of environmental conditions. The addition of pH and oxygen uptake rate (OUR) to the fuzzy rule set could shorten aerated cycle times and further reduce energy consumption (Puig *et al.*, 2006). Ferrer *et al.* (1998) used a similar fuzzy DO controller for a pilot BARDENPHO activated sludge system; showing similar improvement in precision compared to an ON/OFF approach with energy savings of up to 40%. Du *et al.* (2018) developed an RBF NN to adjust cascade control parameters to improve DO controller performance, including significantly reduced variability (67% for dry weather flow, 59–93% for wet weather) and slightly reduced aeration energy (100 kWh/d).

An alternative to adjusting individual DO controllers is to identify the optimum operating strategy given a system-wide model. To accomplish this, Asadi *et al.* (2017) compared MARS, ANN, and RF, among others, of DO in the Detroit Water and Sewerage Department's secondary aeration basins. MARS predicted DO and other effluent water quality variables better than ANN and RF (using MAE and R^2). Using the MARS predictive model, they then compared two sets of weights, one that emphasized the best treated water quality and one that emphasized energy consumption, using SA. When optimizing for best water quality, they showed that it was possible to reduce air flow rate by 30% without compromising treated water quality. However, nutrients were not considered, which is an important driver in aeration requirements and strategy for the majority of WWTP. Additionally, Asadi *et al.* (2017) concluded that more frequent sampling of the influent variables was required for ML models compared to statistical models. This comparison holds when the assumptions made about the shape of the relationships among the variables are true. In general, the simpler assumptions made by statistical models can more accurately fill gaps than ML models when data are sparse. In contrast, at a minimum ML models require examples of conditions in training data for reasonably accurate predictions of similar conditions in testing data.

16.3.3 Anaerobic digestion

The primary functions of AD are the stabilization of solids and reduction of chemical oxygen demand (COD). The secondary function, but critical for decarbonzation, is the production of energy. Fifty to seventy percent of biogas produced from AD is methane (Holubar *et al.*, 2003) and can be used onsite to generate energy or be cleaned, sold, and distributed via natural gas pipelines. To maximize energy reduction in the wastewater treatment process, energy positive processes like AD need to be operated strategically to minimize process upsets. For example, large fluctuations in COD loading can lead to the accumulation of intermediate compounds, which are toxic to critical microbiota within the system. However, AD is one of the most difficult processes to model, monitor, and control (Olsson, 2006), and thus AD is operated conservatively with a high factor of safety to ensure stability. Ultimately, this leads to substantial process inefficiencies, including reduced methane production, higher pumping costs due to the increased number of AD reactors in operation, and higher effluent COD. Due to the complex relationships between control and response variables, which are further decoupled by long

Table 16.3 Examples of data science water applications for AD.

Author(s)	Objective	Method	Configuration	Results
Akbaş *et al.* (2015)	Prediction, optimization	ANN	PSO	Methane percentage +5%, biogas production +64%
Holubar *et al.* (2003)	Prediction, optimization	ANN	One-at-a-time-search	Methane concentration 60–70%
Huang *et al.* (2016)	Prediction, optimization	ANN	GA	Biogas flow $R^2 = 0.91$, MSE = 2.0
Polit *et al.* (2002)	Prediction	FL	Mamdani	Robust under changing load
Turkdogan-Aydınol and Yetilmezsoy (2010)	Prediction	FL	Mamdani	Methane production $R^2 = 0.98$, biogas production $R^2 = 0.98$
Perendeci *et al.* (2009)	Prediction	ANFIS	Lagged, phase	Effluent COD $R^2 = 0.89$, RMSE = 0.10

retention times, DDM of the AD process could provide insight for more efficient operation. Examples from the scientific literature are discussed below and summarized in Table 16.3.

Turkdogan-Aydınol and Yetilmezsoy (2010) developed a multiple input-multiple output (MIMO) FL model to predict biogas production, which outperformed a multiple non-linear regression model. Polit *et al.* (2002) used a mechanistic mass balance model with fuzzy pH and temperature coefficient adjustments to predict biogas production, and this approach had the ability to track gas production under load adjustments better than the mechanistic model alone. Holubar *et al.* (2003) used a hierarchical system of ANN to predict volatile fatty acid (VFA) production and pH followed by a biogas production and composition forecast of an AD during start-up and stabilization. Operating parameters were adjusted by a one-at-a-time search algorithm to simultaneously maximize organic loading rate (OLR) and methane production.

By applying PSO to an ANN model of biogas production, Akbaş *et al.* (2015) identified the operational conditions that produced the highest percent methane and biogas production. Compared to the average of the historical data, the optimum conditions increased the methane fraction in the biogas by 5% and biogas production by 64%. To achieve this, daily averages of sludge loading rate, temperature, pH, total solids, total volatile solids, VFA, alkalinity, solids retention time (SRT), and OLR were used as inputs to prediction models for percent methane and biogas production. Dimension reduction for input variable selection was applied using a boosting tree algorithm (Breiman, 1996), which improved prediction performance.

Both FL and ANN can be used for prediction, but while ANN has been shown to be more precise, FL is able to better handle variability in the inputs and outputs (Kambalimath & Deka, 2020; Özcan *et al.*, 2009). There is extensive research demonstrating the efficiency of ANN to predict biogas production and AD performance (Levstek & Lakota, 2010). Therefore, there is a boom of hybrid fuzzy models, such as ANFIS, to address AD systems (Abrahart *et al.*, 2008). Perendeci *et al.* (2009) showed that an ANFIS model was able to predict effluent COD of a seasonal anaerobic wastewater treatment system, improving performance by adding input variables, including an indicator of whether the system was under start-up or pseudo-steady-state conditions using 10days of historical COD data.

Unlike the case of coagulation in WTP (Section 16.3.2), a WWTP has some control over the organic loading rate, temperature, and SRT depending on the number and size of AD available. To utilize fully DDM for decarbonization, predictive models should be fit for biogas production (both quantity and quality, such as specific methane mass flow rate) using optimization methods such as PSO or GA to identify the ideal operating conditions. Individual models can also be developed for variables that cannot be quickly measured but are critical to understanding performance, such as VFA.

16.4 RECOMMENDATIONS FOR FULL-SCALE IMPLEMENTATION

In 2020, leaders in the water and wastewater industry met to discuss cyber infrastructure for data-driven water systems; identifying knowledge gaps and capable personnel at every step in DDM from data generation to use, application, and presentation (Ren *et al.*, 2020). While there are some working groups (e.g., Smart Water Networks Forum, a UK-based non-profit) and international challenges (e.g., Intelligent Water Systems Challenge, co-sponsored by the Water Research Foundation and Water Environment Federation), there is no single consortium or text that covers the broad range of topics associated with DDM. In the absence of a comprehensive set of recommendations, it currently falls to individual utilities to explore new opportunities. The lack of modern digital infrastructure at most modern WTP and WWTP is the largest hurdle to implementation of DDM for decarbonization. Facilities rarely have the data management procedures in place to broadly and methodically organize databases or consistent protocols to clean inherently noisy data. In order to maintain real-time analyses, a host of programmatic and practical implications must also be considered. For example, the majority of SCADA systems are designed for short-term data storage (maximum of three months) and cascade control loops (primarily feedback). To integrate DDM, either (1) SCADA systems must be upgraded to incorporate historical data and advanced modeling or (2) control strategies must be designed allow for DDM outputs to communicate with the existing data framework. The latter option is the most practical and widely used DDM implementation strategy due to the familiarity of existing staff with and dependability of basic control structures.

Layers of complexity to improve process efficiency can be added after the stability of a DDM system is demonstrated. The stability demonstration must address considerations such as process-wide variability compared to a conventional control strategy (Newhart *et al.*, 2020), and upper and lower limits or other contingencies in the event of data loss or infeasible model predictions. The sequential approach to developing DDM control is an opportunity to develop a robust product in tandem with the operations staff, who could be held legally liable for negligence in the event of an accidental discharge and thus balance exercising caution with their intimate knowledge of full-scale process dynamics. When investigating DDM as a potential solution for decarbonization in water and wastewater treatment, the following factors should be explicitly discussed and defined prior to determining a DDM method and integration strategy:

- goals and key performance indicators (KPI), both plant-wide and process-specific, that address treatment and energy performance (e.g., kWh/MG, gCO_2/MG);
- limiting technological factors, including operational constraints, instrumentation, data management, control system structure, and cybersecurity constraints;
- unknowns, such as impacts on other processes or the rate of adoption of new technology.

Once the project constraints above are identified, DDM method selection begins in earnest. The general steps include:

(1) Identify predictor (input) and response (output) variables that would integrate easily with the existing control strategy and provide substantial benefit.
(2) Develop practical and sound data blending protocols that consider real world implementation (i.e., merge observations when variables have different sampling frequencies). This includes careful consideration for when laboratory data become available.
(3) Perform variable selection to minimize the amount of error introduced into the DDM by irrelevant inputs.
(4) Pilot different modeling frameworks, both simple and advanced, to assess the best candidates for the desired response. If optimization is the ultimate goal, follow the predictive model development with an experiment that lends itself towards identifying the 'best' optimization algorithm for the given problem.

(5) Tune predictive and optimization models by increasing and decreasing model complexity to provide the most accurate performance on testing data, which is not used in model fitting.
(6) Program the model onto a server (using a programming language such as R or Python), which can export data to the data archiving system used at the specific utility. The data archiving system frequently has access to the SCADA system without posing a security risk.
(7) Monitor the stability of the prediction over time, and identify unforeseen contingencies that need to be integrated with the new control strategy prior to full-scale deployment.
(8) Schedule monitoring periods for the model developer, control expert, and operations to simultaneously observe full-scale implementation. These periods should span weeks to thoroughly evaluate different environmental conditions and should then progressively increase in runtime until the operational staff are comfortable with unsupervised operation.
(9) Compare the impact of the new DDM control strategy to the original strategy using pre-established KPIs. If the new DDM control strategy meets or exceeds the KPIs of the original strategy, then the previous steps can be repeated to incorporate additional predictions or rely more directly on the predictions by eliminating layers in a control loop.

16.5 CONCLUSIONS

Interest in integrating DDM into WTP and WWTP is growing rapidly, but utilities are largely overwhelmed by the task. Simultaneous development of good, internal data management protocols and the application of DDM to water and wastewater treatment challenges could dramatically reduce the carbon cost of clean water. Given that water treatment currently consumes 3% of electrical energy generated in the US each year and is projected to increase to 6% due to increasing demand and intensity of treatment processes (i.e., higher quality effluent) (Chaudhry & Shrier, 2010), data-driven process optimization is the proverbial 'low-hanging-fruit' for carbon and cost reduction. There is a large body of scientific literature in which conventional and novel DDM methods are applied to engineered environmental systems like WTP and WWTP; however, the heuristic nature of many ML approaches make declaring any one method the 'best' for a specific process application impossible. Experimentation with an individual utility's datasets using published literature as guidelines is truly the 'best' framework. Fundamentally, given existing machinery and treatment technologies, it is the investment in people and improved operational strategies that will help WTP and WWTP achieve their full treatment potential with the smallest environmental impact.

ACKNOWLEDGEMENTS

This work was supported by the National Science Foundation PFI:BIC Award No: 1632227; the National Science Foundation Engineering Research Center program under cooperative agreement EEC-1028968 (ReNUWIt).

REFERENCES

Abraham A. (2005). Adaptation of fuzzy inference system using neural learning. In: Fuzzy Systems Engineering, N. Nedjah and L. de Macedo Mourelle (eds), Springer, Berlin/Heidelberg, Germany, pp. 53–83.

Abrahart R. J., See L. M. and Solomatine D. P. (2008). Practical Hydroinformatics: Computational Intelligence and Technological Developments in Water Applications. Springer Science & Business Media, Berlin/Heidelberg, Germany.

Akaike H. (1974). A new look at the statistical model identification. *IEEE Transactions on Automatic Control*, **19**(6), 716–723, https://doi.org/10.1109/TAC.1974.1100705

Akbaş H., Bilgen B. and Turhan A. M. (2015). An integrated prediction and optimization model of biogas production system at a wastewater treatment facility. *Bioresource Technology*, **196**, 566–576, https://doi.org/10.1016/j.biortech.2015.08.017

Asadi A., Verma A., Yang K. and Mejabi B. (2017). Wastewater treatment aeration process optimization: a data mining approach. *Journal of Environmental Management*, **203**, 630–639, https://doi.org/10.1016/j.jenvman.2016.07.047

Badruzzaman M., Cherchi C., Oppenheimer J., Gordon M., Bunn S. and Jacangelo J. G. (2014). Implementation of energy and water quality management systems modified with a GHG module. Proceedings of the Annual Conference & Exposition, Boston, MA, USA.

Bai H., Zhu R., An H., Zhou G., Huang H., Ren H. and Zhang Y. (2019). Influence of wastewater sludge properties on the performance of electro-osmosis dewatering. *Environmental Technology*, **40**(21), 2853–2863, https://doi.org/10.1080/09593330.2018.1455744

Barbu M., Vilanova R., Meneses M. and Santin I. (2017). Global evaluation of wastewater treatment plants control strategies including CO2 emissions. *IFAC-PapersOnLine*, **50**(1), 12956–12961, https://doi.org/10.1016/j.ifacol.2017.08.1800

Beale R. and Jackson T. (1990). Neural Computing – An Introduction. Taylor & Francis, UK.

Beheshti Z. and Shamsuddin S. M. (2013). A review of population-based meta-heuristic algorithm. *International Journal of Advances in Soft Computing and its Applications*, **5**, 1–35.

Bello O., Hamam Y. and Djouani K. (2014). Fuzzy dynamic modelling and predictive control of a coagulation chemical dosing unit for water treatment plants. *Journal of Electrical Systems and Information Technology*, **1**(2), 129–143, https://doi.org/10.1016/j.jesit.2014.08.001

Bishop C. M. (1995). Neural Networks for Pattern Recognition. Oxford University Press, Inc., USA.

Biswas W. K. and Yek P. (2016). Improving the carbon footprint of water treatment with renewable energy: a Western Australian case study. *Renewables: Wind, Water, and Solar*, **3**(1), 14, https://doi.org/10.1186/s40807-016-0036-2

Boulesteix A.-L. and Schmid M. (2014). Machine learning versus statistical modeling. *Biometrical Journal*, **56**(4), 588–593, https://doi.org/10.1002/bimj.201300226

Breiman L. (1996). Bagging predictors. *Machine Learning*, **24**(2), 123–140.

Breiman L. (2001). Random forests. *Machine Learning*, **45**(1), 5–32, https://doi.org/10.1023/A:1010933404324

Burnham K. P. and Anderson D. R. (2004). Multimodel inference: understanding AIC and BIC in model selection. *Sociological Methods and Research*, **33**(2), 261–304, https://doi.org/10.1177/0049124104268644

Chaudhry S. and Shrier C. (2010). Energy sustainability in the water sector: challenges and opportunities. Proceedings of Annual Conference and Exposition, 2010 Annual Conference and Exposition of the American Water Works Association, Chicago, IL.

Chen C.-L. and Hou P.-L. (2006). Fuzzy model identification and control system design for coagulation chemical dosing of potable water. *Water Science and Technology: Water Supply*, **6**(3), 97–104, https://doi.org/10.2166/ws.2006.782

Cherchi C., Badruzzaman M., Oppenheimer J., Bros C. M. and Jacangelo J. G. (2015). Energy and water quality management systems for water utility's operations: a review. *Journal of Environmental Management*, **153**, 108–120, https://doi.org/10.1016/j.jenvman.2015.01.051

Corominas L., Garrido-Baserba M., Villez K., Olsson G., Cortés U. and Poch M. (2018). Transforming data into knowledge for improved wastewater treatment operation: a critical review of techniques. *Environmental Modelling and Software*, **106**, 89–103, https://doi.org/10.1016/j.envsoft.2017.11.023

Dellana S. A. and West D. (2009). Predictive modeling for wastewater applications: linear and nonlinear approaches. *Environmental Modelling and Software*, **24**(1), 96–106, https://doi.org/10.1016/j.envsoft.2008.06.002

Du X., Wang J., Jegatheesan V. and Shi G. (2018). Dissolved oxygen control in activated sludge process using a neural network-based adaptive PID algorithm. *Applied Sciences*, **8**(2), 261, https://doi.org/10.3390/app8020261

Eberhart R. and Kennedy J. (1995). A new optimizer using particle swarm theory. Proceedings of the Sixth International Symposium on Micro Machine and Human Science, IEEE, Nagoya, Japan, pp. 39–43.

Ferrer J., Rodrigo M. A., Seco A. and Penya-Roja J. M. (1998). Energy saving in the aeration process by fuzzy logic control. *Water Science and Technology*, **38**(3), 209–217, https://doi.org/10.2166/wst.1998.0210

Fiter M., Güell D., Comas J., Colprim J., Poch M. and Rodríguez-Roda I. (2005). Energy saving in a wastewater treatment process: an application of fuzzy logic control. *Environmental Technology*, **26**(11), 1263–1270, https://doi.org/10.1080/09593332608618596

Flores-Alsina X., Corominas L., Snip L. and Vanrolleghem P. A. (2011). Including greenhouse gas emissions during benchmarking of wastewater treatment plant control strategies. *Water Research*, **45**(16), 4700–4710, https://doi.org/10.1016/j.watres.2011.04.040

Freund Y. and Schapire R. E. (1997). A decision-theoretic generalization of on-line learning and an application to boosting. *Journal of Computer and System Sciences*, **55**(1), 119–139, https://doi.org/10.1006/jcss.1997.1504

Friedman J. H. (1991). Multivariate adaptive regression splines. *Annals of Statistics*, **19**(1), 1–67.

Hering A. S. (2021). Fault isolation, chapter 3. In: Statistical Process Monitoring Using Advanced Data-Driven and Deep Learning Approaches, F. Harrou, Y. Sun, A. S. Hering, M. Madakyaru and A. Dairi (eds), Elsevier, The Netherlands, pp. 71–117, https://doi.org/10.1016/B978-0-12-819365-5.00009-7.

Hassan R., Cohanim B., De Weck O. and Venter G. (2005). A comparison of particle swarm optimization and the genetic algorithm. 46th AIAA/ASME/ASCE/AHS/ASC Structures, Structural Dynamics and Materials Conference, Austin, Texas. American Institute of Aeronautics and Astronautics.

Hastie T. and Tibshirani R. (1999). Generalized Additive Models. Chapman & Hall/CRC, Boca Raton, FL, USA.

Hastie T., Tibshirani R. and Wainwright M. (2015). Statistical Learning with Sparsity: The Lasso and Generalizations. Routledge, United Kingdom.

Helsel D. and Hirsch R. (2002). Statistical methods in water resources. In: Techniques of Water-Resources Investigations of the United States Geological Survey. Hydrologic Analysis and Interpretation, USGS, Reston, VA, pp. 209–218. Available at https://pubs.er.usgs.gov/publication/twri04A3

Holubar P., Zani L., Hager M., Fröschl W., Radak Z. and Braun R. (2003). Start-up and recovery of a biogas-reactor using a hierarchical neural network-based control tool. *Journal of Chemical Technology and Biotechnology*, **78**(8), 847–854, https://doi.org/10.1002/jctb.854

Huang M., Han W., Wan J., Ma Y. and Chen X. (2016). Multi-objective optimisation for design and operation of anaerobic digestion using GA-ANN and NSGA-II. *Journal of Chemical Technology and Biotechnology*, **91**(1), 226–233, https://doi.org/10.1002/jctb.4568

Hüllermeier E. (2015). From knowledge-based to data-driven fuzzy modeling. *Informatik-Spektrum*, **38**(6), 500–509, https://doi.org/10.1007/s00287-015-0931-8

Jackson J. E. (1991). A User's Guide To Principal Components. John Wiley & Sons, Inc.

James G., Witten D., Hastie T. and Tibshirani R. (2013). An Introduction to Statistical Learning. Springer, New York, NY, USA.

Jang J.-S. (1993). ANFIS: adaptive-network-based fuzzy inference system. *IEEE Transactions on Systems, Man, and Cybernetics*, **23**(3), 665–685, https://doi.org/10.1109/21.256541

Jang J.-S. R., Sun C.-T. and Mizutani E. (1997). Neuro-fuzzy and Soft Computing: A Computational Approach to Learning and Machine Intelligence. Prentice Hall, Upper Saddle River, NJ, USA.

Jeppsson U., Pons M. N., Nopens I., Alex J., Copp J. B., Gernaey K. V., Rosen C., Steyer J. P. and Vanrolleghem P. A. (2007). Benchmark simulation model no 2: general protocol and exploratory case studies. *Water Science Technology*, **56**(8), 67–78, https://doi.org/10.2166/wst.2007.604

John G. H., Kohavi R. and Pfleger K. (1994). Irrelevant features and the subset selection problem. In: Machine Learning Proceedings of the Eleventh International Conference, W. W. Cohen and H. Hirsh (eds), Morgan Kaufmann Publishers, San Francisco, CA, pp. 121–129.

Kambalimath S. and Deka P. C. (2020). A basic review of fuzzy logic applications in hydrology and water resources. *Applied Water Science*, **10**(8), 191, https://doi.org/10.1007/s13201-020-01276-2

Kasabov N. K. (1996). Foundations of Neural Networks, Fuzzy Systems, and Knowledge Engineering. MIT Press, Cambridge, MA, USA.

Kazor K., Holloway R. W., Cath T. Y. and Hering A. S. (2016). Comparison of linear and nonlinear dimension reduction techniques for automated process monitoring of a decentralized wastewater treatment facility. *Stochastic Environmental Research and Risk Assessment*, **30**(5), 1527–1544, https://doi.org/10.1007/s00477-016-1246-2

Kebir F. O., Demirci M., Karaaslan M., Ünal E., Dincer F. and Arat H. T. (2014). Smart grid on energy efficiency application for wastewater treatment. *Environmental Progress & Sustainable Energy*, **33**(2), 556–563, https://doi.org/10.1002/ep.11821

Khataee A. R. and Kasiri M. B. (2011). Modeling of biological water and wastewater treatment processes using artificial neural networks. *CLEAN – Soil, Air, Water*, **39**(8), 742–749, https://doi.org/10.1002/clen.201000234

Kira K. and Rendell L. A. (1992). A practical approach to feature selection. In D. Sleeman and P. Edwards (eds), Machine learning proceedings 1992. Morgan Kaufmann, pp. 249–256. Available at https://doi.org/10.1016/B978-1-55860-247-2.50037-1

Kirkpatrick S., Gelatt C. D. and Vecchi M. P. (1983). Optimization by simulated annealing. *Science*, **220**(4598), 671–680, https://doi.org/10.1126/science.220.4598.671

Klanderman M., Newhart K. B., Cath T. and Hering A. S. 2020 Case studies in real-time fault isolation in a decentralized wastewater treatment facility. *Journal of Water Process Engineering*. **38**(2020) 101556, https://doi.org/10.1016/j.jwpe.2020.101556

Kuha J. (2004). AIC and BIC: comparisons of assumptions and performance. *Sociological Methods & Research*, **33**(2), 188–229, https://doi.org/10.1177/0049124103262065

Kurek W. and Ostfeld A. (2013). Multi-objective optimization of water quality, pumps operation, and storage sizing of water distribution systems. *Journal of Environmental Management*, **115**, 189–197, https://doi.org/10.1016/j.jenvman.2012.11.030

Le L. T., Nguyen H., Dou J. and Zhou J. (2019). A comparative study of PSO-ANN, GA-ANN, ICA-ANN, and ABC-ANN in estimating the heating load of buildings' energy efficiency for smart city planning. *Applied Sciences*, **9**(13), 2630, https://doi.org/10.3390/app9132630

Levstek T. and Lakota M. (2010). The use of artificial neural networks for compounds prediction in biogas from anaerobic digestion – a review. *Agricultura*, **7**, 15–22.

Lindtner S., Schaar H. and Kroiss H. (2008). Benchmarking of large municipal wastewater treatment plants treating over 100 000 PE in Austria. *Water Science and Technology*, **57**(10), 1487–1493, https://doi.org/10.2166/wst.2008.214

Ljung L. (2010). Perspectives on system identification. *Annual Reviews in Control*, **34**(1), 1–12, https://doi.org/10.1016/j.arcontrol.2009.12.001

Lundberg S. M. and Lee S.-I. (2017). A unified approach to interpreting model predictions. In I. Guyon, U. V. Luxburg, S. Bengio, H. Wallach, R. Fergus, S. Vishwanathan and R. Garnett (eds), Advances in neural information processing systems. Curran Associates, Inc., Vol. 30, pp. 4765–4774. Available at https://proceedings.neurips.cc/paper/2017/file/8a20a8621978632d76c43dfd28b67767-Paper.pdf

Maier R. M., Pepper I. L. and Gerba C. P. (2009) Environmental microbiology. Academic press. Cambridge, Massachusetts, vol. 397.

Maleki A., Nasseri S., Aminabad M. S. and Hadi M. (2018). Comparison of ARIMA and NNAR models for forecasting water treatment plant's influent characteristics. *KSCE Journal of Civil Engineering*, **22**(9), 3233–3245, https://doi.org/10.1007/s12205-018-1195-z

Manesis S., Sapidis D. and King R. (1998). Intelligent control of wastewater treatment plants. *Artificial Intelligence in Engineering*, **12**(3), 275–281, https://doi.org/10.1016/S0954-1810(97)10002-4

Metropolis N., Rosenbluth A. W., Rosenbluth M. N., Teller A. H. and Teller E. (1953). Equation of state calculations by fast computing machines. *The Journal of Chemical Physics*, **21**(6), 1087–1092, https://doi.org/10.1063/1.1699114

Newhart K. B., Holloway R. W., Hering A. S. and Cath T. Y. (2019). Data-driven performance analyses of wastewater treatment plants: a review. *Water Research*, **157**, 498–513, https://doi.org/10.1016/j.watres.2019.03.030

Newhart K. B., Marks C. A., Rauch-Williams T., Cath T. Y. and Hering A. S. (2020). Hybrid statistical-machine learning ammonia forecasting in continuous activated sludge treatment for improved process control. *Journal of Water Process Engineering*, **37**, 101389, https://doi.org/10.1016/j.jwpe.2020.101389

Newhart K. B., Goldman-Torres J. E., Freedman D. E., Wisdom K. B., Hering A. S. and Cath T. Y. (2021). Prediction of peracetic acid disinfection performance for secondary municipal wastewater treatment using artificial neural networks. *ACS ES&T Water*, **1**(2), 328–338, https://doi.org/10.1021/acsestwater.0c00095

Nielsen M. A. (2015). Neural Networks and Deep Learning. Determination Press, San Francisco, CA, USA.

Nybo P. J., Kallesøe C. S. and Lauridsen K. G. (2017). Method for Operating A Wastewater Pumping Station. US Patent No. 9 719 241. US Patent and Trademark Office, Alexandria, VA, USA.

Odom G. J., Newhart K. B., Cath T. Y. and Hering A. S. (2018). Multistate multivariate statistical process control. *Applied Stochastic Models in Business and Industry*, **34**(6), 880–892, https://doi.org/10.1002/asmb.2333

Olsson G. (2006). Instrumentation, control and automation in the water industry – state-of-the-art and new challenges. *Water Science and Technology*, **53**(4–5), 1–16, https://doi.org/10.2166/wst.2006.097

Oppong G., Montague G. A., O'Brien M., McEwan M. and Martin E. B. (2013). Towards advanced control for anaerobic digesters: volatile solids inferential sensor. *Water Practice and Technology*, **8**(1) 7–17, https://doi.org/10.2166/wpt.2013.002

Özcan F., Atiş C. D., Karahan O., Uncuoğlu E. and Tanyildizi H. (2009). Comparison of artificial neural network and fuzzy logic models for prediction of long-term compressive strength of silica fume concrete. *Advances in Engineering Software*, **40**(9), 856–863, https://doi.org/10.1016/j.advengsoft.2009.01.005

Perendeci A., Arslan S., Tanyolaç A. and Çelebi S. S. (2009). Effects of phase vector and history extension on prediction power of adaptive-network based fuzzy inference system (ANFIS) model for a real scale anaerobic wastewater treatment plant operating under unsteady state. *Bioresource Technology*, **100**(20), 4579–4587, https://doi.org/10.1016/j.biortech.2009.04.049

Polit M., Estaben M. and Labat P. (2002). A fuzzy model for an anaerobic digester, comparison with experimental results. *Engineering Applications of Artificial Intelligence*, **15**(5), 385–390, https://doi.org/10.1016/S0952-1976(02)00091-X

Ross T. J. (2010). Properties of Membership Functions, Fuzzification, and Defuzzification. In Fuzzy Logic with Engineering Applications. John Wiley & Sons Ltd., pp. 89–116. https://doi.org/10.1002/9781119994374.ch4

Prakash A., Shukla N., Shankar R. and Tiwari M. K. (2008). Solving machine loading problem of FMS: an artificial intelligence (AI) based random search optimization approach. In: Handbook of Computational Intelligence in Manufacturing and Production Management, D. Laha and P. Mandal (eds), IGI Global, Hershey, Pennsylvania, USA, pp. 19–43.

Puig S., Corominas L., Traore A., Colomer J., Balaguer M. D. and Colprim J. (2006). An on-line optimisation of a SBR cycle for carbon and nitrogen removal based on on-line pH and OUR: the role of dissolved oxygen control. *Water Science and Technology*, **53**(4–5), 171–178, https://doi.org/10.2166/wst.2006.121

Reeves C. and Rowe J. E. (2002). Genetic Algorithms: Principles and Perspectives: A Guide to GA Theory. Springer Science & Business Media. Boston, MA, USA.

Ren Z., Liner B., Ferguson C., Fisher A., Newhart K., Wang M. and Sharpless C. (2020). Report on NSF Mid-scale Research Infrastructure Workshop for Intelligent Water Systems. NSF Award #CBET 2035032 (online).

Sadatiyan Abkenar S. M., Stanley S. D., Miller C. J., Chase D. V. and McElmurry S. P. (2015). Evaluation of genetic algorithms using discrete and continuous methods for pump optimization of water distribution systems. *Sustainable Computing: Informatics and Systems*, **8**, 18–23, https://doi.org/10.1016/j.suscom.2014.09.003

Sadollah A. (2018). Which membership function is appropriate in fuzzy system? In: Fuzzy Logic Based in Optimization Methods and Control Systems and its Applications, A. Sadollah (ed.), InTech, London, UK, pp. 3–6.

Schmidhuber J. (2015). Deep learning in neural networks: an overview. *Neural Networks*, **61**, 85–117, https://doi.org/10.1016/j.neunet.2014.09.003

Schwarz G. (1978). Estimating the dimension of a model. *The Annals of Statistics*, **6**(2), 461–464, https://doi.org/10.1214/aos/1176344136

Shankar V. K. A., Umashankar S., Paramasivam S. and Hanigovszki N. (2016). A comprehensive review on energy efficiency enhancement initiatives in centrifugal pumping system. *Applied Energy*, **181**, 495–513, https://doi.org/10.1016/j.apenergy.2016.08.070

Shapley L. S. (1951). Notes on the N-Person Game–II: The Value of an N-Person Game. Rand Corporation, Santa Monica, California, USA.

Sharma S., Sharma S. and Athaiya A. (2020). Activation functions in neural networks. *International Journal of Engineering Applied Sciences and Technology*, **04**(12), 310–316, https://doi.org/10.33564/IJEAST.2020.v04i12.054

Shi C. Y. (2011). Mass Flow and Energy Efficiency of Municipal Wastewater Treatment Plants. IWA Publishing, London, UK.

Steinberg D. and Colla P. (1995). CART: tree-structured non-parametric data analysis. San Diego, CA: Salford Systems.

Sutton C. D. (2005). Classification and regression trees, bagging, and boosting. In: Handbook of Statistics. Data Mining and Data Visualization, C. R. Rao, E. J. Wegman and J. L. Solka (eds), Elsevier, The Netherlands, pp. 303–329.

Takagi T. and Sugeno M. (1983). Derivation of fuzzy control rules from human operator's control actions. *IFAC Proceedings Volumes*, **16**(13), 55–60, https://doi.org/10.1016/S1474-6670(17)62005-6

Tchobanoglous G., Stensel H. D., Tsuchihashi R., Burton F. L. and Abu-Orf M. (2014). Fundamentals of chemical coagulation. In: Wastewater Engineering: Treatment and Resource Recovery, G. Bowden, W. Pfrang and Metcalf & Eddy (eds), McGraw-Hill Education, New York, NY, USA, pp. 460–473.

Tibshirani R. (1996). Regression Shrinkage and Selection via the Lasso. *Journal of the Royal Statistical Society. Series B (Methodological)*, **58**(1), 267–288.

Tibshirani R., Saunders M., Rosset S., Zhu J. and Knight K. (2005). Sparsity and smoothness via the fused lasso. *Journal of the Royal Statistical Society: Series B (Statistical Methodology)*, **67**(1), 91–108, https://doi.org/10.1111/j.1467-9868.2005.00490.x

Torregrossa D., Hansen J., Hernández-Sancho F., Cornelissen A., Schutz G. and Leopold U. (2017). A data-driven methodology to support pump performance analysis and energy efficiency optimization in waste water treatment plants. *Applied Energy*, **208**, 1430–1440, https://doi.org/10.1016/j.apenergy.2017.09.012

Traoré A., Grieu S., Puig S., Corominas L., Thiery F., Polit M. and Colprim J. (2005). Fuzzy control of dissolved oxygen in a sequencing batch reactor pilot plant. *Chemical Engineering Journal*, **111**(1), 13–19, https://doi.org/10.1016/j.cej.2005.05.004

Trosset M. W. (2001). What is simulated annealing? *Optimization and Engineering*, **2**(2), 201–213, https://doi.org/10.1023/A:1013193211174

Turkdogan-Aydınol F. I. and Yetilmezsoy K. (2010). A fuzzy-logic-based model to predict biogas and methane production rates in a pilot-scale mesophilic UASB reactor treating molasses wastewater. *Journal of Hazardous Materials*, **182**(1), 460–471, https://doi.org/10.1016/j.jhazmat.2010.06.054

Van Leeuwen J., Chow C. W. K., Bursill D. and Drikas M. (1999). Empirical mathematical models and artificial neural networks for the determination of alum doses for treatment of southern Australian surface waters. *Aqua*, **48**(3), 115–127.

Vrieze S. I. (2012). Model selection and psychological theory: a discussion of the differences between the Akaike information criterion (AIC) and the Bayesian information criterion (BIC). *Psychological Methods*, **17**(2), 228–243, https://doi.org/10.1037/a0027127

Wallace J., Champagne P. and Hall G. (2016). Multivariate statistical analysis of water chemistry conditions in three wastewater stabilization ponds with algae blooms and pH fluctuations. *Water Research*, **96**, 155–165, https://doi.org/10.1016/j.watres.2016.03.046

Wang X., Ratnaweera H., Holm J. A. and Olsbu V. (2017). Statistical monitoring and dynamic simulation of a wastewater treatment plant: a combined approach to achieve model predictive control. *Journal of Environmental Management*, **193**, 1–7, https://doi.org/10.1016/j.jenvman.2017.01.079

Wise B. M. and Gallagher N. B. (1996). The process chemometrics approach to process monitoring and fault detection. *Journal of Process Control*, **6**(6), 329–348, https://doi.org/10.1016/0959-1524(96)00009-1

Wu G.-D. and Lo S.-L. (2008). Predicting real-time coagulant dosage in water treatment by artificial neural networks and adaptive network-based fuzzy inference system. *Engineering Applications of Artificial Intelligence*, **21**(8), 1189–1195, https://doi.org/10.1016/j.engappai.2008.03.015

Wu G.-D. and Lo S.-L. (2010). Effects of data normalization and inherent-factor on decision of optimal coagulant dosage in water treatment by artificial neural network. *Expert Systems with Applications*, **37**(7), 4974–4983, https://doi.org/10.1016/j.eswa.2009.12.016

Yuan M. and Lin Y. (2006). Model selection and estimation in regression with grouped variables. *Journal of the Royal Statistical Society: Series B (Statistical Methodology)*, **68**(1), 49–67, https://doi.org/10.1111/j.1467-9868.2005.00532.x

Zadeh L. A. (1973). Outline of a new approach to the analysis of complex systems and decision processes. *IEEE Transactions on Systems, Man and Cybernetics*, **1100**, 38–45.

Zangooei H., Delnavaz M. and Asadollahfardi G. (2016). Prediction of coagulation and flocculation processes using ANN models and fuzzy regression. *Water Science and Technology*, **74**(6), 1296–1311, https://doi.org/10.2166/wst.2016.315

Zhang Z., Zeng Y. and Kusiak A. (2012). Minimizing pump energy in a wastewater processing plant. *Energy*, **47**(1), 505–514, https://doi.org/10.1016/j.energy.2012.08.048

Zheng F. and Zhong S. (2011). Time series forecasting using a hybrid RBF neural network and AR model based on binomial smoothing. *World Academy of Science, Engineering and Technology*, **75**, 1471–1475.

Zou H. (2006). The adaptive lasso and its oracle properties. *Journal of the American Statistical Association*, **101**(476), 1418–1429, https://doi.org/10.1198/016214506000000735

doi: 10.2166/9781789061796

Chapter 17

Decarbonization policies and water sector opportunities

Jason A. Turgeon[1], Steven A. Conrad[2] and Peter A. Vanrolleghem[3]*

[1]United States Environmental Protection Agency, Boston, MA, USA
[2]Colorado State University, Fort Collins, CO, USA
[3]Université Laval, Quebec, QC, Canada
**Correspondence: turgeon.jason@epa.gov*

17.1 INTRODUCTION

The water sector sits in an unusual place in most economies. The sector, which encompasses drinking water, wastewater treatment/water resource recovery, and stormwater management, is driven by a wide variety of policies. Foremost among these are regulatory mandates to protect water quality and human health. However, many other types of policies impact the sector. These include directives to keep prices low (often below the true cost of water) for reasons of social equity or economic development, inclusion in efforts to make local governments more sustainable, and efforts to make essential services more resilient to natural disasters, pandemics, and security threats.

In this chapter, we will discuss some of the many policies and policy responses that impact efforts to decarbonize the water sector and examine ways in which the sector can respond to sometimes competing or conflicting policy demands. The scope and scale of policymaking varies greatly based on the involvement of readers in the overall process, therefore we wrote this chapter with the following readership in mind: Some readers may be involved in policy making but may not come from an engineering background. Other readers may be engineers or scientists attempting to respond to policy directives in the design and operation of water utilities. Whether the reader is approaching this chapter from a policy development perspective or an operational perspective, we hope to convey a handful of key concepts in this chapter. A unifying thread among these concepts is the need to seek out multiple benefits whenever possible. Decarbonization is often a co-benefit of other policies discussed in this chapter. The web of policy directives demands an integrated approach to resource management.

17.2 CORE CONCEPTS

17.2.1 Concept 1: In the scale of national or global policies the sector's energy use is relatively small, but there are other resources contained in wastewater worth considering

Energy use in the water sector has been a concern almost as long as there has been a water sector. In the ages before electrical pumps and motors made conveying water easy, early engineers in Rome,

Istanbul, and elsewhere went to great lengths to create elaborate aqueduct systems to deliver fresh water hundreds of miles by gravity alone, and to design wastewater collection systems that would also remove used water and rainwater by gravity. Even in the earliest days of the industrial era, steam driven pumping systems and rudimentary aeration systems for activated sludge processes were recognized as enormous energy consumers.

In the United States, the passage of the Clean Water Act in October of 1972 was immediately followed by a July 1973 report titled *Electrical Power Consumption for Municipal Wastewater Treatment*, which found that electrical power consumption for municipal wastewater treatment was about 1% of residential energy consumption, and that consumption was expected to double as more treatment plants were built, with a further increase of 40–50% for tertiary facilities (Smith, 1973). Forty years later, after the construction of many thousands of new publicly owned treatment facilities, a large percentage of which are performing tertiary treatment, the Electric Power Research Institute (EPRI) surveyed the sector and found that municipal wastewater treatment accounted for about 0.8% of total electric demand in the US (drinking water treatment and distribution accounted for an additional 1%) (EPRI/WRF, 2013). Total energy use in 2012 across public water supply and treatment, and municipal wastewater treatment in the US consumed 39.2 and 30.2 billion kilowatt hours (29.2 and 30.2 TWh) electricity respectively, approximately 1.85% of total electricity use in the US (EPRI/WRF, 2013). While electric use in the sector indeed climbed over the decades between these two reports, a combination of increased energy use in other sectors and energy efficiency efforts in the water sector keep the overall percentage of electric consumption in the sector relatively low. At the global level, the International Energy Agency (IEA) estimates that electricity consumption for water and wastewater services is as high as 4% of world electric demand, in part driven by groundwater pumping and desalination in areas without reliable freshwater supply (IEA, 2016).

While these numbers may be high, they are still quite small when placed in the context of overall primary energy use. Primary energy use includes the fuel consumption needed for electrical energy generation, including the wasted heat in fossil electric generation, transmission losses, and including energy for industrial heat, space heat, water heat, and fuels for transportation of all types. Saul Griffith, under contract to the US ARPA-E program, converted the EPRI estimates to primary energy as part of his 'Super Sankey' diagram detailing energy flows across the US. At this scale, energy use is measured in 'quads' or quadrillion BTUs, and the US uses about 100 quads per year for all economic activities. In this metric, municipal water use accounts for 0.13 quads and wastewater for 0.1 quads (Otherlab, 2018). While the law of large numbers says that even one quarter of a quad is still an amazingly large amount of energy, it may be difficult to convince policy makers to focus their attention on the water sector given the scale of the overall energy use of the US.

This puts the water sector in a difficult position if the discussion of decarbonization stays focused on electricity at the facility level. Drinking water and water resource recovery facilities use massive amounts of electricity relative to many consumers, with a real and measurable impact on the environment. In many cities, water and wastewater utilities are the single-largest electricity user (USEPA, 2021b). However, in the context of state, national, and international energy and decarbonization policy decisions, the amount of primary energy the sector uses are too small to warrant sector-specific policies.

Policy makers and utility managers need to consider the larger picture. If the goal is economy-wide decarbonization and environmental protection, the optimal approach may not result in each and every water resource recovery facility being a net-zero energy producer with its own complex and capital intensive distributed electric system. We are not suggesting that water utility managers and policy makers be content to simply sit and wait for the grid to decarbonize. At a minimum, both policy and operations should continue to emphasize major efforts on energy efficiency, including breakthrough technologies and approaches that drastically reduce energy use, regardless of what happens to the grid. Already large urban water and wastewater systems may achieve enough economy of scale to cost-effectively continue to pursue onsite renewable energy production. However, with utility-scale renewable energy prices steeply declining to the point where it is now more cost-effective in many areas to build new solar capacity than operate existing coal-fired power plants (IRENA, 2019), the grid in many countries may be able to deliver carbon-free electricity in sufficient quantities at sufficiently low prices to decarbonize the sector's electric use in the foreseeable future.

However, water resource recovery facilities offer other avenues to decarbonize beyond electricity: organic carbon resources that can be used to displace fossil heating, vehicle fuel, or commodity chemicals; recoverable nutrients that can displace energy-intensive nitrate fertilizers or limited sources of phosphate fertilizers; vast sources of recoverable heat that can be used in district heating well beyond the fenceline of a treatment plant; and recoverable water. These resources, especially when complemented by similar 'waste' streams (also full of recoverable resources) from other industries like food and beverage producers, agriculture, or even waste heat from data centers, may indeed have enough value to society to warrant policies that encourage their recovery and reuse.

To fully decarbonize the wastewater sector, accounting for electric use alone is not enough. The accounting must include N_2O emissions and fugitive methane releases, as well as Scope 2 and 3 emissions such as chemical consumption and fuel used to transport sludge for offsite disposal. An example of this type of accounting conducted by Water UK is in Figure 17.1 below. To address these other sources of emissions, decision-makers need to seek out partnerships, think beyond onsite electricity production, and deliver co-benefits. By expanding the scope of the effort, the water sector can play a meaningful role in decarbonization in areas that go well beyond those impacted by electricity alone. One example of non-electric decarbonization in the water sector is using the biogenic carbon resource in biosolids. Current best practice is to use anaerobic digestion to convert about 50% of that carbon

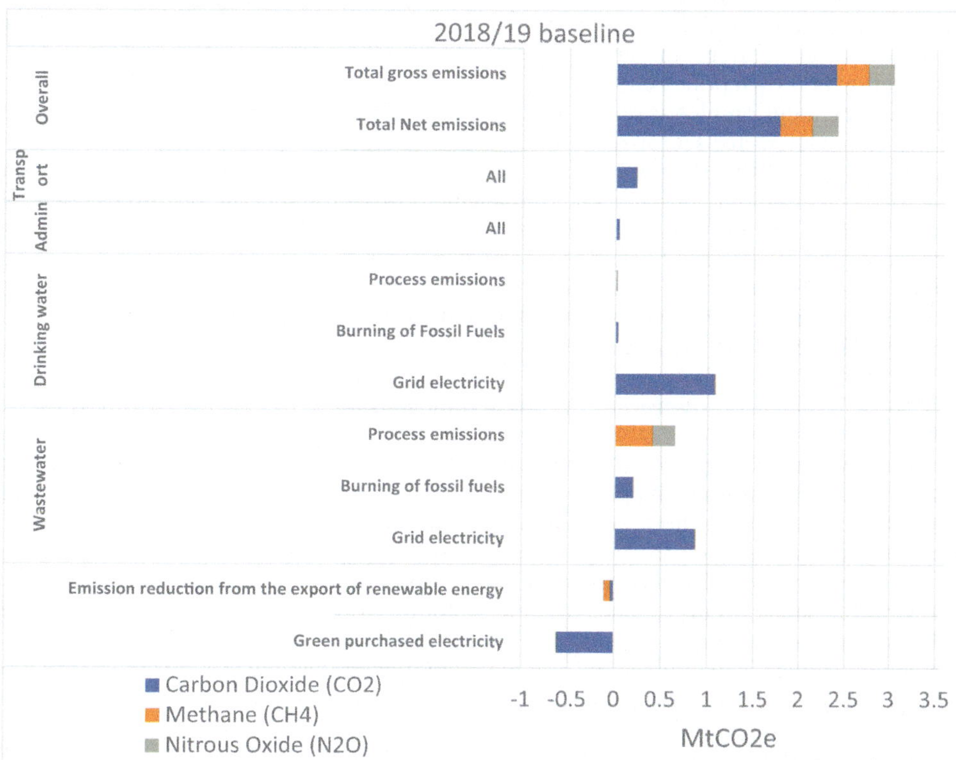

Figure 17.1 Reproduced from Water UK Net Zero 2030 Routemap, published by (Water UK, 2020) with kind permission.

to biogas, which must then go through multiple cleanup steps to be useful as a replacement for fossil natural gas. Once cleaned, the gas can be used for heating fuel, as fuel for combined heat and power generation, as compressed natural gas vehicle fuel, or be injected into the natural gas pipeline network as 'renewable natural gas,' with the end use dictating the cleanup processes. The biosolids are often co-digested with other wet feedstocks such as animal manures or food waste to assist the economics through additional energy production and 'tipping fees,' which are received by digester operators as payment for accepting outside wastes. Given the current low prices for natural gas, the emission concerns related to fugitive methane emissions, the fact that only half the resource is converted to energy, and the high capital and operation costs related to anaerobic digestion, gas cleanup, and onsite energy generation, many utilities have found that this pathway is simply not cost effective, even with co-digestion and tipping fee income, and even in regions that have strong policies supporting digestion and relatively high energy prices, such as Massachusetts (USA).

In a future with ample renewable electricity on the electric grid and concerns about fugitive emissions, some industry observers have suggested that a better path than biogas-specific policy incentives would be to seek out technologies that convert the carbon to a more valuable end product. A variety of technologies are currently under development, from arrested methanogenesis to produce renewable commodity chemical building blocks around which biorefineries can be built (Bhatt *et al.*, 2020) to hydrothermal liquefaction (Chen, 2020). It is beyond the scope of this chapter to go into detail on these technologies, but the most successful policies will be those that encourage the highest and best use of carbon supplies, regardless of feedstock (biosolids, manure, or food waste) and regardless of the technology used to recover the energy resource.

On the other hand, many experts feel that given the thousands of existing digesters in the US and the deep technical expertise the industry has amassed, biogas facilities (to include wastewater digesters, standalone food waste digesters, manure digesters, and facilities that co-digest some mix of those feedstocks) deserve their own policy carve outs. Recent research from the Water Research Foundation reveals an increasing focus on biogas policy and regulations, and, whereas for solar there are multiple market entries, the wastewater sector represents a significant portion of the potential biogas energy market (Kenway *et al.*, 2019). Biogas resource recovery projects are more likely to progress given familiarity with the technology. Further, the management of sector specific resources could not only benefit energy production but could reduce carbon throughout the system. These are reasonable policy differences, and it is possible that the correct answer is simply to accommodate both sides with incentives for both biogas and developing new technologies such as arrested methanogenesis or hydrothermal liquefaction. In the big picture, what really matters is that these wet carbon feedstocks are captured and used to offset fossil fuels.

17.2.2 Concept 2: There is no overarching policy that mandates decarbonization in the water sector globally

Policymakers and water industry professionals have often struggled to develop and respond to policies that encourage decarbonization or its proxy, reduced fossil energy use. This is not for a lack of effort. In 2019, the Water Research Foundation investigated the 'Opportunities and Barriers for Renewable and Distributed Energy Resource Development at Drinking Water and Wastewater Utilities.' As the intricate chart in Figure 17.2 demonstrates, in the US alone there are dozens of policies that seek to impact the water sector's energy use in some fashion. However, far from spurring widespread industry adoption of low-cost on-site renewable energy generation at water and wastewater facilities, the resulting policy matrix is bewilderingly complex and indeed illegible at a normal printed scale.

In Figure 17.3, a zoomed in section of the policy web (highlighted in the red box above) further illustrates the complex policy landscape.

Faced with such overwhelming complexity in the energy policy landscape, which is layered on top of the regulatory mandates to protect public health and the environment, and local policies that aim to keep water/sewer rates low for a variety of reasons, it is unsurprising that the pace of adoption of

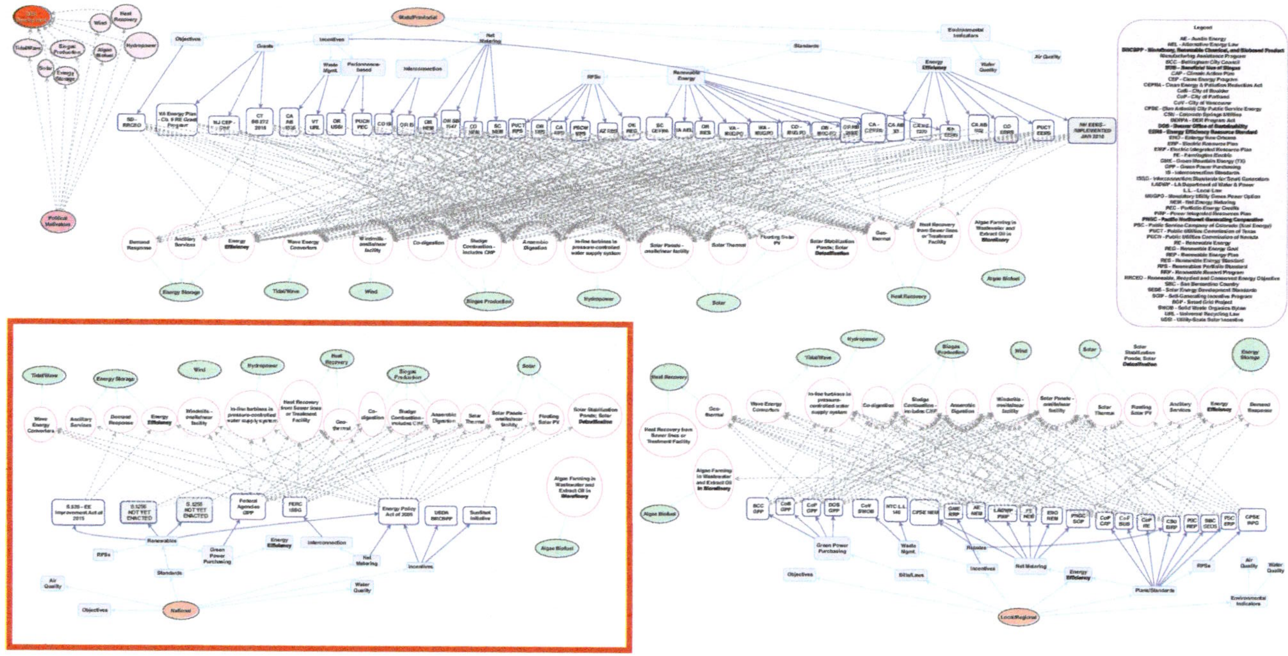

Figure 17.2 Visual Overview of Regulatory Drivers for Distributed Energy Resource Recovery and Renewable Energy applicable to water and wastewater utilities, http://dx.doi.org/10.13140/RG.2.2.24472.01286. Reprinted with permission (Kenway *et al.*, 2019).

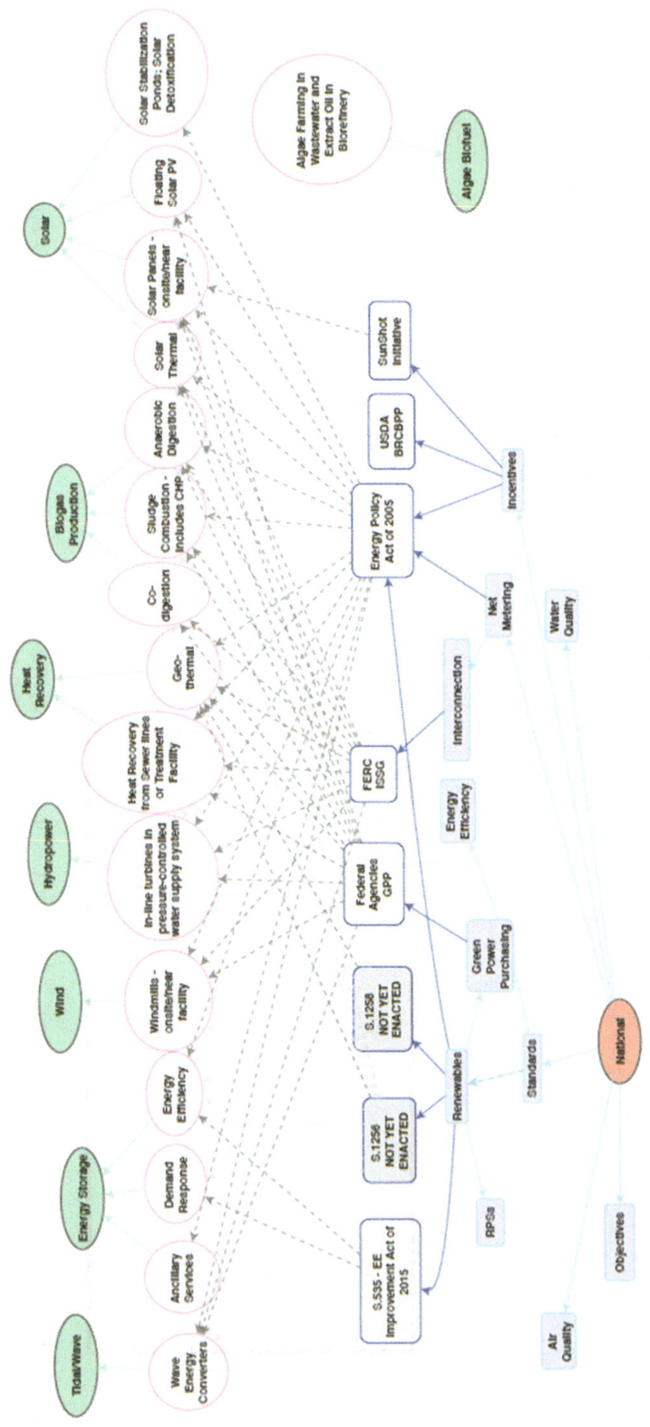

Figure 17.3 Visual Overview of Regulatory Drivers for Distributed Energy Resource Recovery and Renewable Energy applicable to water and wastewater utilities, http://dx.doi.org/10.13140/RG.2.2.24472.01286. Reprinted with permission (Kenway *et al.*, 2019).

onsite distributed renewable energy in the water sector remains stubbornly low despite a decades-long focus on energy use from policy makers and water sector NGOs.

This overlapping array of national, state, and local policies, each of which was created with good intentions, makes it difficult for any single technology or technique to win widespread adoption. This system requires each utility to go through the laborious and expensive process of tailored research and technology analyses and trying to create tailor-made systems that fit their unique mix of policies and other drivers. As a result, only the most motivated utilities have been able to achieve net zero energy, and almost none have become the net energy producers that engineers and scientists agree they could be. Even those that have become net zero, like the Gresham, Oregon, USA, facility, still rely on additional inputs of energy such as solar and imported organic waste (Modern Power Systems, 2019).

Policies can be technology forcing, like regulations mandating numerical pollutant limits. For example, in the UK, the Climate Change Act 2008 set legally binding limits on the total amount of greenhouse gas emissions the nation can emit for a given five-year period (UK Department for Business, Energy & Industrial Strategy, 2021). They can create market forces that drive efficiency, such as the global carbon tax proposed by Canadian Prime Minister Justin Trudeau at COP26 in Glasgow (Tasker, 2021). They can be funding driven, such as the Water Security Grand Challenge from the US Department of Energy (US DOE, 2021), or voluntary and largely unfunded, like the US Environmental Protection Agency's Water Reuse Action Plan (US EPA, 2021a). Or they can combine regulation and funding, like the Massachusetts food waste ban and accompanying funding for development of anaerobic digesters (Massachusetts DEP, 2021).

There are several important takeaway messages from this complex policy matrix. One is simply that policies, whether they be national, regional, state/provincial, or local, are not uniform and can sometimes be contradictory. Another is that there are a variety of forms that policies can take. These include policies that energize creativity by creating opportunities, such as cap-and-trade methods (and even within cap-and-trade, there are different types of policies), and policies like carbon taxes that may achieve similar goals through different mechanisms. Each is shown to drive different but overlapping reasons for renewable energy investments in the sector (Strazzabosco *et al.*, 2020).

17.2.3 Concept 3: Seek out co-benefits with other policy areas

As we write this, the world is facing a series of 'cascading crises,' as described by US President Joe Biden (Biden, 2021). These include but are not limited to the current COVID-19 global pandemic, the climate crisis, growing inequality, and systemic racism, not only in the US but all over the world. Regardless of the reader's stance on politics, it is rational to assume that addressing these cascading crises will be the priority of policy makers around the world for the foreseeable future. Thus, any efforts to create policy that will assist in decarbonizing the water sector will be more successful not only if these efforts articulate how they fit into the larger effort to address these multiple crises, but if these efforts themselves can address multiple big picture policy goals. Although decarbonizing for its own sake does address the climate crisis, any policy proposal or project that pushes decarbonization at the expense of other top-level policy concerns is unlikely to gain traction.

For this reason, the concept of 'co-benefits' is essential both to crafting new policy and to responding to existing policy. Put simply, projects that can meet multiple goals will be more likely to move forward than projects that do not. As an example, in 2015 the government of California, in response to an ongoing drought, mandated an immediate 25% cut in urban water consumption. As an unintended but welcome co-benefit, researchers found that electric use related to lower water consumption was reduced by a staggering 1830 GWh, more than the total reduction from all other investor-owned utility energy-efficiency programs in the state combined during the same time period (Spang *et al.*, 2017). With this information in hand, policy makers may seek to consider advancing water conservation, which has its own well-documented benefits, to reduce energy consumption even in years where drought is not a concern.

As a second example, there is an international focus on addressing systemic racism in the wake of recent incidents in the US and around the world. While this may seem far removed from decarbonizing the water sector, it may have very real impacts on water projects. Traditionally, our industry's facilities, despite their contributions to public health, have been considered a burden to their host communities in terms of odors and localized air emissions, truck traffic, and unsightly facilities barricaded from the public with high concrete walls and chain link fences. Projects that are sited in or near population centers of racial minorities and/or economically distressed people are coming under increased scrutiny for their impacts on those populations. One approach to mitigate these negative impacts is the redesign of water resource recovery facilities to provide public amenities. The Wusong Wastewater Treatment Plant upgrade in Shanghai, China (Figure 17.4), won an 'award of merit' in 2019 from Engineering News-Record for its incorporation of an indoor botanical garden that does double-duty as both a publicly accessible park and integral piece of the treatment train that boosts energy efficiency, controls odors, and reduces the overall physical space needs of the facility (Engineering News Record, 2019). It does not require a great leap of imagination to think that future treatment plants will be required offer similar co-benefits to their host populations, especially when those populations are historically oppressed.

Perhaps one of the greatest areas for potential co-benefits regarding the decarbonization of wastewater comes from capturing the low-grade heat contained in the water. Given the vast quantities of wastewater running under the streets in most urban centers, recovering even a small portion of the thermal energy contained in the wastewater using heat exchangers offers the potential for highly efficient district heating that can offset other, more carbon-intensive, heating sources. Researchers in the UK modeled four treatment facilities and found that recovering thermal energy for district heating offered the potential to reduce carbon emissions by 30–110 kg CO_2e/yr.pop. (Hawley & Fenner, 2012).

One implementation of this technology is in Vancouver, Canada, at the False Creek Neighborhood Energy Utility, where the self-funded project 'eliminates more than 60% of the greenhouse gas pollution associated with heating buildings.' (City of Vancouver, n.d.) However, when developers sought to recreate this approach just a few miles away in Seattle, Washington, USA, they ran into regulatory hurdles that prevented them from tapping into this resource. In response, the local authorities developed a standardized approach to permitting wastewater heat recovery projects, with the goal of providing multiple benefits that include lowering individual building's carbon emissions, giving developers another tool to meet stringent energy codes, and attracting a broader range of tenants, buyers, and investors (Landers, 2021).

Figure 17.4 Wusong WWTF, image courtesy of Organica Water.

Another example of a wastewater heat recovery project with co-benefits beyond decarbonization comes from the town of Avon, Colorado, USA. Here, local regulators were concerned that effluent temperatures from the local wastewater treatment facility were raising the temperature of the receiving water and impairing cold-water fish species. In response to this policy driver, the town incorporated a small district heat system (Figure 17.5). The town's largest municipal energy user is the town recreation center, which includes multiple heated pools and about 3700 square meters of heated space. By using waste heat from the wastewater to heat the pools, provide building heat, and to provide salt-free snow melting on town sidewalks, the town reduces effluent temperatures while simultaneously providing a low-cost heat source to offset fossil fuels used in the town facilities (Strehler *et al.*, 2010). The town purchases wind power to offset the electric demand of the heat pump, ensuring the system is zero carbon. (Avon, Colorado, 2021).

As a practical approach to creating projects that achieve multiple policy goals and co-benefits at the same time, some local governments have embraced the approach of Integrated Resource Management (IRM), sometimes called Integrated Resource Recovery (IRR). This is an interdisciplinary, cooperative project management approach that relies heavily on early stakeholder involvement and an iterative process to continually refine a project, ensuring broad support before it comes up to a vote or similar public approval process (Thurm, 2016). The government of Singapore has used a variation of this approach to integrate multiple objectives, resulting in such showcase facilities as the Marina Barrage, which simultaneously provides 10% of the country's water needs, alleviates flooding, and provides public access with over 15 million visits per year for tourism and recreation (Chye, 2018).

Combining several of these themes, we challenge the reader to envision water resource recovery facilities as a cohesive part of a larger integrated whole. Despite their tremendous public health benefits, our facilities are perceived as being burdensome on local communities and, indeed, are being fought off as tools of institutional oppression. In the US state of New Jersey, legislation passed in

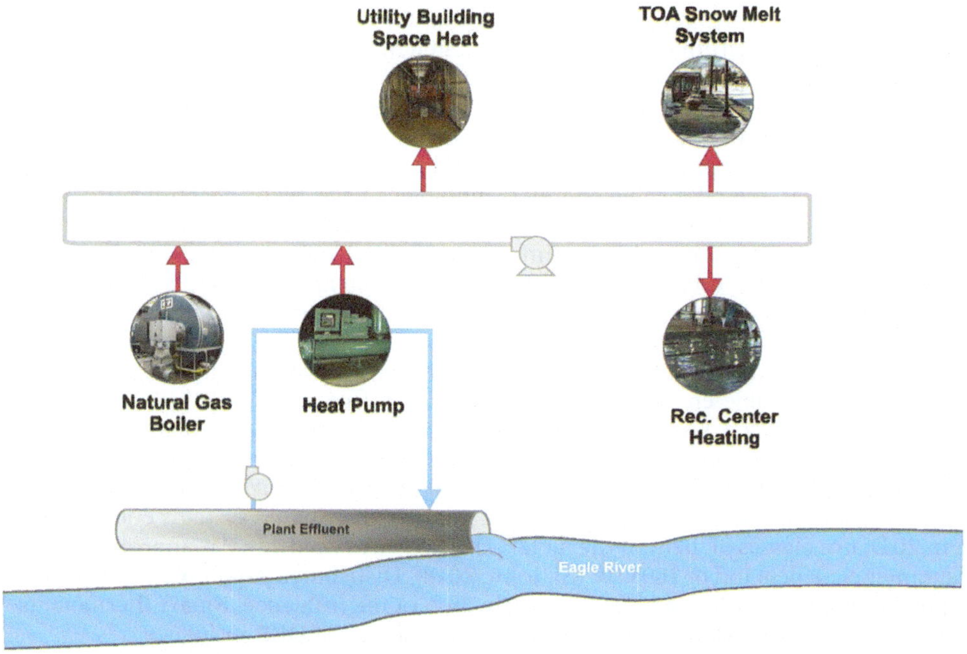

Figure 17.5 Avon, CO, wastewater heat recovery schematic. Image courtesy of Jennifer Strehler/CDM Smith.

2020 requires the state to evaluate the environmental and public health impacts of facilities including sewage treatment plants, sludge incineration facilities, resource recovery facilities, and cogeneration facilities – all of which are part of our current concept of water resource recovery facilities – on overburdened communities (State of New Jersey, 2020).

Imagine, instead of a world where our industry's facilities are considered to be part of a systemic burden placed on disadvantaged communities, a world in which water resource recovery facilities are also viewed as community assets, places that are attractive to live next to and which bring up, not down, their communities by offering beautiful, clean public spaces full of gardens, as in Wusan.

Today, our facilities, despite receiving 5–10 times more chemical and thermal energy in the influent than is required to clean that influent (Water Research Foundation, n.d.), are vast energy consumers. Imagine, instead, a world in which these facilities, through a combination of efficiency and energy recovery, provide low-cost carbon-free power and district heat to their host communities, as in Vancouver, Canada.

Today, the vast majority of facilities simply pass on water cleaned to the bare minimum of regulatory standards to a receiving water, with no attempt made to recover that water to offset the need for new potable water upstream of the treatment plant. Imagine, instead, a world in which water resource recovery facilities treat that water as a resource, as in Singapore.

17.2.4 Concept 4: Water reuse and energy recovery may be more beneficial in a distributed setting where the water and energy are recovered at/near the point of generation

Increasing energy demand for water systems, increasing energy prices, advancement towards green energy and greenhouse gas emissions mitigation goals, and the impacts of climate change, are all driving water and wastewater utilities to seek investment in self-generated energy. Self-generation of energy may reduce costs, improve system resilience and reliability, and reduce GHG emissions. Renewable forms of distributed energy available to utilities can include organic matter in wastewater, hydropower, thermal heat in wastewater, waste heat from converting gas to electricity, solar, and wind. Drawing on all these forms of energy, wastewater treatment plants can potentially generate far more energy than their sites require (Kenway *et al.*, 2019). Collectively, the thermal and organic energy potential contained within wastewater is estimated to be approximately eight times the amount of treatment energy required by the wastewater sector in 2012 and a large portion of this potential remains untapped (Kenway *et al.*, 2019). Further, as utilities can store energy (e.g. as gas or water at elevation), and can often shift their own energy demands through time, they are an important potential element of a more renewables-based future grid.

Distributed resources and renewable energy technologies have evolved significantly over the past ten years with overall distributed energy resource (DER) capacity expected to continue to grow due to innovation lowering costs, along with continuation of federal and state subsidies. This trend is seen particularly with solar and wind renewable technologies, which are becoming more competitive with conventional technologies. For municipal and private water and wastewater utilities there is a unique opportunity to invest in localized resource recovery.

Capacity and generation for DER technologies have been growing steadily over the past ten years. For example, total renewable energy generation has grown from 360 TWh in 2000 to 844 TWh in 2020 in the US (USEIA, 2021). Capacity of renewables increased over 300% to 265 GW between 2000 and 2020. Detailed statistics on the distributed and behind the meter fraction of these total numbers are generally not readily available. However, distributed and small-scale systems generally contribute less than 1% of generation or capacity. During peak periods, distributed systems contribute more. For example, in 2017, five classes of 'behind the meter' DER's contributed 44 GW, approximately 6% of the total US summer peak demand of 769 GW (St. John, 2018). Distributed solar and small-scale combined heat and power (under 50 MW) contributed nearly 80% of the influence. Smart thermostats, electric vehicles and distributed energy storage contributed the balance of DER influence.

Energy generation in the water sector globally is dominated by biogas-based technologies. In the UK the water sector provided 8.5% from renewable sources, with 80% from biogas from anaerobic digestion (Howe, 2009). In the US there are over 14,500 publicly operated wastewater treatment plants (WWTP) treating an average flow of approximately 32,345 million gallons per day (MGD) (Shen *et al.*, 2015) and, of those, 1027 have a capacity above 5 MGD and treat 80% of the wastewater generated. According to Tarallo *et al.* (2015), about 851 trillion British thermal units (BTU) of energy is contained within the wastewater of these 1027 WWTP annually.

Despite a considerable amount of thermal energy carried in the wastewater (2800 megajoules of waste heat per person annually released to the sewer (Larsen *et al.*, 2016), very little is currently utilized. State-level analysis by the University of Queensland (Hivert, 2019) confirmed a total potential of approximately 200,000 GWh/y similar to that estimated by Tarallo (2014). Chemical energy in the form of biogas has been successfully recovered for many years. Biogas is mostly recovered from the process of anaerobic digestion of the sewage sludge, even if there are examples of biogas recovery directly from the wastewater stream (Degarie *et al.*, 2000). Biogas generation from sewage sludge can be enhanced by combining the sludge with an external organic feedstock, this is called co-digestion.

There are now a range of international examples of water and wastewater utilities successfully implementing DER projects using on-site digester gas (biogas), solar-PV, hydro, wind turbines and other renewable sources. Water and wastewater utilities are often good candidates for DER as they can own large amounts of contiguous land, have high (and movable) energy demand, and can provide other types of ancillary grid services. Utilities are also participating in demand response programs by using emergency generators and other energy sources to offset peak grid electricity demands. However, identifying key options, and navigating regulatory requirements, tariff structures, dynamic policy positions, and workforce capacity have been great challenges to invest with certainty.

Lacking any unifying policy for decarbonization and renewable energy adoption in the water sector, adoption has been driven by a multitude of factors from financial opportunities to broad climate mitigation goals. Strazzabosco *et al.* (2020) investigated drivers in renewable energy adoption in the Australian water industry and found reducing energy costs as the most significant factors influencing renewable energy projects. However, the study notes that compulsory greenhouse gas emissions reduction requirements as the most influential policy supporting renewable energy projects. Similar to findings in the US (Kenway *et al.*, 2019), the Australian study questioned the relevance of government financial policy or renewable energy markets as being influential for the water industry, suggesting a broader role for the water industry in decarbonization looking at district level opportunities.

One example of this role is in how Metro Vancouver in Vancouver, Canada is directly injecting cleaned, excess, biomethane from wastewater treatment plants into existing natural gas distribution systems in response to British Columbia's climate action. Discussions about a biogas upgrade project was surrounded by the opportunities to use all elements of liquid waste as resources to reduce Metro Vancouver's corporate greenhouse gas (GHG) emissions, and the region's GHG emissions (Kenway *et al.*, 2019). Metro Vancouver had signed the provincial government's Climate Action Charter in 2007, along with almost all other local governments in British Columbia, which committed Metro Vancouver to be carbon neutral by the 2012 reporting year. The Lulu Island Wastewater Treatment plant (LIWWTP) Green Biomethane Project in Vancouver, British Columbia is a collaboration between Metro Vancouver and natural gas provider FortisBC. The objective of this collaboration is to clean unused biogas to pipeline quality, allowing it to be sold to FortisBC as renewable biomethane. Renewable natural gas (or biomethane) is produced from biogas as a byproduct of anaerobic digestion from the LIWWTP. Routinely, biomethane gas is used to provide heat to the plant's buildings and digesters and excess gas is safely flared into the atmosphere. From 2014 to 2016 the total energy demand and biomethane used by the LIWWTP was about 216 220 and 84 640 GJ, respectively, for all three years and the percentage of heat demand met by biomethane was about 40% for all three years. An interconnection agreement was established between FortisBC and Metro Vancouver to sell excess biomethane from the LIWWTP.

Another example is the management of organics. By routing urban organic waste to regional wastewater treatment plants, the water sector can reduce methane emissions at solid waste facilities and also improve co-digestion and electricity production. For example, the New York City Department of Environmental Protection (NYCDEP) and Waste Management New York have a collaboration to source separate organic waste and preprocessed food scraps in order to improve co-digestion processes at Brooklyn's Newtown Creek Wastewater Treatment Plant. The program is a result of a Mayor Bloomberg's PlaNYC initiative to make New York City to be the most sustainable city in the world. The PlaNYC set goals for an 80% reduction in GHG emissions by 2050, energy-neutral in-city wastewater treatment operations by 2050, maximizing beneficial use while minimizing fugitive emissions of biogas by 2050, and reaching zero waste to landfill by 2030 (City of New York, 2021). In the collaboration, NYCDEP reuses the biogas produced from anaerobic digestion in 13 of its 14 treatment plants. The gas is most often utilized in on-site boilers for heating or for powering equipment. At the Newtown Creek facility, 250 tons of source separated food waste (collected and processed by Waste Management) is injected into the wastewater treatment plant digesters to augment biogas production.

While there are many challenges with DER development, three stand out: (i) integration into the grid; (ii) a need for location-specific knowledge; and (iii) integrated planning. Metering and grid interconnection procedures are shaping the future of the electric grid (IREC/VSI, 2014). Lack of coordination in planning and deployment of DER, as well as lack of adequate management systems, will increase the cost of infrastructure upgrades and reduce the full value of DER as experienced in Germany (EPRI, 2014). An integrated grid and focusing on co-benefits can enable higher penetration of DER, reduce voltage loss and environmental impact, defer capacity upgrade, engage in demand management programs, and improve power system resiliency (EPRI, 2014). However, water and wastewater utilities find barriers emerge as traditional energy utilities can be challenged by DER as a new option to traditional energy utility business models (IREC/VSI, 2014; Willis *et al.*, 2012, 2015).

Key policy and regulatory changes to support DER include minimizing the regulations that small/medium size WWTPs must meet or look for ways to promote cooperation amongst wastewater treatment districts. For example, not all plants would need a digester if there were more interaction around plants. Larger treatment facilities could more actively source organics from smaller plants or other sources for co-digestion. Wastewater utilities could also establish a closer and mutually beneficial relationship with municipalities and other food producers or look into developing a campus environment.

Campus environments would make it easier to manage resource recovery opportunities. For example, WWTP, city hall, landfills, and council properties can be considered all part of the same campus. This would allow for surplus energy produced to be shared on a 'campus grid or micro-grid.' Increasingly, water and wastewater infrastructure are considered critical and therefore given license to operate micro-grids as part of a resiliency policy. Such an approach could be further enhanced by taking a 'whole-community planning' view and work with federal, regional, and local planning agencies, who are more often dependent on local utility services.

Overall, there should be greater organizational support for 'testing-out' DER options before energy and cost crisis drives DER development. Pilot projects could lead to incorporating energy consumption and DER opportunity into planning of plant designs, help establish connections with planning divisions to explore energy reduction opportunities as part of infrastructure and asset management planning (i.e. upgrade and rehab of pump stations, system configurations, incorporate distributed energy such as micro-turbine installations or energy saving techniques, etc.). Investment in personnel, able to work with renewable energy within water or wastewater systems is needed.

Looking forward, water utilities entering the DER market need to consider the market forces, the changing technologies, the changing subsidies, and the changing regulation, but they must also consider others' interests to direct and modify market change. The financial returns on a DER project installed today are likely to change over its operating life. Some jurisdictions will continue subsidies and regulatory policies that support DER. Others will limit the economic advantages of DER.

17.2.5 Concept 5: Change is constant. We need water policy and technological platforms that can more easily adapt to change

Many of the other chapters in this book discuss innovation and efficiency. They look at a wide variety of technologies that could help us get from where the industry is now to some idealized world where we have decarbonized water in the future. The authors of this book, with many hundreds of combined years of experience in academia, government, engineering, and facility management, have come together to try to envision how we can possibly change our industry's entire way of doing business. We do this because, even though the wastewater sector has provided the greatest public health service in human history, we have done so without regard to our part in a growing environmental catastrophe that threatens to negate the work we have done in the last two centuries. And now, we are pivoting to become a part of the solution to this new public health threat.

However, we are constrained by a system that never envisioned this kind of change. Our rules and regulations, our engineering approaches and technologies, our funding mechanisms, our methods of communicating to the public we serve, indeed every facet of our system, were all designed under a set of assumptions that was based in an industrial mindset. Clean water would be made dirty. Dirty water would be collected. And our industry would then make the water clean again, using as much energy and as many chemicals and as much manpower as needed, with little regard to the cost.

To reduce waste and inefficiency, we pursued economies of scale, investing huge amounts in industrialized water treatment facilities designed to last for decades, with the assumption that not much about the way we made the water dirty would change, that populations would either stay stable or grow, that we only needed to make the water clean enough to swim in or fish in and that we knew, by and large, what level of cleanliness that was. These industrial facilities were sited in places that were convenient for designers, and it was tacitly assumed that they would, as all industrial facilities were assumed to do, make life a little unpleasant for the neighbors.

Most of this set of assumptions came from the environmental movement of the late 1960s and early 1970s, which resulted in the creation of public agencies charged with environmental protection in nations around the world. Those agencies then created water quality standards and guidelines for treatment plants. And around the world, these assumptions – that our facilities should be large, expensive, and built to last; that they would have access to unlimited energy and chemical resources; that they should discharge treated water without further reuse; that they need not necessarily be good neighbors – have proven surprisingly durable.

In fact, these assumptions were outdated almost as soon as they were implemented. Before the industrialized world had even finished designing and building its first fleet of wastewater treatment plants, we were already faced with challenges. Our initial policies assumed that pollution came from a pipe, overlooking the contribution that stormwater plays. Our policies assumed that we need only treat for a handful of pollutants, overlooking first the nutrients nitrogen and phosphorus, then failing to anticipate the dangers of personal care products and pharmaceuticals, and most recently being caught off guard by the presence of microplastics and per- and polyfluoroalkyl substances (PFAS) that are problematic in even the most minute quantities. And our founding policies made the disastrous assumption that we could simply tap into a limitless supply of fossil fuels for energy and chemicals forever, with no adverse effects.

So, just as the world faces a series of cascading crises – attacks on science and democracy, the existential threat of climate change, systemic oppression, widening inequality, and a global pandemic that has killed millions of people and disrupted economies around the world, to name but a few – the wastewater treatment industry also faces its own set of cascading crises. We are being asked to simultaneously:maintain aging infrastructure;

- remove an ever-growing list of pollutants to ever-lower levels;
- devise ways to capture and clean up stormwater;

- adapt to a changing climate that threatens to disrupt our industry with a biblical series of floods, droughts, and sea level rise;
- prepare for the eventuality of water reuse;
- rapidly build new facilities in areas where population is growing and maintain older, now oversized facilities in areas where populations have shrunk;
- decarbonize our sector, one of the most energy-intensive of all public services;
- and to do all of the above in an equitable way that does not add to the historic burden placed on communities of color, indigenous peoples, and poor people in countries around the world.

This is no small set of changes from what was originally envisioned in the 1960s, and yet it is not anywhere near a comprehensive list. It does not account for the massive migration we can expect if even mid-range climate predictions come true (Lustgarden, 2020). It does not account for the desperate need for our industry to finally develop some version of water and sanitation systems that work well for the developing world with the resources available to them. It does not account for the impact future pandemics, which may be waterborne, could have on our systems, and it does not account for the unknowns that none of us have yet contemplated.

The intention here is not to depress the reader with an insurmountable list of things that are going wrong. Human beings are endlessly inventive and innovative. In barely a decade, we have gone from having essentially no electric cars on the market to one in which over 50% of the new car sales in Norway are now electric (The Guardian, 2021). In that same period, solar PV has gone from an expensive indulgence for the rich to 'the cheapest electricity in history' (Evans & Gabbatiss, 2020). If the automotive and electrical industries, both of which dwarf the wastewater industry, can pivot and adapt that quickly, there is no reason to think that we cannot.

However, to do so will require us to adopt policies that anticipate constant change. We need policies that allow for flexibility and innovation, in an industry that is famously reluctant to innovate. In 2013, a group of researchers proclaimed that 'there is an innovation deficit in urban water management.' After examining the reasons for this deficit, they arrived at the inescapable conclusion that '(t)o solve current urban water infrastructure challenges, technology-focused researchers need to recognize the intertwined nature of technologies and institutions and the social systems that control change' (Kiparsky *et al.*, 2013).

In other words, technology alone will not get us out of this mess. We need policies and institutions that are willing to support us as we innovate and iterate, and that means challenging our current set of assumptions. Our policies need to be supportive of a broad range of technologies, not pick chosen winners and losers. Facilities may not need to last 40 or more years. They may not need to be massively scaled to be efficient. They must not be bad neighbors. They must be able to be upgraded or replaced easily and cost-effectively as we find new threats to our environment and public health in our used water.

17.3 CONCLUSION AND POLICY SUGGESTIONS

In this chapter, we have highlighted five core concepts that policy makers and water resource recovery professionals can consider as they work to decarbonize the water resource recovery industry. Here, we summarize those concepts and provide policy suggestions where we can.

The first of these concepts is that the water sector may not use enough energy to warrant sector specific energy policy. Despite the terawatt hours of electric use in our drinking water and water resource recovery facilities, they are still a tiny fraction of the overall primary energy used by society. Policy and operations should continue to emphasize major efforts on energy efficiency, including breakthrough technologies and approaches that drastically reduce energy use. However, pushing to make each of the hundreds of thousands of facilities around the world zero net energy through on-site energy production may be a distraction from more productive efforts to decarbonize the larger electric grid. Further, to fully decarbonize the wastewater sector, accounting for electric use alone is not enough. The accounting must include N_2O emissions and fugitive methane releases, as well as Scope 2 and 3 emissions such as chemical consumption and fuel used to transport sludge for offsite disposal. To address these other sources of emissions, decision-makers need to seek out partnerships and deliver co-benefits. These include using the carbon and thermal energy resources contained in wastewater to offset fossil energy use in other sectors. We encourage policy makers to seek out policies that are technology neutral and that foster breakthrough innovation.

The second concept discusses the lack of an overarching policy mandating water sector decarbonization. The water sector has broad policy mandates to protect human health and the environment through the provision of safe drinking water and sewage treatment. However, these policies did not anticipate for the interplay of the energy used to provide these crucial services with the impacts of the energy used to do so. In practice, many efforts to improve public and environmental health in the water sector come at the expense of climate through increased energy and chemical use or by increasing emissions like N_2O. In their current form, policies around water sector energy use and decarbonization are a bewildering and complex web of sometimes contradictory local, state/regional, national, and even international policies. This is a natural follow-on to the first concept, as many of these policies were not designed with the water sector in mind. Although we do not offer any specific policy suggestions in this section, we are optimistic that efforts discussed in the other sections will eventually allow for a clear path to decarbonizing this sector.

The third concept area covers the idea of co-benefits with other policy areas. Put simply, projects that can meet multiple goals will be more likely to move forward than projects that do not. Here, we challenge the reader to envision water resource recovery facilities as a cohesive part of a larger integrated whole. We offer examples of co-benefits including water efficiency projects that also deliver large energy savings, water resource recovery facilities designed to serve as public botanical gardens and recreation facilities that are true community assets, and facilities that provide district heat for their neighbors. This concept area lends itself more to responses to existing policies than to the development of new ones.

The fourth concept area examines the potential to deploy distributed energy resources that take advantage of the unique attributes of water resource recovery facilities. These include receiving carbon and thermal energy resources in the wastewater influent, energy intensive processes that could receive renewable energy 'behind the meter' from onsite generation, and the possibility to integrate with other public sector services in a campus-like setting to share these resources. Policy suggestions include removing barriers to implementing these existing technologies.

The final concept area covers the constantly changing demands placed on the water resource recovery sector and the need for policies that are flexible enough to accommodate these changes. Here, we point to research showing that there is an innovation deficit in the water sector. We call for a rethinking of policies that lock the sector into decades-long investments in large, complex, industrial facilities and instead allow for a more rapid, iterative approach to solving water challenges that will continue to evolve even after we have achieved decarbonization.

17.4 ADDITIONAL RESOURCES

Below we list a few online resources that can be valuable for better understanding the landscape of the related policies and pathways.

Resource recovery from Water: from concept to standard practice. Editors: Ilje Pikaar, Xia Huang, Francesco Fatone, Jeremy S. Guest. https://www.sciencedirect.com/journal/water-research/special-issue/104CRLSTGFT

Mobilizing for a zero carbon America: Jobs, jobs, jobs, and more jobs, A Jobs and Employment Study Report. Saul Griffith, Sam Calisch, Alex Laskey. Rewiring America. July 29, 2020. https://www.ourenergypolicy.org/resources/mobilizing-for-a-zero-carbon-america-jobs-jobs-jobs-and-more-jobs/

Heat Pumps Using Waste Water in Goethenburg, Sweden. Case study on www.celsiuscity.eu, Jan 16, 2020. https://celsiuscity.eu/heat-pumps-using-waste-water-in-gothenburg-sweden/

Water UK Net Zero 2030 Routemap: Unlocking a net zero carbon future (online resource provided by Water UK): https://www.water.org.uk/routemap2030/

Emerging solutions to the water challenges of an urbanizing world. Tove A. Larsen, Sabine Hoffmann,,Christoph Lüthi, Bernhard Truffer, Max Maurer. https://www.science.org/doi/10.1126/science.aad8641

Municipal wastewater sludge as a renewable, cost-effective feedstock for transportation biofuels using hydrothermal liquefaction. Timothy E. Seiple, Richard L. Skaggs, Lauren Fillmore, André M. Coleman. https://doi.org/10.1016/j.jenvman.2020.110852

REFERENCES

Avon, Colorado (2021). Avon's Heat Recovery System. Retrieved from Avon, Colorado town website: https://www.avon.org/926/Avons-Heat-Recovery-System#:~:text=The%20first%2Dof%2Dits%2D,buildings%20in%20the%20town%20core

Bhatt A. H., Ren Z. and Tao L. (2020). Value proposition of untapped wet wastes: carboxylic acid production through anaerobic digestion. *iScience*, **23**(6), 101221, https://doi.org/10.1016/j.isci.2020.101221

Biden J. (2021). President Biden's Full Inauguration Speech. Retrieved from NY Times: https://www.nytimes.com/2021/01/20/us/politics/biden-inauguration-speech-transcript.html

Chen W. T. (2020). A perspective on hydrothermal processing of sewage sludge. *Current Opinion in Environmental Science and Health*, **14**, 63–73. https://doi.org/10.1016/j.coesh.2020.02.008

Chye K. T. (2018). From Resource to Asset: Building a Water-resilient Singapore. Retrieved from Global Infrastructure Initiative: https://www.globalinfrastructureinitiative.com/article/resource-asset-building-water-resilient-singapore

City of New York (2021). OneNYC. Retrieved from City of New York: OneNYC: http://onenyc.cityofnewyork.us/

City of Vancouver (n.d.). False Creek Neighbourhood Energy Utility. Retrieved from https://vancouver.ca/home-property-development/southeast-false-creek-neighbourhood-energy-utility.aspx

Degarie C., Crapper T., Howe B., Burke B. and Mccarthy P. (2000). Floating geomembrane covers for odour control and biogas collection and utilization in municipal lagoons. *Water Science and Technology*, **42**, 291–298, https://doi.org/10.2166/wst.2000.0664

Engineering News Record (2019). Award of Merit, Water/Wastewater: Wusong Wastewater Plant Upgrade – Botanical Garden. Retrieved from Engineering News-Record: https://www.enr.com/articles/47613-award-of-merit-waterwastewater-wusong-wastewater-plant-upgrade_botanical-garden

EPRI (2014). The Integrated Grid: Realizing the Full Value of Central and Distributed Energy Resources. Electric Power Research Institute, Palo Alto.

EPRI/WRF (2013). Electricity Use and Management in the Municipal Water Supply and Wastewater Industries. Electric Power Research Institute, Palo Alto.

Evans S. and Gabbatiss J. (2020). Solar is now 'cheapest electricity in history', confirms IEA. Retrieved from Carbon Brief: https://www.carbonbrief.org/solar-is-now-cheapest-electricity-in-history-confirms-iea#:~:text=Multiple%20Authors,-Simon%20EvansJosh&text=The%20world's%20best%20solar%20power,Agency's%20World%20Energy%20Outlook%202020

Hawley C. and Fenner R. (2012). The potential for thermal energy recovery from wastewater treatment works in southern England. *Journal of Water and Climate Change*, **3**(4), 287–299, https://doi.org/10.2166/wcc.2012.013

Hivert G. (2019). State Analysis of Biogas Opportunities in WWTP. Internal Project Report, University of Queensland, Brisbane.

Howe A. (2009). Evidence: Renewable Energy Potential for the Water Industry. UK Environment Agency, Bristol.

IEA (2016). Water Energy Nexus, Excerpt From the World Energy Outlook. International Energy Agency, Paris.

IREC/VSI (2014). Freeing the Grid 2006–2014. Interstate Reenwable Energy Council and Vote Solar Initiative, Oakland.

IRENA (2019). Renewable Power Generation Costs in 2019. International Renewable Energy Agency, Abu Dhabi. Retrieved from https://www.irena.org/-/media/Files/IRENA/Agency/Publication/2020/Jun/IRENA_Power_Generation_Costs_2019.pdf

Kenway S., Conrad S., Jawad M., Gledhill J., Bravo R., McCalley J. and Howe C. (2019). Opportunities and Barriers for Renewable and Distributed Energy Resource Development at Drinking Water and Wastewater Utilities (No. 4625). Water Research Foundation, Denver.

Kiparsky M., Sedlak D., Thompson J. B. and Truffer B. (2013). The innovation deficit in urban water: The need for an integrated perspective on institutions, organizations, and technology. *Environmental Engineering Science*, **30**(8), 395–408, https://doi.org/10.1089/ees.2012.0427

Landers J. (2021). King County, Washington, promotes sewer heat recovery as renewable energy source. Retrieved from ASCE Source: https://www.asce.org/publications-and-news/civil-engineering-source/civil-engineering-magazine/article/2021/01/king-county-washington-promotes-sewer-heat-recovery-as-renewable-energy-source (Accessed 15 February 2022).

Larsen T. A., Hoffmann S., Luthi C., Truffer B. and Maurer M. (2016). Emerging solutions to the water challenges of an urbanizing world. *Science (New York, N.Y.)*, **352**(6288), 928–933, https://doi.org/10.1126/science.aad8641

Lustgarden A. (2020). The Great Climate Migration. Retrieved from The New York Times: https://www.nytimes.com/interactive/2020/07/23/magazine/climate-migration.html (Accessed 15 February 2022).

Massachusetts DEP (2021). Anaerobic Digestion & Organics Diversion. Retrieved from Massachusetts DEP: https://www.mass.gov/lists/anaerobic-digestion-organics-diversion (Accessed 15 February 2022).

Modern Power Systems (2019). Achieving Energy Net Zero at the Gresham Wastewater Treatment Plant. Retrieved from Modern Power Systems: https://www.modernpowersystems.com/features/featureachieving-energy-net-zero-at-the-gresham-wastewater-treatment-plant-7254009/ (Accessed 15 February 2022).

Otherlab (2018). US Energy Flow Super Sankey. Retrieved from Otherlab: https://www.otherlab.com/blog-posts/us-energy-flow-super-sankey with additional data specific to the water and wastewater sectors retrieved from Otherlab's interactive tool at http://www.departmentof.energy/ (Accessed 15 February 2022).

Shen Y., Linville J. L., Urgun-Demirtas M., Mintz M. M. and Snyder S. W. (2015). An overview of biogas production and utilization at full-scale wastewater treatment plants. *Renewable and Sustainable Energy Reviews*, **50**, 346–362. https://doi.org/10.1016/j.rser.2015.04.129

Smith R. (1973). Electrical Power Consumption for the Wastewater Treatment Industry. Office of Research and Monitoring, National Environmental Research Center. US Environmental Protection Agency, Cincinnati.

Spang E., Holguin A. and Loge F. (2017). The estimated impact of california's urban water conservation mandate on electricity consumption and greenhouse gas emissions. *Environmental Research Letters*, **13**(1), 1–10. https://doi.org/10.1088/1748-9326/aa9b89

State of New Jersey (2020). Governor Murphy Signs Historic Environmental Justice Legislation. Retrieved from Official Site of the State of New Jersey: https://www.nj.gov/governor/news/news/562020/20200918a.shtml (Accessed 15 February 2022).

St. John J. (2018). Distributed Energy Poised for Explosive Growth on the U.S. Grid. Retrieved from Greentech Media: https://www.greentechmedia.com/articles/read/distributed-energy-poised-for-explosive-growth-on-the-us-grid (Accessed 15 February 2022).

Strazzabosco A., Conrad S., Lant P. and Kenway S. (2020). Expert opinion on influential factors driving renewable energy adoption in the water industry. *Renewable Energy*, **162**, 754–765, https://doi.org/10.1016/j.renene.2020.08.054

Strehler J., Vanderburgh S., Parry D. and Rynders T. (2010). Colorado community benefits from installing waste heat recovery system; ASME Paper 90479. Proceedings of ASME 2010 4th International Conference on Energy Sustainability, ASME, Phoenix, AZ.

Tarallo S. (2014). Utilities of the Future: Energy Findings. Water Environment Research Foundation, Alexandria.

Tarallo S., Shaw A., Kohl P. and Eschborn R. (2015). A Guide to Net-Zero Energy Solutions for Water Resource Recovery Facilities. Water Environment Research Foundation, Alexandria.

Tasker J. P. (2021). Trudeau calls for global carbon tax at COP26 summit. Retrieved from CBC News: https://www.cbc.ca/news/politics/trudeau-carbon-tax-global-1.6233936 (Accessed 15 February 2022).

The Guardian (2021). Electric cars rise to record 54% market share in Norway. The Guardian, p. n.a. Retrieved from https://www.theguardian.com/environment/2021/jan/05/electric-cars-record-market-share-norway#:~:text=Battery%20electric%20vehicles%20(BEVs)%20made,Road%20Federation%20(OFV)%20said (Accessed 15 February 2022).

Thurm B. (2016). Toward an Integrated Resource Management strategy? Retrieved from Innovative Governance of Large Urban Systems: https://iglus.org/toward-an-integrated-resource-management-strategy/ (Accessed 15 February 2022).

UK Department for Business, Energy & Industrial Strategy (2021). 2019 UK Greenhouse Gas Emissions, Final Figures. UK Department for Business, Energy & Industrial Strategy, London.
US DOE (2021). About the Water Security Grand Challenge. Retrieved from Energy.gov: https://www.energy.gov/water-security-grand-challenge/water-security-grand-challenge (Accessed 15 February 2022).
USEIA (2021). Annual Energy Outlook 2021 with Projections to 2050. US Energy Information Administration, Washington, DC.
US EPA (2021a). Water Reuse Action Plan. Retrieved from US EPA: https://www.epa.gov/waterreuse/water-reuse-action-plan (Accessed 15 February 2022).
USEPA (2021b). Energy Efficiency for Water Utilities. Retrieved from United States Environmental Protection Agency: Sustainable Water Infrastructure: https://www.epa.gov/sustainable-water-infrastructure/energy-efficiency-water-utilities (Accessed 15 February 2022).
Water Research Foundation (n.d.). Water Research Foundation. Retrieved from Energy Optimization: https://www.waterrf.org/sites/default/files/file/2020-07/4949-EnergyOptimization.pdf
Water UK (2020). Net Zero 2030 Routemap. Available at: https://www.water.org.uk/routemap2030/ (accessed 12 December 2021)
Willis J., Stone L., Durden K., Beecher N., Hemenway C. and Greenwood R. (2012). Barriers to Biogas for Renewable Energy. IWA Publishing, London.
Willis J., Andrews N., Stone L., Cantwell J. and Greenwood R. (2015). Identification of Barriers to Energy Efficiency and Solutions to Promote These Practices. IWA Publishing, London.

doi: 10.2166/9781789061796

Chapter 18

Outlook for the carbon-negative circular water economy

*Glen T. Daigger**

University of Michigan, Ann Arbor, MI, USA
**Correspondence: gdaigger@umich.edu*

18.1 INTRODUCTION

The previous chapters of this book have articulated many of the opportunities currently available to the water sector to decarbonize. Numerous innovators and early adopters are investigating options, conducting trials, and implementing options to increase energy efficiency, reduce carbon footprint, and recover resources from the water cycle. They are doing this in the general absence of regulatory drivers but are doing so based on a combination of practical concerns and organizational and community values. Reduced reliance on outside supplies for critical resources, such as energy, provides utilities with practical advantages, such as increased ability to manage and control operating costs under variable economic conditions. Recovery of products that are valued by customers not only provides revenue that at least partially off-set costs but also provides greater assurance that management options for these residuals will continue to be available. If viewed as wastes, rather than valuable products, opposition to associated management options (such as landfilling) can arise and threaten the ability to manage these residuals. Utilities and communities also reduce carbon emissions and recover resources to reduce their environmental footprint, in conformance with their broader commitment to environmental protection. The knowledge and experienced gained by these innovators and early adopters is essential to better understand what is possible and which of the available options may best fit various situations. The success being achieved by these innovators and early adopters also provides examples and evidence needed by others to subsequently adopt some of the new technologies and approaches being investigated, consistent with the social processes underlying the typical S-curve of adoption of innovations and new technologies (Rogers, 2003).

The momentum for change within the water sector is certainly accelerating, but there is much more to do. Long term, the water sector needs to transform to function effectively as we transition from the current linear economy to a circular one – in fact the water sector can provide leadership for other sectors providing essential public services. Two questions that may be asked are: (1) will a broad range of water sector actors adopt new resource recovery technologies and practices, or only a modest proportion and (2) how can the current transition be accelerated.

18.2 RESOURCE RECOVERY

18.2.1 Historical perspective

Let us begin by addressing some perceptions concerning the water sector. To be clear, the term 'the water sector' includes all entities engaged in managing water to meet the needs of humans and the environment, including water supply, water treatment, wastewater management, residuals management, stormwater, and flood protection. It is perceived by many that the principal concern of the water sector is water service and the local water environment, rather than broader environmental concerns such as climate change. One consequence of this perception is the further belief that water professionals feel justified to use whatever resources are needed, irrespective of their broader environmental impacts, as long as established regulations are complied with. This latter connection to regulations is based on the fact that regulations are generally focused on protection of the health of the population served by a water utility and the local aquatic environment. It is further perceived that the water profession is slow to change in response to evolving external changes. Unfortunately, evidence supporting these perceptions can be easily found by those seeking it. Fortunately, these perceptions do not fully characterize the water sector as it has functioned historically or as it continues to function.

The record clearly demonstrates that the water sector has adapted to societal needs and made significant changes over time, as discussed in Chapter 17. David Sedlak's book *Water 4.0* (Sedlak, 2014) summarizes the historic progression of the water sector, along with future prospects, in an interesting and compelling fashion. Outputs from the International Water Association (IWA) Cities of the Future program provide further evidence. Consider, for example, the well-known depiction of the transition of cities from the initial provision of water supply to water sensitive cities. As indicated in Figure 18.1, urban water management system service delivery functions, depicted in the lower portion of this figure, have progressed in response to the increasingly aspirational socio-political drivers depicted in the upper portion of the figure. Various stages in this progression are given a series of descriptive names. While many cities have yet to progress to the water-sensitive status, water management within individual cities generally develops along the illustrated trajectory.

Examination of the progression of functions presented in Figure 18.1, and comparison to existing regulations, indicates that regulations are not the sole driver for the systems implemented by the water

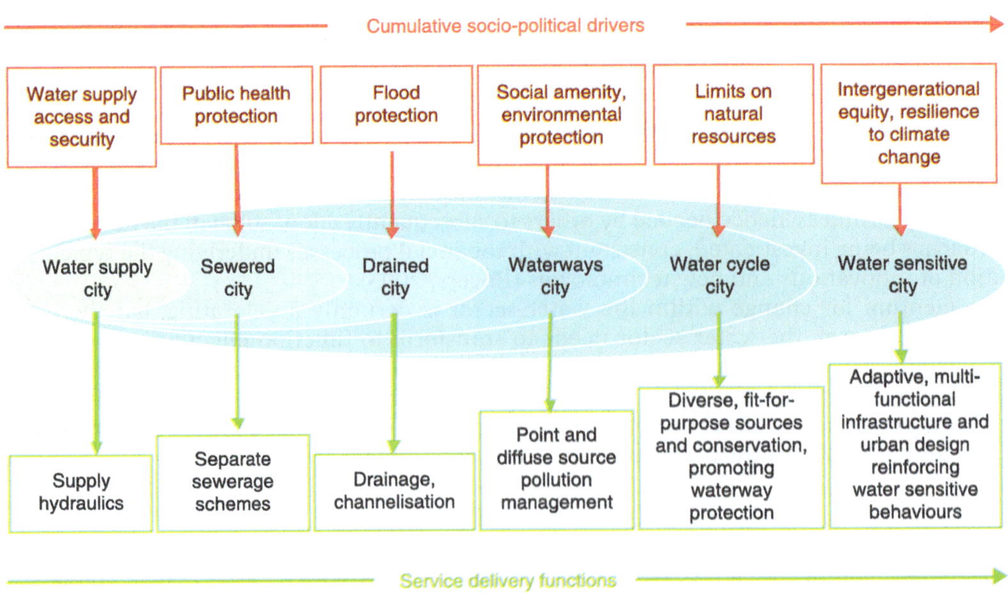

Figure 18.1 Progressive development of water-sensitive cities. From Brown *et al.* (2009).

sector. This is not to discount the role of regulations, when consistently enforced, to promote reliable service provision. This progression of functions is continuing today, as innovative and early adopter water sector utilities (water, wastewater, flood control) have transitioned from one stage to the next in response to a combination of need and opportunity. Broader regulations subsequently followed. This progression from innovator and early adopters to the establishment of regulations will be discussed further in a following section.

The water sector has also acted on broader environmental concerns over the years. Water supply systems have historically been designed and operated to be as energy efficient as possible, for example incorporating hydropower whenever possible. Energy-efficiency has also been a principal concern for wastewater treatment historically, and land application of biosolids which beneficially uses the nutrients and organic matter content of biosolids is a historic and continuing practice in many locations. Water reuse, both indirect and direct, is a historical practice, with rapidly expanding applications in many locations. These resource efficiency and recovery actions have been accomplished, of course, in the context of meeting regulatory requirements. The numerous examples presented in the previous chapters of this book illustrate current progress addressing global energy and environmental issues by the water sector. It is imperative at the present time, however, that efforts such as these be dramatically increased and expanded in scope by the water sector.

18.2.2 Value hierarchies

One tool that can assist the water sector to accelerate implementation of resource recovery is a vision of the journey, similar to the one for water cities illustrated in Figure 18.1. Our friend and colleague Jes LaCour Jansen provides one such depiction (Figure 18.2) that, while applying most specifically to used water, is illustrative for the entire water cycle. Presented as a pyramid, on the right-hand side the value of various types of products is presented in the order of ascending value. Examples of products that can be extracted from the used water stream are arrayed on the left-hand side to illustrate how products of increasing value can be extracted. Water is provided at the base of the pyramid as its recovery is a given. Many of the extractable products are carbon-based and/or can be used to produce energy without the use of fossil fuels, although with some exceptions such as nutrients. Current products,

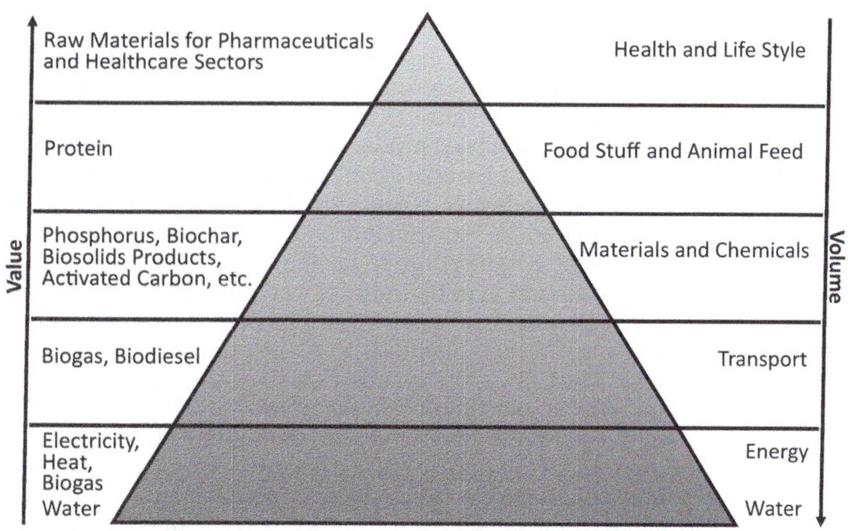

Figure 18.2 Resource recovery pyramid aligned with the recovery of resources from the water cycle. Adapted from a diagram presented by Jes LaCour Jansen based on similar diagrams in van der Hoek *et al.* (2016) and as can be seen at http://www.betaprocess.eu/the-value-pyramid.php.

in addition to water, include biogas, heat, electricity (produced, e.g., through a combined heat and power, CHP, system using biogas as fuel), biosolids, and phosphorus (e.g., as struvite). The range of potential products illustrates, however, the greater diversity of products that can be recovered, along with their increasing value both monetarily and to society. Missing from this pyramid are inorganic substances other than the nutrients nitrogen and phosphorus, such as minerals and metals, which have been demonstrated to offer significant potential value (Westerhoff *et al.*, 2015). Resources can also be extracted from other segments of the urban water cycle, for example the recovery and reuse of coagulants used for water treatment.

An aside. One of the reasons that I like the resource recovery pyramid in Figure 18.2 is that it reminds me of Maslow's Hierarchy of human needs that is a fundamental principal of the science of psychology. Maslow's Hierarchy progresses from basic human needs for survival to those that are more aspirational like personal esteem and self-actualization. I also see that same progression in the progressive development of water sensitive cities presented in Figure 18.1, the only difference being that the vertical progression of Figure 18.2 and the classic presentation of Maslow's Hierarchy is presented horizontally in Figure 18.1. One important message that I take from this is that psychology teaches us that people can generally envision progression from one level of the hierarchy to the next as being possible, but not jumps of two or more levels. Thus, if one is referring to Maslow's Hierarchy when seeking to motivate people to action, one needs to understand where the target audience is at on the Hierarchy and how the actions one wants them to take can elevate them to the next level. Asking them to take actions that will elevate them two levels is often not successful. People can view moving up 'one step' to be possible, but not two or more steps. Does this same psychology apply in the water sector? Do we need to understand where an individual utility is at on the pyramid in Figure 18.2 and focus on encouraging them to 'just take the next step'? I suggest that these are good questions to consider.

As shown in previous chapters, many of the present on-going efforts in the water sector focus on increasing the recovery of carbon in used water through the capture of carbon in the liquid stream and conversion to biogas through anaerobic digestion. These efforts include mainstream processes, such as use of anaerobic membrane bioreactors (AnMBR), or through the capture of carbon and stabilization in anaerobic digesters. As suggested by its lower location on the Figure 18.2 pyramid, this may represent a high quantity but relatively low-value product compared to others. Moreover, as the electrical grid transitions to renewable sources such as solar and wind, the production and use of biogas through a CHP system may represent a decreasing contribution to reducing global environmental impacts. Economic value may also decline over time as the cost to produce electricity decreases, as is occurring in numerous locations as the cost for solar and wind energy systems decline. In short, the water sector will need to 'climb' the pyramid illustrated in Figure 18.2 to continue to add value to society.

Factors other than the inherent value of a subject product can significantly influence the valorization of recovered resources. Consider the case of water. Drinking water produced from 'natural' water resources, such as surface water and groundwater, is often viewed by consumers as the 'gold standard' that other 'water products' are compared to. Non-potable water products may be accepted by users for their desired use, but often only at a reduced price compared to potable water, even if potable water produces no added value. Consideration of the value of water for human consumption is further complicated by the fact that, because water is a human right, its price rarely reflects the true cost of producing and distributing it. While alternate methods to secure the right for those for which the true cost make it unaffordable are available, such as subsidies, they are rarely used. Thus, water may be one of the lowest priced and least appreciated basic requirements. Then there is the issue of the acceptance of potable water reuse, whether indirect or direct, even though this water may be of higher quality than traditional drinking water. Fortunately, this latter situation is changing (although slowly) as potable reuse is becoming increasingly acceptable to the public. One may wonder to what extent the water sector may have contributed to the current situation through the historic practice of 'extolling the virtue' of drinking water and not better educating the public about the entire water cycle. The key point, however, is that influencing perceptions may be as important as the actual practices employed.

18.2.3 Advancing markets and products

Existing resource recovery practices are largely defined by the ease with which relevant markets are available. Water represents an example, since water utilities are already in the business of providing water services to customers. Biogas use represents another. Biogas produced from used water is typically valorized using CHP systems because the electricity produced can generally be fully used on-site. Issues can still arise for the produced electricity with the electric utility serving the facility where the CHP system is located concerning the purchase and sell-back of electricity. This has led all too often in the past to the determination that biogas use by CHP is not economical. CHP systems also produce heat, which can be used on-site. The on-site demand is often not sufficient to fully use the available heat, and thus it is not fully valorized unless access to outside customers such as district heating systems is available. Biogas can also be upgraded to natural gas quality for sale to the local natural gas utility, although difficulties concerning institutional arrangements and pricing can arise again. Upgraded biogas is also used, in some instances, to fuel municipal vehicles, thus simplifying institutional arrangements. Biogas can also serve as feedstock to produce higher value products, such as microbial proteins. As indicated in Figure 18.2, this represents higher-valued uses of the recovered product. In contrast, utilities have extensive experience developing markets for biosolids products, especially in the agricultural community where a majority of biosolids in some locations are reused, but also for a variety of consumer products. While such programs do not generally produce sufficient revenue to off-set costs, they do provide a publicly acceptable (and thus secure) method for biosolids management while producing environmental benefits.

In short, resources not only need to be recovered, but they must be converted into products where sufficient demand exists and with a supporting value chain and business model so that they meet the needs of both the water utility and the customer. This requires that the resource be recovered in a form and quality that meets the specification of the users, in sufficient quantities to attract customers, be available at the times that the customer can use them, and that the economic value proposition for the utility is acceptable. The strategic and operational value of resource recovery, compared to disposal, must be considered when assessing the value to the utility. As mentioned above, if water utilities are producing useful products with secure markets, they are less vulnerable to interruptions that can arise if their residuals are viewed as wastes that must be disposed of. Thus, it is not necessary for resource recovery to be revenue positive for it to have value from both an economic perspective and to ensure continuity of operations. While the water sector has demonstrated the ability to identify useful products and develop the necessary value chains and business models, the necessary skills are not as widespread as needed now and into the future. This points out the need for education and skill building in these areas.

The task of significantly increased resource recovery can be made easier as we adopt alternate water management models that facilitate resource recovery. The rising acceptance of water reuse is leading to alternate water supply models where water reuse and stormwater capture are viewed as integral components of a robust and resilient water supply that can function well under drought conditions. Interest in 'chemical-free' water treatment technologies is also an important preventative by eliminating treatment chemicals and the resulting residuals. An option with great potential is transition to source separation, coupled with resource recovery. We have recently presented work indicating that city-scale urine separation, collection, and conversion to fertilizer products, evaluated over a range of settings (geographical locations, treated effluent nutrient limits, greenhouse gas content of the electrical grid) consistently reduces the life cycle impacts of used water management (Hilton *et al.*, 2020). The biggest objection I hear to source separation is the difficulty retrofitting it into existing infrastructure, although many do not investigate options to do so when buildings are renovated. At the same time, we build new buildings in the old way, thus perpetuating this claimed constraint. The solution is obvious. Include source separation plumbing in new buildings and in building renovations (which always occur) and transition over time.

The production of single cell protein is another example of a significantly different water management approach with inherent resource recovery features. Feedstock products are already being produced from food processing residuals, including single cell protein from wastewater. Efforts are also underway to use algae to produce a variety of products, not only single cell protein but also other, higher-value products. These systems may use traditional open ponds, but the development of photobioreactors is also being pursued. Willy Verstraete and colleagues (Pikaar *et al.*, 2018) have also proposed approaches that can fully use the nutrients in used water to produce high-quality single cell protein. Approaches such as these deserve serious consideration and further development when viewed in the context of the resource recovery pyramid (Figure 18.2).

18.3 ACCELERATING TRANSITIONS

I have already asserted above that the water sector is not only capable of but has transitioned to meet evolving societal needs, for example as illustrated in Figure 18.1. Moreover, resource recovery has always been an accepted and utilized practice by the water sector, but one that must be accelerated. Thus, an important question is how do we accelerate the adoption of new technologies and practices, such as resource recovery, which lead to increased environmental, economic, and societal contributions by the water sector (increased sustainability contribution) and can result in decarbonization of the water sector (the topic of this book).

Let us start with the simple steps. The long lifetime of water sector assets has often been offered as one reason for slow change by the water sector. First, we must recognize that infrastructure assets do not have to be used as originally intended but, rather, can be repurposed to different functions as circumstances change. This is done all the time. I would further submit that, it is not necessarily the long lifetime of water infrastructure, but rather that we plan, design, and construct our infrastructure without fully considering future uses. Consequently, we do not sufficiently consider designs that can be more easily adapted to future uses (Daigger, 2011). I further submit that, if we simply increasingly include flexibility to adapt to future approaches as we plan and design our infrastructure, we can put in place infrastructure that can be more easily adapted (Daigger, 2017). There will be some (or perhaps many) that judge that we already do this, and therefore do not need to do more. I fully agree that future requirements are routinely included in water infrastructure planning and design. I would suggest, however, that this is traditionally done in a manner which either implicitly or explicitly assumes modest change and needed responses compared to the pace and nature of the changes that the water sector must now be preparing itself for. Historically, our approach to future requirements has been quite deterministic, formulating a limited number of future scenarios and associated responses based largely on current technologies and approaches. While this approach has been successful in the past, it does not position water sector utilities very well for the more variable and uncertain future that we face. Formulating responses in terms of current technologies and approaches also furthers 'lock-in' to current approaches.

More aggressively incorporating resource recovery into urban water planning and implementation requires incorporation of a much wider range of external factors than typically included in historical planning processes. Not only must factors such as changing service demands, environmental factors (including climate change), and societal values be considered, but also the uncertain changes resulting from the broader transition to a circular economy. The latter transition will dramatically alter the nature of the products that can be produced, and their economic value. Fortunately, members of our community are developing alternative planning approaches that are better suited to a more rapidly evolving and uncertain future. For example, Malekpour *et al.* (2016) have developed an exploratory planning and implementation approach that focuses on developing a robust plan that performs satisfactorily over a wide range of circumstances, rather than the current predictive approach that focuses on development of an optimal plan that performs the best among others within a defined set of circumstances. Again, as discussed immediately above, approaches exist to implement water sector infrastructure that can be more easily adapted to alternate uses (Daigger, 2017). The approach proposed by Malekpour *et al.* (2016) also includes explicit identification of a wide range of societal needs and explicitly incorporates addressing these needs as an integral part of the process. Approaches such as these focus on maintaining flexibility to adjust to future conditions while also taking steps to learn and incorporate these learnings into the plan as it evolves. Integrated planning and implementation approaches, such as these, need to be further developed and become the norm. Important learning occurs, not only within individual utilities and communities, but also by the profession as a whole. Promoting learning across the entire profession is an essential function of our professional associations.

Time is of the essence, not only because of the need to reduce the environmental footprint of the water sector but also the nature of the change process. O'Callaghan *et al.* (2018) have studied the adoption of new technologies and practices by the water sector, demonstrating that the classic S-curve of adoption applies (mentioned above, but see Rogers (2003) if background in this model is needed). They also confirm the long timeframe for adoption by the water sector. Examining differences in adoption rates of new technologies and innovations, they divide these into needs driven and value driven categories (O'Callaghan *et al.*, 2019). They find that needs driven technologies and innovations are generally adopted faster than value driven ones. Unfortunately, resource recovery and decarbonization changes are often viewed and evaluated as value driven, suggesting the need to change this paradigm to recognize the urgency with which we need to be reducing our environmental footprint. The long timeframe historically experienced within the water sector may also not be inherent but a consequence of past practices and attitudes, as discussed above. Incorporation of evolving planning and implementation approaches, such as described above, coupled with the implementation of more flexible and adaptable infrastructure, may relieve some of the factors historically constraining the rapid adoption of new technologies, approaches, and practices by the water sector.

We also need to more fully understand the processes by which change occurs in the water sector and incorporate this expanded understanding into our efforts to accelerate change. In this regard, I refer to work by Rebekah Brown and colleagues who have studied the evolution of water management practices and the associated technologies, most particularly in Melbourne, Australia (Barron, *et al.*, 2017; Brodnik and Brown, 2018; Brown, *et al.*, 2013; and Brown, 2005). Figure 18.3 illustrates the process as a progressive dialogue between those advocating for the new approach (advocating narrative) and those opposing it (contrasting narrative). This brings to the fore another model of change that I was exposed to many years ago that was presented to me as the 'rule of thirds'. This model suggests that people in general fall into one of three groups in their response to new ideas and concepts. A portion (in this model, represented by one of the 'thirds') like new ideas, another portion is resistant to new ideas (another 'third') and the remainder are uncertain. The model presented in Figure 18.3 is consistent

Figure 18.3 Schematic development of new practices.

Table 18.1 Development of practices.

Transition Phase	Actors	Bridges	Knowledge	Projects	Tools
6. Embedding new practice	Multi-agency coalition	Formalized institution	Next research agenda	Standard practice	Political mandate, coordinating authority, comprehensive regulatory models and tools
5. Policy and practice diffusion	Policy and decision coalition	Science-industry-policy-building	Modeling solutions, capacity building	Numerous industry-led field experiments	Legislative amendments, market offsets, national best-practice guidelines, regulatory models
4. Knowledge dissemination	Informal policy coalition	Science-industry-policy-building	Advanced technological solutions	Major scientific field demonstrations	Best-practice guidelines, targets
3. Shared understanding and issue agreement	Technical solutions coalition	Science-industry-policy	Basic technological solutions	Minor scientific field demonstrations	Draft best-practice guidelines
2. Issue definition	Science leaders	Science-industry	Cause-effect	Laboratory-based and scientific solution prototypes	N/A
1. Issue emergence	Issue activists	N/A	Issue discovery	High profile scientific studies	N/A

with the 'rule of thirds' and frames it as a dialogue between those who welcome new ideas and concepts and those who tend to resist new ideas and concepts. The important take-away from the 'rule of thirds' is that these two groups, those who welcome and those who resist new ideas and concepts, are the actors in the resulting dialogue, but the audience is those who are uncertain. Thus, to create change it is not necessary to convince those who are naturally resistant to new ideas or concepts, but rather those who are uncertain and who tend to not engage in the dialogue, at least not initially. Of course, this is not to suggest that membership in one of these three cohorts is good or bad. Rather, it simply reflects human nature when a sufficient number of people are involved.

Figure 18.3 also illustrates the nature of the dialogue, progressing from the emergence of issue(s), to identification of solutions and their initial implementation, and finally to the development of policies and regulations and embedding the results into emerging practice. Table 18.1 further identifies the principal actors engaged in these dialogues at each stage, highlights the importance of bridging agents and/or institutions to facilitate the dialogue illustrated in Figure 18.3, and the progression of knowledge that underpins progress to refine and define the new ideas and concepts. Pilot and demonstration projects are essential to develop the knowledge needed for the dialogue to progress. Tools to consolidate the knowledge gained at each stage are also listed. An important outcome of reviewing this process is that evidence must first be produced to support policymaking, regulation, and translation into standard practice. This model makes it clear that pilot studies, demonstration projects, and actual applications are essential to provide the evidence needed for policymaking and regulations to subsequently occur. There are some who envision the transition process as beginning with policies and regulations. This is not generally the case, as good policies and regulations need to be based on evidence, and evidence is often needed to develop the consensus required for the development and acceptance of policies and regulations. This emphasizes the importance of the activities already occurring throughout the water profession, as they are essential components of the change process. We must capitalize on these nascent actions though, to consolidate the progress being achieved and accelerate its translation into policies, regulations, and standard practice. Again, this can be an important role for our professional associations.

18.4 THE PATH FORWARD

Water management is a path, not a destination. This has always been and will always be the case, and is set in the context of the broader socio-economic changes occurring. The water sector has demonstrated the ability to adapt to changing circumstances, sometimes retroactively and sometimes proactively. The overall global pace of changes demands that adaptations must be more proactive than in the past. Adoption of the One Water and resource recovery paradigms, championed by leaders of the water profession and increasingly being more broadly adopted, represent the directions needed at this time. While the focus of this book, decarbonization, aligns closely with resource recovery, the companion One Water transition that the profession is also pursuing in tandem also needs to be recognized as these paradigms interact within the water sector as we proactively follow the path before us. Two core concepts underlying One Water are: (1) the concept of a portfolio of solutions and (2) developing solutions that perform well 'in the extreme'. The portfolio concept is illustrated by approaches to water supply in water-short locations which involve a combination of water supplies that function well over a range of hydrologic conditions. The appropriate portfolio for a particular community consists of a combination of traditional water supplies (surface and ground water), coupled with water conservation, rainwater harvesting, reuse, and desalination. The same concept is increasingly being applied to stormwater management, incorporating traditional approaches such as pipes and dikes with natural systems and land use planning. Planning for extremes reflects the fact that the portfolio must function well over a broad range of hydrologic conditions. This means moving beyond the search for an 'optimal', least-cost solution that can be defined only for a specified set of conditions.

The thought process underlying One Water can also serve well as the profession increasingly adopts and implements resource recovery as a core objective. To date, resource recovery beyond water has focused on the same resources generally recovered from the water cycle, namely biogas and biosolids products. and with biogas generally converted into electricity and heat. As illustrated in Figure 18.2, these are relatively low value products. Moreover, as discussed above the value of electricity from both environmental/societal and economic perspectives is likely to decline over time as renewable resources such as solar and wind increasingly displace fossil fuels for energy production. While biogas and traditional biosolids may continue to be desirable products in some instances, a wider range of products must be added to the 'portfolio' routinely available to utilities. Greater flexibility to adjust the resources recovered at any particular time also needs to be built into water management systems so that utilities can adjust to the evolving circular economy and produce products that meet demands in an economic fashion. Building this flexibility into water management systems will require not only infrastructure that can be adjusted more quickly from a functional perspective, but also building professional and institutional capacity into water management institutions. An increased business mentality is an example as while the water profession must continue to maintain water service (quantity and quality), the ability to develop the systems needed to convert recovered resources into products that supply economic needs must become more common.

New planning, management, and implementation paradigms and practices are needed, as described above. We need to transition from our historic approach of periodic planning and adaptation to changing conditions to a more continuous process. We also need to refocus our decision-making from being dominated by cost-effectiveness under a defined set of conditions to place more importance on maintaining flexibility and incorporating learning elements that lead to changes in the plan. All of this is possible.

Not addressed above is the need to accomplish these transformations while also addressing societal issues of equity and inclusion. Here I will simply refer to some of the resources developed by IWA to assist water professionals to see that all citizens of planet earth enjoy the human right to water and sanitation (Bos, 2016; Hirano & Latorre, 2020a, 2020b). Society has committed to achieve this right universally through adoption of water and sanitation as a human right and through expression of the sustainable development goals.

Table 18.2 Excerpts from the table of contents of The leadership challenge (Kouzes & Posner, 2017).

Practice 1: Model the Way

1. Clarify values
   a. Find your voice
   b. Affirm shared values
   c. Take action: Clarify values
2. Set the Example
   a. Live the shared values
   b. Teach others to model the values
   c. Take action: Set the example

Practice 2: Inspire a Shared Vision

1. Envision the future
   a. Imagine the possibilities
   b. Find a common purpose
   c. Take action: Envision the future
2. Enlist Others
   a. Appeal to common ideals
   b. Animate the vision
   c. Take action: Enlist others

Practice 3: Challenge the process

1. Search for opportunities
   a. Seize the initiative
   b. Exercise outsight
   c. Take action: Search for opportunities
2. Experiment and Take Risks
   a. Generate small wins
   b. Learn From experience
   c. Take action: Experiment and take risks

Practice 4: Enable Others to Act

1. Foster collaboration
   a. Create a climate of truest
   b. Facilitate relationships
   c. Take action: Foster collaboration
2. Strengthen Others
   a. Enhance self-determination
   b. Develop competencies and confidence
   c. Take action: Strengthen others

Practice 5: Encourage the Heart

1. Recognize contributions
   a. Expect the best
   b. Personalize recognition
   c. Take action: Recognize contributions
2. Celebrate the Values and Victories
   a. Create a spirit of community
   b. Be personally involved
   c. Take action: Celebrate the values and victories

Good policies and regulations can certainly assist with the necessary on-going transformations. As discussed in Chapter 17 and illustrated in Figures 18.3 and Table 18.1, however, it must be recognized that policies and regulations lag behind emerging practice. Evidence and experience are needed to create the consensus needed for the adoption of policies and regulations, and to form the basis for the development of constructive policies and regulations. Thus, we must always be 'pushing the envelope' to both learn and provide the basis for change. This is a matter of leadership.

My favorite book on leadership is The Leadership Challenge by Kouzes and Posner (2017). I was first introduced to it (the third edition actually) about 25 years ago and have found its content to be very useful, both for myself and when I needed to work with others to increase leadership skills. Table 18.2 summarizes the five core practices Kouzes and Posner have found to be the foundation for good leadership. Kouzes and Posner also emphasize that leadership is not an inherent but rather a learned skill. Leadership by the water profession is truly the core element of the path forward. As indicated in Table 18.2, effective leadership boils down to five core practices. The fact that we have so many leaders who are already practicing this skill makes it clear that the water sector is up to the task. We just need to keep moving forward and use every opportunity to accelerate the process. Together, we can make the water sector more sustainable, resilient, and equitable.

REFERENCES

Barron N. J., Kuller M., Yasmin T., Castonguay A. C., Copa V., Duncan-Horner E., Gimelli F. M., Jamali B., Nielsen J. S., Ng K., Novalia W., Shen P. F., Conn R. J., Brown R. R. and Deletic A. (2017). Towards water sensitive cities in Asia: an interdisciplinary journey. *Water Science and Technology*, **76**(5), 1150–1157. https://doi.org/10.2166/wst.2017.287

Bos R. (2016). Manual of the Human Rights to Safe Drinking Water and Sanitation for Practitioners. IWA Publishing, London.

Brodnik C. and Brown R. (2018). Strategies for developing transformative capacity in urban water management sectors: the case of Melbourne, Australia. *Technological Forecasting & Social Change*, **137**, 147–159. https://doi.org/10.1016/j.techfore.2018.07.037

Brown R. R. (2005). Impediments to integrated urban stormwater management: the need for institutional reform. *Environmental Management*, **36**(3), 455–468. https://doi.org/10.1007/s00267-004-0217-4

Brown R. R., Keath N. and Wong T. H. F. (2009). Urban water management in cities: historical, current and future regimes. *Water Science and Technology*, **59**(5), 847–855. https://doi.org/10.2166/wst.2009.029

Brown R. R., Farrely M. A. and Loorback D. A. (2013). Actors working the institutions in sustainability transitions: the case of Melbourne's stormwater management. *Global Environmental Change*, **20**(2), 287–297.

Daigger G. T. (2011). A practitioner's perspective on the uses and future developments for wastewater treatment modelling. *Water Science and Technology*, **63**(3), 516–526. https://doi.org/10.2166/wst.2011.252

Daigger G. T. (2017). Flexibility and adaptability: essential elements of the WRRF of the future. *Water Practice & Technology*, **12**(1), 156–165. https://doi.org/10.2166/wpt.2017.019

Hilton S. P., Keoleian G. A., Daigger G. T., Zhou B. and Love N. G. (2020). Life cycle assessment of urine diversion and conversion to fertilizer products at the city scale. *Environmental Science & Technology*, **55**(1), 593–603. https://doi.org/10.1021/acs.est.0c04195

Hirano M. and Latorre C. (2020a). Guidelines for Public Participation in the Regulation of Urban Water Services. IWA Publishing, London.

Hirano M. and Latorre C. (2020b). Tools for Public Participation in the Regulation of Urban Water Services. IWA Publishing, London.

Kouzes J. M. and Posner B. Z. (2017). The Leadership Challenge: How to Make Extraordinary Things Happen in Organizations, 6th edn. John Wiley & Sons, Hoboken, New Jersey.

Malekpour S., de Haan F. J. and Brown R. R. (2016). A methodology to enable exploratory thinking in strategic planning. *Technological Forecasting & Social Change*, **105**, 192–202. https://doi.org/10.1016/j.techfore.2016.01.012

O'Callaghan P., Daigger G., Adapa L. and Buisman C. (2018). Development and application of a model to study water technology adoption. *Water Environment Research*, **90**, 563–574. https://doi.org/10.2175/106143017X15054988926479

O'Callaghan P., Adapa L. and Buisman C. (2019). Analysis of adoption rates of needs driven versus value driven innovation water technologies. *Water Environment Research*, **91**, 144–156. https://doi.org/10.1002/wer.1013

Pikaar I., de Vrieze J., Rabaey K., Herrero M., Smith P. and Verstraete W. (2018). Carbon emission avoidance and capture by producing in-reactor microbial biomass based food, feed and slow release fertilizer: potentials and limitations. *Science of the Total Environment*, **644**, 1525–1530. https://doi.org/10.1016/j.scitotenv.2018.07.089

Rogers E. M. (2003). Diffusion of Innovations, 5th edn. Free Press, NY, NY.

Sedlak D. (2014). Water 4.0: The Past, Present, and Future of the World's Mot Vital Resource. Yale University Press, New Haven, Connecticut.

van der Hoek J. P., de Fooij H. and Struker A. (2016). Wastewater as a resource: strategies to recover resources from Amsterdam's wastewater. *Resources, Conservation and Recycling*, **113**, 53–64. https://doi.org/10.1016/j.resconrec.2016.05.012

Westerhoff P., Lee S., Yang Y., Gordon G. W., Hristovski K., Halden R. U. and Herckes P. (2015). Characterization, recovery opportunities, and valuation of metals in municipal sludges from U.S. Wastewater treatment plants nationwide. *Environmental Science & Technology*, **49**, 9479–9488. https://doi.org/10.1021/es505329q

Index

Page numbers in "f" refer to Figures. Page numbers in "t" refer to Tables.

A

Accelerating Transitions, 326–329
Activated Sludge Process, 10, 37, 53–57, 60, 62t, 75, 269, 304
Aeration, 7, 10, 15–16, 23, 39, 48, 53, 55–57, 63, 68, 78, 94, 101–107, 110, 117, 121, 124, 127, 130, 132, 144–145, 147, 152–153, 173, 191–192, 195, 200, 207–208, 212, 214, 267, 290, 293–294t, 304
Algae Membrane Bioreactors, 8, 189–190
Ammonia Oxidizing Bacteria, 99
Anaerobic Co-Digestion, 68, 71
Anaerobic Digestion, 6–7, 11, 16, 23t, 38–39, 52, 58–60, 62–63, 67, 68, 70–72, 75, 78, 87, 91–92, 121t–123, 125, 130, 143–145, 149, 154–157, 169, 174t, 217, 227, 230, 232, 234, 267, 269, 276, 290, 294, 305, 306, 313–314, 324
Anaerobic Membrane Bioreactors, 7–8, 189, 208, 324
Anammox Process, 105

B

Biofiltration, 213–214, 222t, 267
Biogas Production, 7, 11, 39, 58–60, 63, 70–71, 95, 125, 132, 152, 154–155, 192, 232, 276, 295t, 307t, 314
Biogas Upgrading, 72, 143–144, 151, 157–158, 159t–160t, 161
Biomethane, 4, 9, 72–73, 77, 143–144, 152, 157, 158, 161, 313
Biorefinery, 28, 144, 151

C

Capacity Utilization, 7, 51, 55, 60–63
Carbon Capture, 9–10, 20, 22t, 55–56, 63, 69t, 72–73, 84, 85f, 95, 101–102, 106, 121t, 237, 238t, 241
Carbon Dioxide Emissions, 18, 128
Carbon Sequestration and Mitigation, 40
Carbon Valorization, 83–84, 86
Carbon-Water Nexus, 18
Chemical Energy, 8, 51, 126, 131, 135, 145, 313
Chemical Use, 6–8, 51, 58, 60–62, 75–76, 109, 118, 125, 292, 317
Chemically Enhanced Primary Treatment, 56, 103
Circular Water Economy, 1, 12–13, 321
Combined Heat and Power, 4, 6, 38–39, 69t, 72, 208, 217, 254, 306, 312, 324
Constructed Wetlands, 116, 121t, 123, 126, 210–212, 221, 222t
Contaminants of Emerging Concern, 16, 209

D

Data Science Tools, 275, 277–290
Deammonification, 11, 103–104, 108, 132

Decarbonization, 1, 4, 6–9, 11–12, 15–19, 21, 24, 27–29, 33, 51–53, 55, 58, 60, 61–64, 68, 71, 77, 99–100, 102, 106–107, 109–110, 115, 116f, 117, 120, 121t, 123–126, 127t–128t, 130–132, 134–136, 143, 149, 154, 171–173, 174t, 180, 182
Desalination, 8, 16, 33, 39, 187–189, 196–200, 304, 329
Direct Emissions, 2, 5, 35, 37, 39, 45, 61–62, 123, 207
District Heating, 8, 11, 250–251, 254, 305, 310, 325

E

Electrofermentation, 84, 90–93
Electromethanogenesis, 87, 228–229, 238t
Emission Factors, 5, 40, 43–45, 57, 60t
Energy Balance, 67, 68f, 175, 191
Energy Footprints, 2
Energy Frugality, 22t
Energy Neutrality, 11, 68, 69t, 74, 265
Energy Positive, 6, 11, 69t, 74, 191–192, 294
Energy-Water Nexus, 15–16, 18
Enhanced Biological Phosphorus Removal, 120, 122t, 126

F

Forward Osmosis, 8, 189, 192–195, 198–199
Fugitive Emissions, 42t, 58, 306, 314

G

Global Warming, 2–4, 38, 41–42, 75, 77f, 132, 135–136, 173, 211–212
Greenhouse Gas Emissions, 17, 22t, 35, 130, 134, 143, 174t, 177, 200, 207–209, 212, 216–217, 221, 223, 309, 312–313

H

Heat Pumps, 8, 19, 249, 257, 262
Hydraulic Energy, 6, 51
Hydrothermal Liquefaction, 8, 176, 306

I

Indirect Emissions, 2, 35, 39, 61–62, 123
Intergovernmental Panel on Climate Change, 35, 120, 259

L

Land Application, 6–7, 40, 42, 67, 69t, 110, 120, 121t, 122, 171, 174t, 175–177, 179–181, 214, 222t, 323
Landfilling, 8, 171–172, 176–177, 179–180, 217, 321
Life Cycle Assessment, 72, 75–77, 133, 179–180

M

Machine Learning, 11, 276–277
Membrane Biofilm Reactor, 8, 189, 192, 193f
Membrane Distillation, 8, 189, 195, 199
Membrane Systems, 55
Methane Emissions, 3, 6, 38, 45, 48, 59–60, 63, 212, 306, 313
Microalgae, 9–10, 95, 128t, 134, 143–148, 151–152, 154–158, 159t–160t, 190, 214, 218–221
Microbial Electrochemical Technology, 9, 83–95

N

National/Global Policies, 303–306
Natural Treatment Systems, 9, 207–223
Net Zero, 1–13, 37, 40, 136, 208–209, 214, 223, 304, 305f, 309
Nitrite Oxidizing Bacteria, 37, 99, 100f
Nitrite Shunt, 8, 103–108, 127t, 131–132
Nitrous Oxide Emissions, 17, 101–103, 211–212
Nutrient Removal, 10, 62, 75, 125–126, 128t, 131–132, 134, 136, 153, 190, 269, 293

O

OneWater, 1, 2f, 6, 12

P

Partial Denitrification/Anammox, 8, 105, 108–109
Partial Nitritation/Anammox, 8, 104, 127t, 132
Perfluoroalkyl Substances, 16, 315
Phosphorus Management, 6, 8, 113–137
Photobiological Systems, 6, 143–161
Photosynthetic Platforms, 144
Phototrophic Systems, 9, 143–144, 148–149, 150t
Photovoltaics, 39
Pressure Retarded Osmosis, 8, 189, 196
Principal Component Analysis, 281
Process Intensification, 6, 8, 11, 94

Process Mixing, 39, 58, 72
Purple Photosynthetic Bacteria, 9, 143, 153

R

Resource Recovery, 1, 4, 6–12, 46, 51, 62, 67–68, 69t, 73–74, 77–78, 93–94, 99, 117, 121t, 151, 171, 173, 176, 178, 180, 188–195, 218, 263, 267, 270–271, 303–314, 316–318, 321–329
Reverse Osmosis, 8, 39, 136, 187, 189, 194, 198

S

Secondary Treatment, 51, 53, 55–57, 74, 155, 210, 211f
Sludge Management, 7, 337–338, 68, 76, 171, 173–176
Sludge Pretreatment, 70, 72, 179

T

Technoeconomic Assessment, 259
Thermal Energy, 2, 8, 11, 23t, 179, 187, 199, 249–262, 310, 312–313, 317

U

Urban Water Cycle, 2f, 5, 17, 33–48, 224

V

Value Hierarchies, 323–324

W

Wastewater Collection, 1–2, 4, 6, 33–34, 37, 47, 117, 124, 267, 304
Wastewater Pumping, 7, 53–55, 220
Wastewater Treatment, 2, 5–6, 8–11, 16–18, 22t–23t, 33–41, 44–48, 51, 55–56, 61–63, 67, 69t, 70–71, 73–78, 85, 87, 89, 93–95, 102, 110, 115–120, 123–126, 132–136, 143–154, 157, 171, 175, 179–182, 187–200, 209–214, 218–219, 221–223, 260, 265–272, 275–276, 283–288, 293–294, 303–304, 311–315, 323
Water Distribution, 24, 208, 275, 288, 291
Water Frugality, 22t
Water Reclamation, 2, 4, 6, 51, 55, 60t, 61, 69t, 133, 177, 187, 189, 192, 194t, 195, 209, 266
Water Resource Recovery Facilities, 6, 46, 51, 67, 69t, 77, 93, 171, 304–305, 310–312, 316–317
Water Reuse, 35, 38, 61, 73–74, 187, 194, 266–267, 309, 312, 316, 323–325
Water Sector, 1–2, 4–6, 9, 11–13, 15–16, 18, 27–29, 33, 38, 63, 68, 99, 113, 136, 171, 173, 183, 188, 209–212, 272, 303–318, 321–323
Water Treatment, 1, 2, 5–6, 8–9, 11, 15–18, 22t–23t, 33–45, 47–48, 51, 55–56, 61–63, 67, 69t, 70–71, 73–78, 85, 87, 89, 93–95, 102, 110, 115, 115, 120, 123–126, 30–36, 143–144, 150t, 171, 175, 180–182, 181, 189, 200, 209–210, 220–223, 226, 266, 272, 276, 281, 283, 295, 303–304, 815–815, 372–375

IWA Publishing's authorised EU representative for General Product Safety Regulations is Diane D'Arras, 15 rue Duret, 75116 Paris, France, e-mail: safety@iwap.co.uk.

Printed and bound by CPI Group (UK) Ltd, Croydon, CR0 4YY
16/06/2026
02140376-0002